Nonculturable Microorganisms in the Environment

Nonculturable Microorganisms in the Environment

Edited by

Rita R. Colwell
Center of Marine Biotechnology, Columbus Center,
University of Maryland Biotechnology Institute,
Baltimore, Maryland 21202

D. Jay Grimes
Institute of Marine Sciences,
University of Southern Mississippi,
Ocean Springs, Mississippi 39566-7000

ASM PRESS Washington, DC

Cover photo courtesy of Sitthipan Chaiyanan, Tim Maugel, and Rita R. Colwell

American Society for Microbiology
1752 N Street, N.W.
Washington, DC 20036-2084

Library of Congress Cataloging-in-Publication Data

Nonculturable microorganisms in the environment / edited by Rita R. Colwell and D. Jay Grimes.
p. cm.
Includes bibliographical references and index.
ISBN 1-55581-196-5
1. Microbial ecology. 2. Microbiology—Cultures and culture media.
I. Colwell, Rita R., 1934– II. Grimes, D. J. (David J.), 1943–

QR100.N66 2000
579—dc21

00-029291

Printed in the United States of America

CONTENTS

CONTRIBUTORS

J. Scott Angle • College of Agriculture and Natural Resources, University of Maryland, 1201 Symons Hall, College Park, MD 20742

John A. Baross • School of Oceanography, Box 357940, University of Washington, Seattle, WA 98195

Jeffrey J. Byrd • Department of Biology, St. Mary's College of Maryland, St. Mary's City, MD 20686

Rita R. Colwell • Center of Marine Biotechnology, Columbus Center, University of Maryland Biotechnology Institute, 701 E. Pratt St., Baltimore, MD 21202, and Department of Cell and Molecular Biology, University of Maryland, College Park, MD 20742

J. William Costerton • Center for Biofilm Engineering, Montana State University, Bozeman, MT 59717

Jody W. Deming • School of Oceanography, Box 357940, University of Washington, Seattle, WA 98195

T. Martin Embley • Microbiology Group, Department of Zoology, The Natural History Museum, Cromwell Road, London SW7 5BD, United Kingdom

Michel J. Gauthier • INSERM Unité 452, Faculté de Médecine, Avenue de Valombrose, F-06107, Nice Cedex 2, France

D. Jay Grimes • Institute of Marine Sciences, University of Southern Mississippi, P.O. Box 7000, Ocean Springs, MS 39566-7000

Russell T. Hill • Center of Marine Biotechnology, University of Maryland Biotechnology Institute, Columbus Center, 701 East Pratt St., Baltimore, MD 21202

Anwarul Huq • Center of Marine Biotechnology, Columbus Center, University of Maryland Biotechnology Institute, 701 E. Pratt St., Baltimore, MD 21202, and Department of Cell and Molecular Biology, University of Maryland, College Park, MD 20742

Thomas L. Kieft • Department of Biology, New Mexico Institute of Mining and Technology, Socorro, NM 87801

Ivor T. Knight • Department of Biology, James Madison University, Harrisonburg, VA 22807

Kazuhiro Kogure • Ocean Research Institute, University of Tokyo, 1-15-1, Minamidai, Nakano, Tokyo 164, Japan

Mark W. LeChevallier • American Water Works Service Co. Inc., P.O. Box 1770, Voorhees, NJ 08043

Morris A. Levin • University of Maryland Biotechnology Institute, 701 E. Pratt St., Suite 231, Baltimore, MD 21202

Steven E. Lindow • Department of Plant and Microbial Biology, University of California, Berkeley, Berkeley, CA 94720-3102

Kevin C. Marshall • School of Microbiology and Immunology, The University of New South Wales, Sydney, NSW 2052, Australia

Gordon A. McFeters • Microbiology Department and Center for Biofilm Engineering, Montana State University, Bozeman, MT 59717

Aaron L. Mills • Laboratory of Microbial Ecology, Department of Environmental Sciences, Clark Hall, University of Virginia, Charlottesville, VA 22903

Kenneth H. Nealson • Division of Geology and Planetary Sciences, California Institute of Technology, MS 170-25, Pasadena, CA 90125

James D. Oliver • Interdisciplinary Biotechnology Program, Department of Biology, University of North Carolina at Charlotte, 9201 University City Blvd., Charlotte, NC 28223

Irma N. G. Rivera • Center of Marine Biotechnology, Columbus Center, University of Maryland Biotechnology Institute, 701 E. Pratt St., Baltimore, MD 21202, and Department of Microbiology, Biomedical Science Institute, São Paulo University, São Paulo, S.P. Brazil, CEP 05508-900

Frank T. Robb • Center of Marine Biotechnology, University of Maryland Biotechnology Institute, Columbus Center, 701 East Pratt St., Baltimore, MD 21202

Erko Stackebrandt • DSMZ-German Collection of Microorganisms and Cell Cultures, D-38124 Brauschweig, Germany

Minoru Wada • Ocean Research Institute, University of Tokyo, 1-15-1, Minamidai, Nakano, Tokyo 164, Japan

Mark Wilson • Biology Department, The Colorado College, Colorado Springs, CO 80903

PREFACE

Twenty years ago the discovery that *Vibrio cholerae* does not "die off" in the environment was made. We hypothesized that this gram-negative, rod-shaped bacterium possessed the capacity to survive for very long periods of time, i.e., months or years, under conditions adverse to active growth and reproduction. The term "somnicell" was proposed in 1987 to describe this state of being. Subsequently a series of experiments showed that this phenomenon was not unique to *Vibrio cholerae*. Many other gram-negative bacteria were found to undergo the "viable but not culturable" state in response to adverse environmental conditions. Such bacterial species included *Escherichia coli*, *Salmonella enteriditis*, *Campylobacter jejuni*, *Legionella pneumophila*, and *Shigella sonnei*, among others. Clearly, the ability to tolerate the vicissitudes of the environment, notably changes that accompany climate and season, is part of a strategy for survival that is pervasive among bacterial species that are both aquatic in natural habitat and opportunistic human pathogens. It raises a question as to whether the pathogenicity manifested by these bacteria is opportunism, or simply metabolic functions of bacteria that are naturally occurring in the environment and contributing to the stability and sustainability of the environment, with the inadvertent host, the human, suffering because of misplaced consequences of those metabolic functions. It is an interesting question to ponder. Throughout the pages that follow, invited authors describe, from each of their perspectives, the significance and consequences of the viable but nonculturable state.

This book has been long in preparation, but has become ever more timely because of the volume of literature that has appeared describing and characterizing the phenomena of dormancy/nonculturability. Not unexpectedly, there have been those who dispute the existence of any state except viable and reproducing, or dying and/or already dead. For these scientists, there is no resting stage, no cell strategy for survival.

The authors of this book are not evangelical in their writings, but provide evidence as they have gathered it and discuss their results evenly. A general discussion of terminology and very brief history is provided (chapter 1). The morphology of cells in the viable but not culturable state is discussed by Jeffrey Byrd (chapter 2), and the nature of very small, predominantly not yet cultured bacteria in soil and subsurface environments is covered in detail by Tom Kieft (chapter 3), who also discusses the lower size limit for microbial cells.

Membrane bioenergetics is explored by Minoru Wada and Kazuhiro Kogure, who provide an interesting hypothesis for nonculturability in relation to luminous symbionts of marine and estuarine animals (chapter 4).

A major impediment to detecting and characterizing bacteria in the environment when they are not as yet cultured (chapter 5), or are in the viable but not culturable state, has been overcome to a significant extent by advances made in molecular biology. The molecular biology-based methods that have been developed for microbial ecology, e.g., gene probes, PCR, etc., are described in chapters 6 and 16 by Ivor Knight and Jim Oliver, respectively. The fascinating architecture of biofilms and the inherent difficulties of recovering culturable bacteria from natural ecosystems are covered in chapters 8 and 9 by Kevin Marshall and Bill Costerton.

Strategies for survival in the environment, originally described in a very early review by Roszak and Colwell in 1987, have been elegantly and extensively updated, including some new concepts, by Jody Deming and John Baross (chapter 10). The role of as yet uncultured microbial species and of bacteria in the viable but unculturable state in biogeochemistry is detailed in chapter 12. The conditions under which the survival strategy is elicited include biofilm production, chronic or "aseptic" disease, soil and subsurface environments, the deep sea, and aquatic systems. The implications of viable but not culturable cells, dormancy, and as yet uncultured cells for release of genetically engineered microorganisms to the environment are broad-reaching (chapters 13 and 14). These implications include clinical medicine; there are increasingly frequent reports of heart disease and strokes being induced by low-grade, chronic, bacterial infections in which the causative agent cannot be cultured but is detectable by molecular genetic methods (chapter 18).

It is, in fact, the availability of molecular genetic methods during the past decade and their application in microbial ecology that has provided the strongest evidence for the existence of potentially millions of species of bacteria and other microorganisms in microbial ecosystems. These bacteria have been successfully described by their gene sequences when culturing attempts have not succeeded. Stackebrandt and Embley (chapter 5) have defined these "as yet uncultured" microorganisms that may respond in the future to more elaborate and sophisticated culture methods and media. For the moment, legions of bacteria in the Sargasso Sea of the Atlantic Ocean are represented only by their characteristic nucleic acid sequences. Whether all of the Sargasso Sea microorganisms will be cultured, or only some, is purely speculative at this time.

It is clear that agriculture, medicine, environmental science, and biotechnology—to mention only a few of the major fields affected by lack of knowledge of the microorganisms forming their microbial populations of interest—will benefit from recognition and understanding of bacteria that are as yet uncultured, or viable but not culturable, and yet dominant. The authors of the chapters in this book give insight into the array of habitats, terrestrial and aquatic (marine and freshwater), where the "elusive" microorganisms that are viable but not culturable, or are as yet unculturable, reside.

Until culture methods and media are devised that accommodate all microbial species, we will find ourselves debating the "alive versus dead or dying" question and devising the most intricate methods for testing viability. The best evidence will eventually come from whole genome sequencing and parsing of genes to deduce pathways accounting for dormancy, somnicell, and/or nonculturability. The genetic evidence is being gathered, with the full sequences of *V. cholerae*, *E. coli*, and *Shewanella putrefaciens* now available and at least 20 other microbial genomes fully sequenced and dozens more in the planning stage. As the science of microbiology continues to evolve, this cosmic question will certainly be answered, if not from the clues provided by the genes themselves, then certainly from the bacteria in the ice cores of Lake Vostok in Antarctica or the samples collected from future space flights to the frozen depths of Oceanus, or perhaps Mars. The "nanobes," those structures observed in meteorites gathered from the Antarctic and more recently in deep subsurface cores, may provide additional clues leading to enlightenment with respect to the viable but nonculturable state.

It is appropriate to acknowledge those who made this book become reality, who provided the hard work of compiling the manuscripts and gently nudging the authors. Mrs. Norma Brinkley deserves heartfelt thanks for her steadfast commitment to getting the job done, and Gregory Payne has exhibited the patience of Job in seeing this task to completion. A huge debt is owed to Huai Shu Xu, who persevered in his scholarship, despite the most difficult obstacles in his early years, to participate in the discovery and elucidation of the viable but nonculturable *V. cholerae*, and to Nell Roberts, whose keen insight and sheer stubbornness in field observations and data gathering provided a strong foundation for our theories and hypotheses. Finally, our thanks to all the authors for their willingness to write, rewrite, and update, and to our spouses, Dr. Jack H. Colwell, who makes everything possible, and Beverly Grimes, who made it possible again.

Rita R. Colwell
D. Jay Grimes

Nonculturable Microorganisms in the Environment
Edited by R. R. Colwell and D. J. Grimes

Chapter 1

Semantics and Strategies

Rita R. Colwell and D. Jay Grimes

The term "nonculturable" was invoked by Xu et al. in 1982 (47) to describe starved but viable cells in a survival or dormant state of existence, observed to occur after an actively growing culture of *Vibrio cholerae* O1 or *Escherichia coli* was placed in a nutrient-free microcosm, e.g., a saline solution free of nutrient and incubated at low temperature. The cells were grown in a nutrient broth under optimal conditions, harvested by centrifugation, washed, and placed in sterile artificial seawater (15‰ salinity). The cells were enumerated directly, using both acridine orange and fluorescent-antibody staining, and viewed by epifluorescent microscopy. They were also plated using media optimized for their growth. It was determined by direct microscopic observation that the total number of cells did not decrease with time. However, as incubation continued over several days, it was found that colonies no longer formed when the same samples were plated on media that had been optimized for growth of the culture. To determine whether such cells were viable or dead, a microscopic technique developed by Kogure et al. (25) was employed to examine the cultures that yielded no colonies on transfer to solid media and no growth in liquid media. The Kogure et al. (25) method, a direct-viable-count procedure, when applied to these cultures revealed that nearly all of the cells enumerated by acridine orange direct counting (AODC) were metabolically active, even though they could not be recovered on any plating medium or in broth. Hence, the phrase "viable but nonculturable" (VBNC) was coined to describe these bacteria that were apparently in a dormant state.

During the next 18 years, the early studies of Xu et al. (47) were expanded to include a variety of bacterial species and strains studied by many investigators, as the following chapters document. From the results of hundreds of studies published since 1982, it is now more clearly understood that lack of nutrients (40, 44, 47, 48), low temperatures (14, 47), high pressure (3, 46), sharp changes in pH or salinity (41), and other environmental factors can initiate a complex series, or cascade, of cellular events that include changes in cellular morphology, changes in

Rita R. Colwell • Center of Marine Biotechnology, Columbus Center, University of Maryland Biotechnology Institute, 701 E. Pratt St., Baltimore, MD 21202. ***D. Jay Grimes*** • Institute of Marine Sciences, University of Southern Mississippi, P.O. Box 7000, Ocean Springs, MS 39566-7000.

concentration and/or structure of major biopolymers (proteins, membrane lipids, nucleic acids), and cessation of ability to grow on solid or in liquid laboratory media that would otherwise support growth of the bacterial strain employed in the studies. Although such cells have been described as VBNC under defined conditions, several terms have been used to define this state. One such term was "somnicell" (38), implying that the cells are arrested in growth and division, essentially "resting" or "sleeping." The ability to form somnicells, i.e., to enter the VBNC state, is now recognized as a strategy commonly employed by many gram-negative heterotrophic bacteria as well as nonsporulating (non-spore-forming) gram-positive and gram-variable bacteria; that is, it is an inducible capacity to survive under environmental conditions less than optimal for growth and multiplication, a feature particularly valuable for those bacteria whose natural habitat is the aquatic environment. Furthermore, evidence is accumulating for genetic regulation of this state (26, 35, 37).

The term "nonculturable" in the early laboratory studies (47) was considered to be descriptive. Not unexpectedly, however, controversy arose and continues, nearly 20 years later, over whether these cells are alive or dead (2, 29). The opinions range from the vigorous insistence that the cells are dead (21, 24), to puzzlement over the significance of their continuing metabolism and integrity of cellular structure (8), to the position that nonculturability is a stage in the life cycle of cells, i.e., a survival mechanism under adversity (23, 38). At the time Xu et al. (47) conducted their studies, cells labeled as VBNC were nonculturable under the carefully defined conditions of those experiments designed to characterize the VBNC cell. Interestingly, in parallel with the definition and description of VBNC cells, there was a recognition that the vast majority of bacteria in the environment are, in fact, not yet cultured, and this has led to new hypotheses (7). Even today, it is still not possible to cultivate in vitro most bacterial species directly from environmental samples, or after exposure of previously culturable cells to environmental conditions unfavorable for growth and multiplication (7, 10, 11, 33). In a few cases, bacteria that are nonculturable under specific conditions are capable of resuming growth and cell division after specific treatment, such as heat shock or addition of water and/or nutrient (27; K. Amako, personal communication). For several bacterial species, it has been discovered that passage of VBNC cells through an appropriate animal host will induce return of culturability (14, 15, 33, 36, 39). Some investigators have shown that cells can be induced to the culturable state by manipulating in vitro culture conditions (30, 32, 37, 39). Temperature is a significant factor for species such as *Vibrio vulnificus* (33), and heat shock activates others (27; Amako, personal communication). It is theorized that conditions in the natural environment, e.g., in coastal waters, estuaries, or riverine systems, the effects of climate, season, nutrient flux, and related physicochemical factors regularly induce alternating nonculturable or dormant and active growth states in situ. Accordingly, the term somnicell (38) may ultimately prove more descriptive of this phenomenon, more accurately defining the strategy for survival (somnicells being those in a dormant state and/or in a type of microbial hibernation), since the conditions inducing return to active growth remain incompletely understood.

It should also be pointed out that "nonculturable" and "not yet cultured" are terms that have been used to describe bacteria that are presumably functional, i.e., probably not starved, but characteristically metabolizing very slowly and having become highly adapted to oligotrophic conditions, for which no method of laboratory culture has yet been successfully devised (31). For example, *Legionella pneumophila* in the natural environment is very difficult to isolate in culture, even when the numbers of cells present in a given sample are very large (48). Gene probe, fluorescent antibody, and PCR methods frequently yield large numbers ($\geq 10^6$) of *L. pneumophila* in water samples that are otherwise culture negative. In a very early publication describing Legionnaire's disease (28), the etiologic agent was described as nutritionally fastidious and capable of growth only in highly complex laboratory media and in animal models. As information concerning growth requirements of this species was elucidated in numerous and extensive experiments, media were developed for *L. pneumophila* that improved isolation success significantly, with relatively good growth of the species as well. The studies of Hussong et al. (20), and Paszko-Kolva et al. (34) provided good evidence that *L. pneumophila* forms somnicells, that is, cells not growing on routine culture media but remaining viable, even though not culturable.

In the open ocean, this phenomenon appears to be even more widely extant. Novel cold water Archaea were detected in seawater samples collected from depths of 100 to 500 m (9, 12). It is now recognized that these Archaea represent a significant proportion of the total bacterial rRNA in the oceans throughout the world, yet these bacterial species have not been isolated in pure culture, even though easily detected by PCR amplification, characterized, and classified solely on the basis of their rRNA. These Archaea are providing a new understanding of oceanic processes, without having been cultivated. Many other examples of "yet-to-be-cultured" bacteria and Archaea present in the biosphere are known (4–6), frequently from extreme environments, and still await developments in microbial culture technology so that they too can be isolated and phenetically described (45).

Tiedje (42) and Allsopp et al. (1) suggest that, based on a wide variety of evidence, ca. 300,000 to 1 million species of bacteria inhabit the earth, yet *Bergey's Manual* (18) lists only 3,100 species described to date (4, 6, 16). Soil microbial populations alone may comprise 10,000 species per g of soil, based on the heterogeneity of DNA extracted from soil samples (43). Even the most conservative of estimates indicate that millions of bacteria and Archaea have yet to be cultured and inventoried (42). Many of these species may never be isolated in monoculture, i.e., "pure" culture (see chapter 9, this volume).

A varied and rich terminology has been developed to describe physiological and morphological aspects of bacteria that are less active metabolically or have ceased growth, yet maintain detectable metabolic activity. Such terminology includes "dormant," "quiescent," "ultramicrobacterial," "minicell," "cryptically growing," and "slow growth." Dormancy, alone, has been described in a variety of terms, but generally "resting" and "quiescence" are the most frequently applied (17). Most of these terms and their fundamental conceptual basis were discussed by Roszak and Colwell (38), and a more recent listing of terms is provided by Barer and Harwood (2).

"Nonculturable" has sometimes been used to describe bacteria which are injured and therefore no longer capable of initiating growth on otherwise suitable media. This is not a precise application of the term, and the relationship between sublethal injury and dormant or somnicells is discussed by McFeters and LeChevallier (chapter 15).

The microbiological literature abounds with accounts of the appearance of bacterial species in habitats not considered to be a reservoir for those species. Since many bacterial species can make a rapid transition from the nonculturable state to an actively growing and dividing state, the genetic resources of given environments may be far richer than has been suspected. Nonculturable cells, in general, become so small in size they will escape detection using traditional methods (microscopy after collection by filtration), and when conditions return to favoring growth, the appearance of these species in culture may be unexpected. Bacteria in the aquatic environment have been shown to switch between starvation (31) and active growth in the presence of nonlimiting concentrations of nutrients, e.g., the r-strategists, as defined by Hirsch et al. (17), or the copiotrophs of Poindexter (35). As improved understanding of genetic and environmental mechanisms controlling the reversible transitions between active growth and dormancy, or nonculturability, is gained, biogeochemical cycles, microbial ecology, and plant and animal diseases will be better understood, making it possible to develop predictive models. Furthermore, the role of VBNC bacteria in chronic disease is beginning to be recognized (7, 10), as well as their being recognized as a potential hazard for the food industry (19).

The chapters that follow provide both reviews of the literature on the VBNC state in bacteria and information concerning their cellular morphology, physiology, genetics, and ecology. From the discussions provided in each of these chapters, it will be obvious that there is much yet to be learned about bacteria in their natural habitat and how microorganisms have evolved strategies for survival, persistence, and distribution in their environment, either directly in the environment or associated with a plant or animal host. Many aspects, including the epidemiology of opportunistic human-pathogenic VBNC bacteria, remain largely unexplored. Thus, there is much work yet to be done.

REFERENCES

1. **Allsopp, D., R. R. Colwell, and D. L. Hawksworth.** 1995. *Microbial Diversity and Ecosystem Function.* CAB International, Wallingford, U.K.
2. **Barer, M. R., and C. R. Harwood.** 1999. Bacterial viability and culturability. *Adv. Microb. Physiol.* **41:**93–137.
3. **Berlin, D. L., D. S. Herson, D. T. Hicks, and D. G. Hoover.** 1999. Response of pathogenic *Vibrio* species to high hydrostatic pressure. *Appl. Environ. Microbiol.* **65:**2776–2780.
4. **Bull, A. T.** 1991. Biotechnology and biodiversity, p. 203–219. *In* D. L. Hawksworth (ed.), *The Biodiversity of Microorganisms and Invertebrates: Its Role in Sustainable Agriculture.* CAB International, Wallingford, U.K.
5. **Chang, H. R., L. H. Loo, K. Jeyaseelan, L. Earnest, and E. Stackebrandt.** 1997. Phylogenetic relationships of *Salmonella typhi* and *Salmonella typhimurium* based on 16S rRNA sequence analysis. *Int. J. Syst. Bacteriol.* **47:**1253–1254.

6. **Colwell, R. R.** 1996. Microbial biodiversity—global aspects, p. 1–11. *In* R. R. Colwell, U. Simidu, and K. Ohwada (ed.), *Microbial Diversity in Time and Space*. Plenum Press, New York, N.Y.
7. **Costerton, J. W., P. S. Stewart, and E. P. Greenberg.** 1999. Bacterial biofilms: a common cause of persistent infections. *Science* **284:**1318–1322.
8. **Davies, C. M., S. C. Apte, and S. M. Peterson.** 1995. β-D-Galactosidase activity of viable, non-culturable coliform bacteria in marine waters. *Lett. Appl. Microbiol.* **21:**99–102.
9. **DeLong, E. F.** 1992. Archaea in coastal marine environments. *Proc. Natl. Acad. Sci. USA* **89:**5685–5689.
10. **Domingue, G. J., G. M. Ghoniem, K. L. Bost, C. Fermin, and L. G. Human.** 1995. Dormant microbes in interstitial cystitis. *J. Urol.* **153:**1921–1926.
11. **Dupray, E., M. Pommepuy, A. Derrien, M. P. Caprais, and M. Cormier.** 1993. Use of the direct viable count (D.V.C.) for the assessment of survival of *E. coli* in marine environments. *Water Sci. Tech.* **27:**395–399.
12. **Fuhrman, J. A., K. McCallum, and A. A. Davis.** 1992. Novel major archaebacterial group from marine plankton. *Nature* (London) **356:**148–149.
13. **Grimes, D. J.** 1995. Culture collections and nonculturable cells. *USFCC Newsl.* **25:**1–3.
14. **Grimes, D. J., R. W. Atwell, P. R. Brayton, L. M. Palmer, D. M. Rollins, D. B. Roszak, F. L. Singleton, M. L. Tamplin, and R. R. Colwell.** 1986. Fate of enteric pathogenic bacteria in estuarine and marine environments. *Microbiol. Sci.* **3:**324–329.
15. **Hänninen, M.-L., M. Hakkinen, and H. Rautelin.** 1999. Stability of related human and chicken *Campylobacter jejuni* genotypes after passage through chick intestine studied by pulsed-field gel electrophoresis. *Appl. Environ. Microbiol.* **65:**2272–2275.
16. **Hawksworth, D. L., and L. A. Mound.** 1991. Biodiversity databases: the crucial significance of collections, p. 17–29. *In* D. L. Hawksworth (ed.), *The Biodiversity of Microorganisms and Invertebrates: Its Role in Sustainable Agriculture.* CAB International, Wallingford, U.K.
17. **Hirsch, P., M. Bernhard, S. S. Cohen, J. C. Ensign, H. W. Jannasch, A. L. Koch, K. C. Marshall, A. Marin, J. S. Poindexter, S. C. Rittenberg, D. C. Smith, and H. Veldkamp.** 1979. Life under conditions of low nutrient concentrations group report, p. 357–372. *In* M. Shilo (ed.), *Strategies of Microbial Life in Extreme Environments.* Dalem Konferenzen Life Sciences Research Report 13. Verlag Chemie, Weinheim, West Germany.
18. **Holt, J. G. (ed.).** 1984–1989. *Bergey's Manual of Systematic Bacteriology.* The Williams & Wilkins Co., Baltimore, Md.
19. **Huq, A., and R. R. Colwell.** 1996. A microbiological paradox: viable but nonculturable bacteria with special reference to *V. cholerae*. *J. Food Prot.* **59:**96–101.
20. **Hussong, D., R. R. Colwell, M. O'Brien, E. Weiss, A. D. Pearson, R. M. Weiner, and W. D. Burge.** 1987. Viable *Legionella pneumophila* not detectable by culture on agar media. *Bio/Technology* **5:**947–952.
21. **Kaprelyants, A. S., J. C. Gottschal, and D. B. Kell.** 1993. Dormancy in non-sporulating bacteria. *FEMS Microbiol. Rev.* **104:**271–286.
22. **Kaprelyants, A. S., G. V. Mukamolova, and D. B. Kell.** 1994. Estimation of dormant *Micrococcus luteus* cells by penicillin lysis and by resuscitation in cell-free spent culture medium at high dilution. *FEMS Microbiol. Lett.* **115:**347–352.
23. **Keilin, D.** 1959. The Leeuwenhoek Lecture. The problem of anabiosis or latent life; history and current concept. *Proc. R. Soc. London Ser. B* **150:**149–191.
24. **Kell, D. B., H. M. Davey, G. V. Mukamolova, T. V. Votyakova, and A. S. Kaprelyants.** 1995. A summary of recent work on dormancy in nonsporulating bacteria: its significance for marine microbiology and biotechnology. *J. Mar. Biotechnol.* **3:**24–25.
25. **Kogure, K., U. Simidu, and N. Taga.** 1979. A tentative direct microscopic method for counting living marine bacteria. *Can. J. Microbiol.* **25:**415–420.
26. **Kolter, R., D. A. Siegele, and A. Tormo.** 1993. The stationary phase of the bacterial life cycle. *Annu. Rev. Microbiol.* **47:**855–874.
27. **Kondo, K., A. Takade, and K. Amako.** 1994. Morphology of the viable but nonculturable *Vibrio cholerae* as determined by the freeze fixation technique. *FEMS Microbiol. Lett.* **123:**179–184.

28. **McDade, J. E., C. C. Shepard, D. W. Fraser, T. R. Tsai, M. A. Redus, and W. R. Dowdle.** 1977. Legionnaire's disease: isolation of a bacterium and demonstration of its role in other respiratory disease. *N. Engl. J. Med.* **297:**1197–1203.
29. **McDougald, D., S. A. Rice, D. Weichart, and S. Kjelleberg.** 1998. Nonculturability: adaptation or debilitation? *FEMS Microbiol. Ecol.* **25:**1–9.
30. **Morgan, J. A. W., P. A. Cranwell, and R. W. Pickup.** 1991. Survival of *Aeromonas salmonicida* in lake water. *Appl. Environ. Microbiol.* **57:**1777–1782.
31. **Morita, R. Y.** 1999. Feast or famine in the deep sea. *J. Ind. Microbiol. Biotechnol.* **22(4/5):**540–550.
32. **Nilsson, L., J. D. Oliver, and S. Kjelleberg.** 1991. Resuscitation of *Vibrio vulnificus* from the viable but nonculturable state. *J. Bacteriol.* **173:**5054–5059.
33. **Oliver, J. D.** 1993. Formation of viable but nonculturable cells, p. 239–272. *In* S. Kjelleberg (ed.), *Starvation in Bacteria.* Plenum Press, New York, N.Y.
34. **Paszko-Kolva, C., M. Shahamat, and R. R. Colwell.** 1993. Effect of temperature on survival of *Legionella pneumophila* in the aquatic environment. *Microb. Releases* **2:**73–79.
35. **Poindexter, J. S.** 1981. The caulobacters: ubiquitous unusual bacteria. *Microbiol. Rev.* **45:**123–179.
36. **Rahman, I., M. Shahamat, M. A. R. Chowdhury, and R. R. Colwell.** 1996. Potential virulence of viable nonculturable *Shigella dysenteriae* type I. *Appl. Environ. Microbiol.* **62:**115–120.
37. **Ravel, J., R. T. Hill, and R. R. Colwell.** 1994. Isolation of a *Vibrio cholerae* transposon-mutant with an altered viable but nonculturable response. *FEMS Microbiol. Lett.* **120:**57–62.
38. **Roszak, D. B., and R. R. Colwell.** 1987. Survival strategies of bacteria in the natural environment. *Microbiol. Rev.* **51:**365–379.
39. **Roszak, D. B., D. J. Grimes, and R. R. Colwell.** 1984. Viable but non-recoverable stage of *Salmonella enteritidis* in aquatic systems. *Can. J. Microbiol.* **30:**334–338.
40. **Shiba, T., R. T. Hill, W. L. Straube, and R. R. Colwell.** 1995. Decrease in culturability of *V. cholerae* caused by glucose. *Appl. Environ. Microbiol.* **61:**2583–2588.
41. **Tholozan, J. L., J. M. Cappelier, J. P. Tissier, G. Delattre, and M. Federighi.** 1999. Physiological characterization of viable-but-nonculturable *Campylobacter jejuni* cells. *Appl. Environ. Microbiol.* **65:**1110–1116.
42. **Tiedje, J. M.** 1994. Microbial diversity: of value to whom? *ASM News* **60:**524–525.
43. **Torsvik, V., J. Goksoyr, and F. L. Daae.** 1990. High diversity in DNA of soil bacteria. *Appl. Environ. Microbiol.* **56:**782–787.
44. **Vives-Rego, J., R. López-Amorós, and J. Comas.** 1994. Flow cytometric narrow-angle light scatter and cell size during starvation of *Escherichia coli* in artificial seawater. *Lett. Appl. Microbiol.* **19:** 374–376.
45. **Ward, N., F. A. Rainey, B. Goebel, and E. Stackebrandt.** 1995. Identifying and culturing the "unculturables": a challenge for microbiologists, p. 89–112. *In* D. Allsopp, R. R. Colwell, and D. L. Hawksworth (ed.), *Microbial Diversity and Ecosystem Function.* CAB International, Wallingford, U.K.
46. **Welch, T. J., A. Farewell, F. C. Neidhardt, and D. H. Bartlett.** 1993. Stress response of *Escherichia coli* to elevated hydrostatic pressure. *J. Bacteriol.* **175:**7170–7177.
47. **Xu, H.-S., N. Roberts, F. L. Singleton, R. W. Attwell, D. J. Grimes, and R. R. Colwell.** 1982. Survival and viability of nonculturable *Escherichia coli* and *Vibrio cholerae* in the estuarine and marine environment. *Microb. Ecol.* **8:**313–323.
48. **Yamamoto, H., Y. Hashimoto, and T. Ezaki.** 1996. Study of nonculturable *Legionella pneumophila* cells during multiple-nutrient starvation. *FEMS Microbiol. Ecol.* **20:**149–154.

Nonculturable Microorganisms in the Environment
Edited by R. R. Colwell and D. J. Grimes

Chapter 2

Morphological Changes Leading to the Nonculturable State

Jeffrey J. Byrd

Specific morphological changes occur when bacterial cells are introduced into nutrient-depleted environments. In addition, bacteria that are indigenous to these conditions tend to be smaller than bacteria found in nutrient-rich environments. Changes in bacterial morphology during starvation-survival have been previously reviewed (45, 61). The conditions under which bacteria alter their morphology are discussed, with reasons for the changes suggested.

SOILS

It was proposed many years ago that indigenous soil bacteria maintain themselves in a coccoid state (26). Bacteria observed by brightfield microscopy were 0.5 to 0.8 μm in diameter (22, 23). Subsequent examination by transmission electron microscopy revealed that 72% of the soil organisms were less than 0.3 μm in diameter (11). Because small coccoid soil bacteria are difficult to stain with a fluorochrome (73) and the resolution of the brightfield microscope was not maximum, the discrepancy between brightfield and electron microscopy measurements was understandable.

Bacterial cells isolated from soil by density gradient centrifugation and fractionated by filtration were examined for their ability to form colonies (14). Most of the cells able to produce colonies were found to be >0.4 μm prior to growth. Bakken and Olsen (14) also found that cell size for cells <0.4 μm did not increase during in vitro growth. This was not true for *Agromyces ramosus*, which was isolated as a coccobacillus and formed a mycelium when grown in culture, reverting to small rods upon starvation (24). *Agromyces* is notoriously difficult to isolate, requiring extraordinary isolation techniques (implemented over an isolation period of 3 to 4 weeks). *A. ramosus* is but one example of soil microorganisms difficult to isolate. There are many small autochthonous soil bacteria yet to be isolated, because of unknown characteristics of these bacteria.

Jeffrey J. Byrd • Department of Biology, St. Mary's College of Maryland, St. Mary's City, MD 20686.

Analysis of the properties of soil bacteria cultured under starvation conditions has yielded varying results. As mentioned above, *Agromyces* changes from mycelium in culture to a coccoid-like rod when nutrient stressed (24). A size change during culture has also been reported for *Rhizobium leguminosarum* (74, 87) and many deep subsurface endolithic isolates (5). Yet, when rod and coccoid forms of *Arthrobacter* were added to soil (16) or phosphate-buffered saline (18), no change in shape was observed for either form, and both were equally resistant to dessication or starvation. Curiously, the coccoid form of *Arthrobacter* is the predominant morphology manifested in soil. Little change in morphology was observed during starvation in phosphate-buffered saline for *Arthrobacter globiformis*, *Arthrobacter nicotianae*, *Brevibacterium linens*, *Corynebacterium fascians*, *Mycobacterium rhodochorus*, and *Nocardia rosium* (17).

The small cell size of soil organisms may be a function of both low nutrient concentration and the surface structure of soil. Since only the coccoid forms of *Arthrobacter* are detected in soil, there must be conditions or initiators that inhibit or block induction of rod formation. This is especially true if both forms are resistant to desiccation. In addition, *Agromyces ramosus* becomes smaller in cell size when grown on the surface of agar than in broth (24). Thus, surface interaction, along with a reduction in nutrient concentration, may stimulate the smaller size of many microorganisms in a soil environment.

INTERFACES

Interaction of aquatic bacteria with interfaces (air-water or solid-water) stimulates some bacterial species to undergo a decrease in cell size similar to that of soil bacteria. *Vibrio* sp. strain DW1 and *Pseudomonas* sp. strain S9 were shown to "dwarf" upon interaction with an air-water interface (40, 46, 47). When these dwarfed organisms were then placed into a rich nutrient medium, they were observed to increase in size by a factor of 12 prior to exponential growth (46). The induction of smaller sized cells at solid-water interfaces was shown by Humphrey et al. (39) to be associated with marine bacteria possessing a hydrophilic cell surface. Cells with a hydrophobic cell surface did not change morphology at solid-water interfaces. Furthermore, dwarfing was reversibly inhibited by low temperature and low pH.

MARINE SYSTEMS

As in soil, very small bacteria (<0.3 μm) are found in seawater and can be detected when marine samples are examined by direct microscopy (42, 43), epifluorescence microscopy (28, 48, 66, 91, 95, 96), and scanning electron microscopy (86). These findings have led many investigators to study morphological changes of marine bacteria subjected to starvation. While most of the studies reported in the literature deal with multiple-nutrient starvation (e.g., inoculation into artificial seawater) Holmquist and Kjelleberg (37) examined the marine *Vibrio* strain S14 starved for multiple nutrients, as well as carbon, nitrogen, and phosphorus starvation, individually. They found that, even though carbon and multiple-nutrient starvation resulted in small cells, nitrogen and phosphorus starvation yielded long filaments or swollen, large rods, respectively. Therefore, starvation for specific elements may result in very different morphologies, with a carbon- or a multiple-nutrient starvation regime most closely resembling the overall process in the environment.

Multiple-nutrient starvation has been shown to decrease the cell volume of *Vibrio cholerae* and *Vibrio* sp. strain S14 by as much as 85% (12) and 70% (59), respectively. The size reduction for *Vibrio* sp. strain S14 was complete within 24 h, thereby indicating that this process can take place relatively rapidly after starvation begins. This response to starvation is characteristic of many marine isolates, as shown in Table 1.

Thus, bacteria in a nutrient-depleted aquatic environment, in general, occur much smaller in cell size than in a nutrient-rich environment. In 1952, Oppenheimer (72) found that, by using a 0.4-μm-pore-size membrane to filter seawater, 12 culturable bacteria per milliliter were able to pass through the filter. This discovery was confirmed by Anderson and Heffernan in 1965 (9). The search for filterable bacteria was launched and even in fresh water, vibrios were isolated from the filterable population (60). A variety of terms, e.g., filterable bacteria (86), ultramicrobacteria (89), and dwarf cells (39), have been used to describe this group of small bacteria.

To identify and characterize the filterable population, water filtered through 0.45-μm filters was then passed through 0.2-μm filters (86). Those bacteria that were able to pass through the 0.2-μm filter (ultramicrobacteria) were a very different population from those retained by the same filter (86). Isolation of the larger bacteria was accomplished on standard (full-strength) media, but this medium was unsuitable for culture of ultramicrobacteria. Dilute media were necessary for isolation of the ultramicrobacteria (57, 58, 89). Even with specialized treatment, the growth rates were slow, and many generations were required for ultramicrobacteria to increase in size (58, 89). The ultramicrobacterial population included the genera *Vibrio*, *Aeromonas*, *Pseudomonas*, and *Alcaligenes* (58).

Table 1. Selected bacterial species observed to decrease in size during starvation

Organism	Reference(s)
Alteromonas denitrificans	65
Barophile CNPT-31	77
Escherichia coli	35
Flavobacterium	44
Pseudomonas aeruginosa	13
Pseudomonas sp. strain S9	44, 59
Spirillum sp. strain 0114	44
Vibrio anguillarum	64
Vibrio cholerae	12
Vibrio fluvialis	85
Vibrio marinus	32
Vibrio sp. strain Ant-300	7, 62, 63, 66, 68
Vibrio sp. strain DW1	44
Vibrio sp. strain S14	37, 44, 59, 69
Vibrio vulnificus	71

With the finding that bacteria in the marine environment occur predominantly in an ultramicrobacterial state, it can be assumed that marine bacteria brought into laboratory culture should be able to reenter the ultramicrobacterial state. Studies examining starvation survival of marine *Vibrio* sp. strain Ant-300 (6, 63, 66) have shown that Ant-300 can achieve a cell volume of 0.05 μm^3 (62), with 50% of the cells able to pass through a 0.4 μm-membrane within 3 weeks of starvation (66). By 6 weeks, the bacteria became ultramicrocells (6), illustrating that bacterial species comprising the natural flora of seawater retain the ability to enter the ultramicrobacterial state.

MISCELLANEOUS ENVIRONMENTS

Some bacterial species that do not fit into the above categories also show morphological change when starved. One example of a group of bacteria that changes morphology during exposure to limiting environmental conditions is the genus *Campylobacter*. *Campylobacter* spp. occur as rods, spirals, and S-shape during active growth. As active growth of culture decreases or if conditions of the culture change, the cell morphology changes to a coccoid shape (76, 84). *Campylobacter* spp. starved in either natural stream water (80) or physiological saline solution (15) undergo conversion to the coccoid state.

The enteric bacterium *Escherichia coli* reacts to starvation in a manner similar to that of marine bacteria. Grossman et al. (35) showed that *E. coli* reduces its mean cell diameter from 2.72 to 2.14 μm after 60 min under conditions of amino acid starvation. Therefore, with the change in size and morphology of *E. coli* and *Campylobacter* being comparable to that of bacteria whose normal habitat is the natural environment, enteric bacteria may also employ this as a survival strategy upon introduction into nutrient-deplete environments.

HOW SIZE REDUCTION OCCURS

Two mechanisms can be used to explain how bacteria reduce their size during starvation. One is that bacteria dividing in a limiting environment do not increase in size prior to cell division. For this to occur, the bacteria must replicate their genome prior to exposure to starvation conditions. The marine bacterium *Vibrio* sp. strain Ant-300 and the fish pathogen *Yersinia ruckeri* were shown to contain between two and six genomes per cell (67, 88), providing enough genomic material to divide under stressful conditions. Reduction division is observed in the laboratory as an increase in culturable cell number during starvation. After reductive division, the cells continue to decrease in size (30, 46). Bacterial species known to undergo reduction division upon starvation are listed in Table 2.

Another mechanism of reducing cell size during exposure to starvation conditions is gradual size reduction without cell division, observed by Oliver et al. (71) for *Vibrio vulnificus* under certain conditions. They found that when *V. vulnificus* was incubated in artificial seawater at 5°C, the cells began to decrease in size and change cell morphology from rods to cocci without increase in cell number. Along with reduction in size, the cells became undetectable, i.e., nonculturable. This was

Table 2. Bacterial species demonstrating reductive division upon starvation

Organism	Reference
Acinetobacter sp.	41
Pseudomonas aeruginosa	20
Pseudomonas sp.	44, 47, 49
Rhizobium leguminosarum	44
Spirillum sp. strain 0114	44
Vibrio cholerae	12
Vibrio sp. strain Ant-300	6, 8, 30, 66–68
Vibrio sp. strain DW1	46, 47, 80
Vibrio sp. strain S14	44, 69
Vibrio vulnificus	71

not the case when the cells were incubated at room temperature. Instead, the cells underwent reduction division and remained culturable. Under such conditions, direct reduction of size was reported for *Enterobacter agglomerans* and a *Pseudomonas* sp. which produced small vesicles released from the outer membrane (59). It was proposed that release of membrane vesicles was a method for allowing reduction in size.

As with temperature, nutrient conditions play a role in reductive division. Eberl et al. (29) found that *Pseudomonas putida* KT2442 did not go through reductive division when exposed to phosphate starvation but demonstrated reductive division under carbon starvation. Change in Mn(II) concentration was shown by Lin et al. (56) to play a role in reductive division in *Deinococcus radiodurans*. If small amounts of Mn(II) were present upon entry into stationary phase, *Deinococcus radiodurans* proceeded to undergo three rounds of reductive division. Therefore, the presence or absence of reduction depends upon the nutrient conditions the bacterium sees during entry into stationary phase.

MEMBRANE AND EXTRACELLULAR CHANGES

As bacteria respond to starvation conditions, certain cell membrane changes occur. The most notable is a shrinkage of the cytoplasmic membrane, as has been reported for *Escherichia coli* (75), *Vibrio* sp. strain Ant-300 (66), *Vibrio* sp. strain S14 (59), and dwarfed soil bacteria (11). However, membrane shrinkage was not observed for *Pseudomonas* sp. strain S9 and *Enterobacter agglomerans* DW101 which, when examined after exposure to starvation conditions for 24 h, showed no change in the cytoplasmic membrane (59). Since the time period of starvation was only 24 h and the shrinkage reported for *E. coli* and *Vibrio* strain Ant-300 took place after 1 week, the length of time for starvation may have been insufficient to detect the change in the latter case.

Along with changes in the cytoplasmic membrane, there are changes in proteins found in the periplasmic space. Holmquist and Kjelleberg (38) found three dominant carbon-starvation-induced periplasmic proteins produced by *Vibrio* sp. strain S14 which were not produced during nitrogen or carbon starvation. However, starvation-specific proteins in the periplasmic space were found in *Vibrio* sp.

strain DW1 (1). Therefore, production of periplasmic proteins may be species specific.

Additionally, production of extracellular polysaccharides is species variable during starvation. For example, it has been reported that there is a drastic reduction in production of exopolysaccharide for *Vibrio* sp. strain Ant-300 (68) and *Klebsiella pneumoniae* (53, 54). In contrast, Wrangstadh et al. (93) found that during periods of energy and nutrient starvation, the marine isolate *Pseudomonas* sp. strain S9 produced an exopolysaccharide. The polysaccharide was found to influence detachment of the bacterium from surfaces and possibly play a role in releasing the bacterium during starvation conditions to scavenge for nutrients. This mechanism was also proposed as a possible starvation-survival strategy for benthic marine cyanobacteria (31) and the oil-degrading *Acinetobacter calcoaceticus* (82). In conclusion, membrane and exterior cell structure alterations contribute to the morphological changes observed in bacteria exposed to limiting environmental conditions.

GENETICS OF ROUND CELL FORMATION

The rounded morphology observed in stationary phase has been attributed to stationary-phase induction of the *bolA* gene in *E. coli*. The *bol*A codes for a small (13.5K) protein that regulates the production of the penicillin-binding protein PBP6 (2), a carboxypeptidase involved in septum formation. Regulation of PBP6 by *bolA* has been proposed to be at the level of transcription (2), with a fourfold increase in PBP6 seen during stationary phase (19). For the morphogene *bolA* to function, an active *ftsZ* gene product must be present in the cell (3). The *ftsZ* gene product is an important protein in the early stages of cell division. Consequently, overexpression of *ftsZ* will cause the production of minicells in *E. coli* (90).

The stationary-phase specificity of *bolA* in *bolA* mutants was determined. The *bolA* mutants are able to grow and divide normally during exponential growth; therefore, it would appear that *bolA* is not essential for cell division under elevated nutrient concentrations (2). However, overexpression of *bolA* will result in the formation of round cells in exponentially growing cells, even in the absence of PBP5 and PBP6. To demonstrate a direct link to stationary-phase production of *bolA*, its regulation was studied, and it was found that *bolA* is regulated by a novel sigma factor (σ^S), required for transcription of many stationary-phase genes (51, 52, 94). In *rpoS* mutants the cells were rod shaped during exponential phase but became filamentous to short rod shaped during stationary phase. In contrast, the wild-type cells were coccobacillary during stationary phase. Because of the role that *rpoS* plays in reductive division, regulators of *rpoS*, such as *dnaK* (78, 79), will affect the occurrence of reductive division. Osmotic stress will similarly stimulate certain *rpoS*-dependent genes (36), indicating that morphology clearly is influenced by environmental conditions.

Reductive division is also regulated by genes responsible for activating damage control defense mechanisms. One gene that fits into this category and is required for reductive division to occur is ArcC (70). It was proposed that ArcA functions in the defense against oxygen radicals produced during stationary phase (70) and that if it is not present, the cells do not go through reductive division.

SMALL CELLS AND PREDATION

The increased numbers of smaller sized bacteria found to occur in response to changes in environmental conditions may be determined by ecological factors other than simply decreased nutrient availability. Wilkner et al. (92) examined predation of *E. coli* minicells by protozoa and found that ca. 27 to 100% of the total population was ingested. Since then, many studies have established size preference of bacteria by grazing protozoans. Flagellates have been shown to preferentially ingest larger bacteria in both fresh (25, 83) and estuarine (33) water. Such predation resulted in a decrease overall in the average bacterioplankton cell size (10, 50). However, if predation pressure is removed from the actively growing bacterioplankton populations, the average bacterial cell size increased by a factor of at least 2 (4, 55). Although there was size preference demonstrated for flagellates also, the results were not as dramatic as for the ciliates, which had only a slight preference for larger bacteria (33). In addition, ciliates did show a preference for specific genera of bacteria.

As mentioned above, rod- and coccoid-shaped forms of *Arthrobacter* are equally resistant to desiccation in soil (16), but it is coccoid-shaped bacteria that mainly occur in soil. It is possible that rod morphology may be more attractive to bacterial predators, such as *Agromyces* (21). Byrd et al. (21) found that *Agromyces ramosus* was present in soil in the coccoid-rod form, until stimulated by the presence of nutrient, prey species, or a bacterial predator. Upon stimulation, *A. ramosus* produced mycelia in the direction of the stimulator. In the case of stimulation towards a predator, *Agromyces* was subsequently killed by the predator; this was not observed for nonstimulated coccoid cells.

PRACTICAL APPLICATIONS OF SMALL, STARVED BACTERIA

A practical application of small, starved bacteria is in petroleum microbiology. It has been shown that *Klebsiella pneumoniae* (53, 54) and *Pseudomonas* sp. strain FC3 (27), originally isolated from oil well water, decreased in size under starvation conditions in phosphate-buffered saline. Along with the decrease in size, the cells also reduced production of a glycocalyx. It has been proposed that these starved bacteria can be injected into rock and sandpack pores. Because of their size, the cells will penetrate deeper than actively growing bacteria. Upon addition of nutrient, the cells resuscitate and grow to a larger size, thereby plugging holes in the rock. Thus, the process allows enhanced recovery of oil trapped in oil-bearing substrata by sealing those areas already depleted of oil.

CONCLUSIONS

The change in size and shape of various bacteria subjected to starvation conditions, along with the finding that bacteria in environmental samples, in general, are much reduced in size, leads to the hypothesis that this smaller size is the natural state of bacteria. *E. coli* cells subjected to amino acid starvation decrease cell volume by 45%, in comparison to cells growing in a nutrient-rich medium (34).

When nutrients are returned to these small, starved cells, they are able to divide, even when they are 30% smaller than the original cells (34, 81). Thus, cells exposed to limiting conditions for growth in the natural environment may replicate at smaller cell volumes then predicted from studies conducted using pure cultures grown under nutrient-rich conditions. One theory proposed to explain this reduced cell size is that the reduction increases the surface-to-volume ratio, thereby aiding cells in sequestering available nutrients in low-nutrient environments (6). In addition to a lower surface-to-volume ratio, the smaller size may enhance survival by protecting against predation and may function as a response mechanism upon interaction with surfaces and at interfaces. Most likely, all of these factors play a role in the morphological changes occurring during growth-limiting conditions and comprise the totality of the species survival strategy. When in the small morphology state, the cells are more difficult to isolate into laboratory culture and appear therefore to have adapted to their environmental surroundings.

Acknowledgments. The preparation of this chapter was supported in part by a Faculty Development Grant from St. Mary's College of Maryland. I thank Charles Kaspar, Lisa Steele, and Valarie Miller for helpful review of the manuscript.

REFERENCES

1. **Albertson, N. H., G. W. Jones, and S. Kjelleberg.** 1987. The detection of starvation-specific antigens of two marine bacteria. *J. Gen. Microbiol.* **133:**2225–2232.
2. **Aldea, M., T. Garrido, C. Hernandez-Chico, M. Vicente, and S. R. Kushner.** 1989. Induction of growth-phase-dependent promoter triggers transcription of *bolA*, an *Escherichia coli* morphogene. *EMBO J.* **8:**3923–3931.
3. **Aldea, M., C. Hernandez-Chico, A. G. de la Campa, S. R. Kushner, and M. Vicente.** 1988. Identification, cloning, and expression of *bolA*, an *ftsZ*-dependent morphogene of *Escherichia coli. J. Bacteriol.* **170:**5169–5176.
4. **Ammerman, J. W., J. A. Fuhrman, A. Hagstrom, and F. Azam.** 1984. Bacterioplankton growth in seawater. I. Growth kinetics and cellular characteristics in seawater cultures. *Mar. Ecol. Prog. Ser.* **18:**31–39.
5. **Amy, P. S., C. Durham, D. Hall, and L. Haldeman.** 1993. Starvation-survival of deep subsurface isolates. *Curr. Microbiol.* **26:**345–352.
6. **Amy, P. S., and R. Y. Morita.** 1983. Starvation-survival patterns of sixteen freshly isolated open-ocean bacteria. *Appl. Environ. Microbiol.* **45:**1109–1115.
7. **Amy, P. S., C. Pauling, and R. Y. Morita.** 1983. Recovery from nutrient starvation by a marine *Vibrio* sp. *Appl. Environ. Microbiol.* **45:**1685–1690.
8. **Amy, P. S., C. Pauling, and R. Y. Morita.** 1983. Starvation-survival processes of a marine vibrio. *Appl. Environ. Microbiol.* **45:**1041–1048.
9. **Anderson, J. I. W., and W. P. Heffernan.** 1965. Isolation and characterization of filterable marine bacteria. *J. Bacteriol.* **90:**1713–1718.
10. **Andersson, A., U. Larsson, and A. Hagstrom.** 1986. Size-selective grazing by a microflagellate on pelagic bacteria. *Mar. Ecol. Prog. Ser.* **33:**51–57.
11. **Bae, H. C., E. H. Cota-Robles, and L. E. Casida, Jr.** 1972. Microflora of soil as viewed by transmission electron microscopy. *Appl. Microbiol.* **23:**637–648.
12. **Baker, R. M., F. L. Singleton, and M. A. Hood.** 1983. Effects of nutrient deprivation of *Vibrio cholerae*. *Appl. Environ. Microbiol.* **46:**930–940.
13. **Bakhrouf, A., M. Jeddi, A. Bouddabous, and M. J. Gauthier.** 1989. Evolution of *Pseudomonas aeruginosa* cells towards a filterable stage in seawater. *FEMS Microbiol. Lett.* **59:**187–190.
14. **Bakken, L. R., and R. A. Olsen.** 1987. The relationship between cell size and viability of soil bacteria. *Microb. Ecol.* **13:**103–114.

15. **Beumer, R. R., J. de Vries, and F. M. Rombouts.** 1992. *Campylobacter jejuni* nonculturable coccoid cells. *Int. J. Food Microbiol.* **15:**153–163.
16. **Boylen, C. W.** 1973. Survival of *Arthrobacter crystallopoietes* during prolonged periods of extreme dessication. *J. Bacteriol.* **113:**33–37.
17. **Boylen, C. W., and M. H. Mulks.** 1978. The survival of coryneform bacteria during periods of prolonged nutrient starvation. *J. Gen. Microbiol.* **105:**323–334.
18. **Boylen, C. W., and J. L. Pate.** 1973. Fine structure of *Arthrobacter crystallopoietes* during long-term starvation of rod and spherical stage cells. *Can. J. Microbiol.* **19:**1–5.
19. **Buchanan, C. E., and M. O. Sowell.** 1982. Synthesis of penicillin-binding protein 6 by stationary-phase *Escherichia coli. J. Bacteriol.* **151:**491–494.
20. **Byrd, J. J., H.-S. Xu, and R. R. Colwell.** 1993. Viable but nonculturable bacteria in drinking water. *Appl. Environ. Microbiol.* **57:**875–878.
21. **Byrd, J. J., L. R. Zeph, and L. E. Casida, Jr.** 1985. Bacterial control of *Agromyces ramosus* in soil. *Can. J. Microbiol.* **31:**1157–1163.
22. **Casida, L. E., Jr.** 1965. Abundant microorganisms in soil. *Appl. Microbiol.* **13:**327–334.
23. **Casida, L. E., Jr.** 1971. Microorganisms in unamended soil as observed by various forms of microscopy and staining. *Appl. Microbiol.* **21:**1040–1045.
24. **Casida, L. E., Jr.** 1977. Small cells in pure cultures of *Agromyces ramosus* and in natural soil. *Can. J. Microbiol.* **23:**214–216.
25. **Chrzanowski, T. H., and K. Simek.** 1990. Prey-size selection by freshwater flagellated protozoa. *Limnol. Oceanogr.* **35:**1429–1436.
26. **Conn, H. J.** 1948. The most abundant groups of bacteria in soil. *Bacteriol. Rev.* **12:**257–273.
27. **Cusack, F., S. Singh, C. McCarthy, J. Grieco, M. de Rocco, D. Nguyen, H. Lappin-Scott, and J. W. Costerton.** 1992. Enhanced oil recovery—three-dimensional sandpack simulation of ultramicrobacteria resuscitation in reservoir formation. *J. Gen. Microbiol.* **138:**647–655.
28. **Daley, R. J., and J. E. Hobbie.** 1975. Direct count of aquatic bacteria by a modified epifluorescent technique. *Limnol. Oceanogr.* **20:**875–881.
29. **Eberl, L., M. Givskov, C. Sternberg, S. Moller, G. Christiansen, and S. Molin.** 1996. Physiological responses of *Pseudomonas putida* KT2442 to phosphate starvation. *Microbiology* **142:**155–163.
30. **Faquin, W. C., and J. D. Oliver.** 1984. Arginine uptake by a psychrophilic marine *Vibrio* sp. during starvation-induced morphogenesis. *J. Gen. Microbiol.* **130:**1331–1335.
31. **Fattom, A., and M. Shilo.** 1985. Production of emulcyan by *Phormidium* J-1: its activity and function. *FEMS Microbiol. Ecol.* **31:**3–9.
32. **Felter, R. A., R. R. Colwell, and G. B. Chapman.** 1969. Morphology and round body formation in *Vibrio marinus. J. Bacteriol.* **99:**326–335.
33. **Gonzalez, J. M., E. B. Sheer, and B. F. Sheer.** 1990. Size-selective grazing on bacteria by natural assemblages of estuarine flagellates and ciliates. *Appl. Environ. Microbiol.* **56:**583–589.
34. **Grossman, N., and E. Z. Ron.** 1989. Apparent minimal size required for cell division in *Escherichia coli. J. Bacteriol.* **171:**80–82.
35. **Grossman, N., E. Z. Ron, and C. L. Woldringh.** 1982. Changes in cell dimension during amino acid starvation of *Escherichia coli. J. Bacteriol.* **152:**35–41.
36. **Hengge-Aronis, R., R. Lange, N. Henneberg, and D. Fischer.** 1993. Osmotic regulation of *rpoS*-dependent genes in *Escherichia coli. J. Bacteriol.* **175:**259–265.
37. **Holmquist, L., and S. Kjelleberg.** 1993. Changes in viability, respiratory activity and morphology of the marine *Vibrio* sp. strain S14 during starvation of individual nutrients and subsequent recovery. *FEMS Microbiol. Ecol.* **12:**215–224.
38. **Holmquist, L., and S. Kjelleberg.** 1993. The carbon starvation stimulon in the marine *Vibrio* sp. S14 (CCUG15956) includes three periplasmic space protein responders. *J. Gen. Microbiol.* **139:**209–215.
39. **Humphrey, B., S. Kjelleberg, and K. C. Marshall.** 1983. Responses of marine bacteria under starvation conditions at a solid-water interface. *Appl. Environ. Microbiol.* **45:**43–47.
40. **Humphrey, B. A., and K. C. Marshall.** 1984. The triggering effect of surfaces and surfactants on heat output, oxygen consumption and size reduction of a starving marine *Vibrio. Arch. Microbiol.* **140:**166–170.

41. **James, G. A., D. R. Korber, D. E. Caldwell, and J. W. Costerton.** 1995. Digital image analysis of growth and starvation responses of a surface-colonizing *Acinetobacter* sp. *J. Bacteriol.* **177:**907–915.
42. **Jannasch, H. W.** 1955. Zur Okologie ser zymogenen planktischen Bacterienflora naturlicher Gewasser. *Arch. Mikrobiol.* **23:**146–180.
43. **Jannasch, H. W.** 1958. Studies on planktonic bacteria by means of a direct membrane filter method. *J. Gen. Microbiol.* **18:**609–620.
44. **Kjelleberg, S., and M. Hermansson.** 1984. Starvation-induced effects on bacterial surface characteristics. *Appl. Environ. Microbiol.* **48:**497–503.
45. **Kjelleberg, S., M. Hermansson, P. Marden, and G. W. Jones.** 1987. The transient phase between growth and nongrowth of heterotrophic bacteria with emphasis on the marine environment. *Annu. Rev. Microbiol.* **41:**25–49.
46. **Kjelleberg, S., B. A. Humphrey, and K. C. Marshall.** 1982. Effect of interfaces on small, starved marine bacteria. *Appl. Environ. Microbiol.* **43:**1166–1172.
47. **Kjelleberg, S., B. A. Humphrey, and K. C. Marshall.** 1983. Initial phases of starvation and activity of bacteria at surfaces. *Appl. Environ. Microbiol.* **46:**978–984.
48. **Kogure, K., U. Simidu, and N. Taga.** 1979. A tentative direct microscopic method for counting living marine bacteria. *Can. J. Microbiol.* **24:**415–420.
49. **Kurath, G., and R. Y. Morita.** 1983. Starvation-survival physiological studies of a marine *Pseudomonas* sp. *Appl. Environ. Microbiol.* **45:**1206–1211.
50. **Kuuppo-Leinikki, P.** 1990. Protozoan grazing on planktonic bacteria and its impact on bacterial population. *Mar. Ecol. Prog. Ser.* **63:**227–238.
51. **Lange, R., and R. Hengge-Aronis.** 1991. Growth phase-regulated expression of *bol*A and morphology of stationary-phase *Escherichia coli* cells are regulated by the novel sigma factor σ^s. *J. Bacteriol.* **173:**4474–4481.
52. **Lange, R., and R. Hengge-Aronis.** 1991. Identification of a central regulator of stationary-phase gene expression of *Escherichia coli*. Mol. Microbiol. **5:**49–59.
53. **Lappin-Scott, H. M., F. M. Cusack, F. A. Macleod, and J. W. Costerton.** 1988. Nutrient resuscitation and growth of starved cells in sandstone cores—a novel approach to enhance oil recovery. *Appl. Environ. Microbiol.* **54:**1373–1382.
54. **Lappin-Scott, H. M., F. M. Cusack, F. A. Macleod, and J. W. Costerton.** 1988. Starvation and nutrient resuscitation of *Klebsiella pneumoniae* isolated from oil well waters. *J. Appl. Bacteriol.* **64:** 541–550.
55. **Larsson, U., and A. Hagstrom.** 1982. Fractionated phytoplankton primary production, exudate release, and bacterial production in a Baltic eutrophication gradient. *Mar. Biol.* **67:**57–70.
56. **Lin, C. L., C. S. Lin, and S. T. Tan.** 1995. Mutations showing specificity for normal growth or Mn(II)-dependent post-exponential-phase cell division in *Deinococcus radiodurans*. *Microbiology* **141:**1707–1714.
57. **MacDonell, M. T., and M. A. Hood.** 1982. Isolation and characterization of ultramicrobacteria from a Gulf Coast estuary. *Appl. Environ. Microbiol.* **43:**556–571.
58. **MacDonell, M. T., and M. A. Hood.** 1984. Ultramicrovibrios in Gulf Coast estuarine water: isolation, characterization and incidence, p. 551–562. *In* R. R. Colwell (ed.), *Vibrios in the Environment.* John Wiley and Sons, Inc., New York, N.Y.
59. **Marden, P., A. Tunlid, K. Malmcrona-Friberg, G. Odham, and S. Kjelleberg.** 1985. Physiological and morphological changes during short term starvation of marine bacterial isolates. *Arch. Microbiol.* **142:**326–332.
60. **Martin, A., Jr.** 1963. A filterable *Vibrio* from fresh water. *Proc. Pa. Acad. Sci.* **36:**174–178.
61. **Morita, R. Y.** 1985. Starvation and miniaturisation of heterotrophs, with special emphasis on maintenance of the starved viable state, p. 111–130. *In* M. M. Fletcher and G. D. Floodgate (ed.), *Bacteria in Their Natural Environments.* Academic Press, London, United Kingdom.
62. **Moyer, G. L., and R. Y. Morita.** 1989. Effect of growth rate and starvation-survival on cellular DNA, RNA, and protein of a psychrophilic marine bacterium. *Appl. Environ. Microbiol.* **55:**2710–2716.
63. **Moyer, G. L., and R. Y. Morita.** 1989. Effect of growth rate and starvation-survival on the viability and stability of a psychrophilic marine bacterium. *Appl. Environ. Microbiol.* **55:**1122–1127.

64. **Nelson, D. R., Y. Sadlowski, M. Eguchi, and S. Kjelleberg.** 1997. The starvation-stress response of *Vibrio* (*Listonella*) *anguillarum*. *Microbiology* **143:**2305–2312.
65. **Nissen, H.** 1987. Longterm starvation of a marine bacterium, *Alteromonas denitrificans*, isolated from a Norwegian fjord. *FEMS Microbiol. Ecol.* **45:**173–183.
66. **Novitsky, J. A. and R. Y. Morita.** 1976. Morphological characterization of small cells resulting from nutrient starvation of a psychrophilic marine vibrio. *Appl. Environ. Microbiol.* **32:**617–622.
67. **Novitsky, J. A., and R. Y. Morita.** 1977. Survival of a psychrophilic marine vibrio under long-term nutrient starvation. *Appl. Environ. Microbiol.* **33:**635–641.
68. **Novitsky, J. A., and R. Y. Morita.** 1978. Possible strategy for the survival of marine bacteria under starvation conditions. *Marine Biol.* **48:**289–295.
69. **Nystrom, T., and S. Kjelleberg.** 1989. Role of protein synthesis in the cell division and starvation induced resistance to autolysis of a marine *Vibrio* during the initial phase of starvation. *J. Gen. Microbiol.* **135:**1599–1606.
70. **Nystrom, T., C. Larsson, and L. Gustafsson.** 1996. Bacterial defense against aging: role of the *Escherichia coli* ArcA regulator in gene expression, readjusted energy flux and survival during stasis. *EMBO J.* **15:**3219–3228.
71. **Oliver, J. D., L. Nilsson, and S. Kjelleberg.** 1991. Formation of nonculturable *Vibrio vulnificus* cells and its relationship to the starvation state. *Appl. Environ. Microbiol.* **57:**2640–2644.
72. **Oppenheimer, C. H.** 1952. The membrane filter in marine microbiology. *J. Bacteriol.* **64:**783–786.
73. **Postma, J., and H. J. Altemuller.** 1990. Bacteria in thin soil sections stained with fluorescent brightner Calcofluor White M2R. *Soil Biol. Biochem.* **22:**89–96.
74. **Postma, J., J. D. van Elsas, J. M. Govaert, and J. van Veen.** 1988. The dynamics of *Rhizobium leguminosarum* biovar *trifolii* introduced into soil as determined by immunofluorescence and selective plating techniques. *FEMS Microbiol. Ecol.* **53:**251–260.
75. **Reeve, C. A., P. S. Amy, and A. Martin.** 1984. Role of protein synthesis in the survival of carbon-starved *Escherichia coli* K-12. *J. Bacteriol.* **160:**1041–1046.
76. **Rhodes, H. E.** 1954. The illustration of the morphology of *Vibrio fetus* by electron microscopy. *Am. J. Vet. Res.* **15:**630–633.
77. **Rice, S. A., and J. D. Oliver.** 1992. Starvation response of the marine barophile CNPT-3. *Appl. Environ. Microbiol.* **58:**2432–2437.
78. **Rockabrand, D., T. Aurgher, G. Korinek, K. Livers, and P. Blum.** 1995. An essential role for the *Escherichia coli* DnaK protein in starvation-induced thermotolerance, H_2O_2 resistance, and reductive division. *J. Bacteriol.* **177:**3695–3703.
79. **Rockabrand, D., K. Livers, T. Austin, R. Kaiser, D. Jensen, R. Burgess, and P. Blum.** 1998. Roles of DnaK and RpoS in starvation-induced thermotolerance of *Escherichia coli*. *J. Bacteriol.* **180:**846–854.
80. **Rollins, D. M., and R. R. Colwell.** 1986. Viable but nonculturable stage of *Campylobacter jejuni* and its role in survival in the natural aquatic environment. *Appl. Environ. Microbiol.* **52:**531–538.
81. **Ron, E. Z., N. Grossman, and C. E. Helmstetter.** 1977. Control of cell division in *Escherichia coli* effect of amino acid starvation. *J. Bacteriol.* **129:**569–573.
82. **Rosenberg, E., N. Kaplan, O. Pines, M. Rosenberg, and D. Gutnick.** 1983. Capsular polysaccharides interfere with adherence of *Acinetobacter calcoaceticus* to hydrocarbon. *FEMS Microbiol. Lett.* **17:**157–160.
83. **Simek, K., and T. H. Chrzanowski.** 1992. Direct and indirect evidence of size-selective grazing on pelagic bacteria by freshwater nanoflagellates. *Appl. Environ. Microbiol.* **58:**3715–3720.
84. **Smibert, R. M.** 1978. The genus *Campylobacter*. *Annu. Rev. Microbiol.* **32:**673–709.
85. **Smigielski, A. J., B. J. Wallace, and K. C. Marshall.** 1989. Changes in membrane functions during short-term starvation of *Vibrio fluvialis* strain NCTC 11328. *Arch. Microbiol.* **151:**336–347.
86. **Tabor, P. S., K. Ohwada, and R. R. Colwell.** 1981. Filterable marine bacteria found in the deep sea: distribution, taxonomy and response to starvation. *Microb. Ecol.* **7:**67–83.
87. **Thorne, S. H., and H. D. Williams.** 1997. Adaptation to nutrient starvation in *Rhizobium leguminosarum* bv. Phaseoli: Analysis of survival, stress resistance, and changes in macromolecular synthesis during entry to and exit from stationary phase. *J. Bacteriol.* **179:**6894–6901.

88. **Thorsen, B. K., O. Enger, S. Norland, and K. A. Hoff.** 1992. Long-term starvation survival of *Yersinia ruckeri* at different salinities studied by microscopical and flow cytometric methods. *Appl. Environ. Microbiol.* **58:**1624–1628.
89. **Torella, F., and R. Y. Morita.** 1981. Microcultural study of bacterial size changes and microcolony formation by heterotrophic bacteria in seawater. *Appl. Environ. Microbiol.* **41:**518–527.
90. **Ward, J. E., Jr., and J. Lutkenhaus.** 1985. Overproduction of Fts Z induces minicell formation in *E. coli. Cell* **42:**941–949.
91. **Watson, S. W., T. J. Novitsky, H. L. Quinby, and F. W. Valois.** 1977. Determination of bacterial number and biomass in the marine environment. *Appl. Environ. Microbiol.* **33:**940–946.
92. **Wikner, J., A. Andersson, S. Normark, and A. Hagstrom.** 1986. Use of genetically marked minicells as a probe in measurement of predation on bacteria in aquatic environments. *Appl. Environ. Microbiol.* **52:**4–8.
93. **Wrangstadh, M., P. L. Conway, and S. Kjelleberg.** 1988. The role of an extracellular polysaccharide produced by the marine *Pseudomonas* sp. S9 in cellular detachment during starvation. *Can. J. Microbiol.* **35:**309–312.
94. **Zambrano, M. M., D. A. Siegele, M. Almiron, A. Tormo, and R. Kolter.** 1993. Microbial competition *Escherichia coli* mutants that take over stationary phase cultures. *Science* **259:**1757–1760.
95. **Zimmerman, R.** 1977. Estimation of bacterial numbers and biomass by epifluorescence microscopy and scanning electron microscopy, p. 103–120. *In* G. Rheinheimer (ed.), *Microbial Ecology of a Brackish Water Environment.* Springer-Verlag, New York, N.Y.
96. **Zimmerman, R., and L.-A. Meyer-Reil.** 1974. A new method for fluorescence staining of bacterial populations on membrane filter. *Kiel. Meeresforsch.* **30:**24–27.

Noncultivable Microorganisms in the Environment
Edited by R. R. Colwell and D. J. Grimes

Chapter 3

Size Matters: Dwarf Cells in Soil and Subsurface Terrestrial Environments

Thomas L. Kieft

Soils are perhaps the most challenging of natural environments to characterize microbiologically. The difficulties arise from the spatial heterogeneity, temporal variability, and multiphase nature of soil environments, and from the concomitant high diversity of microorganisms, most of which cannot be cultured. While the situation is not unique to soils, the problem is especially acute in soils because the nonculturable (or at least noncultured by current techniques) fraction of the community is generally very large. Comparisons of dilution plate counts of culturable microorganisms with direct microscopic counts of total microorganisms indicate that only 0.001 to 4% of microorganisms in soils can be cultured on organic growth media (32, 50, 142). Information regarding the nature of these uncultivated microbes has come from microscopic studies (e.g., 8), indirect characterization by activity measurements (e.g., 7), and molecular approaches, such as amplification, cloning, and sequencing of genes encoding small subunit RNA genes (e.g., 118).

Bacteria make up the largest proportion of individual microorganisms in soil environments (1, 147) and can comprise the largest component of the biomass, as well (73). Microscopic studies have shown that most of these bacteria are extremely small (8, 9, 15). Bae et al. (8) found as much as 72% of soil bacteria to be small coccoid cells with diameters less than 0.3 μm, and they termed these "dwarf" cells. So, what is the nature of these undersized, Lilliputian cells? Do they represent bacterial populations that are distinct from those that can be readily cultured, or are they merely different morphological forms of commonly cultured, normal-sized bacteria? Starvation of laboratory-cultured bacteria commonly leads to reduction in cell size (33, 138), and so it may be that these small cells are starved, miniaturized versions of culturable bacteria. However, there is evidence that at least some bacterial populations in soils are intrinsically undersized, i.e., they do not grow in size in response to nutrient amendment (11, 12, 84, 85).

Thomas L. Kieft • Department of Biology, New Mexico Institute of Mining and Technology, Socorro, NM 87801.

Very small forms of bacteria are also commonly found in marine and freshwater systems and, in the case of marine systems, have received extensive study. Torrella and Morita (177) coined the term "ultramicrobacteria" for cells less than 0.3 μm and found that some responded to nutrient addition by increasing in cell number and size while others multiplied but retained their small size. The term "dwarf" has been used to describe the cells resulting from starvation-induced miniaturization or "dwarfing" of marine bacteria (45, 104). Miniaturization of cells causes an increase in the surface-to-volume ratio, which enhances the organisms' ability to glean energy-rich nutrients from dilute solution. Depending on the particular strain of bacteria, dwarfing may be preceded by a brief period of fragmentation, the process whereby bacteria undergo cell division without growth, thereby increasing cell numbers and decreasing cell size (104, 139). Fragmentation has also been termed reductive division, and in the case of rapid shift-down from a condition of active growth, it may represent completion of ongoing cell division (140). Most of the bacteria in which fragmentation and dwarfing have been studied could be described as copiotrophic, i.e., capable of rapid growth in the presence of high concentrations of energy-rich organic substrates (102). Each of the terms "dwarf" and "ultramicrobacteria" (and the synonym "ultramicrocells" [132]) has been used to describe very small bacteria in both soil and aquatic environments. Thus, there is no clear distinction between these terms, and they are used interchangeably here; however, "dwarf" may be preferable, since it could be applied to unusually small Archaea, should they exist. The term "filterable bacteria" has been used to refer to very small bacteria, specifically those that can pass through membrane filters (0.2 or 0.45-μm-pore-size), and these are roughly synonymous with ultramicrobacteria and dwarf cells (82, 170, 178). The term "nannobacteria" has also been coined in reference to nanometer-scaled (0.01- to 0.2-μm-diameter) objects that have been observed in geologic materials (51). These, along with the related "nanofossils," have generated considerable controversy (see "What is the Lower Size Limit for Microbial Cells?" below). "Nanobacteria" (87, 88), "nano-organisms," and "nanobes" (181) present more appropriate spellings for claims of submicrometer-sized bacterial forms. Given this superabundance of terms, there may be a need for a single, unifying term for all unusually diminutive bacteria; in keeping with recent trends, I offer "volumetrically challenged microorganisms." The term "mini-cell" is not used here, as it generally designates the undersized, DNA-less products of uneven cell division in a mutant strain of *Escherichia coli* (47).

Recent studies of subsurface terrestrial environments have extended our concept of the biosphere to depths considerably below surface soils. Microbial communities have been discovered in subsurface environments underlying surface soils, and (as in soils) the indigenous microorganisms have been found to include dwarf cells or ultramicrobacteria. Diverse communities of microorganisms have been found in a variety of deep subsurface environments, including shallow aquifers in Oklahoma (17, 20, 153), coastal plain sediments of the southeastern United States (14, 16, 36, 55, 58, 97, 154, 156), various aquifers in Europe (79, 107, 108, 148, 150), rocks and sediments of the thick unsaturated zones in arid and semiarid regions of the western United States (27, 39, 67, 68, 90, 93, 94 and T. L. Kieft, unpublished data), and Russian permafrost soils (185). It has even been proposed that life first arose in the subsurface and that the biosphere extends to great depth across widespread areas of the earth and possibly other planets (64, 174). As in soil communities, the majority of subsurface bacteria cannot be cultured (88), and the in situ cell sizes of most are quite small (20, 162, 183). Anyone who has performed direct microscopic counts of bacteria from subsurface environments knows the experience of struggling to see minuscule coccoid cells that barely exceed the limits of resolution for light microscopy. For recent reviews of subsurface microbiology, see Chapelle (35), Amy and Haldeman (4), and Fredrickson and Fletcher (57).

Since the concentrations of organic substrates are extremely low in subsurface environments (at least those not subjected to groundwater pollution), starvation-induced dwarfing likely contributes to the presence of small cells. However, intrinsically small ultramicrobacteria may also comprise a significant proportion of subsurface communities. The study of dwarf cells in subsurface environments is complicated by the fact that the total numbers of bacteria are frequently orders of magnitude lower than in surface environments and by the fact that culturable cells commonly comprise an even lower proportion of the total numbers of cells than they do in soils (144).

The nature of very small, predominantly noncultured bacteria in soil and subsurface environments has recently been the subject of intense investigation and debate. This chapter reviews the physical and chemical characteristics that influence nutrient availability in soils and subsurface environments, the responses of soil microorganisms to nutrient deprivation, the characteristics of dwarf cells as they relate to nutrient limitation, and the potential uses of ultramicrobacteria. The focus is mainly on pristine soil and subsurface environments, i.e., those not physically disturbed or chemically contaminated, as these undisturbed environments are most likely to contain very small, nutrient-stressed cells. Because the majority of small noncultured cells in soils and subsurface environments are vegetative forms of bacteria, this chapter addresses primarily these forms. Specialized cells adapted for starvation and desiccation survival (i.e., endospores, actinomycete spores, bacterial cysts, etc.) are not covered here. Reviews of nutrient stress, small bacteria, and related topics include those of Dawes (44), Williams (182), Bakken and Olsen (11), Lappin-Scott and Costerton (112, 113), Morita (129–132), England et al. (49), and Koch (105).

SOIL AND SUBSURFACE ENVIRONMENTS: CHEMICAL AND PHYSICAL FACTORS

Nutrient Sources, Concentrations, and Fluxes

A number of features of the soil environment directly influence the growth and survival of microorganisms. Foremost among these is the availability of energy, especially organic substrates for heterotrophic microorganisms. Organic carbon is generally the factor most limiting to microorganisms in soil environments (46, 131). Although organic matter makes up a significant component of most soils, comprising 0.2 to 10% by weight in most mineral soils and as much as 80% in peat soils (172), the majority is relatively refractory to microbial metabolism and is unevenly distributed within the soil. Carbon inputs to soils are primarily plant material deposited as litter on the surface and as root material (exudates and dead roots) below ground. A portion of this carbon (~10%) is in the form of simple sugars, amino acids, amino sugars, and organic acids, all of which are soluble and readily metabolized by microorganisms (147). The remainder occurs as polymers, including cellulose (15 to 60%), hemicellulose (10 to 30%), lignin (5 to 30%), and protein (2 to 15%). These plant polymers are metabolized by soil microorganisms at various rates; lignin is metabolized most slowly. Much of the soil organic matter derived from plants, animals, and microorganisms is transformed by a combination of microbiological and nonbiological processes into humus. Humus is a highly stable complex of heteropolymeric substances comprising humic acids, fulvic acids, and humin. These humic compounds are relatively stable and only very slowly degraded by microorganisms in soils. The half-life of stabilized soil organic matter (i.e., the time required for half of the soil organic matter to be biologically degraded in situ) has been estimated to be 2,000 years (147).

Deep subsurface terrestrial environments are generally even more nutrient limiting than surface soils. Organic carbon availabilities are extremely low, with groundwater dissolved organic carbon (DOC) concentrations typically ranging from 0.1 to 10 mg liter^{-1} (61). DOC values measured in Atlantic coastal plain aquifers in South Carolina ranged from 0.3 to 0.4 mg liter^{-1} (134, 145). Total organic carbon (TOC) in cored subsurface material from the same aquifers ranged from 49 mg kg^{-1} to 15 g kg^{-1}. TOCs in deep lacustrine (lake), paleosol (buried soil), and fluvial (river) sediments in south-central Washington State ranged from 210 to 16,000 mg kg^{-1} (134). Although the chemical composition of subsurface organic carbon has not been well characterized, we can surmise that it is predominantly humic in nature. The organic matter in these environments can be surface soil organic matter that has been transported by groundwater flow and/or organic matter that was entrained with the mineral matter at the time of geological deposition. In either case, the organic matter is mostly humus and therefore poorly metabolized. The organic carbon that is not degraded in surface soils (and is therefore available for transport to deeper layers) is thus relatively recalcitrant. This DOC is further acted upon by microorganisms as it is transported along groundwater flow paths, resulting in decreased overall concentrations and increased proportions of recalcitrant humic compounds. Organic carbon that was buried along with subsurface geological material thousands to millions of years ago has also been acted upon by microorganisms in many cases, leaving predominantly humic compounds. Murphy et al. (135) characterized particulate organic carbon in the Middendorf aquifer and found that it was composed almost entirely of humic and fulvic acids and humin. Localized accumulations were found in the form of lignite, and these were hypothesized to serve as substrates for microbial activity. In considering nutrient availability, one must also consider nutrient flux (i.e., the concentration of nutrients per unit time). In subsurface environments, nutrient flux is a function of the groundwater flow rate as well as the concentration of nutrients in the water. The flow rates of groundwater through aquifers can be extremely slow. For example, the age of the water in the Atlantic coastal plain Middendorf aquifer (i.e., the time since recharge) ranges from near zero in a borehole near the recharge zone to 11,500 years in a borehole that was approximately 85 km along the flowpath from the recharge zone (134). Rates of flow in fine-grained aquitards and in deep unsaturated systems are even slower. Thus, nutrient fluxes in deep subsurface environments are often infinitesimal.

Substances other than soluble or particulate organic carbon can serve as electron donors for microbial metabolism. Gaseous energy sources (e.g., volatile organics, H_2, H_2S, NH_3, CH_4) can occur in soils and unsaturated subsurface environments; however, they have so far received little attention as potential energy sources in these environments. Volatile organic acids can be produced by fermentation in anaerobic aquitards (128, 134) and these can diffuse to overlying unsaturated zones as well as to nearby aquifers. Inorganic sources of energy can also occur in the subsurface. It has been reported that abiologically generated H_2 in basaltic subsurface environments can support autotrophic methanogens (148, 174). Chemolithotrophs, including hydrogen oxidizers, sulfur oxidizers, and nitrifiers, have been detected in deep subsurface environments (58); however, they appear to constitute only a small fraction of the total community, at least among the culturable microorganisms. Exceptions to this can occur in some disturbed subsurface environments, e.g., neutrophilic iron oxidizers (e.g., *Leptothrix* and *Gallionella*) can proliferate in the vicinity of a well screen when oxygen is introduced into a ferrous-iron-rich aquifer (35).

Factors other than the availability of organic carbon substrates may limit microbial activities. Availability of inorganic nutrients, such as nitrogen and phosphorous, may be limiting (145). Availability of terminal electron acceptors (O_2, NO_3^-, Fe^{3+}, SO_4^{2-}, etc.) can also limit microbial activities. A sequential depletion of electron acceptors occurs as water moves along a flow path such as in the Middendorf aquifer (134). Physical constraints can also limit access to nutrients (see discussion below).

The slow to nonexistent nutrient fluxes in deep subsurface environments cause the microorganisms to be extremely nutrient stressed. The persistence of microorganisms can be explained either by slow metabolism of available organic matter, primarily relatively recalcitrant humic compounds, or by very slow rates of endogenous metabolism, i.e., persistence in a quiescent state. Rates of microbial respiration (CO_2 production) in subsurface environments have been estimated by groundwater modeling techniques to be in the range of 10^{-3} to 10^{-6} mmol liter^{-1} year^{-1} (36, 134, 135). This is comparable to the rates in the deep ocean, where microorganisms also occur primarily as nutrient-starved ultramicrobacteria. Average generation times for subsurface bacteria have been estimated to be in the range of hundreds to thousands of years, with the slowest rates of metabolism and the longest generation times occurring in confining clay layers (aquitards) and thick unsaturated zones of arid regions (95, 144, 155).

A final point to consider when comparing nutrient availabilities in various environments is the duration of nutrient deprivation. In aquatic environments, microbes may be thought of as cycling between nutrient-rich "zones of proliferation" and nutrient-poor "zones of quiescence" (117). The period of cycling is typically weeks to months, although recycling times as long as 1,000 years may occur in the deep antarctic ocean (117, 139). In soils, starvation may last for days to months, typically until the next rainfall event redistributes nutrients and stimulates microbial proliferation. In contrast to these surface environments, conditions in the subsurface are relatively static, and opportunities for microbial dispersal are considerably less. In some isolated subsurface environments with negligible groundwater flow, the indigenous bacteria may be subject to nutrient deprivation lasting many thousands of years.

Water Availability

Surface soils undergo fluctuations in water content, ranging from nearly saturated conditions to extreme desiccation. A portion of the soil microbial community dies during each drying and wetting cycle (98). The remaining viable microorganisms presumably have adaptations for desiccation survival; some of these are obvious and well characterized, e.g., fungal and actinomycete spores, bacterial cysts, and bacterial endospores. Vegetative bacterial cells also survive soil desiccation (21, 159, 164); however, their mechanisms of desiccation tolerance are poorly understood. One strategy may be to produce an extracellular polysaccharide matrix that retains water, thus maintaining a relatively moist microhabitat surrounding the cell (2, 163). Desiccated bacteria may also accumulate intracellular compatible solutes in much the same way that they do in response to solute water stress (69). Moisture loss in porous media also indirectly causes nutrient stress. The decrease in water content in soils causes the rates of solute diffusion to decline to nearly zero (146). The diminishing thickness of water films on soil solids also causes a sharp decline in microbial mobility. Thus, the microorganisms become physically separated from nutrients as the soil dries. Filamentous actinomycetes and fungi may gain access to nutrients by bridging air-filled voids, but nonfilamentous organisms cannot. Thus, nonfilamentous forms must either sporulate or survive as vegetative cells. Survival of vegetative cells requires endogenous metabolism, which in turn generally causes the cells to diminish in size. Desiccation of surface soils has been found to cause a decrease in the average cell size of bacteria (173). It can also cause some bacteria to enter a viable but nonculturable state (151).

Water availability can also be a controlling factor in deep subsurface environments, particularly in unsaturated zones; however, the conditions are relatively constant and seldom as extreme as in surface soils. Unsaturated (vadose) zones lie above the water table and are unsaturated with respect to water (i.e., they contain air-filled voids as well as water). They range in thickness from 1 or 2 m in mesic environments to hundreds of meters in arid and semiarid environments. Matric water potentials in vadose zones are typically between 0 and −0.5 MPa (91), conditions that are not severe and thus do not cause desiccation stress. Instead, as discussed above for surface soils, decreases in water content cause decreases in solute diffusion and also curtail microbial movement. Solute diffusion rates in unsaturated zones are typically less than half those of saturated systems (90) and microbial motility has been shown to be severely restricted at matric water potentials less than (more negative than) −0.1 MPa (65). Thus, the indirect effect of the moderately low matric water potentials of unsaturated zones is a decrease in the potential for transport of microorganisms through these zones and inaccessibility of nutrients.

Microorganisms exist in many unsaturated subsurface rocks and sediments, but their numbers are generally lower than in aquifers (27, 39, 68, 90, 91). Given the general inaccessibility of nutrients in unsaturated zones, the microorganisms are even more severely nutrient-starved than in saturated environments, and this in turn should be evidenced by small size and nonculturability of microorganisms. Indeed, the majority of vadose zone bacteria appear very small when viewed in freshly collected samples by direct microscopy (T. L. Kieft, unpublished data), and total microorganisms exceed culturable microorganisms by at least three orders of magnitude (39, 67, 90). In some arid and semiarid environments, the recharge rates through the unsaturated zone are practically nil. For example, viable bacteria found in unsaturated subsurface sediments of south-central Washington State appear to have persisted for 13,000 years, despite a groundwater recharge rate of approximately 1 $\mu m\ year^{-1}$ (56, 90, 121). At another site in eastern Washington state, bacteria appear to have survived under unsaturated conditions in sediments that are a million years old with groundwater that is 1,200 years old (94). Given the slow nutrient fluxes in these unsaturated environments, survival is presumably by endogenous metabolism, which may lead to cell miniaturization.

Soil Texture, Structure, and Mineralogy

The texture (grain-size distribution) of deep subsurface environments has repeatedly been found to be correlated with the numbers and activities of microorganisms (14, 36, 58, 154, 156, 171). In saturated environments, coarse-textured sandy aquifers generally contain higher numbers of microorganisms than fine-textured aquitards. This can be attributed to differences in the potential for transport of microbes into these subsurface sediments and to differences in nutrient fluxes. Sandy aquifers tend to have higher rates of groundwater flow and thus generally have higher nutrient fluxes than fine-textured sediments. Fredrickson et al. (59) measured pore throat sizes and microbial abundance and activities in shale and in sandstone cores collected at a site in New Mexico. They found more rapid rates of potential microbial activity (^{14}C-substrate mineralization and $^{35}SO_4^{2-}$ reduction) in samples with pore throats larger than 0.2 μm than in those with smaller pore throats. Relatively large, interconnected pore throats apparently allow greater nutrient diffusion and thus more sustained metabolic activity. The fine-grained shales with small pore throat diameters also contained bacteria, but these required long incubation times to be resuscitated. These bacteria were likely small, dormant forms, perhaps surviving in this Cretaceous shale since its deposition. Onstott et al. (143) found evidence of bacteria in a deep (2,800 m), hot (76°C), natural gas-bearing formation with micrometer-sized pores and pore throats of $<0.04\ \mu m$. This means that the bacteria are surviving while remaining trapped within pores. Bacteria have also been found surviving in fine-grained lacustrine sediments at a site in south-central Washington State, where clay-rich lacustrine sediments had higher numbers and activities of microorganisms than nearby coarse-textured fluvial sands (60, 92). In this particular case, the higher numbers in the lacustrine sediments may reflect the high organic carbon content, the large microbial communities that likely existed in the original lake sediments, and the potential for clay minerals to enhance survival of bacteria.

Clay minerals exert a variety of influences on microorganisms in porous media, including soil. One such influence is the diminished availability of energy sources for microorganisms when organic matter sorbs to clay minerals (175). This negative effect is counteracted by positive influences. There is a growing body of evidence (reviewed by England, ref. 49) that clay minerals protect bacteria from various stresses in soil, including starvation, desiccation, and protozoan predation. Bushby and Marshall (29) found that montmorillonite decreased the susceptibility of *Rhizobium leguminosarum* to desiccation, and they attributed this to sorption of water away from the bacterial cells by the montmorillonite. They hypothesized that cells survive better if cellular water contents are diminished. Direct coating of bacteria with montmorillonite may be involved in this protection from starvation and desiccation (49, 122). Addition of other clay minerals including bentonite, kaolinite, vermiculite, and illite has also been shown to protect bacteria from starvation and desiccation stress in soil (75, 76, 176). The presence of clay minerals influences the distribution of pore sizes within the soil; pores with neck diameters ≤ 6 μm are thought to be protective microhabitats that exclude predatory protozoa (74). This effect may be greatest in the presence of swelling clay minerals such as bentonite, montmorillonite, and vermiculite (49, 76). Hassink et al. (73) found that the bacterial biomass in soil was positively correlated with the volume of pores in the 0.2 to 1.2 μm range. Since bacteria are thought to be on average one-third smaller than the pores they occupy (100), this finding is consistent with the majority of soil microbes being dwarf sized.

STARVATION RESPONSES IN SOIL AND SUBSURFACE MICROORGANISMS

Nearly all bacteria show at least some ability to survive periods of nutrient deprivation. The ability to survive nutrient deprivation has been best studied in marine bacteria, for which four patterns of starvation survival have been identified (6, 130). In the first of these, an initial increase in the number of cells (due to fragmentation) is followed by a decline in cell numbers to a steady-state number of cells that remain viable for months. This pattern has been observed in a psychrophilic vibrio (138), a *Pseudomonas* sp. (111), and freshly isolated strains (6). In the second pattern, the number of cells declines immediately but eventually stabilizes at a population size at which the remaining bacteria maintain viability indefinitely. This is essentially the same as the first pattern except that the fragmentation stage does not occur. This pattern has been observed in many bacteria including freshly isolated strains (6). Most bacteria tested so far show one of these patterns. In the third pattern, also observed in fresh isolates (6), the population increases immediately in response to starvation (presumably due to fragmentation), and remains at this level indefinitely. Given a long enough period of incubation, these might eventually complete the first pattern. In a curious fourth pattern, the number of viable cells remains at the initial, prestarvation level indefinitely despite nutrient deprivation. Again given enough time, these cells might show one of the other patterns. This fourth pattern has been observed in marine nitrifiers (130).

Cellular changes in marine bacteria, particularly marine vibrios, undergoing starvation have been well documented. Concomitant with changes in numbers and size brought about by fragmentation and dwarfing, the chemical composition of starving cells undergoes changes due to endogenous metabolism. The typical pattern is that cellular lipids, carbohydrates, and energy-storage compounds (e.g., poly-beta-hydroxybutyrate) are rapidly utilized, followed by nucleic acids and proteins (81, 11, 133). The greatest variability in these patterns appears to be the extent to which intracellular RNA and DNA are consumed. However, in the majority, intracellular contents of both RNA and DNA decline to a steady-state level after a few days to weeks of energy deprivation. Marine bacteria are also known to alter their membrane phospholipid fatty acids (PLFAs) in response to starvation stress. These stress-induced changes in PLFA profiles include increases in the ratios of saturated to unsaturated fatty acids (66, 161), increases in the ratios of *trans-* to *cis-*monoenoic fatty acids (66), and increases in the mole percent of cyclopropyl fatty acids (66). These changes in PLFAs presumably stabilize membrane structure in the face of stress; they may also be associated with the dwarfing response.

Starved marine bacteria frequently alter their surface characteristics to increase sorption at solid-liquid interfaces, thereby increasing access to nutrients that have sorbed to those same surfaces (45, 83, 103, 104). This bacterial sorption to surfaces is mediated initially by increases in cell surface hydrophobicity and can be made irreversible by production of extracellular polysaccharides. Since soil and subsurface microorganisms are sorbed to surfaces throughout most of their life cycles, starvation conditions may not elicit the same changes in cell surface characteristics. There may, in fact, be a selective advantage for starved bacteria in soils and subsurface environments to migrate away from surfaces so that they are free to move through the soil solution by diffusion, advection, and/or chemotaxis, thereby increasing their probability of encountering a higher nutrient microsite. As examples, starvation of a subsurface *Klebsiella pneumoniae* resulted in a decrease in the polysaccharide coating of cells (116), and an actinomycete isolate decreased extracellular polysaccharides in response to starvation survival (77). The subsurface isolates of Amy et al. (3) also apparently lost some of their subsurface polysaccharide during starvation. Loss of extracellular polymeric substances may enhance dispersal in subsurface microorganisms; additionally, degradation of polymeric substances may provide energy for survival.

Given this extensive background of information on starvation in marine bacteria, one can ask whether starved terrestrial bacteria undergo the same morphological and metabolic changes. Although there have been far fewer studies, evidence suggests that at least some of the same patterns occur in starved soil and subsurface bacteria. The vegetative bacterial cell types that have been best studied for their starvation survival characteristics are *Arthrobacter* spp. and related coryneform bacteria. Bacteria of the genus *Arthrobacter* are among the most common in soils and may constitute one-half or more of the culturable bacteria (30). The arthrobacters, along with several related high G+C gram-positive bacteria, are also among the most commonly isolated bacteria from subsurface environments (5, 68). The reasons for their dominance in soils and subsurface environments appear to be their ability to function as oligotrophs (i.e., to utilize high-affinity uptake systems under low nutrient conditions), their abilities to survive prolonged periods of nutrient deprivation, and their desiccation tolerance. Arthrobacters are often pleomorphic: in laboratory batch cultures, they are rod shaped during exponential growth and metamorphose into cocci during stationary phase. Under conditions of nutrient limitation and at slow growth rates in chemostat culture, they occur only as cocci. These conditions somewhat mimic those in soils where arthrobacters are thought to occur exclusively as cocci.

It has been reported that when *Arthrobacter crystallopoietes* is cultured in a nutrient-rich liquid medium in the laboratory and is then shifted to a nonnutrient buffer, the starving cells retain their rod shape (22, 25). Rod-shaped cells of *A. crystallopoietes* subjected to starvation underwent no outward changes in cell morphology; however, internal changes occurred: glycogen stores were depleted, numbers of ribosomes declined, the volume of the nucleoplasm increased, and the number of vesicular membranes increased. This lack of transformation of starved rods into cocci might be explained by the fact that starvation conditions were different from those occurring in soil. It may also be that some arthrobacters undergo dwarfing under starvation conditions, while others do not. Kieft et al. (99) found that when a surface soil strain and a closely related subsurface strain of *Arthrobacter* sp. were starved in sediment microcosms, the cells diminished in volume about 10-fold. If arthrobacters miniaturize in situ in response to nutrient deprivation, they could constitute a major fraction of the dwarf bacteria in soil and subsurface environments.

Arthrobacters resort to endogenous metabolism when challenged by nutrient depletion, as do most other bacteria. Their capacity for surviving extremely prolonged starvation appears to be due in part to their habit of profoundly decelerating this metabolism of endogenous substrates. Boylen and Ensign (22) found that 100% of starved *A. crystallopoietes* cells remained culturable in nonnutrient buffer for 30 days and that the cells decreased their rate of endogenous metabolism 80-fold within the first 2 days of starvation. Other coryneform soil bacteria (*Brevibacterium* sp., *Corynebacterium* sp., and *Mycobacterium* sp.) showed similar patterns of starvation survival and reductions in endogenous metabolism (24). The levels of endogenous metabolism observed in these organisms are far below those of any other vegetative bacterial cells. *A. crystallopoietes* is also extremely resistant to starvation in desiccated soil, and it further reduces its rate of endogenous metabolism under these conditions. Boylen (21) found that 50% of cells remained viable in air-dried soils for at least 6 months and that endogenous metabolism (measured as production of $^{14}CO_2$ by ^{14}C-labeled cells) diminished to a rate at which it could be projected that 50% of cell carbon would remain after incubation for 12 years.

Long-term survival is also facilitated by a store of energy-rich endogenous substrates. *A. crystallopoietes* can accumulate up to 40% of its dry weight as a glycogen-like polyglucose under conditions of excess carbon (23). This polyglucose is preferentially metabolized during starvation; protein and RNA also serve as endogenous substrates (23, 166). The arthrobacters appear not to undergo major changes in membrane PLFA profiles in response to starvation (97, 99, 109).

Gram-negative bacteria such as *Pseudomonas* spp. are also commonly cultured from soil; however, their abilities to survive starvation and desiccation are generally thought to be inferior to those of the arthrobacters. In laboratory tests of survival comparing strains of *Pseudomonas* and *Arthrobacter*, the arthrobacters maintain a higher percent viability (or at least culturability) (37, 164). However, these organisms do persist in soils and subsurface environments, albeit in lower numbers than the gram-positive bacteria. Pseudomonads and related genera are also commonly isolated from the deep subsurface. Although less is known of how gram-negative bacteria survive in soils and subsurface environments, the mechanisms are presumably similar to those of their aquatic relatives and to other vegetative cells in soil, i.e., endogenous metabolism of storage products and other cellular metabolites. Kieft et al. (96, 99) found that subsurface *Pseudomonas* spp. evinced changes in membrane PLFAs that are typical in other stressed gram-negative bacteria (i.e., an increase in the ratio of saturated to unsaturated fatty acids, increases in the ratios of *trans*- to *cis*-monoenoic fatty acids, and increases in the ratios of cyclopropyl fatty acids to their monoenoic precursor molecules). These similarities in PLFA alterations suggest that subsurface terrestrial pseudomonads actively respond to starvation stress in a manner similar to their aquatic counterparts.

Amy et al. (3) tested six deep subsurface bacterial strains isolated from volcanic tuff for their ability to survive starvation and observed virtually the same pattern observed for many marine heterotrophic bacteria. Five of the six isolates increased the number of culturable cells (presumably through fragmentation) during the first 2 days of starvation but then declined to a steady-state level at which the remaining cells remained viable indefinitely. Herman and Costerton (77) reported fragmentation of filaments during the starvation response of a subsurface actinomycete. All six of the subsurface strains of Amy et al. (3) became miniaturized during starvation. Dwarfing has also been documented in starved subsurface *K. pneumoniae* (116), *Pseudomonas* spp., *Acinetobacter calcoaceticus*, *Flavobacterium* sp., *Achromobacter* sp., *Alcaligenes* spp. (170), and *Arthrobacter fluorescens* (99). Thus, dwarfing is a prevalent consequence of, and adaptation to, nutrient deprivation in porous media.

CHARACTERISTICS OF SOIL AND SUBSURFACE DWARF CELLS

Morphological and Metabolic Characteristics

Much of what is known of dwarf cells in soils and subsurface environments is based on microscopic studies of their size and morphology. Various techniques involving blending soil slurries followed by centrifugation have been developed for physically separating bacterial cells from soil for characterization (8–10, 15, 50). Using transmission electron microscopy to view cells eluted from a surface soil, Bae et al. (8) found that 72% were less than 0.3 μm in diameter (dwarf cells), 26% were between 0.31 and 0.5 μm, and 2% were between 0.51 and 0.9 μm. None was greater than 0.9 μm in diameter. A few of the apparent dwarf cells were as small as 0.08 μm in diameter, a size that is below the lower limit for resolution of light microscopy. Laboratory-grown soil bacterial cells typically have a larger size distribution, with the minimum size being at least 0.5 μm and the average being 1.0 μm or greater. Bae et al. (8) also observed unusual morphologies, including myxobacteria- and *Azotobacter*-like cysts, thick electron-dense periplasmic spaces in gram-negative bacteria, and large intracellular electron-transparent areas suggestive of storage products (e.g., polyphosphate, PHB, and glycogen). The presence of intracellular storage granules within dwarf cells is consistent with these cells surviving by frugal metabolism of endogenous substrates. Ghiorse and Balkwill (63) observed similar morphological features in bacteria from a shallow aquifer in Louisiana. Further examination of soil bacteria by freeze-etching transmission electron microscopy (9) and by scanning electron microscopy (11, 12) revealed that many soil bacteria are rod shaped rather than coccoid, but that most can still be considered dwarfs, based on their cell volumes. Bakken and Olsen defined dwarfs as cells with volumes less than 0.07 μm^3 (11, 12). Bae and Casida (9) showed that when air-dried soil is moistened, the percentage of dwarf bacteria transiently decreases (i.e., average cell size increases), without an increase in the overall number of cells. As has been well documented for many soils, addition of moisture increases nutrient availability, resulting in a burst of respiratory activity. Bae and Casida's work showed that at least some of the dwarf cells in soil respond to nutrients by an increase in cell size. As the soil was further incubated, presumably causing nutrient availability to wane, the percentage of dwarfs again increased. When glucose and ammonium were added along with moisture, the increase in average cell size was even more pronounced.

Experiments in which dwarf cells have been extracted from soil in relatively pure form have been especially useful. Bakken and Olsen (11, 12) separated bacteria from soil by a blending, gradient centrifugation procedure developed by Bakken (10) and then fractionated these purified cells into different size categories by sequentially filtering them through membrane filters of successively smaller pore size. Though filtration is an imperfect means of separating the bacteria into different size fractions, it does establish a maximum size limit for each category. By quantifying size distributions, total numbers, and culturability of bacteria, Bakken and Olsen found that the culturability of cells was proportional to their size: only 0.2% of cells with diameters less than 0.4 μm were culturable, whereas 30 to 40% of cells with diameters larger than 0.6 μm were culturable. The second important finding was that most of the culturable dwarf cells did not increase in diameter when they multiplied. This was shown by microscopic examination of bacteria in colonies from the dilution plate counts and by a microcolony incubation technique similar to that used by Torella and Morita (177) to determine the responses of individual dwarf cells to nutrient amendment. In other words, a significant proportion of the dwarf cells or ultramicrobacteria in soil consists of intrinsically small bacteria, rather than starved, miniaturized forms of larger bacteria. These inherently small bacteria have been termed "natural ultramicrobacteria" (167) and "true ultramicrobacteria" (84). Bakken and Olsen (13) also quantified DNA in the bacteria in different size fractions and found the amount per cell to be the same for all fractions and also the same as in cultured bacteria. This shows that even though most dwarf cells are nonculturable, they have retained DNA contents that are sufficient to represent one full genome per cell. More recently, Lindahl et al. (119)

showed that small soil bacteria (<0.4 μm diameter) have enough membrane lipids (measured as PLFAs) to maintain intact cell membranes. PLFA profiles of these small cells were different from those of larger cells. PLFAs indicated a higher proportion of gram-positive cells and also that they appear to be under greater nutrient stress than larger-sized bacteria from the same soil, as indicated by PLFA stress ratios, e.g., increased ratio of cyclopropyl fatty acids to their monoenoic precursors.

Although dwarf cells in soil appear to dominate numerically, they may be in the minority in terms of total cell biovolume or biomass, comprising only 10% (11, 12). Bakken and Olsen (11, 12) therefore concluded that the majority of heterotrophic carbon flow through soil is mediated by the larger bacteria that are more amenable to isolation and characterization. Similarly, Baath (7) found that large-sized soil bacteria showed higher thymidine and leucine incorporation rates than bacteria that passed through a 0.4-μm filter; and Christensen et al. (38) measured an increase in large-sized cells (>0.18 μm^3 volume) when soil extract was added as a nutrient, whereas the number of small-sized cells (<0.065 μm^3 volume) did not increase. While these studies indicate that dwarf cells are less metabolically active, they may carry out unique reactions in soil, possibly metabolizing gaseous substrates or recalcitrant humic compounds. They may be responsible for the slow turnover of recalcitrant humic matter in soils. As such, they may fit the description of autochthonous microorganisms in soil, and their unique metabolic capabilities may have practical applications. Winogradsky (184) considered the autochthonous soil bacteria to be predominantly small cocci.

Relatively little is currently known regarding dwarf cells in deep subsurface environments. They have yet to be purified and characterized in the same manner as surface soil dwarfs. This is due in part to the relatively few subsurface environments that have been sampled for microbiology, the frequently low numbers of microbes found in these environments, and the relatively short time that the deep subsurface has been under investigation. Future efforts in this area should be valuable.

Phylogeny of Dwarf Cells

Molecular techniques, particularly small subunit rRNA-based approaches, now enable identification of microorganisms regardless of whether they can be cultivated in vitro. This has resulted in the identification of new, previously unrecognized taxonomic groups of soil microorganisms, most of which remain resistant to cultivation (19, 86, 112, 118). These novel, uncultivated microbes include members of the *Archaea* (19, 86) as well as the *Bacteria.* In most cases, the phylogenetic data give few hints regarding the phenotypes of these microorganisms, and thus it is possible that they include dwarf forms. A few small soil bacteria have been cultivated and then identified by 16S rRNA gene sequencing. Iizuka et al. (84) isolated three groups of dwarf bacteria from a Japanese soil and classified them based on 16S rRNA gene sequencing. The bacteria were rod-shaped cells with widths ranging from 0.35 to 0.50 μm and cell volumes ranging from 0.07 to 0.18 μm^3. All isolates showed stable dwarf sizes, i.e., they did not enlarge in response to nutrients. They were all capable of relatively rapid growth in vitro with doubling times of less than 6 h. Two groups were gram negative, with their closest phylogenetic relatives being *Xanthomonas campestris* and *Pseudomonas lemoignei*; the third group consisted of high G+C gram-positives with sequences similar to *Rathayibacter tritici* and *Microbacterium lacticum.* It is perhaps surprising to find these intrinsically small bacteria to be relatively easily cultivated copiotrophs that are related to well-characterized, normal-sized bacteria. Other small bacteria show more unusual phylogenies and are generally more resistant to cultivation. Janssen et al. (85) isolated three strains of the *Verrocomicrobiales* bacterial lineage from microcosms containing anoxic rice paddy soil. These rice paddy isolates have a stable, small cell size: width, 0.33 to 0.37 μm; length, 0.47 to 0.49 μm; cell volume, 0.03 to 0.04 μm^3. They are obligately fermentative gram-negative bacteria. 16S rRNA gene sequence analysis showed these isolates to be most closely related to *Verrucomicrobium spinosum*, an aquatic strain that was the only previously cultivated member of this lineage. Population densities of members of this lineage were 1.2×10^5 to 7.3×10^5 cells g (dry weight) soil^{-1}, as estimated by an MPN (most probable number) culture method. More recently, *Verrucomicrobium* spp. have been quantified in pasture soils using primers specific for this genus and an MPN-PCR method (141). In this study, the verrucomicrobia were estimated at 5×10^6 bacteria g (dry weight) of soil^{-1}, and their DNA was estimated to comprise 0.2% of total soil DNA. More studies of this type will be needed before generalizations can be made about the classification of dwarf microorganisms in soil.

16S rRNA-based molecular characterization have now been used to characterize deep subsurface microbial communities. For example, Ekendahl et al. (48), Pedersen et al. (149), and Pedersen (148) have used 16S rRNA sequencing to characterize uncultured bacteria in deep granitic groundwater from sites in Sweden. They found several sequences that do not match those of known species of bacteria as well as sequences indicating described species. Unusual archaeal and bacterial sequences have also been reported from a paleosol (34). Reports of 16S rRNA sequences from other subsurface environments are no doubt forthcoming, and likely they will also unearth novel sequences denoting novel bacteria. As in surface soil communities, some of these novel sequences will no doubt be derived from dwarf cells.

WHAT IS THE LOWER SIZE LIMIT FOR MICROBIAL CELLS?

Interest in submicrometer-sized bacterial forms intensified with the publication by McKay et al. (127) of putative evidence from a Martian meteorite of past life on Mars. One of their multiple lines of evidence was the finding of small ovoid features, 20 to 100 nm in diameter, visible in scanning electron micrographs, as well as an oblong, tubelike segmented form approximately 20 nm in diameter and 0.5 μm long (89). The authors identified these as probable fossil microorganisms and cited Folk's report of "nannobacteria" in sedimentary carbonates on Earth (51) as Earthly analogs for minute Martian forms. Folk has claimed that nannobacteria range from 10 to 200 nm in diameter, that they are responsible for mineral precipitation in many environments, and that they comprise the majority of the Earth's biomass (52)! Idiosyncratic spelling aside, the evidence supporting Folk's "nannobacteria" as microbial cells, living or fossilized, has been scant, limited primarily to scanning electron micrographs of nanometer-scaled objects viewed in a variety of geological materials (51, 53, 153). These nannobacteria and "nanofossils" have been assailed as mineral surface irregularities and as artifacts of metal coating for electron microscopy (26). Another possible explanation is that they are fossilized fragments of bacteria (101). Nonetheless, the paper of McKay et al. (127) and Folk's claims of living cells with diameters 50 nm or less (53) precipitated a lively debate regarding the lower size limits of bacterial cells.

Smallness confers selective advantages, the obvious one being high surface-to-volume ratio for efficient transport of metabolites into and out of cells. Other possible advantages include protection from predation (105), ability to occupy small pore spaces, and the opportunity for transport through fine-grained porous media. In the absence of opposing selective pressures, microorganisms would evolve ever smaller forms and disappear altogether. Several factors have been suggested to constrain the smallest possible size of prokaryotes. Foremost is the need for a cell to contain enough macromolecules (DNA, RNA, ribosomes, proteins, etc.) for metabolism and reproduction or, as Koch (105) stated, a cell must be "large enough to house the total amount of needed stuff." Individual prokaryotic ribosomes are 20–25 nm in diameter (137), and active cells contain hundreds or more. Theoretically, a cell could get by with less stuff if the macromolecules could function more efficiently, e.g., if the genome contained fewer redundancies, and if enzymes were more efficient (lower k_s, higher V_{max}). However, more than 3 billion years of evolution appears not to have optimized genome size or enzyme efficiency. Moreover, most cells need enough stuff to survive and function when conditions change or, in Koch's terms, cells need a "catastrophe kit" to handle nutritional deprivation and other environmental adversity. Cell differentiation into spores and other resting forms requires extra genes, including those encoding regulatory systems. Likewise, abilities to use multiple substrates and to function at different temperatures, pH, etc., all add to the genetic load. This applies to most microorganisms; however, some obligately parasitic microorganisms, especially intracellular ones, have a relatively streamlined lifestyle requiring minimal catabolism and have a correspondingly reduced genome and undersized cells.

Qualitatively, a minimum size to contain necessary stuff is undeniable; however, the quantitative question still remains: how much cellular stuff is required and how small can it be packaged? Koch (105) gives a range of 0.02 to 0.6 μm^3 for volumes of living cells. The 0.02-μm^3 cell volume corresponds to a 0.337-μm diameter sphere. Psenner and Loferer (158) suggested that at least a 0.3-μm diameter sphere (0.014 μm^3) is required to contain enough macromolecules for life. Using similar considerations, Maniloff (123) put the lower limit at 0.14-μm diameter (0.0014-μm^3 volume for a sphere). Nealson (136, 137) noted the requirement for solutes: metabolically active cells require a sufficient concentration of metabolites in the cytosol. If living cells require micromolar to millimolar concentrations of these metabolites, cells smaller than about 0.1 μm in diameter would have too few molecules for life. For example, a 0.05-μm diameter sphere containing a 10-mM metabolite solution (high for most bacteria) will have only a few hundred molecules per cell; at a more typical cellular concentration, 10 μM, such a sphere would be unlikely to contain even one solute molecule (137)! These arguments become even more compelling when one considers that a portion of the volume of living cells is occupied by a surrounding lipid membrane, with typical thickness of approximately 8 nm. The smallest of Folk's nannobacteria (54) would have no space for cytoplasm. The lower size limit for microbial cells has yet to be defined, and ever more infinitesimal forms will likely continue to be discovered; however, it is unlikely that they will fall much below 0.2 μm in diameter. As for extraterrestrial microbes, they might conceivably be subject to somewhat different constraints on cell size; however, they would still presumably be limited by solution chemistry and thus could not be much smaller than their Earthly counterparts.

TRANSPORT OF STARVED AND DWARF MICROORGANISMS

Studies of the transport of bacteria through porous media such as soils and subsurface environments are motivated by efforts to protect groundwater aquifers from contamination with pathogens, to deliver bacteria to chemically contaminated aquifers for bioremediation, to inject bacteria into petroleum reservoirs to enhance oil recovery, and to understand the origins of bacteria in pristine deep subsurface environments. Characteristics of bacteria that influence their transport through porous media include size, cell surface characteristics (e.g., hydrophobicity, charge, presence of extracellular polysaccharides), growth rate, survival, motility, and chemotaxis (31, 62, 70, 72, 54, 125, 153, 160, 168, 181). Each of these bacterial characteristics is phenotypically plastic and can change with the nutrient status of the cells. Given this situation, one can ask whether small bacteria are transported more rapidly through porous media and whether nutrient deprivation influences bacterial transport. Unfortunately, answering these questions is difficult because (i) size usually does not vary independently from other cell characteristics; (ii) bacteria are quite variable, both within and among populations, in their responses to starvation conditions; and (iii) nonbiological factors affecting bacterial transport (e.g., ionic strength of the porewater, texture and mineralogy of the porous medium, porosity and tortuosity of the porous medium, and groundwater velocity), are highly variable.

As discussed above, nutrient deprivation can lead to bacterial dwarfing in soil and subsurface environments. One might expect the influence of cell miniaturization on transport to be somewhat predictable, but so far this has not been the case. Size-exclusion effects may cause large bacteria to move through porous media faster than smaller cells. This effect has been demonstrated using carboxylated microspheres of different sizes as surrogate bacteria in a field transport experiment (72). The same investigators found that bacteria can travel through the subsurface faster than a conservative solute tracer, and they also attributed this to a size-exclusion effect (72). However, other investigators have found no effect of cell size on the retention times of bacteria during transport through porous media (54).

Besides speed of transport, one can also consider the percentage of cells that becomes trapped within the porous medium and the influence of bacterial size and nutrient deprivation on this percent retention. While considerations of straining (i.e., the trapping of particles by smaller-diameter pore throats) would suggest that a higher percentage of larger cells should be retained within a porous medium during transport, this appears not to be an important factor (70, 72). In comparing transport of two bacterial strains that differed primarily in cell size through columns of sand, Fontes et al. (54) found that although the retention times were not significantly different between the two strains, a higher percentage of the small cells was retained in the columns. Similarly, Gannon et al. (62) found a higher percent retention of small bacteria than large bacteria during transport through soil, and Camper et al. (31) found that a higher percentage of small starved *Pseudomonas fluorescens* cells was retained in a porous medium column compared to larger, nonstarved cells of the same strain. This higher rate of retention of small cells may be explainable by colloid filtration theory, which predicts higher probabilities of collision between small particles and solid collector surfaces, and thus higher filtration efficiencies, for small particles compared to large particles. Harvey and Garabedian (71) measured the size frequency distributions of indigenous bacteria along a flow path within a subsurface contamination plume and found that the average size of the bacteria increased with distance down the flow path, indicating that small cells were more likely to be retained than large cells. These data are consistent with a transport model based on colloid filtration theory.

Changes in the surface chemistry of bacteria under starvation conditions and the effects of these changes in surface chemistry on bacterial transport are extremely important but largely unpredictable. Gannon et al. (62) found no correlation between hydrophobicity, surface charge, or the presence of capsules and transport of 19 different strains of bacteria through soil. As discussed above, many copiotrophic marine bacteria increase sorption at liquid-solid interfaces under starvation conditions by becoming relatively hydrophobic and, in some cases, by generating sticky extracellular polysaccharides. Some soil and subsurface bacteria may follow this pattern, thereby diminishing their transport during nutrient-poor conditions and enhancing transport under nutrient-rich conditions (72). However, starvation of other terrestrial bacteria may result in their being less surface active and thus more readily transported through porous media. Applications of this phenomenon are addressed below. At the present time, our understanding of bacterial transport through porous media is primarily phenomenological, and much of the experimental data appear contradictory. Biological factors are often confounding, and nonbiological factors are seldom standardized among experiments. Our inability to culture most indigenous dwarf bacteria further hinders efforts in this area.

APPLICATIONS OF DWARF CELLS

The potential for ultramicrobacteria to be transported through porous media such as petroleum reservoirs and groundwater aquifers has suggested several applications. J. W. Costerton and his colleagues have been pioneers in proposing, testing, and promoting innovative uses of ultramicrobacteria (termed "UMB" by this group) (113). They have patented the use of starved, miniaturized, copiotrophic bacteria for selectively plugging portions of oil-bearing substrata for microbially enhanced oil recovery (40, 41). Secondary recovery of oil from petroleum reservoirs is generally attempted by water flooding, i.e., injection of water through an injection well to displace the oil, which is then pumped from the reservoir through a production well. This process is hampered by "fingering," which is the preferential flow of water through highly permeable "thief" zones that typically have little residual oil. Selective plugging is the process of plugging the highly permeable zones in order to enhance the water sweep through low permeability zones. The use of bacteria with large amounts of extracellular polysaccharide capsular material has been proposed for selective plugging, and it has been demonstrated that such bacteria can be effective in forming a biofilm that greatly reduces the permeability of a porous medium (169). UMB in the form of starved copiotrophs are especially attractive for selective plugging purposes because they can penetrate deeply in the highly permeable zone and can then be resuscitated with nutrient addition, thus forming a biofilm that extends throughout the porous medium. Lappin-Scott et al. (116) demonstrated that a *K. pneumoniae* isolated from oil well water had the requisite physiological characteristics for deep penetration and selective plugging: it underwent fragmentation and dwarfing, it lost extracellular polysaccharide when nutrient starved, it maintained viability during long-term starvation, and small starved cells could be rapidly resuscitated by nutrient addition to form large cells with copious extracellular polysaccharide. In subsequent tests of this organism in sintered glass (122) and sandstone cores (115), UMB penetrated uniformly throughout the cores, whereas nonstarved normal cells penetrated only short distances; the starved cells could be resuscitated by nutrient addition to generate biofilms that greatly reduced permeabilities. Similar results have been obtained in a three-dimensional model system using a starved *Pseudomonas* sp. (43) and in sandstone cores using thermophilic sulfate-reducing bacteria derived from a petroleum reservoir (18). Thus, laboratory experiments suggest that this is a promising technology. Field tests of the method in petroleum reservoirs have been conducted, with reported UMB-created plugs extending more than 200 m through the subsurface (42).

Costerton and others have envisioned more applications of UMB in which starved cells could be transported through the subsurface and then stimulated by addition of organic nutrients. One such application is the production of subsurface "biobarriers" (114). A biobarrier consisting of a thick biofilm of bacteria could be used to retard subsurface transport of hazardous solutes, e.g., heavy metals, organic pollutants, and acid mine drainage. If the UMB can also transform the pollutants in a useful way (e.g., by mineralizing an organic pollutant), the biobarrier could eliminate the pollutant while containing it. Ultramicrobacteria (either starved forms of larger bacteria or intrinsically small bacteria) that are capable of biodegrading environmental contaminants could be useful for bioremediating relatively inaccessible contaminated aquifers. Bioaugmentation efforts in which bacteria are injected into contaminated aquifers often fail because the bacteria do not penetrate deeply. Biodegradative bacteria that are selected for small size and lack of sorption to aquifer solids could be especially effective.

VIABLE BUT NONCULTURABLE AND INTACT BUT NONVIABLE CELLS

The viable-but-nonculturable-cells phenomenon has been investigated primarily in bacteria of public health importance in aquatic environments, e.g., *Vibrio cholerae*, *Vibrio vulnificus*, *Campylobacter jejuni*, *E. coli*, and others (126, 165). Though they have been less studied in soil, there is evidence that soil microorganisms (introduced and indigenous) can be in a viable but nonculturable state. Turpin et al. (179) reported that the culturability *Salmonella enterica* serovar Typhimurium applied to nonsterile soil declined rapidly over a period of days while the number of viable cells remained constant. Not surprisingly, these cells also miniaturized into small cocci during the same time period. Though their experiment did not provide unequivocal evidence that the cells remained viable (e.g., by the direct viable count method of Kogure et al. [106]), the data do suggest that enteric pathogens applied to soils (e.g., in sewage sludge) may remain infectious, though undetectable by standard culturing methods. Pedersen and Leser (161) reported that a strain of *Enterobacter cloacae*, a rhizosphere copiotroph, entered a viable but nonculturable state when introduced to soil and leaf surfaces. Clearly, microorganisms introduced into soil may not be monitorable by standard culture methods.

The reasons for nonculturability of most soil and subsurface microorganisms are numerous. Certainly some cells fail to replicate in or on growth media because they are dead or moribund. These can be considered intact but nonviable. Others are viable but not cultured by current methods. For many of these microbes, we have incomplete understanding of their nutritional needs and therefore have not devised a suitable growth medium. Others may be viable but injured and in need of unknown resuscitation methods. The nutritional needs of some microorganisms may be so closely tied to other organisms that they cannot be cultured in isolation on dilution plates. Dilute, low-nutrient media generally yield higher counts of culturable microorganisms than media with higher nutrient concentrations. However, even the standard low-nutrient media may be too rich, causing a hypothesized "substrate-accelerated death" (157). This is a standard litany of excuses that microbiologists give for being able to culture so few of the microbes in most natural environments, and a combination of some or all of these explanations is undoubtedly correct. Evidence suggests that a large proportion of the indigenous dwarfs or ultramicrobacteria in soil fall into the viable but nonculturable category.

In deep subsurface environments, the total number of microorganisms may exceed the number of culturable cells by 1 to 6 orders of magnitude, with 3 to 4 orders being most typical. This raises the question of how many of the nonculturable cells are viable. Brockman et al. (28) used total extractable PLFAs, as a measure of the total viable cells in subsurface environments and converted PLFA concentration to cell numbers based on the average PLFA content per cell. PLFAs, are thought to be a good measure of living cells because they are de-phosphorylated rapidly by phosphatases when cells die. By comparing total cell data from acridine orange direct counts, PLFA-derived cell numbers, and numbers of culturable cells, the proportions of viable but nonculturable cells and the number of intact but nonviable cells were estimated. It was concluded from these calculations that a majority of nonculturable cells is viable in relatively high-nutrient-flux subsurface environments such as the Atlantic coastal plain aquifer sediments, while intact but nonviable cells dominate in vadose zone sediments and rocks. Diglyceride fatty acids (DGFA) have been used as a measure of dead cells, because phosphatases are thought to remove phosphate groups from membrane lipids immediately upon cell death (162). Kieft et al. (94) found DGFA:PLFA ratios as high as 100 in deep vadose zone sediments (compared to 0.001 in near-surface soils), indicating that dead cells greatly outnumber live cells in the deeper sediments.

SUMMARY AND CONCLUSIONS

Most bacteria in soils and subsurface terrestrial environments are small, dwarf forms, and most bacteria in these same environments are not culturable by current methods. Obviously there is a large overlap between these groups. Dwarf cells include starved, miniaturized forms that may be much larger when cultured, as well as intrinsically small forms that appear never to increase their size in response to nutrient amendment. Dwarf bacteria of the first type may be the products of starvation/miniaturization patterns similar to those observed in marine copiotrophic bacteria. Dwarf bacteria of the second type may be more numerous in situ and may represent previously unknown taxonomic groups of bacteria. Practical applications of starved/miniaturized forms of culturable bacteria are currently being tested in the areas of microbially enhanced oil recovery and in bioremediation of subsurface contamination. Molecular approaches, particularly 16S rRNA-based techniques, are now beginning to provide information on the identities of the noncultured forms of bacteria in soil and other environments. In some cases, knowledge of the phylogenetic relationships of uncultured bacteria to characterized bacteria may provide clues to their metabolism and even to techniques for culturing them. Some may have potential for mediating novel and useful biochemical reactions, e.g., catabolism of recalcitrant organic pollutants or production of pharmaceuticals. In other cases, possibly most cases, dwarf bacteria will continue to resist cultivation and will yield to characterization only by indirect means.

REFERENCES

1. **Alexander, M.** 1977. *Introduction to Soil Microbiology,* 2nd ed. John Wiley and Sons, New York, N.Y.
2. **Allison, S. M., and J. L. Prosser.** 1991. Survival of ammonia oxidizing bacteria in air-dried soil. *FEMS Microbiol. Lett.* **79:**65–68.
3. **Amy, P. S., C. Durham, D. Hill, and D. L. Haldeman.** 1993. Starvation-survival of deep subsurface isolates. *Curr. Microbiol.* **26:**345–352.
4. **Amy, P. S., and D. L. Haldeman (ed.).** 1997. *The Microbiology of the Terrestrial Subsurface.* CRC Press, Boca Raton, Fla.
5. **Amy, P. S., D. L. Haldeman, D. Ringelberg, D. H. Hall, and C. Russell.** 1992. Comparison of the identification systems for study of water and endolithic bacterial isolates from the subsurface. *Appl. Environ. Microbiol.* **58:**3367–3373.
6. **Amy, P. S., and R. Y. Morita.** 1983. Starvation-survival patterns of sixteen freshly isolated open ocean bacteria. *Appl. Environ. Microbiol.* **45:**1109–1115.
7. **Baath, E.** 1994. Thymine and leucine incorporation in soil bacteria with different cell size. *Microb. Ecol.* **27:**267–278.

8. **Bae, H. C., E. H. Cota-Robles, and L. E. Casida.** 1972. Microflora of soil as viewed by transmission electron microscopy. *Appl. Microbiol.* **23:**637–648.
9. **Bae, H. C., and L. E. Casida.** 1973. Responses of indigenous microorganisms to soil incubations as viewed by transmission electron microscopy of cell thin sections. *J. Bacteriol.* **113:**1462–1473.
10. **Bakken, L. R.** 1985. Separation and purification of bacteria from soil. *Appl. Environ. Microbiol.* **49:**1482–1487.
11. **Bakken, L. R., and R. A. Olsen.** 1986. Dwarf cells in soil—a result of starvation of "normal" bacteria, or a separate population? p. 561–566. *In* F. Megusar and M. Gantar (ed.), *Current Perspectives in Microbial Ecology. Proceedings of the Fourth International Symposium on Microbial Ecology, Ljubljana, Yugoslavia.* Slovene Society for Microbiology, Ljubljana, Yugoslavia,
12. **Bakken, L. R., and R. A. Olsen.** 1987. The relationship between cell size and viability of soil bacteria. *Microb. Ecol.* **13:**103–114.
13. **Bakken, L. R., and R. A. Olsen.** 1989. DNA-content of soil bacteria of different cell size. *Soil Biol. Biochem.* **21:**789–793.
14. **Balkwill, D. L.** 1989. Numbers, diversity, and morphological characteristics of aerobic chemoheterotrophic bacteria in deep subsurface sediments from a site in South Carolina. *Geomicrobiol. J.* **7:**33–52.
15. **Balkwill, D. L., and L. E. Casida.** 1973. Microflora of soil as viewed by freeze-etching. *J. Bacteriol.* **114:**1319–1327.
16. **Balkwill, D. L., J. K. Fredrickson, and J. M. Thomas.** 1989. Vertical and horizontal variations in the physiological diversity of the aerobic chemoheterotrophic bacterial microflora in deep southeast coastal plain deep subsurface sediments. *Appl. Environ. Microbiol.* **55:**1058–1065.
17. **Balkwill, D. L., and W. C. Ghiorse.** 1985. Characterization of subsurface bacteria associated with two shallow aquifers in Oklahoma. *Appl. Environ. Microbiol.* **50:**580–588.
18. **Bass, C. J., R. A. Davey, P. F. Sanders, and H. M. Lappin-Scott.** 1993. Isolation and core penetration of starved and vegetative cultures from North Sea oil systems, abstr. F-16. *In Programme and Abstracts of the 1993 International Symposium on Subsurface Microbiology.* Bath, United Kingdom.
19. **Bintrim, S. B., T. J. Donohue, J. Handelsman, G. P. Roberts, and R. M. Goodman.** 1997. Molecular phylogeny of Archaea from soil. *Proc. Natl. Acad. Sci. USA* **94:**277–282.
20. **Bone, T. L., and D. L. Balkwill.** 1988. Morphological and cultural comparison of microorganisms in surface soil and subsurface sediments at a pristine study site in Oklahoma. *Microb. Ecol.* **16:**49–64.
21. **Boylen, C. W.** 1973. Survival of *Arthrobacter crystallopoietes* during prolonged periods of extreme desiccation. *J. Bacteriol.* **113:**33–37.
22. **Boylen, C. W., and J. C. Ensign.** 1970. Long-term starvation survival of rod and spherical cells of *Arthrobacter crystallopoietes. J. Bacteriol.* **103:**569–577.
23. **Boylen, C. W., and J. C. Ensign.** 1970. Intracellular substrates for endogenous metabolism during long-term starvation of rod and spherical cells of *Arthrobacter crystallopoietes. J. Bacteriol.* **103:** 578–587.
24. **Boylen, C. W., and M. H. Mulks.** 1978. The survival of coryneform bacteria during periods of prolonged nutrient starvation. *J. Gen. Microbiol.* **105:**323–334.
25. **Boylen, C. W., and J. L. Pate.** 1973. Fine structure of *Arthrobacter crystallopoietes* during long-term starvation of rod and spherical stage cells. *Can. J. Microbiol.* **19:**1–5.
26. **Bradley, J. P., R. P. Harvey, and H. Y. McSween, Jr.** 1997. No 'nanofossils' in Martian meteorite. *Nature* **390:**455–456.
27. **Brockman, F. J., T. L. Kieft, J. K. Fredrickson, B. N. Bjornstad, S. W. Li, W. Spangenburg, and P. E. Long.** 1992. Microbiology of vadose zone paleosols in south-central Washington State. *Microb. Ecol.*. **23:**279–301.
28. **Brockman, F. J., D. B. Ringelberg, D. C. White, J. K. Fredrickson, D. L. Balkwill, T. L. Kieft, T. J. Phelps, and W. C. Ghiorse.** 1993. Estimates of intact but nonviable and viable but nonculturable microorganisms in subsurface sediments from six boreholes located in wet and dry climatic regions of the United States, abstr. B-32. *In Programme and Abstracts of the 1993 International Symposium on Subsurface Microbiology.* Bath, United Kingdom.

29. **Bushby, H. V. A., and K. C. Marshall.** 1977. Water status of *Rhizobia* in relation to their susceptibility to desiccation and their protection by montmorillonite. *J. Gen. Microbiol.* **99:**19–27.
30. **Cacciari, I., and D. Lippi.** 1987. Arthrobacters: successful arid soil bacteria. A review. *Arid Soil Res. Rehab.* **1:**1–30.
31. **Camper, A. K., J. T. Hayes, P. J. Sturman, W. L. Jones, and A. B. Cunningham.** 1993. Effects of motility and adsorption rate coefficient on transport of bacteria through saturated porous media. *Appl. Environ. Microbiol.* **59:**3455–3462.
32. **Casida, L. E.** 1965. Abundant microorganism in soil. *Appl. Microbiol.* **13:**327–334.
33. **Casida, L. E.** 1977. Small cells in pure cultures of *Agromyces ramosus* and in natural soil. *Can. J. Microbiol.* **23:**214–216.
34. **Chandler, D. P., F. J. Brockman, T. J. Bailey, and J. K. Fredrickson.** 1998. Phylogenetic diversity of Archaea and Bacteria in a deep subsurface paleosol. *Microb. Ecol.* **936:**37–50.
35. **Chapelle, F. H.** 1993. *Ground-Water Microbiology and Geochemistry*. Wiley, New York, N.Y.
36. **Chapelle, F. H., and D. R. Lovley.** 1990. Rates of microbial metabolism in deep coastal plain aquifers. *Appl. Environ. Microbiol.* **56:**1865–1874.
37. **Chen, M., and M. Alexander.** 1973. Survival of soil bacteria during prolonged desiccation. *Soil Biol. Biochem.* **5:**213–221.
38. **Christensen, H., R. A. Olsen, and L. R. Bakken.** 1995. Flow cytometric measurements of cell volumes and DNA contents during culture of indigenous soil bacteria. *Microb. Ecol.* **29:**49–62.
39. **Colwell, F. S.** 1989. Microbiological comparison of surface soil and unsaturated subsurface soil from a semiarid high desert. *Appl. Environ. Microbiol.* **55:**2420–2423.
40. **Costerton, J. W. F., F. Cusack, and F. A. MacLeod.** 31 January 1989. Microbial process for selectively plugging a subterranean formation. U.S. patent 4,800,959.
41. **Costerton, J. W. F., F. Cusack, T. J. Cyr, S. A. Blenkinsopp, and C. P. Anderson.** 29 December 1992. Microbial manipulation of surfactant-containing foams to reduce subterranean formation permeability. U.S. patent 5,174,378.
42. **Costerton, J. W.** 1993. Ultramicrobacteria and biofilms—mode of growth is pivotal in the subsurface environment, abstr. B-1. *In Programme and Abstracts of the 1993 International Symposium on Subsurface Microbiology.* Bath, United Kingdom.
43. **Cusack, F., S. Singh, C. McCarthy, J. Grieci, M. de Rocco, D. Nguyen, H. Lappin-Scott, and J. W. Costerton.** 1992. Enhanced oil recovery: three-dimensional sandpack simulation of ultramicrobacteria resuscitation in reservoir formation. *J. Gen. Microbiol.* **138:**647–655.
44. **Dawes, E. A.** 1985. Starvation, survival and energy reserves, p. 43–79. *In* M. Fletcher and G. D. Floodgate (ed.), *Bacteria in Their Natural Environments.* Academic Press, London, United Kingdom.
45. **Dawson, M. P., B. A. Humphrey, and K. C. Marshall.** 1981. Adhesion: a tactic in the survival strategy of a marine vibrio during starvation. *Curr. Microbiol.* **6:**195–199.
46. **Dommergues, Y. R., L. W. Belser, and E. L. Schmidt.** 1978. Limiting factors for microbial growth and activity in soil. *Adv. Microb. Ecol.* **2:**49–104.
47. **Donachie, W. D., and A. C. Robinson.** 1987. Cell division: parameter values and the process, p. 1578–1593. *In* F. C. Neidhardt, J. L. Ingraham, K. B. Low, B. Magasanik, M. Schaecter, and H. E. Umbarger (ed.), *Escherichia coli and Salmonella typhimurium: Cellular and Molecular Biology.* American Society for Microbiology, Washington, D.C.
48. **Ekendahl, S., J. Arlinger, F. Stahl, and K. Pedersen.** 1994. Characterization of attached bacterial populations in deep granitic groundwater from the Stripa research mine by 16S rRNA gene sequencing and scanning electron microscopy. *Microbiology* **140:**1575–1583.
49. **England, L. S., H. Lee, and J. T. Trevors.** 1993. Bacteria survival in soil: effects of clays and protozoa. *Soil Biol. Biochem.* **25:**525–531.
50. **Faegri, A., V. Lid Torsvik, and J. Goksoyr.** 1977. Bacterial and fungal activities in soil: separation of bacteria and fungi by a rapid fractionated centrifugation technique. *Soil Biol. Biochem.* **9:**105–112.
51. **Folk, R. L.** 1993. SEM imaging of bacteria and nannobacteria in carbonate sediments and rocks. *J. Sed. Petrol.* **63:**990–999.
52. **Folk, R. L.** 1996. In defense of nannobacteria. *Science* **276:**1777.
53. **Folk, R. L., and F. L. Lynch.** 1997. Nannobacteria are alive on Earth as well as Mars, p. 406–419. *In* R. B. Hoover (ed.), *Proceedings of SPIE, Instruments, Methods, and Missions for the Investigation*

of Extraterrestrial Microorganisms, San Diego, CA, 27 July–1 August. SPIE, Vol. 3111. The International Society for Optical Engineering, Bellingham, Wash.

54. **Fontes, D. E., A. L. Mills, G. M. Hornberger, and J. S. Herman.** 1991. Physical and chemical factors influencing transport of microorganisms through porous media. *Appl. Environ. Microbiol.* **57:** 2473–2481.
55. **Fredrickson, J. K., D. L. Balkwill, J. M. Zachara, S. W. Li, F. J. Brockman, and M. A. Simmons.** 1991. Physiological diversity and distributions of heterotrophic bacteria in deep cretaceous sediments of the Atlantic coastal plain. *Appl. Environ. Microbiol.* **57:**402–411.
56. **Fredrickson, J. K., F. J. Brockman, B. N. Bjornstad, P. E. Long, S. W. Li, J. P. McKinley, J. V. Wright, J. L. Conca, T. L. Kieft, and D. L. Balkwill.** 1993. Microbiological characterization of pristine and contaminated deep vadose sediments from an arid region. *Geomicrobiol. J.* **11:**95–107.
57. **Fredrickson, J. K., and M. Fletcher (ed.).** *Subsurface Microbiology and Biogeochemistry,* in press. John Wiley & Sons, New York, N.Y.
58. **Fredrickson, J. K., T. R. Garland, R. J. Hicks, J. M. Thomas, S. W. Li, and K. M. McFadden.** 1989. Lithotrophic and heterotrophic bacteria in deep subsurface sediments and their relation to sediment properties. *Geomicrobiol. J.* **7:**53–66.
59. **Fredrickson, J. K., J. P. McKinley, B. N. Bjornstad, P. E. Long, D. B. Ringelberg, D. C. White, L. R. Krumholz, J. M. Suflita, F. S. Colwell, R. M. Lehman, and T. J. Phelps.** 1997. Pore-size constraints on the activity and survival of subsurface bacteria in a late Cretaceous shale-sandstone sequence, northwestern New Mexico. *Geomicrobiol. J.* **14:**183–202.
60. **Fredrickson, J. K., J. P. McKinley, S. A. Nierzwicki-Bauer, D. C. White, D. B. Ringelberg, S. A. Rawson, S.-M. Li, F. J. Brockman, and B. N. Bjornstad.** 1995. Microbial community structure and biogeochemistry of Miocene subsurface sediments: implications for long-term microbial survival. *Mol. Ecol.* **4:**619–626.
61. **Freeze, R. A., and J. A. Cherry.** 1979. *Groundwater.* Prentice-Hall, Englewood Cliffs, N.J.
62. **Gannon, J. T., V. B. Manilal, and M. Alexander.** 1991. Relationship between cell surface properties and transport of bacteria through soil. *Appl. Environ. Microbiol.* **57:**190–193.
63. **Ghiorse, W. C., and D. L. Balkwill.** 1983. Enumeration and morphological characterization of bacteria indigenous to subsurface sediments. *Dev. Ind. Microbiol.* **24:**213–224.
64. **Gold, T.** 1992. The deep, hot biosphere. *Proc. Natl. Acad. Sci. USA* **89:**6045–6049.
65. **Griffin, D. M.** 1981. Water potential as a selective factor in the microbial ecology of soils, p. 141–151. *In* J. F. Parr, W. R. Gardner, and L. F. Elliott (ed.), *Water Potential Relations in Soil Microbiology.* Soil Science Society of America, Madison, Wis.
66. **Guckert, J. B., M. A. Hood, and D. C. White.** 1986. Phospholipid ester-linked fatty-acid profile changes during nutrient deprivation of *Vibrio cholerae*: increases in the *trans*/*cis* ratio and proportions of cyclopropyl fatty acids. *Appl. Environ. Microbiol.* **52:**794–801.
67. **Haldeman, D. L., and P. S. Amy.** 1993. Bacterial heterogeneity in deep subsurface tunnels at Rainier Mesa, Nevada, Test Site. *Microb. Ecol.* **25:**183–194.
68. **Haldeman, D. L., P. S. Amy, D. Ringelberg, and D. C. White.** 1993. Characterization of the microbiology within a 21 m^3 section of rock from the deep subsurface. *Microb. Ecol.* **26:**145–159.
69. **Harris, R. F.** 1981. Effect of water potential on microbial growth and activity, p. 23–97. *In* J. F. Parr, W. R. Gardner, and L. F. Elliott (ed.), *Water Potential Relations in Soil Microbiology.* Soil Science Society of America, Madison, Wis.
70. **Harvey, R. W.** 1991. Parameters involved in modeling movement of bacteria in groundwater, p. 89–114. *In* C. J. Hurst (ed.), *Modeling the Environmental Fate of Microorganisms.* American Society for Microbiology, Washington, D.C.
71. **Harvey, R. W., and S. P. Garabedian.** 1991. Use of colloid filtration theory in modeling movement of bacteria through a contaminated sandy aquifer. *Environ. Sci. Technol.* **25:**175–185.
72. **Harvey, R. W., L. H. George, R. L. Smith, and D. R. LeBlanc.** 1989. Transport of microspheres and indigenous bacteria through a sandy aquifer: results of natural- and forced-gradient tracer experiments. *Environ. Sci. Technol.* **23:**51–56.
73. **Hassink, J., L. A. Bouwman, K. B. Zwart, and L. Brusaard.** 1993. Relationship between habitable pore space, soil biota, and mineralization rates in grassland soils. *Soil Biol. Biochem.* **25:**47–55.
74. **Heijnen, C. E., C. Chenu, and M. Robert.** 1993. Micromorphological studies on clay-amended and unamended loamy sand, relating survival of introduced bacteria and soil structure. *Geoderma* **56:**195–297.

75. **Heijnen, C. E., J. D. van Elsas, P. J. Kuikman, and J. A. van Veen.** 1988. Dynamics of *Rhizobium leguminosarum* biovar *trifolii* introduced into soil; the effect of bentonite clay on predation by protozoa. *Soil Biol. Biochem.* **20:**483–488.
76. **Heijnen, C. E., and J. A. van Veen.** 1991. A determination of protective microhabitats for bacteria introduced into soil. *FEMS Microbiol. Ecol.* **85:**73–80.
77. **Herman, D. C., and J. W. Costerton.** 1993. Starvation-survival of a *p*-nitrophenol-degrading bacterium. *Appl. Environ. Microbiol.* **59:**340–343.
78. **Hicks, R. J., and J. K. Fredrickson.** 1989. Aerobic metabolic potential of microbial populations indigenous to deep subsurface environments. *Geomicrobiol. J.* **7:**67–77.
79. **Hirsch, P., and E. Rades-Rohlkohl.** 1983. Microbial diversity in a groundwater aquifer in northern Germany. *Dev. Ind. Microbiol.* **24:**213–224.
80. **Hirsch, P., and E. Rades-Rohlkohl.** 1988. Some special problems in the determination of viable counts of groundwater microorganisms. *Microb. Ecol.* **16:**99–113.
81. **Hood, M. A., J. B. Guckert, D. C. White, and F. Deck.** 1986. Effect of nutrient deprivation on lipid carbohydrate, DNA, RNA, and protein levels in *Vibrio cholerae*. *Appl. Environ. Microbiol.* **52:** 788–793.
82. **Hood, M. A., and M. T. MacDonell.** 1987. Distribution of ultramicrobacteria in a gulf coast estuary and induction of ultramicrobacteria. *Microb. Ecol.* **14:**113–127.
83. **Humphrey, B., S. Kjelleberg, and K. C. Marshall.** 1983. Responses of marine bacteria under starvation conditions at a solid-water interface. *Appl. Environ. Microbiol.* **45:**43–47.
84. **Iizuka, T., S. Yamanaka, T. Nishiyama, and A. Hiraishi.** 1998. Isolation and phylogenetic analysis of aerobic copiotrophic ultramicrobacteria from urban soil. *J. Gen. Appl. Microbiol.* **44:**75–84.
85. **Janssen, P. H., A. Schuhmann, E. Morschell, and F. A. Rainey.** 1997. Novel anaerobic ultramicrobacteria belonging to the *Verrucomicrobiales* lineage of bacterial descent isolated by dilution culture from anoxic rice paddy soil. *Appl. Environ. Microbiol.* **63:**1382–1388.
86. **Jurgens, G., K. Lindstrom, and A. Saano.** 1997. Novel group within the kingdom *Crenarcheota* from boreal forest soil. *Appl. Environ. Microbiol.* **63:**803–805.
87. **Kajander, E. O., and N. Ciftcioglu.** 1998. Nanobacteria: an alternative mechanism for pathogenic intra- and extracellular calcification and stone formation. *Proc. Natl. Acad. Sci. USA* **95:**8274–8279.
88. **Kajander, E. O., I. Kuronen, and N. Ciftcioglu.** 1996. Fatal (fetal) bovine serum: discovery of nanobacteria. *Mol. Biol. Cell* **7**(Suppl.)**:**517.
89. **Kerr, R. A.** 1996. Ancient life on Mars? *Science* **273:**864–866.
90. **Kieft, T. L., P. S. Amy, F. J. Brockman, J. K. Fredrickson, B. N. Bjornstad, and L. L. Rosacker.** 1993. Microbial abundance and activities in relation to water potential in the vadose zones of arid and semiarid sites. *Microb. Ecol.* **26:**59–78.
91. **Kieft, T. L., and F. J. Brockman.** Vadose zone microbiology. *In* J. K. Fredrickson and M. Fletcher (ed.), *Subsurface Microbiology and Biogeochemistry*, in press. John Wiley & Sons, New York, N.Y.
92. **Kieft, T. L., J. K. Fredrickson, J. P. McKinley, B. N. Bjornstad, S. A. Rawson, T. J. Phelps, F. J. Brockman, and S. M. Pfiffner.** 1995. Microbiological comparisons within and across contiguous lacustrine, paleosol, and fluvial subsurface sediments. *Appl. Environ. Microbiol.* **61:**749–757.
93. **Kieft, T. L., W. P. Kovacik, Jr., D. B. Ringelberg, D. C. White, D. L. Haldeman, P. S. Amy, and L. E. Hersman.** 1997. Factors limiting to microbial growth and activity at a proposed high-level nuclear repository, Yucca Mountain, Nevada. *Appl. Environ. Microbiol.* **63:**3128–3133.
94. **Kieft, T. L., E. M. Murphy, D. L. Haldeman, P. S. Amy, B. N. Bjornstad, E. V. McDonald, D. B. Ringelberg, D. C. White, J. O. Stair, R. P. Griffiths, T. C. Gsell, W. E. Holben, and D. R. Boone.** 1998. Microbial transport, survival, and succession in a sequence of buried sediments. *Microb. Ecol.* **36:**336–348.
95. **Kieft, T. L., and T. J. Phelps.** 1997. Life in the slow lane: activities of microorganisms in the subsurface, p. 137–163. *In* P. S. Amy and D. L. Haldeman (ed.), *The Microbiology of the Terrestrial Subsurface.* CRC Press, Boca Raton, Fla.
96. **Kieft, T. L., D. B. Ringelberg, and D. C. White.** 1994. Changes in ester-linked phospholipid fatty acid profiles of subsurface bacteria during starvation and desiccation in a porous medium. *Appl. Environ. Microbiol.* **60:**3292–3299.
97. **Kieft, T. L., and L. L. Rosacker.** 1991. Application of respiration- and adenylate-based soil microbiological assays to deep subsurface terrestrial sediments. *Soil Biol. Biochem.* **23:**563–568.

98. **Kieft, T. L., E. Soroker, and M. K. Firestone.** 1987. Microbial biomass response to a rapid increase in water potential when dry soil is wetted. *Soil Biol. Biochem.* **19:**119–126.
99. **Kieft, T. L., E. Wilch, K. O'Connor, D. B. Ringelberg, and D. C. White.** 1997. Survival and phospholipid fatty acid profiles of surface and subsurface bacteria in natural sediment microcosms. *Appl. Environ. Microbiol.* **63:**1531–1542.
100. **Kilbertus, G.** 1980. Etude des microhabitats contenus dans les aggregats du sol. Leur relation avec la biomasse bacterienne et la taille des prokaryotes presents. *Rev. Ecol. Biol. Sol* **17:**543–557.
101. **Kirkland, B. L., F. L. Lynch, M. A. Rahnis, R. L. Folk, I. J. Molineux, and R. J. McLean.** 1999. Alternative origins for nannobacteria-like objects in calcite. *Geology* **27:**347–350.
102. **Kjelleberg, S.** 1984. Effects of interfaces on survival mechanisms of copiotrophic bacteria in low-nutrient habitats, p. 151–159. *In* M. J. Klug and C. A. Reddy (ed.), *Current Perspectives in Microbial Ecology. Proceedings of the Third International Symposium on Microbial Ecology.* Michigan State University, East Lansing, 7–12 August 1983. American Society for Microbiology, Washington, D.C.
103. **Kjelleberg, S., B. A. Humphrey, and K. C. Marshall.** 1982. Effect of interfaces on small, starved marine bacteria. *Appl. Environ. Microbiol.* 43:1166–1172.
104. **Kjelleberg, S., B. A. Humphrey, and K. C. Marshall.** 1983. Initial stages of starvation and activity of bacteria at surfaces. *Appl. Environ. Microbiol.* 46:978–984.
105. **Koch, A. L.** 1997. What size should a bacterium be? A question of scale. *Annu. Rev. Microbiol.* **50:**317–348.
106. **Kogure, K., U. Simidu, and N. Taga.** 1979. A tentative direct microscopic method for counting living marine bacteria. *Can. J. Microbiol.* **25:**415–420.
107. **Kolbel-Boelke, J., E.-M. Anders, and A. Nehrkorn.** 1988. Microbial communities in the saturated groundwater environment II: diversity of bacterial communities in a pleistocene sand aquifer and their in vitro activities. *Microb. Ecol.* **16:**31–48.
108. **Kolbel-Boelke, J., B. Tienken, and A. Nehrkorn.** 1988. Microbial communities in the saturated groundwater environment I: methods of isolation and characterization of heterotrophic bacteria. *Microb. Ecol.* **16:**17–29.
109. **Kostiw, L. L., C. W. Boylen, and B. J. Tyson.** 1973. Lipid composition of growing and starving cells of *Arthrobacter crystallopoietes*. *J. Bacteriol.* **111:**102–111.
110. **Kramer, J. G. and F. L. Singleton.** 1992. Variations in rRNA content of marine *Vibrio* spp. during starvation survival and recovery. *Appl. Environ. Microbiol.* **58:**201–207.
111. **Kurath G., and R. Y. Morita.** 1983. Starvation-survival physiological studies of a marine *Pseudomonas* sp. *Appl. Environ. Microbiol.* **45:**1206–1211.
112. **Kuske, C. R., S. M. Barns, and J. D. Busch.** 1997. Diverse uncultivated bacterial groups from soils of the arid southwestern United States that are present in many geographic regions. *Appl. Environ. Microbiol.* **63:**3614–3621.
113. **Lappin-Scott, H. M., and J. W. Costerton.** 1990. Starvation and penetration of bacteria in soils and rocks. *Experientia* **46:**807–812.
114. **Lappin-Scott, H. M., and J. W. Costerton.** 1992. Ultramicrobacteria and their biotechnological applications. *Curr. Opin. Biotechnol.* **3:**283–285.
115. **Lappin-Scott, H. M., F. Cusack, and J. W. Costerton.** 1988. Nutrient resuscitation and growth of starved cells in sandstone cores: a novel approach to enhanced oil recovery. *Appl. Environ. Microbiol.* **54:**1373–1382.
116. **Lappin-Scott, H. M., F. Cusack, F. MacLeod, and J. W. Costerton.** 1988. Starvation and nutrient resuscitation of *Klebsiella pneumoniae* isolated from oil well waters. *J. Appl. Bacteriol.* **64:**541–549.
117. **Lewis, D. L., and D. K. Gattie.** 1991. The ecology of quiescent microbes. *ASM News* **57:**27–32.
118. **Liesack, W., and E. Stackebrandt.** 1992. Occurrence of novel groups of the domain *Bacteria* as revealed by analysis of genetic material isolated from an Australian terrestrial environment. *J. Bacteriol.* **174:**5072–5078.
119. **Lindahl, V., A. Frostegard, L. Bakken, and E. Baath.** 1997. Phospholipid fatty acid composition of size fractionated indigenous soil bacteria. *Soil Biol. Biochem.* **29:**1565–1569.
120. **Linder, K., and J. D. Oliver.** 1989. Membrane fatty acid and virulence changes in the viable but nonculturable state of *Vibrio vulnificus*. *Appl. Environ. Microbiol.* **55:**2837–2842.

121. **Long, P. E., S. A. Rawson, E. Murphy, B. Bjornstad.** 1992. Hydrologic and geochemical controls on microorganisms in subsurface formations, p. 49–71. *In* Pacific Northwest Laboratory Annual Report for 1991 to the DOE Office of Energy Research, Part 2, Environmental Sciences (PNL 8000, Pt. 2). Pacific Northwest Laboratories, Richland, Wash.
122. **MacLeod, F. A., H. M. Lappin-Scott, and J. W. Costerton.** 1988. Plugging of a model rock system by using starved bacteria. *Appl. Environ. Microbiol.* **54:**1365–1372.
123. **Maniloff, J.** 1997. Nannobacteria: size limits and evidence. *Science* **276:**176.
124. **Marshall, K. C.** 1968. Interactions between colloidal montmorillonite and cells of *Rhizobium* species with different ionogenic surfaces. *Biochim. Biophys. Acta* **156:**179–186.
125. **McInerney, M. J.** 1991. Use of models to predict bacterial penetration and movement within a subsurface matrix, p. 115–135. *In* C. J. Hurst (ed.), *Modeling the Environmental Fate of Microorganisms.* American Society for Microbiology, Washington, D.C.
126. **McKay, A. M.** 1992. Viable but non-culturable forms of potentially pathogenic bacteria in water. *Lett. in Appl. Microbiol.* **14:**129–135.
127. **McKay, D. S., E. K. Gibson, K. L. Thomas-Keprta, H. Vali, C. S. Romanek, S. J. Clemett, X. D. F. Chillier, C. R. Maechling, and R. N. Zare.** 1996. Search for past life on Mars: possible relic biogenic activity in Martian meteorite ALH84001. *Science* **273:**924–930.
128. **McMahon, P. B., and F. H. Chapelle.** 1991. Microbial production of organic acids in aquitard sediments and its role in aquifer geochemistry. *Nature* **349:**233–235.
129. **Morita, R. Y.** 1982. Starvation-survival of heterotrophs in the marine environment. *Adv. Microb. Ecol.* **6:**171–198.
130. **Morita, R. Y.** 1985. Starvation and miniaturization of heterotrophs, with special emphasis on maintenance of the starved viable state, p. 111–130. *In* M. Fletcher and G. D. Floodgate (ed.), *Bacteria in their Natural Environments.* Academic Press, London, United Kingdom.
131. **Morita, R. Y.** 1988. Bioavailability of energy and its relationship to growth and starvation survival in nature. *Can. J. Microbiol.* **34:**436–441.
132. **Morita, R. Y.** 1990. The starvation survival state of microorganisms in nature and its relation to the bioavailability of energy. *Experientia* **46:**813–817.
133. **Moyer, C. L., and R. Y. Morita.** 1989. Effect of growth rate and starvation-survival on cellular DNA, RNA, and protein of a psychrophilic marine bacterium. *Appl. Environ. Microbiol.* **55:**2710–2716.
134. **Murphy, E. M., J. A. Schramke, J. K. Fredrickson, H. W. Bledsoe, A. J. Francis, D. S. Sklarew, and J. C. Linehan.** 1992. The influence of microbial activity and sedimentary organic carbon on the isotope geochemistry of the Middendorf aquifer. *Water Resources Res.* **28:**723–740.
135. **Murphy, E. M., and J. A. Schramke.** 1998. Estimation of microbial respiration rates in groundwater by geochemical modeling constrained with stable isotopes. *Geochim. Cosmochim. Acta* **62:** 3395–3406.
136. **Nealson, K. H.** 1997. Nannobacteria: size limits and evidence. *Science* **276:**176.
137. **Nealson, K. H.** 1997. Sediment bacteria: who's there, what are they doing, and what's new? *Annu. Rev. Earth Planetary Sci.* **25:**403–434.
138. **Novitsky, J. A., and R. Y. Morita.** 1976. Morphological characteristics of small cells resulting from nutrient starvation of a psychrophilic marine vibrio. *Appl. Environ. Microbiol.* **32:**616–622.
139. **Novitsky, J. A., and R. Y. Morita.** 1977. Survival of a psychrophilic marine vibrio under long-term nutrient starvation. *Appl. Environ. Microbiol.* **33:**635–641.
140. **Nystrom, T., N. H. Albertson, K. Flardh, and K. Kjelleberg.** 1990. Physiological and molecular adaptation to starvation and recovery from starvation by the marine *Vibrio* sp. S14. *FEMS Microbiol. Ecol.* **74:**129–140.
141. **O'Farrell, K. A., and P. H. Janssen**. 1999. Detection of Verrucomicrobia in a pasture soil by PCR-mediated amplification of 16S rRNA genes. *Appl. Environ. Microbiol.* **65:**4280–4284.
142. **Olsen, R. A., and L. R. Bakken.** 1987. Viability of soil bacteria: optimization of plate-counting technique and comparison between total counts and plate counts within different size groups. *Microb. Ecol.* **13:**59–74.
143. **Onstott, T. C., T. J. Phelps, F. S. Colwell, D. Ringelberg, D. C. White, D. R. Boone, J. P. McKinley, T. O. Stevens, P. E. Long, D. L. Balkwill, T. Griffin, and T. Kieft.** 1998. Observations

pertaining to the origin and ecology of microorganisms recovered from the deep subsurface of Taylorsville Basin, Virginia. *Geomicrobiol. J.* **15:**353–385.

144. **Onstott, T. C., T. J. Phelps, T. L. Kieft, F. S. Colwell, D. L. Balkwill., J. K. Fredrickson, and F. J. Brockman.** 1999. A global perspective on the microbial abundance and activity in the deep subsurface, p. 489–500. *In* J. Seckbach (ed.), *Enigmatic Microorganisms and Life in Extreme Environments.* Kluwer Academic Publishers, Dordrecht, The Netherlands.
145. **Palumbo, A. V., P. M. Jardine, J. F. McCarthy, and B. R. Zaidi.** 1991. Characterization and bioavailability of dissolved organic carbon in deep subsurface and surface waters, p. 2-57–2-68. *In* C. B. Fliermans and T. C. Hazen (ed.), *Proceedings of the First International Symposium on Microbiology of the Deep Subsurface.* WSRC Information Services, Aiken, N.C.
146. **Papendick, R. I., and G. S. Campbell.** 1981. Theory and measurement of water potential, p. 1–22. *In* J. F. Parr, W. R. Gardner, and L. F. Elliott (ed.), *Water Potential Relations in Soil Microbiology.* Soil Science Society of America, Madison, Wis.
147. **Paul, E. A., and F. E. Clark.** 1989. *Soil Microbiology and Biochemistry.* Academic Press, San Diego, Calif.
148. **Pedersen, K.** 1997. Microbial life in deep granitic rock. *FEMS Microbiol. Rev.* **20:**399–414.
149. **Pedersen, K., J. Arlinger, S. Ekendahl, and L. Hallbeck.** 1996. 16S rRNA gene diversity of attached and unattached bacteria in boreholes along the access tunnel to the Aspo hard rock laboratory, Sweden. *FEMS Microbiol. Ecol.* **19:**249–262.
150. **Pedersen, K., and S. Ekendahl.** 1990. Distribution and activity of bacteria in deep granitic groundwaters of southeastern Sweden. *Microb. Ecol.* **20:**37–52.
151. **Pedersen, J. C., and C. S. Jacobsen.** 1993. Fate of *Enterobacter cloacae* JP120 and *Alcaligenes eutrophus* AEO106(pRO101) in soil during water stress: effects on culturability and viability. *Appl. Environ. Microbiol.* **59:**1560–1564.
152. **Pedersen, J. C., and T. D. Leser.** 1992. Survival of *Enterobacter cloacae* on leaves and in soil detected by immunofluorescence microscopy in comparison with selective plating. *Microb. Releases* **1:**95–102.
153. **Pedone, V. A., and R. L. Folk.** 1996. Formation of aragonite cement by nannobacteria in the Great Salt Lake, Utah. *Geology* **24:**763–765.
154. **Phelps, T. J., D. B. Hedrick, D. Ringelberg, C. B. Fliermans, and D. C. White.** 1989. Utility of radiotracer activity measurements for subsurface microbiology studies. *J. Microbiol. Methods* **9:** 15–27.
155. **Phelps, T. J., E. M. Murphy, S. M. Pfiffner, and D. C. White.** 1994. Comparison between geochemical and biological estimates of subsurface microbial activities. *Microb. Ecol.* **28:**335–349.
156. **Phelps, T. J., E. G. Raione, D. C. White, and C. B. Fliermans.** 1989. Microbial activities in deep subsurface sediments. *Geomicrobiol. J.* **7:**79–91.
157. **Postgate, J. R., and J. R. Hunter.** 1962. The survival of starved bacteria. *J. Gen. Microbiol.* **29:** 233–263.
158. **Psenner, R., and M. Loferer.** 1997. Nannobacteria: size limits and evidence. *Science* **276:**176–177.
159. **Rattray, E. A. S., J. I., Prosser, L. A. Glover, and K. Killham.** 1992. Matric water potential in relation to survival and activity of a genetically modified microbial inoculum in soil. *Soil Biol. Biochem.* **24:**421–425.
160. **Reynolds, P. J., P. Sharma, G. E. Jenneman, and M. J. McInerney.** 1989. Mechanisms of microbial movement in subsurface materials. *Appl. Environ. Microbiol.* **55:**2280–2286.
161. **Rice, S. A., and J. D. Oliver.** 1992. Starvation response of the marine barophile CNPT-3. *Appl. Environ. Microbiol.* **58:**2432–2437.
162. **Ringelberg, D. B., S. Sutton, and D. C. White.** 1997. Biomass, bioactivity, and biodiversity: microbial ecology of the deep subsurface: analysis of ester linked phospholipid fatty acids. *FEMS Microbiol. Rev.* **20:**371–377.
163. **Roberson, E. B., and M. K. Firestone.** 1992. Relationship between desiccation and exopolysaccharide production in a soil *Pseudomonas* sp. *Appl. Environ. Microbiol.* **58:**1284–1291.
164. **Robinson, J. B., P. O. Salonius, and F. E. Chase.** 1965. A note on the differential response of *Arthrobacter* spp. and *Pseudomonas* spp. to drying in soil. *Can. J. Microbiol.* **11:**746–748.

165. **Roszak, D. B., and R. R. Colwell.** 1987. Survival strategies of bacteria in the natural environment. *Microbiol. Rev.* **51:**365–379.
166. **Scherer, C. G., and C. W. Boylen.** 1977. Macromolecular synthesis and degradation in *Arthrobacter* during periods of nutrient deprivation. *J. Bacteriol.* **132:**584–589.
167. **Schut, F., M. Jansen, T. M. P. Gomez, J. C. Gottschall, H. Harder, and R. A. Prins.** 1995. Substrate uptake and utilization by a marine ultramicrobacterium. *Microbiology* **141:**351–361.
168. **Sharma, P. K., M. J. McInerney, and R. M. Knapp.** 1993. In situ growth and activity and modes of penetration of *Escherichia coli* in unconsolidated porous materials. *Appl. Environ. Microbiol.* **59:**3686–3694.
169. **Shaw, J. C., B. Bramhill, N. C. Wardlaw, and J. W. Costerton.** 1985. Bacterial fouling in a model core system. *Appl. Environ. Microbiol.* **49:**693–701.
170. **Shirey, J. I., and G. K. Bissonnette.** 1991. Detection and identification of groundwater bacteria capable of escaping entrapment on 0.45-μm-pore-size membrane filters. *Appl. Environ. Microbiol.* **57:**2251–2254.
171. **Sinclair, J. L., and W. C. Ghiorse.** 1989. Distribution of aerobic bacteria, protozoa, algae, and fungi in deep subsurface sediments. *Geomicrobiol. J.* **7:**15–31.
172. **Smith, J. L., R. I. Papendick, D. F. Bezdicek, and J. M. Lynch.** 1993. Soil organic matter dynamics and crop residue management, p. 65–94. *In* F. B. Metting (ed.), *Soil Microbial Ecology, Applications in Agricultural and Environmental Management.* Marcel Dekker, New York, N.Y.
173. **Soroker, E., L. J. Waldron, and M. K. Firestone.** 1987. Effects of solute vs. matric water potential on microbial activity and survival in soil, p. 189. *In Agronomy Abstracts,* American Society of Agronomy, Madison, Wis.
174. **Stevens, T. O., and J. P. McKinley.** 1996. Lithoautotrophic microbial ecosystems in deep basalt aquifers. *Science* **270:**450–454.
175. **Stotzky, G.** 1986. Influence of soil mineral colloids on metabolic processes, adhesion, and ecology of microbes and viruses, p. 305–428. *In* P. M. Huang and M. Schnitzer (ed.), *The Interaction of Soil Minerals with Natural Organics and Microbes*, SSSA Spec. Publ. No. 17. Soil Science Society of America, Madison, Wis.
176. **Stutz, E., G. Kahr, and G. Defago.** 1989. Clays involved in suppression of tobacco black rot by a strain of *Pseudomonas fluorescens. Soil Biol. Biochem.* **21:**361–366.
177. **Torrella, F., and R. Y. Morita.** 1981. Microcultural study of bacterial size changes and microcolony and ultramicrocolony formation by heterotrophic bacteria in seawater. *Appl. Environ. Microbiol.* **41:**518–527.
178. **Tuckett, J. D., and W. E. C. Moore.** 1959. Production of filterable particles by *Cellovibrio gilvus. J. Bacteriol.* **77:**227–229.
179. **Turpin, P. E., K. A. Maycroft, C. L. Rowlands, and E. M. H. Wellington.** 1993. Viable but non-culturable salmonellas in soil. *J. Appl. Bacteriol.* **74:**421–427.
180. **Uwins, P. J. R., R. I. Webb, and A. P. Taylor.** 1998. Novel nano-organisms from Australian sandstones. *Am. Mineral.* **83:**1541–1550.
181. **Wan, J., J. L. Wilson, and T. L. Kieft.** 1994. Influence of the gas-water interface on transport of microorganisms through unsaturated porous media. *Appl. Environ. Microbiol.* **60:**509–516.
182. **Williams, S. T.** 1985. Oligotrophy in soil: fact or fiction?, p. 81–110. *In* M. Fletcher and G. D. Floodgate (ed.), *Bacteria in Their Natural Environments.* Academic Press, London, United Kingdom.
183. **Wilson, J. T., J. F. McNabb, D. L. Balkwill, and W. C. Ghiorse.** 1983. Enumeration and characterization of bacteria indigenous to a shallow water-table aquifer. *Ground Water* **24:**225–233.
184. **Winogradsky, S.** 1925. Etudes sur la microbiologie du sol. I. Sur la methode. *Ann. Inst. Pasteur* **39:**299–354.
185. **Zvyagintsev, D. G., D. A. Gilichinskii, S. A. Blagodatskii, E. A. Varob'eva, G. M. Klebnikova, A. A. Arkhangelov, and N. N. Kudryavtseva.** 1985. Survival time of microorganisms in permanently frozen sedimentary rocks and buried soils. *Microbiology* (USSR) **54:**131–136.

Nonculturable Microorganisms in the Environment
Edited by R. R. Colwell and D. J. Grimes

Chapter 4

Membrane Bioenergetics in Reference to Marine Bacterial Culturability

Minoru Wada and Kazuhiro Kogure

Understanding of microbial ecology has been expanded by means of isolation, cultivation, and characterization of microorganisms found in natural environments. However, there are still many bacteria which are metabolically active but remain in the so-called "nonculturable" state in nature (44). This state was originally recognized because of the difference observed between direct viable counts and conventional viable counts in seawater (17). Since then, ecological and practical importance of viable but nonculturable cells have been investigated and debated with particular emphasis on *Vibrio cholerae* and other gram-negative marine bacteria (3, 4, 14, 20, 43, 44). The general features of cells entering the nonculturable state can be summarized as reduction in size (27), decrease in macromolecular synthesis (28), and changes in composition of the cell wall and/or membrane (20, 26). However, little has been investigated with respect to the changes in bioenergetic state.

This chapter focuses on the significance of the primary sodium pump which has been found in species of *Vibrio* with regard to culturability. It provides possible scenarios for what happens in the respiratory electron transport system in the bacterium-animal associations, namely, luminescent symbiosis.

MEMBRANE BIOENERGETICS OF MARINE BACTERIA AND NONCULTURABLE STATE

Ecological Significance of Membrane Bioenergetics

The mechanism of bacterial ATP synthesis is generally explained by the chemosomotic theory (23), in which establishment of a proton gradient across the membrane plays a central role. The respiratory chain is coupled to proton translocation, and the resultant potential energy (proton motive force [pmf]) can drive

Minoru Wada and Kazuhiro Kogure • Ocean Research Institute, University of Tokyo, 1-15-1, Minamidai, Nakano, Tokyo 164, Japan.

a variety of energy-linked processes. The primary sodium pump was first discovered in a marine bacterium, *Vibrio alginolyticus*, by Tokuda and Unemoto (34). In addition to the primary proton pump, this bacterium possesses the primary sodium pump which functions at alkaline pH. This respiratory chain translocates sodium ions as a direct result of electron transport. A similar respiratory chain has been found in other halophilic strains (18). The sodium motive force (smf) is used for various cellular functions, including solute transport (35), pH regulation (34), and flagellar rotation (1, 15).

In the marine environment, smf has a distinct advantage over pmf. Without the primary sodium pump, the cell must use the Na^+/H^+ antiporter to generate a sodium gradient across the membranes. Since the pH of seawater is relatively high, the cell cannot use a chemical gradient of protons (ΔpH), but has to depend solely on electrical potential. Therefore, the Na+/H+ antiporter system cannot be electroneutral, i.e., the extrusion of one mole of sodium ion requires exchange of more than one mole of protons. If the cells possess the primary sodium pump, it is possible to generate a sodium gradient directly without any of the loss stated above (37).

This situation has been shown for growth rate and yield of a mutant (Nap 1) of *V. alginolyticus* grown in a synthetic seawater medium amended with glucose (500 mg/liter) as a sole carbon and energy source. This mutant lacks the primary sodium pump (33) but has maintained a primary proton pump. In the wild strain, the sodium pump functions at alkaline pH, and the proton pump at acidic pH. Therefore, it would be expected that Nap 1 would have a greater energetic problem at alkaline pH. Table 1 shows growth rate and yield at different concentrations of sodium at pH 6.5 and 8.5. At pH 6.5, the growth rate of Nap 1 is comparable or even slightly higher than that of the wild strain, whereas the rate was lower at pH 8.5. Growth

Table 1. Growth rate and yield of *V. alginolyticus* at different concentrations of sodium chloride[a]

pH	NaCl concn (mM)	Growth rate (μm)			Growth yield[b]		
		Wild type	Nap 1	Nap 1/wild type	Wild type	Nap 1	Nap 1/wild type
6.5	400	0.44	0.31	0.7	59	35	0.59
	200	0.35	0.42	1.2	65	38	0.58
	100	0.26	0.32	1.2	64	39	0.61
8.5	400	0.49	0.33	0.67	64	31	0.48
	200	0.41	0.34	0.83	71	34	0.48
	100	0.25	0.15	0.60	70	21	0.30

[a]Cells of *V. alginolyticus* were pregrown in a synthetic seawater medium containing the following: NaCl (400 mM), 23.4 g; KCl, 0.8 g; $MgSO_4 \cdot 7H_2O$, 4.0 g; $CaCl_2$, 1.2 g; KBr, 100 mg; $SrCl_2$, 26 mg; H_3BO_3, 20 mg; polypeptone (Japan Pharmaceutical Co., Tokyo, Japan), 1 g; yeast extract (Difco), 0.2 g in 1 liter of distilled water. When necessary, pH was adjusted with 50 mM morpholineethanesulfonic acid (MES) (pH 6.5), *N*-2-hydroxyethylpiperazine-*N'*-2-ethanesulfonic acid (HEPES) (pH 7.5), or Tricine (pH 8.5). After preculture, the cells were harvested and washed twice with the same medium without organic supplements at different concentrations of sodium chloride. Cells were reinoculated into 1 liter of the culture medium amended with glucose (500 mg/liter) as a sole carbon and energy source and incubated at 20°C with shaking. Growth was monitored spectrophotometrically at 600 nm. During the mid-exponential growth, a portion of the culture was passed through a glass fiber filter (GF/F, Whatman) to determine the amount of bacterial cellular carbon with a CHN analyzer (MT-2, Yanagimoto Co.) and to determine the glucose concentration of the filtrates by the glucose oxidase method. Incorporation of carbon into cells of *V. alginolyticus* was estimated from the decrease in glucose concentration in the medium.

[b]Percent cellular carbon per carbon incorporated.

yields of Nap 1 were generally lower than those of the wild type, and this was more prominent at alkaline pH and lower sodium concentration. This result has been interpreted that the cells have more difficulty in establishing a sodium gradient, due to the lack of the primary sodium pump at alkaline pH.

Can Membrane Bioenergetics Be Related to the Nonculturable State?

Considering the oligotrophic condition of seawater, bioenergetic economy should be critical for survival. Since the electrochemical gradient of proton or sodium ions across the membrane plays a central role in the energetics of bacteria, survival can be regarded as a process whereby cells maintain the necessary gradient with the least supplementation of organic solutes.

When a wild-type strain of *V. alginolyticus* is incubated in a synthetic seawater medium without organic supplement, the viability gradually decreases with time, and Fig. 1 shows this phenomenon in the presence of a sodium concentration of

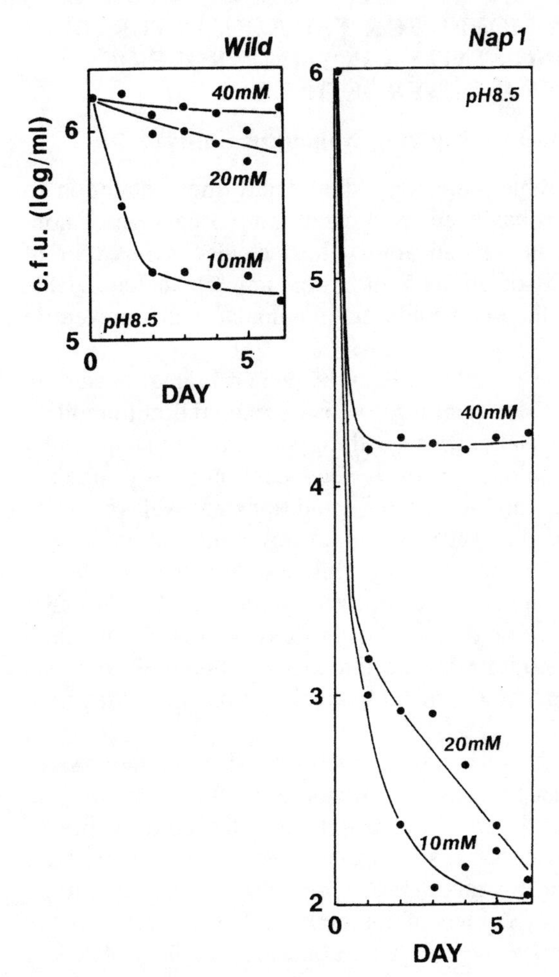

Figure 1. Decrease in viability of *V. alginolyticus* in synthetic seawater at different concentrations of sodium chloride. Cells of *V. alginolyticus* were pregrown, harvested, and washed as described in Table 1, footnote *a*. The washed cells were inoculated into the medium without organic supplement and incubated at 20°C with shaking. Viable counts were obtained by using 1/5-strength Zobell 2216E medium prepared with natural seawater.

40 mM or less. The lower sodium concentration makes it more difficult for cells to establish or utilize smf. This situation is much more serious for Nap 1, which lacks the primary sodium pump. As is shown in Fig. 1, culturability decreases sharply, depending on sodium concentration. Microscopic observation clarified that there were still "cells" present even after incubation for 5 days under this condition. It is probable that there is a threshold in sodium or proton gradient, below which cells have difficulty in maintaining viability. A quantitative study needs to be done to verify this hypothesis.

A second possible reason for nonculturability is that if the primary sodium pump cannot function properly, the cell might have difficulty in handling adverse conditions, such as osmotic stress or the limited nutrient availability in the marine environments. This could be due to injury, a functional problem, or gene expression. Thus, the inhibited cells could remain viable but enter into a nonculturable state. This case can be induced for luminous bacteria and is discussed below.

LUMINOUS SYMBIOSIS AS A MODEL FOR THE NONCULTURABLE AND/OR SLOW GROWING STATE: A POSSIBLE ROLE OF MEMBRANE BIOENERGETICS

Can Symbiosis Be Useful in Studying Nonculturability?

When considering the nonculturable state, one should pay more attention to animal-microbe interactions, since it has been recognized that (i) passage of nonculturable enteric marine bacteria through an animal host results in recovery of culturability (3) and (ii) symbiotic associations with marine animals include viable but nonculturable bacteria, such as the endosymbiotic chemoautotrophs present in marine invertebrates (5, 16).

Although we do not yet know exact mechanisms by which animal-bacterium associations function in favor of either regaining or losing bacterial culturability, the associations should provide bacterial partners with opportunities to switch their metabolic processes. In general, symbiotic growth is dependent on energy supplied by the host organism. Environmental and nutritional conditions are well regulated by the host, and symbionts are forced to survive in a narrow range of conditions. If the integrity of the association becomes higher, this situation may make the isolation more difficult and can be referred to as obligate symbiosis. On the other hand, bacteria facultatively associated with their host would be adapted to more variable conditions which are likely to impose selective pressures on the symbionts to maintain metabolic versatility, generally making the isolation and cultivation easier.

Considering that the obligate symbiosis can be derived from the facultative symbiosis with higher degree of dependence, comparative studies between obligate and facultative bacterial symbionts should gain more insight into the culturability of marine heterotrophic bacteria. In this context, symbiosis between luminous bacteria and fish or cephalopods is promising for investigation, since the bacterium-animal associations provide a wide range of degree of integration. The majority of the luminescent symbiotic bacteria seem to be facultative and are readily isolated in pure culture (11). Symbionts are predominantly members of the genera *Vibrio* and *Photobacterium* that require sodium for growth (2, 32) and/or respiratory chain activities (10). It is therefore probable that the symbiotic relationship can be described in connection with the primary sodium pump. The luminescent symbionts of the flashlight fishes and the deep-sea anglerfishes appear to be obligate, because attempts to cultivate have failed. However, 16SrRNA analysis showed that the symbionts are phylogenetically related to species in the genus *Vibrio* (13) and supported the hypothesis that the symbionts lost metabolic versatility that allows growth in nonsymbiotic environment, after a long period of obligate symbiosis. Thus, comparative studies of the membrane energetics between the laboratory-cultured symbionts and the symbionts in the host light organs are of particular interest.

Slow Growth and the Nonculturable State of Luminous Ectosymbionts: the Bacterial Respiratory Chain Versus the Luciferase System as an Alternative Terminal Oxidase

Luminous bacteria are usually maintained as a pure culture in tubules of the host light organ, which is connected to the exterior to allow constant release of growing cells into the surrounding seawater (11). Although it is not yet clear how their growth is sustained, the symbionts should derive their nutrient from their host and in turn emit light for the host. The physiology and morphology of luminous bacteria maintained in the light organ differ significantly from those cultured in the laboratory (6, 11, 30). Slow growth and a high level of luminescence are phenomena characteristic of the symbiotic state, whereas isolated bacteria in laboratory culture grow faster and emit less light in artificial media. The host may regulate several parameters, including lowering osmolarity and limiting oxygen and iron (7, 12, 25).

From a bioenergetic point of view, the bacterial luciferase system can be seen as an alternative terminal oxidase which competes with the respiratory cytochrome oxidases (9, 10, 21, 24, 38, 42). Thus, the relative transfer of electrons, either to the respiratory chain or to the luciferase system, seems to be highly relevant to maintenance of symbiosis. Although the role of electron transfer in symbiosis has yet to be clarified, Makemson and Hastings (22) suggested that the luciferase of *Vibrio harveyi* can function as an energy-generating system and allow substantial growth without depending on cytochrome oxidases. If this is the case for symbiotic bacteria, the physiological changes described above would be explained; a relative increase in electron transfer to the luciferase system should enhance light production but reduce the efficiency of energy production, thereby resulting in slow growth and high luminescence. Studies with *V. harveyi* suggest that the sodium-driven respiratory chain couples to the luciferase system at the NADH:quinone oxidoreductase site (39, 40). Thus, it is possible that the electron shunt to luciferase in symbionts results in the generation of a sodium gradient across the cell membrane. Considering the slow growth that occurs inside the light organ, the electrochemical gradient generated during symbiosis might be smaller than required for growth in artificial media.

To explain nonculturability, one could expand the hypothesis stated above. If electron shunting is irreversible in the symbiont, the bacterial cells will be unable to generate membrane potential necessary to sustain their population outside the host light organ and therefore fail to adapt to environmental and nutritional changes imposed during cultivation.

The Nonculturable but Infective Luminous Symbiont: the Primary Sodium Pump versus the Primary Proton Pump

Another example of nonculturability among the luminous symbionts was recently described by Lee and Ruby (19). They reported, for the first time, that symbiotic *Vibrio fischeri* could become nonculturable immediately after being expelled into seawater out of the host light organ. However, these free *V. fischeri* cells remained infective for axenic *Euprymna scolopes* larvae. Upon establishing a bacterium-squid symbiosis, the luminous bacteria can infect axenic larvae and proliferate dramatically, until a maximum cell density is reached within the host tubules (31). At present, the reason that the symbiont loses its culturability outside the host is not clear (19).

One intriguing hypothesis is that upon entry into the squid light organ, the bacteria turn from the sodium-driven state to a proton-driven state. Whether the bacteria use the sodium pump or the proton pump is essentially dependent on pH (36). As the pH of an animate surface is likely to be shifted to acidic (1), it is very likely that the pH inside the light organ is lower than that of seawater, thus allowing the symbionts to employ the proton pump. Various cellular mechanisms, including solute transport or maintenance of osmotic equilibrium, could become much more dependent on the proton pump during symbiosis. Once expelled into seawater, the cells would be exposed to drastic changes in nutrition, osmotic strength, and oxygen availability, which could be adverse to the bacterial growth unless the cells are able to respond immediately by switching to the sodium pump. Considering the fact that the cells were in a slow growing state inside the light organ (30), it appears likely that a substantial portion of the cells fail to switch the respiratory chain during the short period of time in which they are exposed to the stresses in seawater, and ultimately they enter into the nonculturable state. Their ability to infect, on the other hand, might be less influenced by environmental change or may easily be restored upon entering the host light organ.

Data obtained for other symbiotic *V. fischeri* isolates appear to be consistent and should be considered with this notion. Table 2 shows the effects of protonophore, carbonyl cyanide *m*-chlorophenylhydrazone (CCCP), on luminescence of *V. fischeri* MJ-1, which was isolated from the light organ of Monocentrid fish (29), and a type strain of *V. harveyi* ATCC 14126. When 10 μM CCCP was added to MJ-1, luminescence was considerably inhibited at both pH 6.5 and 8.5, whereas luminescence of *V. harveyi* was severely inhibited at acidic pH but not at alkaline pH. As described in a previous work, if luminous bacteria possess primary sodium pumps, cells could emit CCCP-resistant luminescence at alkaline pH (39). Although CCCP collapses the proton gradient across the cell membrane, cells can create the sodium gradient through the sodium pump and thereby keep the consistent cellular cationic

Table 2. Effect of CCCP on the luminescence of *V. fischeri* MJ-1 and *V. harveyi* ATCC 14126

pH	CCCP (μM)	Luminescence[a]	
		V. fischeri MJ-1	*V. harveyi* ATCC 14126
6.8	0	100	100
	10	<0.1	4
8.5	0	100	100
	10	31	176

[a]Bacterial strains cultured in LBS (8) were harvested at mid-log phase. After washing twice with 25-mM Tris-HCl (pH 6.8 or 8.5), 30-mM $MgSO_4 \cdot 7H_2O$, 10-mM KCl, and 0.3-M NaCl, cells were resuspended in the same buffer. The cell suspension contained about 0.4 mg of protein per ml. Luminescence was measured with a photometer 10 min after CCCP addition and is shown as percentage of light output without CCCP in each strain at different pH values.

condition which allows the luciferase reaction to continue. On the other hand, at acidic pH where only a proton-driven respiratory chain is functional, cells become entirely vulnerable to the protonophore, and this significantly reduces the luminescent activity. Assuming that CCCP affects and decreases the proton gradient across the cell membrane to a similar extent in either of the two luminous vibrios, sensitivity to CCCP found in MJ-1 at alkaline pH may indicate (i) *V. fischeri* MJ-1 does not have a primary sodium pump similar to that of *V. harveyi* and *V. alginolyticus* or (ii) *V. fischeri* MJ-1 has a primary sodium pump, but it is not active enough to compensate the perturbation of cation distribution across cell membrane caused by protonophore at alkaline pH. Since NADH oxidase activity of the membrane fraction of *V. fischeri* MJ-1 was enhanced specifically by sodium (41), the latter case is likely and consistent with the above hypothesis. As described early in this chapter for a sodium pump defective mutant of *V. alginolyticus* (Fig. 1), malfunction of the primary sodium pump in *V. fischeri* MJ-1 at alkaline pH could result in significant loss of culturability even in a short period of incubation in seawater. Although there are no experimental data available for symbiotic *V. fischeri* in *E. scolapes*, the nonculturability observed by Lee and Ruby (19) may be partly explained by assuming similar incomplete switching of the sodium pump of the symbiotic *V. fischeri.*

CONCLUSIONS

When examining the role of the bacterial physiological state, with respect to culturability of bacteria isolated from the marine environment, the following bioenergetic viewpoint should be considered. First, the primary sodium pump, which is directly coupled to the respiratory chain, should provide a distinct advantage over the primary proton pump for growth and survival in natural seawater. Second, in the case of a bacterium-animal symbiosis, electron transfer to an alternative respiratory system may affect the physiological state of the bacterial symbionts. Further characterization of the respiratory chain and additional bioenergetic information will be useful to substantiate these hypotheses.

Acknowledgments. We thank R. R. Colwell and D. J. Grimes for critical reading and helpful comments on the manuscript. We thank Hajime Tokuda and Paul V. Dunlap for providing us with strains of *V. alginolyticus* 138-2 and its mutant Nap 1 and *V. fischeri* MJ-1, respectively.

REFERENCES

1. **Atsumi, T., L. McCarter, and Y. Imae.** 1992. Polar and lateral flagellar motors of marine Vibrio are driven by different ion-motive forces. *Nature* **355:**182–184.
2. **Baumann, P., L. Baumann, M. Woolkalis, and S. Bang.** 1983. Evolutionary relationships in Vibrio and Photobacterium: a basis for a natural classification. *Annu. Rev. Microbiol.* **37:**369–398.
3. **Colwell, R. R., P. R. Brayton, D. J. Grimes, D. B. Roszak, S. A. Huq, and L. M. Palmer.** 1985. Viable but non-culturable *Vibrio cholerae* and related pathogens in the environment: implications for the release of genetically engineered microorganisms. *Bio/Technology* **3:**817–820.
4. **Colwell, R. R., M. L. Tamplin, P. R. Brayton, A. L. Gauzens, B. D. Tall, D. Herrington, M. M. Levine, S. Hall, A. Huq, and D. A. Sack.** 1990. Environmental aspects of *Vibrio cholerae* in transmission of cholera, p. 327–343. *In* R. B. Sack and Y. Zinnaka (ed.), *Advances on Cholera and Related Diarrheas*, vol. 7. KTK Scientific, Tokyo, Japan.
5. **Distel, D. L.** 1998. Evolution of chemoautotrophic endosymbioses in bivalves. *BioScience* **48:**277–286.
6. **Dunlap, P. V.** 1984. Physiological and morphological state of the symbiotic bacteria from light organ of ponyfish. *Biol. Bull.* **176:**410–425.
7. **Dunlap, P. V.** 1985. Osmotic control of luminescence and growth in *Photobacterium leiognathi* from ponyfish light organs. *Arch. Microbiol.* **141:**44–50.
8. **Dunlap, P. V.** 1989. Regulation of luminescence by cyclic AMP in *cya*-like and *crp*-like mutants of *Vibrio fischeri*. *J. Bacteriol.* **171:**1199–1202.
9. **Grogan, D. W.** 1984. Interaction of respiration and luminescence in a common marine bacterium. *Arch. Microbiol.* **137:**159–162.
10. **Hastings, J. W., and K. H. Nealson.** 1977. Bacterial bioluminescence. *Annu. Rev. Microbiol.* **31:** 549–595.
11. **Hastings, J. W., J. C. Makemson, and P. V. Dunlap.** 1987. How are growth and luminescence regulated independently in light organ symbionts? *Symbiosis* **4:**3–24.
12. **Haygood, M. G., and K. H. Nealson.** 1985. The effect of iron on the growth and luminescence of the symbiotic bacterium *Vibrio fischeri*. *Symbiosis* **1:**39–51.
13. **Haygood, M. G., and D. L. Distel.** 1993. Bioluminescent symbionts of flashlight fishes and deep-sea anglerfishes form unique lineages related to the genus *Vibrio*. *Nature* **363:**154–156.
14. **Hoff, K. A.** 1989. Survival of *Vibrio anguillarum* and *Vibrio salmonicida* at different salinities. *Appl. Environ. Microbiol.* **55:**1775–1786.
15. **Imae, Y., and T. Atsumi.** 1989. Na+-driven bacterial flagellar motors. *J. Bioenerg. Biomembr.* **21:** 705–716.
16. **Jannasch, H. W., and D. C. Nelson.** 1984. Recent progress in the microbiology of hydrothermal vents, p. 170–176. *In* M. J. King and C. A. Reddy (ed.), *Current Perspectives in Microbial Ecology.* American Society for Microbiology, Washington, D.C.
17. **Kogure, K., U. Simidu, and N. Taga.** 1979. A tentative direct microscopic method for counting living marine bacteria. *Can. J. Microbiol.* **25:**415–420.
18. **Kogure, K.** 1998. Bioenergetics of marine bacteria. *Curr. Opin. Biotechnol.* **9:**278–282.
19. **Lee, K-H., and E. G. Ruby.** 1995. Symbiotic role of the viable but nonculturable state of *Vibrio fischeri* in Hawaiian coastal seawater. *Appl. Environ. Microbiol.* **61:**278–283.
20. **Linder, K., and J. D. Oliver.** 1989. Membrane fatty acid and virulence changes in the viable but nonculturable state of *Vibrio vulnificus*. *Appl. Environ. Microbiol.* **55:**2837–2842.
21. **Makemson, J. C.** 1986. Luciferase-dependent oxygen consumption by bioluminescent vibrios. *J. Bacteriol.* **165:**461–466.
22. **Makemson, J. C., and J. W. Hastings.** 1986. Luciferase-dependent growth of cytochrome-deficient *Vibrio harveyi*. *FEMS Microbiol. Ecol.* **38:**79–85.
23. **Mitchell, P.** 1961. Coupling of phosphorylation to electron and hydrogen transfer by a chemosmotic type of mechanism. *Nature* **191:**144–148.
24. **Nealson, K. H., T. Platt, and J. W. Hastings.** 1970. The cellular control of the synthesis and activity of the bacterial luminescent system. *J. Bacteriol.* **104:**313–322.
25. **Nealson, K. H., and J. W. Hastings.** 1977. Low oxygen is optimal for luciferase synthesis in some bacteria: ecological implications. *Arch. Microbiol.* **112:**9–16.

26. **Oliver, J. D.** 1993. Formation of viable but nonculturable cells, p. 239–272. *In* S. Kjelleberg (ed.), *Starvation in Bacteria.* Plenum, New York, N.Y.
27. **Rollins, D. M., and R. R. Colwell.** 1986. Viable but nonculturable stage of *Campylobacter jejuni* and its role in survival in the natural aquatic environment. *Appl. Environ. Microbiol.* **52:**531–538.
28. **Roth, W. G., M. P. Leckie, and D. N. Dietzler.** 1988. Restoration of colony-forming activity in osmotically stressed *Escherichia coli* by betain. *Appl. Environ. Microbiol.* **54:**3142–3146.
29. **Ruby, E. G., and K. H. Nealson.** 1976. Symbiotic association of *Photobacterium fischeri* with the marine luminous fish *Monocentris japonica*: a model of symbiosis based on bacterial studies. *Biol. Bull.* **151:**574–586.
30. **Ruby, E. G., and L. M. Asato.** 1993. Growth and flagellation of *Vibrio fischeri* during initiation of the sepiolid squid light organ symbiosis. *Arch. Microbiol.* **159:**160–167.
31. **Ruby, E. G., and M. J. McFall-Ngai.** 1992. A squid that glows in the light: development of an animal-bacterial mutualism. *J. Bacteriol.* **174:**4865–4870.
32. **Singleton, F. L., R. Attwell, S. Jangi, and R. R. Colwell.** 1982. Effects of temperature and salinity on *Vibrio cholerae* growth. *Appl. Environ. Microbiol.* **44:**1047–1058.
33. **Tokuda, H.** 1983. Isolation of *Vibrio alginolyticus* mutants defective in the respiration-coupled Na+ pump. *Biochem. Biophys. Res. Commun.* **114:**113–118.
34. **Tokuda, H., T. Nakamura, and T. Unemoto.** 1981. Potassium ion is required for the generation of pH-dependent membrane potential and ΔpH by the marine bacterium *Vibrio alginolyticus. Biochemistry* **20:**4198–4203.
35. **Tokuda, H., M. Sugasawa, and T. Unemoto.** 1982. Role of Na+ and K+ in α-aminoisobutyric acid transport by the marine bacterium *Vibrio alginolyticus. J. Biol. Chem.* **257:**788–794.
36. **Tokuda, H., and T. Unemoto.** 1981. A respiration-dependent primary sodium extrusion system functioning at alkaline pH in the marine bacterium *Vibrio alginolyticus. Biochem. Biophys. Res. Commun.* **102:**256–271.
37. **Tokuda, H., and T. Unemoto.** 1985. The Na+-motive respiratory chain of marine bacteria. *Microbiol. Sci.* **2:**65–71.
38. **Ulitzur, S., A. Reinhertz, and J. W. Hastings.** 1981. Factors affecting the cellular expression of bacterial luciferase. *Arch. Microbiol.* **137:**159–162.
39. **Wada, M., K. Kogure, K. Ohwada, and U. Simidu.** 1992. Coupling between the respiratory chain and the luminescent system of *Vibrio harveyi. J. Gen. Microbiol.* **138:**1607–1611.
40. **Wada, M., H. Tokuda, K. Kogure, and K. Ohwada.** 1994. The membrane fraction of *Vibrio harveyi* as a possible site of in vivo luminescence, p. 560–563. *In* A. K. Campbell, L. J. Kricka, and P. E. Stanley (ed.), *Bioluminescence and Chemiluminescence*: *Fundamentals and Applied Aspects.* John Wiley & Sons, Chichester, United Kingdom.
41. **Wada, M., and P. V. Dunlap.** 1997. Molecular cloning of the respiratory NADH dehydrogenase (NDH-2) from *Vibrio fischeri*, abstr. I-60, p. 331. *In Abstracts of the 97th General Meeting of the American Society for Microbiology 1997.* American Society for Microbiology, Washington, D.C.
42. **Watanabe, T., N. Mimura, A. Takimoto, and T. Nakamura.** 1975. Luminescence and respiratory activities of *Photobacterium phosphoreum. J. Biochem.* **77:**1147–1155.
43. **Wolf, P. W., and J. D. Oliver.** 1992. Temperature effects on the viable but nonculturable state of *Vibrio vulnificus. FEMS Microbiol. Ecol.* **101:**33–39.
44. **Xu, H.-S., N. Roberts, F. L. Singleton, R. W. Attwell, D. J. Grimes, and R. R. Colwell.** 1982. Survival and viability of nonculturable *Escherichia coli* and *Vibrio cholerae* in the estuarine and marine environment. *Microb. Ecol.* **8:**313–323.

Nonculturable Microorganisms in the Environment
Edited by R. R. Colwell and D. J. Grimes

Chapter 5

Diversity of Uncultured Microorganisms in the Environment

Erko Stackebrandt and T. Martin Embley

With the ratification of the Convention on Biological Diversity by the signature nations, the importance of biological diversity for maintaining the life-sustaining systems of the biosphere has entered a legislative level. The preamble of this convention highlights the possibilities of significant reduction or loss of biological diversity at source caused by human activities, and it admits a "general lack of information and knowledge regarding biological diversity" and "the urgent need to develop scientific, technical and institutional capacities to provide the basic understanding upon which to plan and implement appropriate measures." Nowhere is the lack of knowledge more acute than for uncultured microbial diversity. There is now a widespread appreciation among microbiologists that cultured microorganisms represent a very small, and not necessarily ecologically important fraction of natural microbial diversity. In fact, the microbiologists (e.g., Winogradsky and Beijerinck) who were most responsible for developing the selective enrichment approach were already aware of its limitations.

It can perhaps be considered coincidence that during the several years of discussion that preceded the Convention on Biological Diversity, molecular methods were developed that enabled a first glimpse of the true microbial diversity in natural habitats. This chapter will summarize the achievements of "molecular microbial ecology" and discuss some of the strengths and weaknesses of the methods of quantifying the metabolically active part of a microbial community. As the full range of environments subjected to molecular analysis cannot be covered in this chapter, only certain examples are selected which were investigated at an early stage after the introduction of these molecular methods in microbial ecology.

Erko Stackebrandt • DSMZ-German Collection of Microorganisms and Cell Cultures, 38124 Braunschweig, Germany. ***T. Martin Embley*** • Microbiology Group, Department of Zoology, The Natural History Museum, Cromwell Road, London SW7 5BD, United Kingdom.

MOLECULAR CHARACTERIZATION OF PURE MICROBIAL CULTURES

To understand the dramatic progress in molecular microbial ecology, one needs to consider that almost all the techniques used were developed by taxonomists over the last 25 years or so for the classification and identification of pure cultures. Information from these studies also provides the context whereby the identities or relationships of uncultured taxa may be inferred and their diversity appreciated. For over 100 years microbiologists had to rely exclusively on phenotypic characters for describing taxa. Numerical taxonomy of unweighed characters (59) was the most advanced form of this phenetic approach for determining close relationships among microorganisms. Additional useful characters could be determined by analyzing episemantic characters, such as lipids, lipopolysaccharides, cell wall structures, and base composition of DNA (27). These properties are useful in that they are exclusive, i.e., two organisms that differ in one or more of these characters are unlikely to be very closely related. The relative distance of any relationship and the phylogenetic position of taxa cannot, however, be reliably determined using these characters.

This goal was first achieved when it became possible to indirectly determine sequence similarity by DNA-DNA hybridization (54). Single-stranded DNA from two strains showing less than 15% base mispairing will reassociate to a measurable extent to form a DNA hybrid (65). The melting point (T_m) of the hybrid should also be determined as a second parameter of relationship, as the stability of the hybrid increases with the increasing relatedness of the strains investigated. Values of greater than 70% DNA similarity correlate with less than 2% differences in the thermal stability of duplexes (30). As measured with experimentally introduced mispairings, thermal stabilities have been estimated to decrease from 1 to 2.2% for each percent mispairing (65). Considering that these experiments have been performed on short stretches and not on complete genomes one can nevertheless argue that organisms that share 70% and higher DNA similarities share at least 96% DNA sequence identity. Different formats have been developed in the past to measure DNA relatedness by hybridization, and good congruency exists between these formats, at least among highly related strains and species.

The inability of DNA-DNA hybridization to unravel relationships among more distantly related strains was quickly recognized and stimulated a search for methods measuring broader taxonomic divergence. When the conserved character of rRNA species and their genes was demonstrated, the determination of rRNA cistron similarities (8, 31) extended the range of relationships which could reliably be inferred to the genus and family level. This approach, in combination with DNA hybridization and phenotypic characterization, produced some major taxonomic rearrangements, especially among the proteobacteria. These studies allowed microbial taxonomists to appreciate for the first time just how biochemically and morphologically versatile even relatively closely related bacteria could be. The main disadvantages of the hybridization approaches were their inability to determine actual sequences and the fact that the data were not cumulative.

It is now possible to sequence directly rRNA, rDNA, and essentially any other gene or protein. For taxonomists, the sequencing revolution began in the mid-1970s with the sequencing of short 16S rRNA oligonucleotides by the cataloging method (62). Currently, the combined application of PCR, commercially available sequencing kits and vectors, and automated sequencing devices allows the rapid determination of the phylogenetic position of any culturable and most uncultured microorganisms. The sequences generated are cumulative so that existing and new data from cultured or uncultured microorganisms can be easily integrated into a single phylogenetic analysis.

SYMBIOSES

The use of molecular sequences has facilitated the identification of symbiotic microorganisms which often cannot be cultured, although it must be said that serious attempts to do so are seldom performed. Phylogenetic analysis of the sequences obtained have clearly demonstrated that symbioses between eukaryotes and prokaryotes have formed repeatedly and that the symbionts themselves are phylogenetically diverse. Thus the ability to colonize, or to be colonized, is common throughout the tree of life. In cases where the microorganism is abundant and easily purified or when its rRNA could be separated from that of the host, the identities of symbionts could be addressed prior to the development of more sophisticated molecular methods. An example is the case of the photosynthetic *Prochloron didemni*, which lives within its ascidian host (57). If, however, the symbiont occurs in small numbers or in the presence of other microbes, its nucleic acids must be amplified by PCR and sequenced, and the symbiont sequence must be verified within its eukaryotic matrix by oligonucleotide probing.

Most of the symbionts recognized so far come from the bacterial domain (such as *Holospora*; *Epulopiscium*; *Buchnera*; symbionts of *Nasonia*, *Dysmicoccus*, and *Pseudococcus*; and others), but the ability to colonize anaerobic protozoa appears to be widespread phylogenetically among methanogenic archaea (14). All the symbiont sequences are novel, but given the small number of sequences for cultured archaea, this is hardly surprising. The basis of the interaction appears to be hydrogen transfer from host to endosymbiont, which can use the gas for methanogenesis. Anaerobic habitats contain large communities of phylogenetically diverse eukaryotes (15, 20) which feed on the rich and varied prokaryote populations. Many of these eukaryotes form more or less permanent functional consortia with different bacteria as endo- or ectosymbionts (16, 18, 19, 21, 35). Some rather elegant experiments have identified physiologies for the bacteria in a small number of cases, e.g., the sulphur-oxidizing ectosymbiont on *Kentrophorus* (19) and the phototroph in *Strombidium* (18), but the phylogenetic relationships, stabilities, and specificity of most of these interactions are completely unknown.

The ability to fix dinitrogen occurs in many different branches of the phylogenetic tree for prokaryotes (13, 74). Among proteobacteria the ability to fix nitrogen is particularly common, and it is found in members of the genera *Azospirillum*, *Azotobacter*, *Bradyrhizobium*, and *Rhizobium*, to name but a few. Judged by the distribution of this ability mapped on the proteobacterial rRNA sequence-based tree, it appears that it can be lost or gained over relatively short phylogenetic distances (74). Thus it is difficult to predict from phylogenetic position alone whether an environmental sequence, which groups with sequences from characterized diazotrophs, actually belongs to a bona fide nitrogen-fixing strain. Some of the phylogenetic instability of this phenotype may be due to the fact that in some taxa the nif genes are borne on plasmids. However, there are insufficient comparative data to say if this is the case in all taxa or even if the systems which fix nitrogen in the phylogenetically different groups of proteobacteria are always the same, or if some or all components have independent evolutionary origins. Although no eukaryotic organism has ever convincingly been shown to fix nitrogen, they frequently benefit from symbioses with diazotrophic bacteria. The symbiotic association between legumes and members of the genera *Rhizobium* and *Bradyrhizobium* is the single most beneficial association between plant and bacterium in agriculture (13). Even here, our appreciation of symbiont diversity is far from complete. In a recent investigation Oyaizu et al. (46) were able to detect at least 16 genetically distinct groups among rhizobia colonizing different legumes. Nitrogen-fixing symbioses are not restricted to plants. The gland of Deshayes in invertebrate shipworms contains large numbers of bacteria. In some cases, these have been isolated and shown to possess the ability to digest cellulose and fix nitrogen (69). Subsequent molecular experiments have nicely complemented these earlier studies to confirm by in situ probing that the isolates are the endosymbionts and that the same group of previously unrecorded gamma-proteobacteria have a worldwide distribution in taxonomically different members of the host group (12).

The comparison of phylogenetic trees of two partners in order to study the long-term history of an association is a well-established tool (in zoology and botany) for studying its evolution, diversity, and ecology (6). The method is increasingly being used to study relationships where one or both partner(s) are microbial. Thus, comparison of methanogen symbiont and anaerobic ciliate host phylogenies demonstrates little evidence of parallel cladogenesis among the taxa sampled so far (14). This suggests that the ability to harbor endosymbionts or to colonize anaerobic hosts may prove to be phylogenetically widespread and that it will occur in most anaerobic habitats where protozoa and methanogens co-occur (e.g., 22). In contrast, the ability to act as primary endosymbionts of different aphids appears to be restricted to a single lineage among the gamma proteobacteria; the *Buchnera aphidicola* complex (43) (which also contains endosymbionts of the tsetse fly [45]), and the carpenter ant *Camponotus* (56). The symbioses appear to be obligate, and the topology of the symbiont tree is completely concordant with the host phylogeny based upon morphology (42). These results imply a single original infection in a common ancestor of the aphids which have so far been studied. The complete congruence between the host and primary endosymbiont trees also allowed the fossil and biogeographic time points for the aphid phylogeny to be used to calibrate the 16S rRNA clock of the very closely related (ca. 8% or less sequence divergence) symbionts to approximately 2 to 4% fixed substitutions per 100 million years (41, 42). In contrast to the phylogenetic uniformity of aphid primary endosymbionts, published data for morphology and heat shock proteins suggest that secondary endosymbionts of aphids may represent a more diverse assemblage (25, 41).

THE SUCCESSFUL ERA OF CLONE LIBRARIES TO ANALYZE MICROBIAL DIVERSITY

Although Norman Pace and colleagues (48) first suggested that rRNA sequences could be used to characterize natural communities without the need to culture, it was not until much later that the first experiments using 16S rRNA sequences to characterize natural communities were published (26, 69). The delay was largely due to the need to develop robust and simple technologies whereby 16S rRNA sequences could be recovered from complex mixtures of environmental nucleic acids and then individually sequenced. Since then, there have been a number of studies (only the earliest, the "trendsetters," are mentioned) which have used 16S rRNA or 16S rDNA sequences to explore microbial diversity in the marine environment (9, 10, 23, 24, 55), within a fresh water thermal spring system (71, 72), and within soil samples (36, 37, 60). A recent summary of the impact of culture-independent studies on the emerging phylogenetic view of bacterial diversity (29) suggests that the number of identifiable bacteria divisions has more then tripled to about 40 due to the molecular analysis of environmental microbial communities.

Marine Environment

Over 70% of the Earth's surface is covered by salt water and most of the biomass and biogeochemical activity occurring therein can be attributed to picoplankton (17). Marine prokaryotes are considered to be the major primary producers and heterotrophic consumers in these systems (24). However, little is known of the genetic structure or diversity of these very large communities. The molecular studies published so far have looked mainly at oligotrophic subsurface marine microbial communities in the Pacific and Atlantic (4, 9, 23, 24, 26, 55). Despite some methodological differences, many of the sequences recovered have belonged to the same previously unrecorded lineages. For example, representatives of two alpha proteobacterial lineages (SAR 11 and SAR 83 [see Fig. 1]) have so far been detected in samples taken from the photic zone of the Atlantic Sargasso Sea (24, 26), from surface waters in the Santa Barbara Channel (10), and in samples from the North Pacific (23, 24, 55). The same cyanobacteria/prochlorophyte-like sequences (SAR 7 cluster) have also been recovered from both oceans (4, 24, 55), and the presence of members of these clusters has been verified in most of subsequent studies in different oceans. These results suggest that members of some prokaryote lineages are transoceanic and (given the small size of the samples processed) may represent clades which are abundant and important in marine environments. The only published probing data for the presence of the SAR 11 cluster showed that rRNA from this group represented approximately 12% of the probeable prokaryotic rRNA from Sargasso Sea samples, and a small but detectable fraction of the rRNA from other marine samples (26). At the moment we can only speculate what physiologies these cosmopolitan lineages possess. The success by which this can be done depends to some extent on the number of sequences from cultured and phenotypically characterized organisms, the nearness of environmental clones to them, and whether or not the new sequences fall within phenotypically homogenous clades of cultured taxa. It is probably safe to assume that the SAR 7 cluster are oxygenic phototrophs

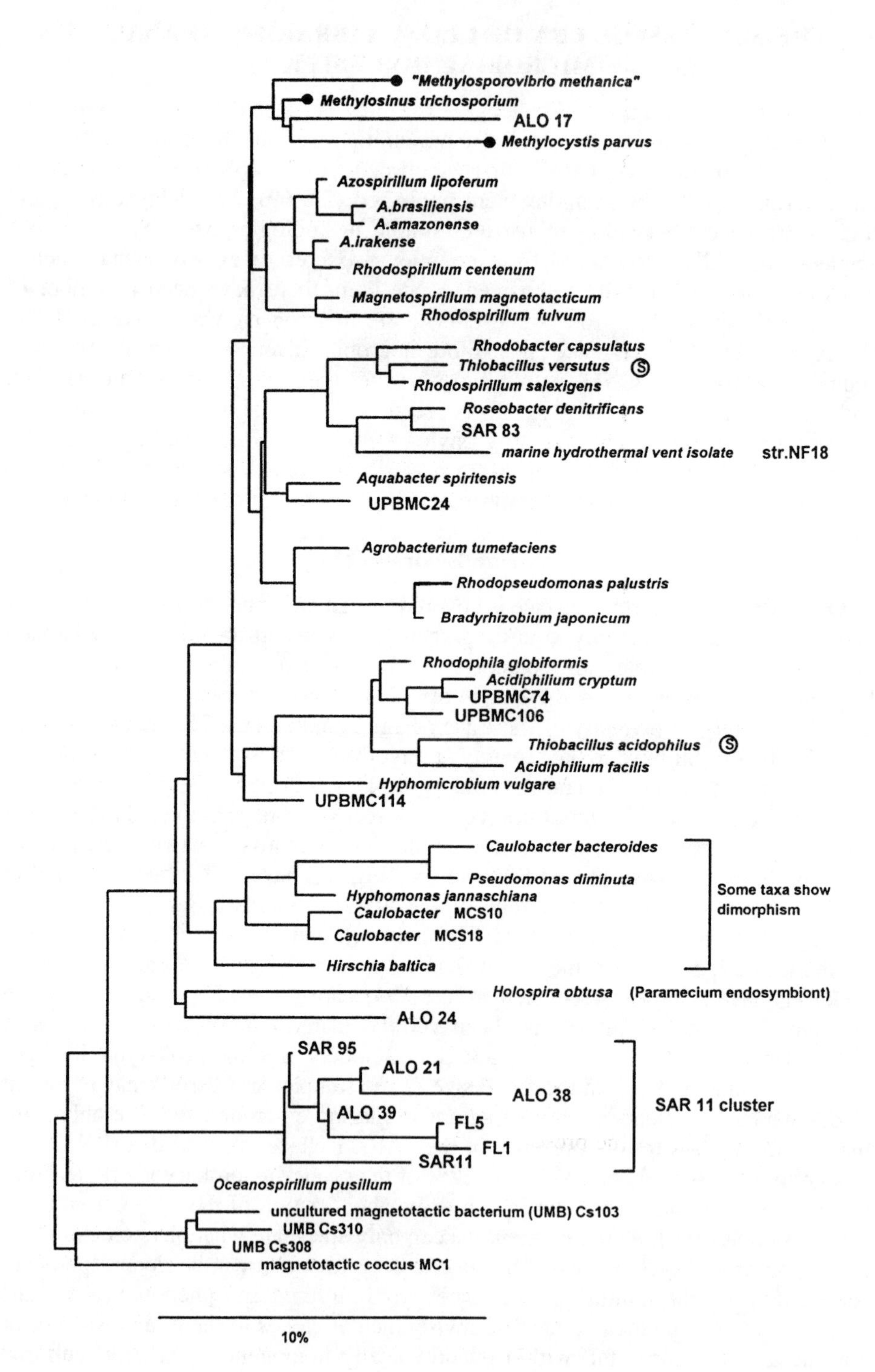
"Methylosporovibrio methanica"
Methylosinus trichosporium
ALO 17
Methylocystis parvus
Azospirillum lipoferum
A.brasiliensis
A.amazonense
A.irakense
Rhodospirillum centenum
Magnetospirillum magnetotacticum
Rhodospirillum fulvum
Rhodobacter capsulatus
Thiobacillus versutus
Rhodospirillum salexigens
Roseobacter denitrificans
SAR 83
marine hydrothermal vent isolate str.NF18
Aquabacter spiritensis
UPBMC24
Agrobacterium tumefaciens
Rhodopseudomonas palustris
Bradyrhizobium japonicum
Rhodophila globiformis
Acidiphilium cryptum
UPBMC74
UPBMC106
Thiobacillus acidophilus
Acidiphilium facilis
Hyphomicrobium vulgare
UPBMC114
Caulobacter bacteroides
Pseudomonas diminuta
Hyphomonas jannaschiana
Caulobacter MCS10
Caulobacter MCS18
Hirschia baltica
Some taxa show dimorphism
Holospira obtusa (Paramecium endosymbiont)
ALO 24
SAR 95
ALO 21
ALO 38
ALO 39
FL5
FL1
SAR11
SAR 11 cluster
Oceanospirillum pusillum
uncultured magnetotactic bacterium (UMB) Cs103
UMB Cs310
UMB Cs308
magnetotactic coccus MC1
10%

since they are closely related to cultured bacteria which possess this phenotype and no nonphotosynthetic member of the cyanobacterial clade has yet been recorded. However, the phenotype of the SAR 11 cluster has to be considered completely unknown; cultured members of the alpha-*Proteobacteria* (*Roseobacter*, *Sulfitobacter, Ruegeria*) and gamma-*Proteobacteria* (*Alteromonas*, *Pseudoalteromonas*) are among the most phenotypically diverse of prokaryotes, and organisms which possess similar rRNA sequences may display morphological dimorphism, chemoorganotrophy, aerobic and anaerobic phototrophism, nitrogen fixation, or magnetotactic behavior (Fig. 1).

As well as these ubiquitous lineages, single or small numbers of (so far) geographically restricted but phylogenetically diverse lineages have also been detected in marine samples. These include a new and very deep branching bacterial lineage, a sequence which is moderately related to cultured flavobacteria, a gram-positive-like sequence, and a small number of single sequences, or clusters of clones, which form deep branches among or close to the *Proteobacteria* (10, 24, 55). These data, from a restricted number of experiments, suggest that the oceans contain a large number of diverse and previously unrecorded microorganisms, a finding that has been verified in many subsequent studies of the marine environment. It is currently impossible to assign a role to most of these environmental lineages.

Particulate organic material ("marine snow") is the site of significant microbial activity in the marine environment. The results of microscopic examination and metabolic studies have predicted that the microbial communities associated with such particles are likely to be different from those found in open water (1, 7). In a preliminary study, DeLong et al. (10) have compared the communities associated with phytodetrital particles and open water, off the coast of California in the Santa Barbara channel. The two types of samples contained markedly different lineages. Many of the coastal open-water clones resembled lineages found in previous studies of free-living marine communities, e.g., members of the SAR 11 cluster were recovered (FL clones in Fig. 1). In contrast, most of the aggregate associated sequences were new moderate- to deep-branching lineages related to *Cytophaga*, *Flavobacteria* and relatives, or to the *Planctomyces*, bacterial groups which contain many examples of surface-associated adaptations.

Hot Spring Environments

The microbial communities associated with hot spring microbial mats have been subjected to detailed study using a variety of techniques (5, 68). As a result, there

Figure 1. Phylogenetic tree of selected alpha-proteobacteria showing some of the phenotypes represented within this group. The tree is based upon only 220 bases of 16S rRNA sequences from the 5′ end of the molecule, hence some deeper relationships are probably unreliable. Sequences for environmental clones are from Schmidt et al. (55) (ALO), DeLong et al. (10) (FL), Britschgi and Giovannoni (4) (SAR), and Liesack and Stackebrandt (37) (MC). All other sequences were obtained from the Ribosomal Database Project (39). The tree was constructed using Neighbour Joining (54) and the Jukes and Cantor (33) correction. The scale bar represents an estimated 10% substitution.

were a number of cultures which could be sequenced for their rRNA prior to beginning population analysis using molecular methods. The mat community therefore provided a nice comparison of microbial composition as assessed by selective culture and the molecular approach. The results of the sequence investigations published so far have clearly demonstrated that a phylogenetically diverse community inhabits even this "simple" system (52). None of the recovered sequences closely resembles sequences from cultured taxa previously isolated from the spring environment. In particular, small subunit rRNA sequences from *Synechoccous lividus* or *Chloroflexus aurantiacus*, both of which were thought to be important in the formation of the mat community, have not yet been recovered.

Using a PCR primer pair, consisting of one "universal" and one archaeal-specific primer, the analysis of a hot spring located in the Mud Volcano area ("Jim's Black Pool") revealed an unexpected large number of distinct archaeal sequences, specifically related to sequences from *Crenarchaeota* (3). A few sequences showed high similarities with 16 rRNA sequences of cultivated *Archaea*, such as *Desulfurococcus mobilis*, *Pyrobaculum islandicum*, and *Thermofilum pendens*, without, however, being identical with any. Several clone sequences could not be related to those of described crenarchaeal species, and each of these novel lineages diverges from the crenarchael stem closer to its root than do the sequences of the cultivated species. The archaeal sequences from the hot spring environment are not closely related to the novel archael sequences retrieved from the marine environment (see below) which appear to possess a similar position intermediate to the crenarchaeota and the euryarchaeota. These studies indicate that the bifurcation of the domain Archaea is somewhat blurred by the position of the novel lineages and that not only the phylogenetic but probably also the physiological diversity of the archaeae is significantly larger than reflected by the few cultured representatives.

Marine Archaea

Cultured archaea have been isolated from what are frequently termed "extreme" environments, and thermophilic, acidophilic, alkalophilic, halophilic, anaerobic, and sulfur-based phenotypes occur throughout the archaeal tree. It was rather surprising, therefore, when two groups of 16S rRNA genes from the archaeal domain were recovered from water samples taken from geographically distant oxic coastal and deep ocean samples (10, 23, 24). One of the groups of related sequences falls within the *Euryarchaota* with the cell-wall-less *Thermoplasma acidophilum* as a distant relative. The other group of sequences are particularly intriguing since they form a deep and long branch close to the *Crenarchaota* at the base of the archaeal tree. The depth and degree of divergence of this previously unknown and cryptic lineage is similar to that which occurs between the two kingdoms of *Archaea*. The precise placement of these organisms is problematic since their rRNA genes do not show the base compositional bias (63 to 67 mol% G+C) found in the basal archaeal lineages, most of which are thermophilic (73). The increased amount of guanine and cytosine in thermophile rRNAs is thought to improve ribosome stability at high temperature. The environmental sequences contain a G+C content of between 51 and 54%; which suggests that they are not thermophilic, and their presence in coastal waters suggests that some may be aerobic. These speculations apart, nothing is known of their phenotype or roles in the ocean. That they are ubiquitous in marine environments is further demonstrated by the discovery of a phylogenetically diverse assemblage of related sequences in the midgut contents of a marine Holothurian living at a depth of 4,800 meters in the North Atlantic Abyss (40).

Soil Environment

Soil constitutes an extremely heterogeneous habitat with complex physicochemical and biological characteristics. Most soil microbes are thought to exist at low metabolic activities or as resting stages because (apart from decaying organic material or plant roots) readily available nutrients are often absent. In addition, many sites in the complex pore space are not accessible by microbes due to size restrictions (28). In contrast to the extensive work on different marine sites, determination of microbial biodiversity in soil is less advanced. Early studies on rumen microorganisms (63) demonstrated the potential difficulties in the isolation of PCR amplifiable DNA due to the presence of interfering material such as humic acids. The presence of these compounds in soil, the problems of quantitative cell recovery, and the presence of residual DNA from dead cells are all factors which have hindered study of soil microorganisms using molecular methods. The studies that have been carried out have clearly demonstrated that soil contains a rich and varied flora. For example, Torsvik and colleagues (64) estimated, from the renaturation kinetics of extracted genomic DNA, that a deciduous soil sample contained approximately 4,000 completely different bacterial genomes.

One of the soil samples that has been investigated thoroughly by sequencing and probing of 16S rRNA genes originated from a subtropical, moderately acidophilic and forested environment in South Queensland, Australia. In contrast to expectation, hardly any sequences in the library originated from the genus *Streptomyces*, members of which were isolated in large numbers from the same soil sample. Moreover, the two library sequences were different from the sequences obtained from the isolates. The reasons for the low representation of *Streptomyces* in the library are still unclear; it may indicate that the streptomycetes were mainly in spore form and failed to lyse. Alternatively, it may indicate that streptomycetes are actually a minor (numerically) component of the soil flora but selective isolation exaggerates their numbers. Of the other groups represented in the clone library only one, albeit the largest, was closely related to sequences from cultivated organisms; nitrogen-fixing members of the alpha-2 subgroup of proteobacteria which includes *Bradyrhizobium* and *Azorhizobium* (36, 37). All the other clones represented novel groups that were only remotely related to known taxa. Several of these constituted two sublines near the base of the actinomycetes, where they grouped with the iron-oxidizing strain TH3 or *Atopobium*, respectively (37, 60). Members of this group have now been detected in many different soil samples, wastewater treatment plants, and even in the marine environment (51). Others formed individual sublines within the radiation of *Planctomycetales* (Fig. 2) and *Verrucomicrobiales*, or they formed novel lineages within the *Acidobacterium-Holophaga* phylum. Additional individual sequences were found to branch off within the

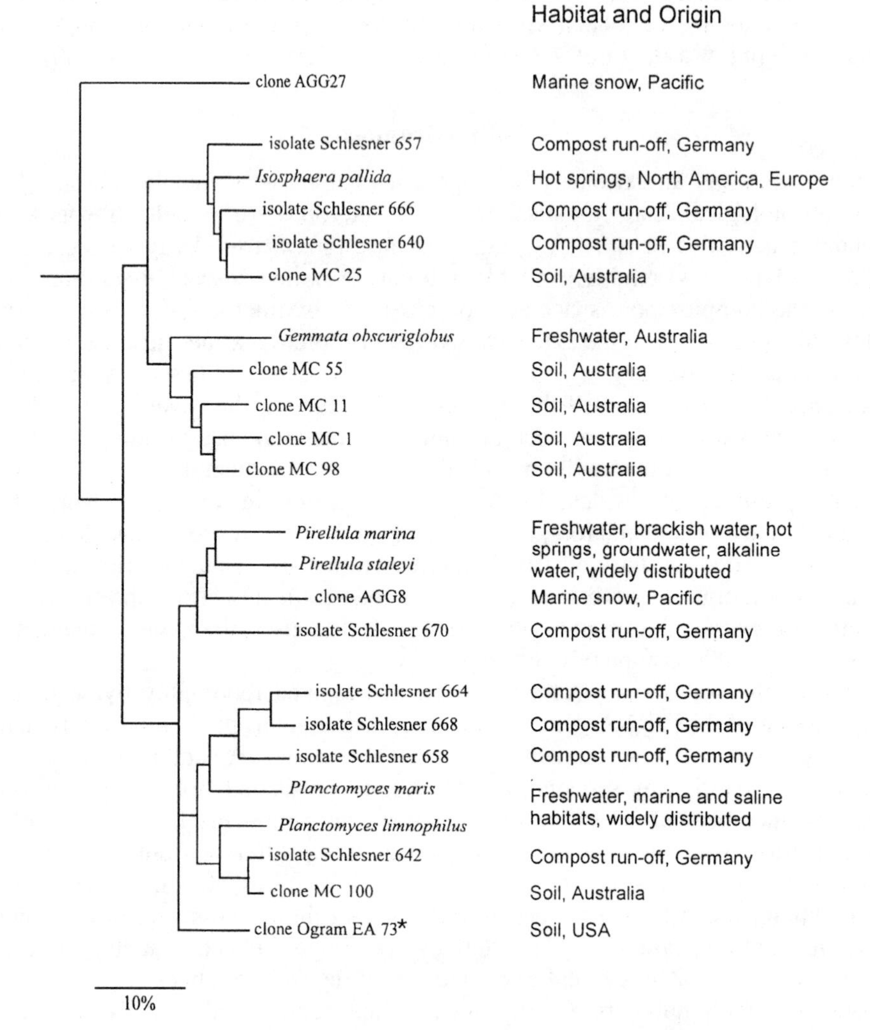

Figure 2. Phylogenetic tree of members of the order *Planctomycetales* and habitat and origin of isolates and clones. The tree is based on about 900 bases of 16S rRNA sequences from the 5′ end of the molecule. Sequences for environmental clones are from DeLong et al. (10) (AGG), Liesack and Stackebrandt (37) (MC) and Bollinger and Ogram, unpublished (EA). The asterisk indicates that the clone most likely represents a chimera, composed of sequences from members of *Pirellula* and/or *Planctomyces*. All other sequences are available from EMBL (37). Pairwise evolutionary distances were computed using the correction of Jukes and Cantor (33). Distance matrix analysis was done according to DeSoete (11). The scale bar represents an estimated 10% substitution.

Bacillus-Clostridium subphylum or were related to *Acidiphilium* and *Thiobacillus acidophilus* within the alpha-1 subclass of *Proteobacteria.*

INTERPRETING MOLECULAR DATA FOR THE ASSESSMENT OF DIVERSITY

The diversity uncovered by 16S rRNA sequence analysis of microbial communities is most often visualized by reference to a tree diagram of relationships. Trees are very informative and can be used to illustrate the number, phylogenetic distribution, and degree of divergence of new lineages, as well as their relationships to cultured taxa. Most environmental analyses have uncovered a heterogeneous mixture of deep and shallow branching lineages, very few of which have shown close relationships to cultured taxa. How does one interpret this tremendous diversity in familiar terms? Is each new sequence a new species? The bacterial species is currently the only taxonomic rank in microbiology which is defined in phylogenetic terms (70). The borderline at which strains are said to belong to a (geno)species is 70% DNA-DNA similarity in binding assays. This admittedly subjective cutoff value corresponds to an estimated 97.5% sequence similarity between two genomes (30). Considering the total bases in the genome of *E. coli* to be approximately 4×10^6, then 4% difference equates to about 1.6×10^5 nucleotide differences (not taking into account the possibility that genome rearrangements are likewise a source of decrease in DNA similarities). This divergence could easily account for the differences in phenotype observed among strains of some species. With the advent of molecular sequence analysis, it is likely that at some stage the DNA hybridization component in the species definition may be replaced by some agreed threshold 16S rDNA similarity value. The reasons are obvious as more taxonomists perform sequence analysis than do DNA hybridization experiments, and because sequences are more direct measurements and less prone to inaccuracies. A correlation plot of these two phylogenetic parameters is not linear, and for highly related organisms the resolving power of DNA hybridization is higher than 16S rRNA sequence analysis (2, 61). However, it is clear that if two organisms share less than about 98% 16S rRNA sequence similarity, it is unlikely that they would be assigned to the same species by DNA-DNA binding. This correlation can help when trying to assess the molecular biodiversity revealed by environmental clones in terms of species composition and numbers.

Prior to trying to interpret what sequence divergence of environmental clones means in terms of species, it is worth emphasizing that most environmental sequences are incomplete. Indeed, some published analyses are based upon as few as 120 nucleotides or 8% of the complete sequence. If these short stretches sample the highly variable regions of 16S rRNA sequence, then dissimilarity will be exaggerated; if the more conserved regions are sampled, then it will be underestimated. In several studies the sequences are more than 900 bases long, and in these analyses the dissimilarity values are probably more reliable. Excluding the groups of highly related clone sequences (some of which emerge because of *rrn* operon microheterogeneity [49]), most clones identified in several environmental studies demonstrate less than 98% sequence similarity to reference sequences or to each other, and therefore most of these probably represent novel species. Some of them may also represent new genera and a few, from their depth of branching and absence of signatures, new phyla.

As previously mentioned, one of the most important findings from the environmental analyses is the lack of absolute 16S rDNA sequence identity between a cultured strain and any clone sequence, regardless of the origin of the clone library. In some cases clone sequences were identical even when they originated from different locations, such as different marine environments, and several clones from this habitat have been identified as being almost identical. At least in the marine environment, some bacterial groups are ubiquitous and highly successful. As the relevant organisms are not in culture it is impossible to say if these small differences in rRNA sequence are reflected by phenotypic differences.

LIMITATIONS OF THE MOLECULAR METHODS TO ACCURATELY ASSESS MICROBIAL DIVERSITY

All of the nucleic-acid-based procedures which are used to analyze natural communities depend on the unbiased isolation of more or less pure and undegraded nucleic acid (66). An obvious point to consider immediately are the different susceptibilities of microbial cells to lysis, which may mean that some fragile cells may lyse during capture, while particularly resistant cells may not lyse at all. Lysis can be monitored by microscopy with relatively pure samples of microbial cells from marine samples. In heterogeneous matrices such as soil, it is probably safer to opt for one of the harsher mechanical procedures such as bead beating. However, this may result in badly degraded DNA which needs to be size fractionated prior to PCR analysis.

If the diversity of the natural community is to be estimated using rRNA sequences, a number of options are available for isolating them from total nucleic acids. The strengths and weaknesses of the different methods have recently been discussed (66). Here we will only comment on potential biases—problems which may affect our picture of sequence diversity when using the most popular method; PCR-based selective enrichment of rRNA genes. It is probably worth noting that primer design depends on published sequences which may not adequately represent the sequence diversity of uncultured organisms. This is probably less of a problem when using universal sequences which occur in all three domains of life, than with lineage-specific primers. However, it has been demonstrated that even Universal PCR primers may interact more favorably with some templates than others (50). The latter include templates from extreme thermophiles whose G+C rich sequences may denature less readily than sequences from mesophiles. The inclusion of formamide or acetamide (50) to facilitate template denaturation may help to avoid preferential amplification of templates with low melting temperatures.

The formation of chimeric PCR products has been reported in a number of studies (34, 38, 47). These recombinant sequences are generated by PCR from two different templates in different cycles to make a single product. One can predict that genes such as rRNA, which contain stretches of very conserved sequences interspersed with more variable regions, will be particularly prone to jumping between templates. Any hybrid sequences which are produced will appear to represent novel lineages in a phylogenetic analysis and give a distorted picture of biodiversity within the original community. All environmental sequences should be checked for secondary structure abnormalities over their entire length. Chimeric products can also be detected by analyzing the relationships inferred when different parts of a sequence are analyzed. The program CHECK_CHIMERA provided by the Ribosomal Database Project (39) can be used for this purpose. It is difficult to estimate how reliably clone libraries reflect natural microbial diversity because sampling of the library is often limited and there is often no replication (of any step) to assess reproducibility.

NEW DIRECTIONS

The limitations of the presently applied techniques to assess and quantify biodiversity stress the need for a better understanding of factors influencing and perhaps biasing extraction of nucleic acids, PCR amplification, and cloning of nucleic acids. Baseline studies in all these areas are required. The first 10 years of molecular ecology can be considered an extremely successful period as it proved the existence of a suspected but hitherto unknown and uncultured diversity among natural communities. However, we would admit to little progress when we merely continue along the same strategies. Indeed, the scope of modern microbial ecology has been extended by the introduction of novel techniques and concepts. It has been recognized that microbiologists can only benefit from this diversity when they obtain pure cultures. Ecological studies can only be performed meaningfully when the occurrence of taxa is quantifiable and the physiological role of individual cells in the ecosystems is known and measurable. Demands for in situ conservation can only be answered when these prerequisites are fulfilled.

Once the composition of a community has been explored using rDNA sequences, it is important to try to gain an estimate of the abundance of different lineages. The most direct way of doing this is to design probes to the extracted sequences. These can be used in a variety of formats to assess the relative abundance of the sequence, or by in situ whole cell hybridization using fluorescent probes, to count single cells (67). The occurrence and distribution of cells in matrices such as soil and rhizosphere habitats can be visualized using fluorescent probes and confocal scanning laser microscopy. Examples of these technologies and their application to uncultured microbes are discussed elsewhere in this volume (see Chapter 8). Provided gene-specific primers are available, in situ detection of PCR-amplified DNA and cDNA may eventually be possible to detect the presence of mRNA from genes of interest in single cells. Thus it may be possible to investigate possible functions and activities of uncultured microorganisms at the cellular level. However, it will first be necessary to understand more about the structural diversity of environmentally important enzymes and their genes.

A few methodological approaches have been described recently that may be applied and further developed to gain greater insight into the various facets of microbial ecology and diversity. One is the determination of the complexity of microbial populations by denaturating gradient gel electrophoresis (DGGE) or temperature gradient gel electrophoresis (TGGE). This molecular approach can be used to analyze the complexity and the dynamics of microbial populations via the separation of PCR-amplified 16S rDNA fragments in polyacrylamide gels using a linear increasing gradient of denaturant or temperature. Appropriate PCR fragments of rRNA are obtained using two primers, one of which contains a GC-rich clamp. The resolution of the method is sufficiently high for the denaturing gradient gel electrophoresis to separate DNA fragments of the same length but with different base-pair sequences. One can thus obtain a fingerprint of PCR products which reflects (with the caveats previously discussed) the occurrence of different rRNA genes in the original sample. The phylogenetic affiliation of the predominant members of the population can be inferred following sequence analysis of the predominant bands. Additional information on the species composition can be obtained using taxon-specific oligonucleotide probes. This approach is also well suited for monitoring the fate of particular microorganisms in complex populations after environmental perturbations (44) and will lead to the development of rapid monitoring of taxa of different ranks in environmental samples via micro- or nano-array technologies. Another method is the use of flow sorting of microorganisms to analyze the activity of bacterial cells depending upon their biovolume, following incubation with tritiated leucine and stained with SYTO 13 (58). Bacterial communities falling into different size-classes according to their average side-angled-scattered light can then be analyzed using the DGGE and TGGE methods to determine the genetic diversity within each class.

In addition to methods which facilitate an appreciation of complexity or diversity, methods also need to be developed to specifically monitor the presence of single cells or fractions of the population. It has recently been shown that magnetic beads can be used to purify single-stranded DNA for sequencing studies. It is also possible to bind specific oligonucleotides, RNA, or DNA fragments to their surface. These could then be used for the selective recovery of complementary sequences, even from matrices of complex composition. In cases where antibodies for specific subgroups of microbes are available, these cells could be directly isolated using iron-coupled antibodies and the magnetic cell separation technique.

Micromanipulators may be used for selective isolation of individual cells from agar plates or from an environmental sample. Since these cells may be unculturable under the artificial, hostile conditions provided, one would like at least to determine their phylogenetic affiliation and provide sufficient DNA for future molecular studies. Protocols for whole genome amplification of individual eukaryotic cells have been described using a mixture of random oligonucleotides. There is no a priori reason why the same techniques could not be applied to prokaryotes as well.

The Convention on Biological Diversity specifically addresses the "in-situ conservation of ecosystems and natural habitats and the maintenance and recovery of viable populations of species in their national surroundings" and ex situ measures, preferably in the country of origin. It is obvious that the latter two notions pose different problems to each of the biological disciplines. Ex situ conservation has a long tradition through the establishment of botanical and zoological gardens and microbial culture collections, and breeding and propagation programs are available for most of the cultivated forms. Ex situ preservation techniques for certain microorganisms are highly advanced and prerequisite for the maintenance of prokaryotes and yeasts. In contrast, in situ conservation programs constitute a challenge of unknown dimension. In situ conservation of animal and plants may be achieved, given maintenance of the climatic and geological intactness of the ecosystem. Loss of individuals can be monitored and measures taken to keep the balance within the ecosystem. As judged from the present knowledge about the activities, importance, distribution, and diversity of microorganisms in the environment, in situ conservation of defined microbiological species or consortia does not appear possible. Programs have been initiated, such as the All Taxa Biodiversity Inventory, that aim at the cataloging of all living forms in a defined area, but again the outcome of such activities for inventorying microorganisms is highly speculative. The term in situ could also be interpreted as the conservation of samples of natural sites under conditions developed for pure cultures, i.e., freeze-drying, liquid nitrogen, freezing under glycerin, and the like. As pointed out above, the main problem with any activities concerning in situ preservation is the enormous lack of knowledge of how to unravel the biodiversity of microorganisms. The order of magnitude of life forms destroyed by conventional preservation methods within an ecosystem is absolutely unknown because of the lack of information on what to preserve in the first place.

Consequently, in order to maintain a natural sample in the laboratory, the sample needs to be analyzed for the full range of majority and minority populations, and their physiological activities, together with the determination of the physical structure and chemical composition of the sample. Microorganisms thrive in a microworld determined by physical and chemical parameters that differ from the surrounding parameters of the macroworld measured routinely. Electrochemical microsensors have been developed for measuring oxygen, nitrate, sulfide, ammonium, NO, and the pH in the environment. They are equipped with tips with diameters not much larger than bacterial cells. Oxygen gradients and hence the consumption of oxygen by the microorganisms can be measured routinely. Similarly it is possible to measure sulfate respiration and (though not directly) nitrate respiration. According to Jørgensen et al. (32), sensors for carbon dioxide and chemical compounds are presently under development. The use of microsensors in biofilms has revealed that oxygen consumption occurs only within levels thinner than 1 millimeter and is therefore not limited by diffusion constraints. It has also revealed that within a thin biofilm different biochemical activities such as oxygen respiration, sulfid oxidation, denitrification, sulfate reduction, and methane production occur simultaneously and in partially separated layers. It is a challenge to molecular ecological study to assign these physiological activities to culturable microorganisms.

REFERENCES

1. **Alldredge, A. L., and M. W. Silver.** 1988. Characteristics, dynamics, and significance of marine snow. *Prog. Oceanogr.* **20:**41–82.
2. **Amann, R. I., C. Lin, R. Key, L. Montgomery, and D. A. Stahl.** 1992. Diversity among Fibrobacter isolates: towards a phylogenetic classification. *Syst. Appl. Microbiol.* **15:**23–31.
3. **Barns, S. M., R. E. Fundyga, M. W. Jeffries, and N. R. Pace.** 1994. Remarkable archaeal diversity detected in a Yellowstone National Park hot spring environment. *Proc. Natl. Acad. Sci. USA* **91:** 1609–1613.

4. **Britschgi, T., and S. J. Giovannoni.** 1991. Phylogenetic analysis of a natural marine bacterioplankton population by rRNA gene cloning and sequencing. *Appl. Environ. Microbiol.* **57:**1707–1713.
5. **Brock, T.** 1978. *Thermophilic Organisms and Life at High Temperatures.* Springer Verlag, New York, N.Y.
6. **Brooks, D. R., and D. A. McLennan.** 1991. *Phylogeny, ecology and behavior.* The University of Chicago Press, Chicago, Ill.
7. **Caron, D. A., P. G. Davis, L. P. Madin, and J. M. Sieburth.** 1982. Heterotrophic bacteria and bacteriverous protozoa in oceanic macroaggregates. *Science* **218:**795–797.
8. **DeLey, J., and J. DeSmedt.** 1975. Improvements of the membrane filter method for DNA:rRNA hybridization. *Antonie Leeuwenhoek J. Microbiol. Serol.* **36:**461–474.
9. **DeLong, E. F.** 1992. Archaea in coastal marine environments. *Proc. Natl. Acad. Sci. USA* **89:**5685–5689.
10. **DeLong, E. F., D. G. Franks, and A. L. Alldredge.** 1993. Phylogenetic diversity of aggregate-attached vs. free-living marine bacterial assemblages. *Limnol. Oceanogr.* **38:**924–934.
11. **DeSoete, G.** 1983. A least squares algorithm for fitting additive trees to proximity data. *Psychometrica* **48:**621–626.
12. **Distel, D. L., E. F. DeLong, and J. B. Waterbury.** 1991. Phylogenetic characterization and in situ localization of the bacterial symbiont of shipworms (*Teredinidae*:Bivalvia) by using 16S rRNA sequence analysis and oligodeoxynucleotide probe hybridization. *Appl. Environ. Microbiol.* **57:**2376–2382.
13. **Eady, R. R.** 1992. The dinitrogen fixing bacteria, p. 534–554. *In* A. Balows, H. G. Trüper, M. Dworkin, W. Harder, and K.-H. Schleifer (ed), *The Prokaryotes*, 2nd ed. Springer Verlag, New York, N.Y.
14. **Embley, T. M., and B. J. Finlay.** 1994. The use of rRNA sequences to unravel the relationships between anaerobic ciliates and their methanogen endosymbionts. *Microbiology* **140:**225–235.
15. **Esteban, G., B. J. Finlay, and T. M. Embley.** 1993. New species double the diversity of anaerobic ciliates in a Spanish lake. *FEMS Microbiol. Lett.* **109:**93–100.
16. **Esteban, G., B. E. Guhl, K. J. Clarke, T. M. Embley, and B. J. Finlay.** 1993. *Cyclidium porcatum* n.sp.: free-living anaerobic scuticociliate containing a stable complex of hydrogenosomes, eubacteria and archaeobacteria. *Eur. J. Protistol.* **29:**262–270.
17. **Fenchel, T.** 1988. Marine plankton food chains. *Annu. Rev. Ecol. System.* **19:**19–38.
18. **Fenchel, T., and C. Barnard.** 1993. A purple protist. *Nature* **362:**300.
19. **Fenchel, T., and B. J. Finlay.** 1989. *Kentrophorus*: a mouthless ciliate with a symbiotic kitchen garden. *Ophelia* **30:**75–93.
20. **Fenchel, T., and B. J. Finlay.** 1990. Anaerobic free-living protozoa: growth efficiencies and the structure of anaerobic communities. *FEMS Microbiol. Ecol.* **74:**269–276.
21. **Fenchel, T., T. Perry, and A. Thane.** 1977. Anaerobiosis and symbiosis with bacteria in free-living ciliates. *J. Protozool.* **24:**154–163.
22. **Finlay, B. J., G. Esteban, K. J. Clarke, A. G. Williams, T. M. Embley, and R. P. Hirt.** 1994. Some rumen ciliates have endosymbiotic methanogens. *FEMS Microbiol. Lett.* **117:**157–162.
23. **Fuhrman, J. A., K. McCallum, and A. A. Davis.** 1992. Novel major archaebacterial group from marine plankton. *Nature* **356:**148–149.
24. **Fuhrman, J. A., K. McCallum, and A. A. Davis.** 1993. Phylogenetic diversity of subsurface marine microbial communities from the Atlantic and Pacific oceans. *Appl. Environ. Microbiol.* **59:**1294–1302.
25. **Fukatsu, T., and H. Ishikawa.** 1993. Occurrence of chaperonin 60 and chaperonin 10 in primary and secondary bacterial symbionts of aphids: implications for the evolution of an endosymbiotic system in aphids. *J. Mol. Evol.* **36:**568–577.
26. **Giovannoni, S. J., T. B. Britschgi, C. L. Moyer, and K. G. Field.** 1990. Genetic diversity in Sargasso Sea bacterioplankton. *Nature* **345:**60–63.
27. **Goodfellow, M., and A. G. O'Donnell (eds).** 1993. *Handbook of New Bacterial Systematics.* Academic Press, London, United Kingdom.
28. **Hartmann, A.** 1994. Towards soil microbial community structure analysis. *Abstr. Moderne Analyseverfahren für die Bestimmung der Artenvielfalt von Mikroorganismen in natürlichen Standorten.* Analytica 94, Munich, Germany.

29. **Hugenholtz, P., B. M. Goebel, and N. R. Pace.** 1998. Impact of culture-independent studies on the emerging phylogenetic view of bacterial diversity. *J. Bacteriol.* **180:**4765–4774.
30. **Johnson, J. J.** 1973. The use of nucleic acid homologies in the taxonomy of anaerobic bacteria. *Int. J. Syst. Bacteriol.* **23:**308–315.
31. **Johnson, J. L., and B. S. Francis.** 1975. Taxonomy of the clostridia: ribosomal ribonucleic acid homologies among the species. *J. Gen. Microbiol.* **95:**229–244.
32. **Jørgensen, B. B.** 1994. Determination of the physical-chemical environment. *Abstr. Moderne Analyseverfahren für die Bestimmung der Artenvielfalt von Mikroorganismen in natürlichen Standorten.* Analytica 94, Munich, Germany.
33. **Jukes, T. H., and C. R Cantor.** 1969. Evolution of protein molecules, p. 21–132. *In* H. N. Munro (ed.), *Mammalian Protein Metabolism.* Academic Press, New York, N.Y.
34. **Kopczynski, E. D., M. M. Bateson, and D. M. Ward.** 1994. Recognition of chimeric small-subunit ribosomal DNAs composed of genes from uncultivated microorganisms. *Appl. Environ. Microbiol.* **60:**746–748.
35. **Lee, J. J., A. T. Soldo, W. Reisser, M. J. Lee, K. W. Jeon, and H. D. Görtz.** 1985. The extent of algal and bacterial endosymbioses in protozoa. *J. Protozool.* **32:**391–402.
36. **Liesack, W., and E. Stackebrandt.** 1992. Unculturable microbes detected by molecular sequences and probes. *Biodiv. Conserv.* **1:**250–262.
37. **Liesack, W., and E. Stackebrandt.** 1992. Occurrence of novel groups of the domain *Bacteria* as revealed by analysis of genetic material isolated from an Australian terrestrial environment. *J. Bacteriol.* **174:**5072–5078.
38. **Liesack, W., H. Weyland, and E. Stackebrandt.** 1991. Potential risks of gene amplification by PCR as determined by 16S rDNA analysis of a mixed culture of strict barophilic bacteria. *Microb. Ecol.* **21:**191–198.
39. **Maidak, B. L., G. J. Olsen, N. Larsen, R. Overbeek, M. J. McCaughey, and C. R. Woese.** 1997. The ribosomal database project. *Nucleic Acids Res.* **25:**109–111.
40. **McInerny, J. O., M. Wilkinson, J. W. Patching, T. M. Embley, and R. Powell.** 1995. Recovery and phylogenetic analysis of novel archaeal rRNA sequences from a deep-sea deposit feeder. *Appl. Environ. Microbiol.* **61:**1646–1648.
41. **Moran, N., and P. Baumann.** 1994. Phylogenetics of cytoplasmically inherited microorganisms of arthropods. *Trends Ecol. Evol.* **9:**15–20.
42. **Moran, N. A., M. A. Munson, P. Baumann, and H. Ishikawa.** 1993. A molecular clock in endosymbiotic bacteria is calibrated using the insect hosts. *Proc. R. Soc. London Ser. B* **253:**167–171.
43. **Munson, M. A., P. Baumann, and M. G. Kinsey.** 1991. *Buchnera* gen. nov. and *Buchnera aphidicola* sp. nov. designation for a phylogenetic taxon consisting of the primary endosymbionts of aphids. *Int. J. Syst. Bacteriol.* **41:**566–568.
44. **Muyzer, G., E. C. DeWaal, and A. G. Uitterlinden.** 1993. Profiling of complex microbial populations by denaturing gradient gel electrophoresis of polymerase chain reaction amplified genes coding for 16S rRNA. *Appl. Environ. Microbiol.* **59:**695–700.
45. **O'Neill, S. L., H. Gooding, and S. Aksoy.** 1993. Phylogenetically distant symbiotic microorganisms reside in *Glossina* midgut and ovary tissues. *Med. Vet. Entomol.* **7:**377–383.
46. **Oyaizu, H., N. Naruhashi, and T. Gamou.** 1992. Molecular methods of analyzing bacterial diversity: the case of Rhizobia. *Biodiv. Conserv.* **1:**237–249.
47. **Paabo, S., D. M. Irwin, and A. C. Wilson.** 1990. DNA damage promotes jumping between templates during enzymatic amplification. *J. Biol. Chem.* **265:**4718–4721.
48. **Pace, N. R., D. A. Stahl, D. J. Lane, and G. J. Olsen.** 1985. The analysis of natural microbial communities by ribosomal RNA sequences. *Microb. Ecol.* **9:**1–56.
49. **Pukall, R., O. Päuker, D. Buntefuß, G. Ulrichs, P. Lebaron, L. Bernhard, T. Guindulain, J. Vives-Rego, and E. Stackebrandt.** 1998. High sequence diversity of *Alteromonas*-related cloned and cellular 16S rDNAs from a Mediterranean seawater mesocosm experiment. *FEMS Microb. Ecol.* **28:**335–344.
50. **Reysenbach, A-L., L. J. Giver, G. S. Wickham, and N. R. Pace.** 1992. Differential amplification of rRNA genes by polymerase chain reaction. *Appl. Environ. Microbiol.* **58:**3417–3418.

51. **Rheims, H., C. Spröer, F. A. Rainey, and E. Stackebrandt.** 1996. Molecular biological evidence for the occurrence of uncultured members of the actinomycete line of descent in different environments and geographical locations. *Microbiology* (Reading) **142:**2863–2870.
52. **Ruff-Roberts, A. L., J. G. Kuenen, and D. M. Ward.** 1994. Distribution of cultivated and uncultivated cyanobacteria and chloroflexus-like bacteria in hot spring microbial mats. *Appl. Environ. Microbiol.* **60:**697–704.
53. **Saitou, N., and M. Nei.** 1987. The neighbour joining method: a new method for constructing phylogenetic trees. *Mol. Biol. Evol.* **4:**406–425.
54. **Schleifer, K. H., and E. Stackebrandt.** 1983. Molecular systematics of prokaryotes. *Annu. Rev. Microbiol.* **37:**143–187.
55. **Schmidt, T. M., E. F. DeLong, and N. R. Pace.** 1991. Analysis of a marine picoplankton community by 16S rRNA gene cloning and sequencing. *J. Bacteriol.* **173:**4371–4378.
56. **Schröder, D., H. Deppisch, M. Obermeyer, G. Krohne, E. Stackebrandt, B. Hölldobler, W. Goebel, and R. Gross.** 1996. Intracellular endosymbiotic bacteria of *Camponotus* species (carpenter ants): systematics, evolution and ultrastructural characterization. *Mol. Microbiol.* **21:**479–489.
57. **Seewaldt, E., and E. Stackebrandt.** 1982. Partial sequence of 16S ribosomal RNA and the phylogeny of *Prochloron. Nature* **295:**618–620.
58. **Servais, P., C. Courties, P. Lebaron, and M. Troussilier.** 1999. Coupling bacterial activity measurements with cell sorting by flow cytometry. *Microb. Ecol.* **38**:180–189.
59. **Sneath, P. H. A., and R. R. Sokal.** 1973. *Numerical Taxonomy: the Principles and Practice of Numerical Classification.* Freeman, San Francisco, Calif.
60. **Stackebrandt, E., W. Liesack, and B. M. Goebel.** 1993. Bacterial diversity in a soil sample from a subtropical Australian environment as determined by 16S rDNA analysis. *FASEB J.* **7:**232–236.
61. **Stackebrandt, E., and B. M. Goebel.** 1994. A place for DNA-DNA reassociation and 16S rRNA sequence analysis in the present species definition in bacteriology. *Int. J. Syst. Bacteriol.* **44:**846–849.
62. **Stackebrandt, E., W. Ludwig, and G. E. Fox.** 1985. 16S ribosomal RNA oligonucleotide cataloguing, p. 75–107. *In* G. Gottschalk (ed), *Methods in Microbiology.* Academic Press, London, United Kingdom.
63. **Stahl, D. A., B. Flesher, H. R. Mansfield, and L. Montgomery.** 1988. Use of phylogenetically-based hybridization probes for studies of ruminal microbial ecology. *Appl. Environ. Microbiol.* **54:** 1079–1884.
64. **Torsvik, V., J. Goksoyr, and F. L. Daae.** 1990. High diversity in DNA of soil bacteria. *Appl. Environ. Microbiol.* **56:**782–787.
65. **Ullman, J. S., and B. J. McCarthy.** 1973. The relationship between mismatched base pairs and the thermal stability of DNA duplexes. *Biochim. Biophys. Acta* **294:**416–424.
66. **Von Wintzingerode, F., U. Göbel, and E. Stackebrandt.** 1997. Determination of microbial diversity in environmental samples: pitfalls of PCR-based rRNA analysis. *FEMS Microbiol. Rev.* **21:**213–229.
67. **Wagner, M., R. I. Amann, H. Lemmer, and K. H. Schleifer.** 1993. Probing activated sludge with oligonucleotides specific for proteobacteria: inadequacy of culture-dependent methods for describing microbial community structure. *Appl. Environ. Microbiol.* **59:**1520–1525.
68. **Ward, D. M., R. Weller, and M. M. Bateson.** 1990. 16S rRNA sequences reveal numerous uncultured microorganisms in a natural community. *Nature* **345:**63–65.
69. **Waterbury, J. B., C. B. Calloway, and R. D. Turner.** 1983. A cellulolytic nitrogen-fixing bacterium cultured from the gland of Deshayes in shipworms (Bivalvia: Teredinidae). *Science* **221:**1401–1403.
70. **Wayne, L. G., D. J. Brenner, R. R. Colwell, P. A. D. Grimont, O. Kandler, M. I. Krichevsky, L. H. Moore, W. E. C. Moore, R. G. E. Murray, E. Stackebrandt, M. P. Starr, and H. G. Trüper.** 1987. Report of the ad hoc committee on reconciliation of approaches to bacterial systematics. *Int. J. Syst. Bacteriol.* **37:**:463–464.
71. **Weller, R., M. M. Bateson, B. K. Heimbuch, E. D. Kopczynski, and D. M. Ward.** 1992. Uncultivated cyanobacteria, Chloroflexus-like inhabitants, and spirochaete-like inhabitants of a hot spring cyanobacterial mat. *Appl. Environ. Microbiol.* **58:**3964–3969.

72. **Weller, R., J. W. Weller, and D. M. Ward.** 1991. 16S rRNA sequences of uncultivated hot spring cyanobacterial mat inhabitants retrieved as randomly primed cDNA. *Appl. Environ. Microbiol.* **57:** 1146–1151.
73. **Winker, S., and C. R. Woese.** 1991. A definition of the domains Archaea, Bacteria and Eucarya in terms of small subunit ribosomal RNA characteristics. *Syst. Appl. Microbiol.* **14:**305–310.
74. **Young, J. P. W.** 1992. Classification of nitrogen fixing organisms, p. 43–86. *In* G. Stacey, R. H. Burris, and H. Evans (ed.), *Biological Nitrogen Fixation*. Chapman and Hall, New York, N.Y.

Nonculturable Microorganisms in the Environment
Edited by R. R. Colwell and D. J. Grimes

Chapter 6

Molecular Genetic Methods for Detection and Identification of Viable but Nonculturable Microorganisms

Ivor T. Knight

The problems associated with culture-based methods for detection and identification of microorganisms in clinical and environmental samples have motivated the development of alternative methods which do not require cultivation of the target organisms. Molecular genetic methods, which specifically target microbial nucleic acids, have become important tools for the identification of both cultured and uncultured microorganisms. The sensitivity of these methods approaches that of culture-based methods and it is arguable that the specificity of these methods generally exceeds that of culture-based methods. Indeed, molecular genetic methods have become a powerful set of tools used by many investigators to detect and identify culturable microorganisms in their viable but nonculturable state, as well as organisms which have yet to be brought into culture.

Since the molecular methods described below can be used to detect and identify microorganisms without the need to cultivate them, the distinction between having detected culturable or nonculturable organisms often cannot be made. Obviously, the only way to determine the relative proportion of culturable and nonculturable target organisms present in a sample is to conduct a parallel analysis using culture techniques. This is often not the aim of investigators who are using these techniques. More often the aim is to detect or identify specific microorganisms regardless of the ability to cultivate them. Furthermore, molecular genetic methods for direct detection (without cultivation) have been used with equal success to detect microbial cells in the viable but nonculturable state and the cells of organisms that have never been successfully cultivated. Although this distinction is an important one, it can make the discussion of molecular methods unnecessarily cumbersome. For this reason I will borrow a term from the literature (26) and refer to organisms in a viable but nonculturable state and previously uncultured organisms as "unculturable."

Ivor T. Knight • Department of Biology, James Madison University, Harrisonburg, VA 22807.

The aims of this chapter are to survey molecular genetic methods in use and to review their application to the detection of unculturable microorganisms. There is no doubt that this rapidly developing area will produce technological advances which will expand the frontiers of our understanding of the viable but nonculturable state and the role of uncultured organisms in natural communities.

HYBRIDIZATION PROBES AND DNA AMPLIFICATION

Increasingly, the tools of molecular biology have been used to investigate the role of nonculturable and uncultured organisms in microbial communities. Applications include detection of pathogenic bacteria in environmental and clinical samples (6, 9, 20, 23, 24, 37); examination of microbial diversity in agricultural, industrial, and natural environments (2, 3, 10, 11, 28, 31, 49); and discovery of previously unidentified taxa (39, 40, 45). The primary methods used in these applications have been hybridization with DNA or RNA probes and DNA amplification using PCR. There has been significant development in both methodologies in recent years, and it is useful to review these methodologies as they have been applied to the detection and identification of unculturable organisms.

Hybridization Probes

Hybridization probes are single-stranded nucleic acids (DNA or RNA), which have been chemically or radioactively labeled, and are used to detect complementary target DNA or RNA in a heterogeneous mixture of nucleic acids. In hybridization assays DNA or RNA probes form a stable, double-stranded structure with the target nucleic acid via hydrogen bonding between complementary bases. Those conditions which affect hydrogen bonding, such as temperature and ionic strength, influence specificity of the reaction. A detailed treatment of the theory and general application in microbial ecology is found elsewhere (22).

The number of different formats used for hybridization assays has expanded considerably since E. M. Southern first demonstrated the use of the Southern blot (47). Most of these formats (including Southern and Northern blots, colony blots, and dot blots) are variations of the mixed-phase format, in which test DNA or RNA is bound to a solid phase (usually a membrane or paper) and hybridization occurs with the probe in a liquid phase. Discussion of the theory and practice of mixed-phase hybridization in general is provided by Meinkoth and Wahl (30). DNA and RNA probes have been used in mixed-phase assays to directly detect target organisms in bulk extracts of nucleic acids from a variety of sample types. Methods for obtaining relatively unbiased bulk nucleic acid extracts directly from complex environmental and clinical samples have been developed in a number of laboratories (18, 32, 35, 46, 50). The nucleic acid extracts are applied directly to a hybridization membrane, in the case of dot and slot blot formats, or are digested with restriction endonucleases and separated by gel electrophoresis before being transferred to a hybridization membrane, in the case of the Southern blot format. The application of this strategy to the direct detection of microorganisms in a variety of environmental samples has been discussed in several reviews (5, 16, 22, 42).

Quantitative use of dot blot or slot blot hybridization assays can provide estimates of the numbers of bacterial cells in a sample by correlating the hybridization signal strength with the amount of target DNA applied to the membrane (21). In a study of estuarine water samples collected in New York Harbor, high *Salmonella* populations were estimated by quantitative dot blot hybridization with a DNA probe in samples where no *Salmonella* could be cultivated (23). Further studies of *Salmonella* inoculated into laboratory microcosms showed that the hybridization signal in dot blots did not decline significantly over a time when the culturable cell population declined from 10^5 to 10^1 CFU/ml (21). Similarly, quantitative dot blot hybridization with an *eltB* probe was used to detect viable but nonculturable enterotoxigenic *Escherichia coli* in ground water (8). These studies show that dot blot hybridization of bulk DNA extracts can be used to detect unculturable bacteria.

In Situ Hybridization

An exciting development in the use of alternative formats for hybridization probes is in situ hybridization, a method in which fluorescently labeled DNA or RNA probes are hybridized with target nucleic acids in whole, permeabilized cells. The application of this method to the detection of single microbial cells by using rRNA-targeted probes in combination with epifluorescent microscopy was developed in Norman Pace's laboratory (12, 14) and is widely used. Amann et al. (4) described the use of flow cytometry to enumerate the fluorescently labeled cells. In situ hybridization with fluorescently labeled probes is analogous to a fluorescent-antibody staining method and can be used to detect and identify both culturable and unculturable cells. Multiple fluorescently labeled probes, each with a different emission wavelength, can be applied to single samples to visualize high genetic diversity (2). The technique has proven useful for detecting uncultured bacterial endosymbionts (3), identifying nonculturable magnetotactic bacteria (49), in situ enumeration of individual species in biofilms (27), and identifying unculturable bacteria in activated sludge (55).

One potential drawback of using fluorescent in situ hybridization with rRNA-targeted probes for detecting cells in a viable but nonculturable state is the relatively low sensitivity of the method. Single cells detected using this method must have abundant rRNA molecules which provide enough targets for hybridization of multiple probe molecules, in order that the fluorescence is multiplied to the level of detection (12). Cells with very low ribosome copy number may not possess enough probe hybridization sites to be detected using conventional rRNA-targeted oligonucleotide probes. For example, low cellular rRNA concentration is the likely reason that fluorescent oligonucleotide probes poorly detect slow-growing organisms such as bacterioplankton in natural aquatic environments samples (15). If a consequence of the transition from a culturable state to a nonculturable state is a reduction in ribosome number, then conventional rRNA-directed probes may not be sensitive enough to detect the nonculturable forms. Probes with multiple fluorescent labels, such as those used by Trebesius et al. (53), or the use of in situ target amplification prior to hybridization (see "Conclusions," below) are strategies which may be usefully applied to this problem.

DNA Amplification

PCR, the method of amplifying regions of DNA using a thermostable DNA polymerase and pairs of oligonucleotide primers, was introduced in 1988 (41) and has become widely used by microbiologists for detecting and identifying both culturable and unculturable organisms. The specificity of the method is similar to that of hybridization probes because DNA amplification depends upon hybridization of the oligonucleotide primers with their target sequences, but the sensitivity is much higher because the target region is amplified as much as a millionfold. Furthermore, enough copies of the target region are produced to permit more detailed analyses such as nucleotide sequence determination. This powerful technique has been used to identify uncultured pathogens in clinical samples (39, 40, 45), to more fully characterize the microbial diversity in natural environments by identifying unculturable bacteria (10, 11, 31, 49), and to detect culturable bacteria in the viable but nonculturable state (17, 25, 36).

The detection, identification, and classification of uncultured organisms have been advanced by amplifying and sequencing phylogenetically informative portions of rRNA genes present in DNA extracted directly from clinical and environmental samples. This strategy employs PCR primers which anneal to highly conserved regions in eubacterial or archaeal rRNA genes and prime amplification of the less conserved sequence between the primer annealing sites. The amplification products are cloned, and individual clones (each of which carries a single amplification product) are selected for DNA sequence determination. The sequence can then be aligned with homologous sequences in other bacteria to determine the taxonomic position of the uncultured organism. This strategy has proved useful for identification of unculturable pathogens in clinical samples, including the causative agents of Whipple's disease (40) and bacillary angiomatosis (39). The details of this approach are reviewed elsewhere (38, 44).

The same strategy has been applied to the problem of characterizing the unculturable component of natural microbial populations. This approach is used to characterize microbial diversity in a variety of habitats. In a study of *Archaea* in coastal marine environments, *Archaea*-specific primers were used to recover ribosomal DNA (rDNA) representing two lineages of uncultured *Archaea* (10). In another study, eubacterial primers were used to amplify rDNA in nucleic acids extracted from aggregate-attached and free-living marine bacteria in marine water samples. The amplification products were characterized by restriction fragment length polymorphism (RFLP) analysis and by nucleotide sequencing in order to compare the phylogenetic diversity of the bacteria present in the two environments (11). Amplification products can also be characterized by denaturing gradient gel electrophoresis (DGGE), a technique which separates same-sized PCR products according to their stability in an increasing gradient of denaturing agents (1, 33). Individual PCR products in the gel can then be identified by Southern hybridization or removed from the gel so that their nucleotide sequences can be determined (e.g., see reference 13).

The ease of use of the PCR and the universality of the eubacterial and archaeal primers make broad-spectrum amplification an attractive approach to identifying unculturable bacteria. The use of broad-spectrum primers, however, can produce artifactual amplification products which are derived from contaminating bacterial DNA in the PCR reagents. Commercial preparations of thermostable polymerases have been shown to be a source of contamination (43), and thus the standard precautions for preventing contamination in the laboratory are ineffective. The use of negative control reactions, where no template DNA is added to the PCR, is a reasonable way to detect such contamination, and Meier et al. (29) have suggested that treatment of reagents with 8-methoxypsoralen and long-wave UV light is an effective method of eliminating contaminating DNA from reagents. Bias in amplification of mixtures of rRNA genes has been demonstrated (48) and, although PCR amplification of rRNA genes can be used to characterize microbial diversity in natural environments, quantitative analyses should be interpreted cautiously.

DNA amplification employing primers specific for the target organism, or group of target organisms, has been used for direct detection in a variety of sample types. Using rRNA sequence data, primers directed to group-specific regions of rRNA genes can be designed which will selectively amplify rDNA from the target group (24, 56). After PCR the identity of the amplification products is usually confirmed by Southern hybridization with a group-specific probe.

PCR amplification of genes other than rRNA genes is frequently used for direct detection of specific organisms in heterogeneous samples. The most common approach to detecting pathogenic bacteria using PCR is to design primers for amplification of genes encoding pathogenic determinants (9, 17, 20, 21, 54). This approach is not only highly specific but also correlates detection of an organism with its potential for pathogenicity. Other strategies for specific detection include (i) amplification of functionally specific genes, such as *luxA* for detection of *Vibrio fischeri* (25) and methane monooxygenase genes for detection of methanotrophic bacteria (28); (ii) amplification of genes encoding enzymes which are diagnostic for an organism or group of organisms, such as *lacZ* for detection of coliform bacteria (7) and *uid* for detection of *E. coli* (6); and (iii) amplification of randomly cloned genome fragments which have been shown by hybridization studies to be specific for the target organism (51, 52).

Amplification of Targets in Nonculturable Cells: Viable and Nonviable

Specific amplification of DNA targets in bulk DNA extracts from environmental and clinical samples permits detection of specific organisms or groups of related organisms without the need to cultivate them. Since DNA recovery procedures do not discriminate between culturable and unculturable forms of the target organisms, all cells with intact amplification targets will be detected. Furthermore, bulk DNA recovery and amplification by PCR would not be expected to discriminate between viable and dead cells with intact amplification targets. Consequently, few studies have examined the relative efficacy of their PCR procedures for detecting culturable and unculturable forms. PCR-based, direct detection methods are generally developed and tested using culturable cells, either in pure culture or by using natural samples which have been spiked with dilutions of freshly cultivated cells. Some studies have specifically addressed the detection of unculturable forms, however, and show that both viable and nonviable unculturable cells are detected using PCR-based assays (8, 9, 17, 19, 25).

The fact that PCR methods can readily detect organisms in the viable but non-culturable state makes it a useful method for directly determining the presence of target organisms in samples. A study by Lee and Ruby (25) describes a PCR detection approach to showing that nonculturable *V. fischeri* is abundant in seawater and further showed that the nonculturable forms are viable and capable of producing symbiotic infections.

The fact that targets in nonviable cells can also be amplified means that caution should be exercised when interpreting the results of PCR-based detection (19). If the goal of a PCR-based assay is to detect only viable organisms, in both the culturable and the unculturable state, then additional methods for assessing viability must be applied. One possibility is to amplify DNA from mRNA targets rather than genomic DNA targets. Since cellular mRNA is much less stable than DNA, the likelihood of finding intact amplification targets in nonviable cells would be much lower. Amplification of mRNA targets involves treatment with reverse transcriptase to produce a copy DNA (cDNA) which then serves as the target for amplification by PCR. Prerequisites for success of this strategy would include choosing a target that is constitutively expressed and finding a good method for extracting intact mRNA from the sample of interest (see, e.g., ref. 18).

CONCLUSIONS

Use of molecular genetic techniques for detection and identification of unculturable organisms involves principally application of hybridization probes and DNA amplification by PCR. New molecular genetic methods are constantly in development, and therefore more powerful applications of current methods and appearance of new methods can be expected. In situ PCR, a technique for amplifying low copy hybridization targets in intact cells (34) has been employed to detect viral targets in infected eukaryotic cells and could be applied to amplification of low copy targets in bacteria prior to detection with hybridization probes. Confocal laser microscopy, in combination with fluorescence-based hybridization assays, also provides a more sensitive method for detecting and identifying unculturable organisms.

An appreciation of the power of molecular genetic techniques should be balanced by an understanding that they are limited to detecting and identifying the nucleic acid component of the organism. The knowledge that the genome of a microorganism is present in a sample is merely the beginning of understanding the status and role of that microorganism in the environment from which the sample was collected.

REFERENCES

1. **Abrams, E. S., and V. P. Stanton.** 1992. Use of denaturing gradient gel electrophoresis to study conformational transitions in nucleic acids. *Methods Enzymol.* **212:**71–104.
2. **Amann, R., J. Snaidr, M. Wagner, W. Ludwig, and K. Schleifer.** 1996. In situ visualization of high genetic diversity in a natural microbial community. *J. Bacteriol.* **178:**3496–3500.

3. **Amann, R., N. Springer, W. Ludwig, H. Gortz, and K. Schleifer.** 1991. Identification *in situ* and phylogeny of uncultured bacterial endosymbionts. *Nature* **351:**161–164.
4. **Amann, R. I., B. J. Binder, R. J. Olson, S. W. Chisholm, R. Devereux, and D. A. Stahl.** 1990. Combination of 16S rRNA-targeted oligonucleotide probes with flow cytometry for analyzing mixed microbial populations. *Appl. Environ. Microbiol.* **56:**1919–1925.
5. **Atlas, R. M., G. Sayler, R. S. Burlage, and A. K. Bej.** 1992. Molecular approaches for environmental monitoring of microorganisms. *BioTechniques* **12:**706.
6. **Bej, A. K., J. L. DiCesare, L. Haff, and R. M. Atlas.** 1991. Detection of *Escherichia coli* and *Shigella* spp. in water by using the polymerase chain reaction and gene probes for *uid*. *Appl. Environ. Microbiol.* **57:**1013–1017.
7. **Bej, A. K., R. J. Steffan, J. DiCesare, L. Haff, and R. M. Atlas.** 1990. Detection of coliform bacteria in water by polymerase chain reaction and gene probes. *Appl. Environ. Microbiol.* **56:**307–314.
8. **Bogert, A. P., and I. T. Knight.** 1995. Detection of enterotoxigenic *E. coli* in ground water using DNA hybridization and PCR, abstr. N-76, p. 345. *In Abstracts of the 95th General Meeting of the American Society for Microbiology 1995.* American Society for Microbiology, Washington, D.C.
9. **Brauns, L. A., M. C. Hudson, and J. D. Oliver.** 1991. Use of the polymerase chain reaction in detection of culturable and nonculturable *Vibrio vulnificus* cells. *Appl. Environ. Microbiol.* **57:**2651–2655.
10. **Delong, E. F.** 1992. Archaea in coastal marine environments. *Proc. Natl. Acad. Sci. USA* **89:**5685–5689.
11. **DeLong, E. F., D. G. Franks, and A. L. Alldredge.** 1993. Phylogenetic diversity of aggregate-attached vs. free-living marine bacterial assemblages. *Limnol. Oceanogr.* **38:**924–934.
12. **DeLong, E. F., G. S. Wickham, and N. R. Pace.** 1989. Phylogenetic stains: ribosomal RNA-based probes for the identification of single cells. *Science* **243:**1360–1363.
13. **Ferris, M. J., G. Muyzer, and D. M. Ward.** 1996. Denaturing gradient gel electrophoresis profiles of 16S rRNA-defined populations inhabiting a hot spring microbial mat community. *Appl. Environ. Microbiol.* **62:**340–346.
14. **Giovannoni, S. J., E. F. DeLong, G. J. Olsen, and N. R. Pace.** 1988. Phylogenetic group-specific oligodeoxynucleotide probes for identification of single microbial cells. *J. Bacteriol.* **170:**720–726.
15. **Hicks, R. E., R. I. Amann, and D. A. Stahl.** 1991. Dual staining of natural bacterioplankton with 4′,6-diamidino-2-phenylindole and fluorescent oligonucleotide probes targeting kingdom-level 16S rRNA sequences. *Appl. Environ. Microbiol.* **58:**2158–2163.
16. **Holben, W. E., and J. M. Tiedje.** 1988. Applications of nucleic acid hybridization in microbial ecology. *Ecology* **69:**561–568.
17. **Islam, M. S., M. K. Hasan, M. A. Miah, G. C. Sur, A. Felsenstein, M. Venkatesan, R. B. Sack, and M. J. Albert.** 1993. Use of the polymerase chain reaction and fluorescent-antibody methods for detecting viable but nonculturable *Shigella dysenteriae* type 1 in laboratory microcosms. *Appl. Environ. Microbiol.* **59:**536–540.
18. **Jeffrey, W. H., S. Nazaret, and R. Von Haven.** 1994. Improved method for recovery of mRNA from aquatic samples and its application to detection of *mer* expression. *Appl. Environ. Microbiol.* **60:**1814–1821.
19. **Josephson, K. L., C. P. Gerba, and I. L. Pepper.** 1993. Polymerase chain reaction detection of nonviable bacterial pathogens. *Appl. Environ. Microbiol.* **59:**3513–3515.
20. **Khan, A. A., and C. E. Cerniglia.** 1994. Detection of *Pseudomonas aeruginosa* from clinical and environmental samples by amplification of the exotoxin A gene using PCR. *Appl. Environ. Microbiol.* **60:**3739–3745.
21. **Knight, I. T., J. DiRuggiero, and R. R. Colwell.** 1991. Direct detection of enteropathogenic bacteria in estuarine water using nucleic acid probes. *Water Sci. Technol.* **24:**261–266.
22. **Knight, I. T., W. E. Holben, J. M. Tiedje, and R. R. Colwell.** 1991. Nucleic acid hybridization techniques for detection, identification and enumeration of microorganisms in the environment, p. 65–91. *In* M. Levin, R. J. Seidler, and M. Rogul (ed.), *Microbial Ecology: Principles, Methods and Application to Environmental Biotechnology*. McGraw-Hill, Inc., New York, N.Y.
23. **Knight, I. T., S. Shults, C. W. Kaspar, and R. R. Colwell.** 1990. Direct detection of *Salmonella* spp. in estuaries by using a DNA probe. *Appl. Environ. Microbiol.* **56:**1059–1066.

24. **Lawrence, L. M., and A. Gilmour.** 1994. Incidence of *Listeria* spp. and *Listeria monocytogenes* in a poultry processing environment and in poultry products and their rapid confirmation by mulitplex PCR. *Appl. Environ. Microbiol.* **60:**4600–4604.
25. **Lee, K.-H., and E. G. Ruby.** 1995. Symbiotic role of the viable but nonculturable state of *Vibrio fischeri* in Hawaiian coastal seawater. *Appl. Environ. Microbiol.* **61:**278–283.
26. **Liesack, W., and E. Stackebrandt.** 1992. Unculturable microbes detected by molecular sequences and probes. *Biodiversity Conserv.* **1:**250–262.
27. **Manz, W., U. Szewzyk, P. Ericsson, R. Amann, K.-H. Schleifer, and T.-A. Stenström.** 1993. In situ identification of bacteria in drinking water and adjoining biofilms by hybridization with 16S and 23S rRNA-directed fluorescent oligonucleotide probes. *Appl. Environ. Microbiol.* **59:**2293–2298.
28. **McDonald, I. R., E. M. Kenna, and J. C. Murrell.** 1995. Detection of methanotrophic bacteria in environmental samples with the PCR. *Appl. Environ. Microbiol.* **61:**116–121.
29. **Meier, A., D. H. Persing, M. Finken, and E. C. Böttger.** 1993. Elimination of contaminating DNA within polymerase chain reaction reagents: implication for a general approach to detection of uncultured pathogens. *J. Clin. Microbiol.* **31:**646–652.
30. **Meinkoth, J., and G. Wahl.** 1984. Hybridization of nucleic acids immobilized on solid supports. *Anal. Biochem.* **38:**267–284.
31. **Mirza, M. S., D. Hahn, S. V. Dobritsa, and A. T. L. Akkermans.** 1994. Phylogenetic studies on uncultured *Frankia* populations in nodules of *Datisca cannabina. Can. J. Microbiol.* **40:**313–318.
32. **Morè, M. I., J. B. Herrick, C. Silva, W. C. Ghiorse, and E. L. Madsen.** 1994. Quantitative cell lysis of indigenous microorganisms and rapid extraction of microbial DNA from sediment. *Appl. Environ. Microbiol.* **60:**1572–1580.
33. **Muyzer, G., E. C. De Waal, and A. G. Uitterlinden.** 1993. Profiling of complex microbial populations by denaturing gradient gel electrophoresis analysis of polymerase chain reaction-amplified genes coding for 16S rRNA. *Appl. Environ. Microbiol.* **59:**695–700.
34. **Nuovo, G. J.** 1994. In situ detection of PCR-amplified DNA and cDNA: a review. *J. Histotechnol.* **17:**235–246.
35. **Ogram, A., G. S. Sayler, and T. Barkay.** 1987. The extraction and purification of microbial DNA from sediments. *J. Microbiol. Methods* **7:**57–66.
36. **Oyofo, B. A., and D. M. Rollins.** 1993. Efficacy of filter types for detecting *Campylobacter jejuni* and *Campylobacter coli* in environmental water samples by polymerase chain reaction. *Appl. Environ. Microbiol.* **59:**4090–4095.
37. **Oyofo, B. A., S. A. Thornton, D. H. Burr, T. J. Trust, O. R. Pavlovskis, and P. Guerry.** 1992. Specific detection of *Campylobacter jejuni* and *Campylobacter coli* by using polymerase chain reaction. *J. Clin. Microbiol.* **30:**2613–2619.
38. **Relman, D. A.** 1993. The identification of uncultured microbial pathogens. *J. Infect. Dis.* **168:**1–8.
39. **Relman, D. A., J. S. Loutit, T. M. Schmidt, S. Falkow, and L. S. Thompkins.** 1990. The agent of bacillary angiomatosis: an approach to the identification of uncultured pathogens. *N. Eng. J. Med.* **323:**1573–1580.
40. **Relman, D. A., T. M. Schmidt, R. P. MacDermott, and S. Falkow.** 1992. Identification of the uncultured bacillus of Whipple's disease. *N. Engl. J. Med.* **327:**293–301.
41. **Saiki, R. K., D. H. Gelfand, S. Stoffel, S. J. Scharf, R. Higuchi, G. T. Horn, K. B. Mullis, and H. A. Erlich.** 1988. Primer-directed enzymatic amplification of DNA with a thermostable DNA polymerase. *Science* **239:**487–491.
42. **Sayler, G. S., and A. C. Layton.** 1990. Environmental application of nucleic acid hybridization. *Annu. Rev. Microbiol.* **44:**625–648.
43. **Schmidt, T. M., B. Pace, and N. R. Pace.** 1991. Detection of DNA contamination in Taq polymerase. *BioTechniques* **11:**176–177.
44. **Schmidt, T. M., and D. A. Relman.** 1994. Phylogenetic identification of uncultured pathogens using ribosomal RNA sequences. *Methods Enzymol.* **215:**205–222.
45. **Solnick, J. V., J. O'Rurke, A. Lee, B. J. Paster, F. E. Dewhirst, and L. S. Tompkins.** 1993. An uncultured gastric spiral organism is a newly identified *Helicobacter* in humans. *J. Infect. Dis.* **168:** 379–385.

46. **Somerville, C. C., I. T. Knight, W. L. Straube, and R. R. Colwell.** 1989. Simple, rapid method for direct isolation of nucleic acids from aquatic environments. *Appl. Environ. Microbiol.* **55:**548–554.
47. **Southern, E. M.** 1975. Detection of species specific sequences among DNA fragments separated by gel electrophoresis. *J. Mol. Biol.* **98:**503–517.
48. **Suzuki, M. T., and S. J. Giovannoni.** 1996. Bias caused by template annealing in the amplification of mixtures of 16S rRNA genes by PCR. *Appl. Environ. Microbiol.* **62:**625–630.
49. **Spring, S., R. Amann, W. Ludwig, K.-H. Schleifer, and N. Peterson.** 1992. Phylogenetic diversity and identification of nonculturable magnetotactic bacteria. *Syst. Appl. Microbiol.* **15:**116–122.
50. **Steffan, R. J., J. Goksøyr, A. K. Bej, and R. M. Atlas.** 1988. Recovery of DNA from soils and sediments. *Appl. Environ. Microbiol.* **54:**2908–2915.
51. **Thiem, S. M., M. L. Krumme, R. L. Smith, and J. M. Tiedje.** 1994. Use of molecular techniques to evaluate the survival of a microorganism injected into an aquifer. *Appl. Environ. Microbiol.* **60:** 1059–1067.
52. **Thierry, D., C. Chureau, C. Aznar, and J.-L. Guesdon.** 1992. The detection of Mycobacterium tuberculosis in uncultured clinical specimens using the polymerase chain reaction and a nonradioactive DNA probe. *Mol. Cell. Probes* **6:**181–191.
53. **Trebesius, K., R. Amann, W. Ludwig, K. Mühlegger, and K.-H. Schleifer.** 1994. Identification of whole fixed bacterial cells with nonradioactive 23S rRNA-targeted polynucleotide probes. *Appl. Environ. Microbiol.* **60:**3228–3235.
54. **Victor, T., R. Du Toit, J. Van Zyl, A. J. Bester, and P. D. Van Helden.** 1991. Improved method for the routine identification of toxigenic *Escherichia coli* by DNA amplification of a conserved region of the heat-labile toxin A subunit. *J. Clin. Microbiol.* **29:**158–161.
55. **Wagner, M., R. Erhart, W. Manz, R. Amann, H. Lemmer, D. Wedi, and K.-H. Schleifer.** 1994. Development of an rRNA-targeted oligonucleotide probe specific for the genus *Acinetobacter* and its application for in situ monitoring in activated sludge. *Appl. Environ. Microbiol.* **60:**792–800.
56. **Wang, R.-F., W.-W. Cao, H. Wang, and M. G. Johnson.** 1993. A 16S rRNA-based DNA probe and PCR method specific for *Listeria ivanovii*. *FEMS Microbiol. Lett.* **106:**85–92.

Nonculturable Microorganisms in the Environment
Edited by R. R. Colwell and D. J. Grimes

Chapter 7

Environmental Parameters Associated with the Viable but Nonculturable State

Michel J. Gauthier

In 1985, Colwell and coworkers introduced the term "viable but nonculturable (VBNC) bacterial cells" to distinguish particular cells that could not form colonies on solid media but maintained metabolic activity and the ability to elongate after the administration of nutrients. The evolution of bacteria towards a VBNC state in natural environments (or under experimental conditions that mimic environmental ones) is now well established and documented (87, 101, 120). This state is obviously of high interest to our general understanding of microbial ecology. It is also of special concern when considering release in the environment of bacterial pathogens or indicators such as fecal coliforms, which may escape detection via routine bacteriological procedures and/or modify their virulence.

This review aims to analyze what is presently known (or assumed) on the possible involvement of physicochemical and biological environmental parameters in the nonculturable response. This first necessitates defining precisely the limits of the VBNC concept. As discussed elsewhere, several steps have been acknowledged between growth and death of bacteria in natural environments. According to Oliver (101), we would define a VBNC bacterium as a metabolically active bacterial cell that crossed a threshold in this way, for known or unknown reasons, and became unable to multiply in or on a medium normally supporting its growth. It must be emphasized that such a definition has, as a matter of fact, a limited ecological sense since it restricts the VBNC concept to those bacteria which are capable of growing on bacteriological media in the laboratory at a given phase of their life cycles. The definition is certainly suitable for copiotrophs such as enteric bacteria but presently leaves out most of the species which live in oligotrophic environments and remain unable to grow under laboratory conditions. Otherwise stated, the discussion of parameters related to the VBNC state will be limited in this review to those studies that clearly assess the VBNC state of bacteria through a joint enumeration of total culturable cells (using plate counts on solid media or MPN in liquid media) and

Michel J. Gauthier • INSERM Unité 452, Faculté de Médecine, Avenue de Valombrose, F-06107 Nice Cedex 2, France.

of total living cells using any noncultural method allowing the determination of cell viability (101, 120). These include: formation of microcolonies (42), elongation test in the presence of yeast extract and nalidixic acid or cephalexin (59, 75, 76), rhodamine 123 staining (50, 92), respiration and INT dehydrogenase assay (71, 141), thymidine incorporation (108, 123), CTC-DAPI double staining (15), or Gfp-tagging and Lux-tagging methods (24). There is however a considerable number of studies examining survival rates of bacteria by relating evolution of their culturability in natural environments or in laboratory microcosms. Some of them, although inexplicit regarding the VBNC state, bring interesting views on the possible role of physicochemical factors as inducers of the nonculturable response of bacteria. These data are also included and discussed in this chapter.

BACTERIA AND ENVIRONMENTS CONCERNED WITH THE VBNC STATE

Table 1 lists bacteria reported to enter the VBNC state and summarizes the environmental conditions under which they have been shown to evolve towards this state.

Most of the bacteria reported to enter the VBNC state are gram-negative species belonging to the gamma subclass of the Proteobacteria branch (129), except for *Rhizobium*, *Agrobacterium*, and *Helicobacter-Campylobacter* species, related to the alpha and epsilon subclasses, respectively. This probably reflects the dominance in the gamma subclass of species with high sanitary significance (enteric bacteria, water-borne human pathogens) (87) and of the heterotrophic culturable copiotrophs most commonly encountered in natural aquatic environments (vibrios, aeromonads, pseudomonads, alteromonads, xanthomonads). As previously pointed out by Oliver (101), the fact that most of these bacteria are gram negative could simply result from the generally higher resistance of gram-positive bacteria to the antibiotics (nalidixic acid, cephalexin) employed in the most frequently used direct-viable-count assay ground on elongation of cells in the presence of yeast extract (75, 76). Within the gram-positive species tested by Byrd et al. (14) in drinking water, *Streptococcus faecalis* and *Micrococcus flavus* exhibited a rapid culturability decline (7 days at 25°C), whereas *Bacillus subtilis* did not lose culturability over 25 days. It is however noteworthy that whether these bacteria enter the VBNC state was not assessed since neither viability nor spore formation was determined. No strictly anaerobic species has yet been analyzed in this way. Nevertheless, one should admit that the list of species presently known to enter the VBNC state better reflects the interest of microbiologists for particular species of higher sanitary or ecological meaning than the actual occurrence of this survival process within natural bacterial communities.

ENVIRONMENTAL FACTORS RELATED TO THE NONCULTURABLE RESPONE IN BACTERIAL

Available data that demonstrate the evolution of bacteria towards the VBNC state (Table 1) mainly come from in vitro experiments carried out with cultures or

Table 1. Bacterial species reported to enter the VBNC state, for which one or several environmental parameters were associated with the nonculturable response

Species or group of species	Environment[a]	Parameters associated with the VBNC state[b] (when known)	Source or reference
Aeromonas salmonicida	LW	T, NS	Morgan et al. (90–92)
Agrobacterium tumefaciens		BA	Alexander et al. (2)
Alcaligenes eutrophus	S	DE	Pedersen and Jacobsen (110)
Arthrobacter crystallopoietes	BFW	NS	Ensign (39)
Campylobacter coli	DiW	T	Jacob et al. (64)
Campylobacter jejuni	CM, FW, RW	T, NS, A, HOT	Rollins and Colwell (118); Medema et al. (88); Beumer et al. (10); Jones et al. (67); Boucher et al. (11); Harvey and Leach (54)
Cytophaga allerginae	A	T, NS, DE	Heidelberg et al. (56)
Enterobacter cloacae	S, LS	L, DE	Pedersen and Leser (109); Pedersen and Jacobsen (110)
Enterococcus faecalis	FW, LW, SW	T, L, NS	Barcina et al. (7); Gonzalez et al. (50); Smith et al. (127); Lleo et al. (80)
Escherichia coli	CM, BFW, FW, LW, RW, WW, BW, SW, SED	T, VL, UVL, TO, S, O, NS, HA, BA	Xu et al. (139); Carrillo et al. (17); Grimes and Colwell (53); Flint (41); Roth et al. (122); Lopez-Torres et al. (81); Linder and Oliver (79), Barcina et al. (6, 7); Byrd and Colwell (13); Davies and Evison (33); Gonzalez et al. (50); Arana et al. (5); Kaprelyants and Kell (70); Duncan et al. (37); Gourmelon et al. (52); Smith et al. (127); Mason et al. (84); Pommepuy et al. (113); Dukan et al. (36); Wang and Doyle (136)
Helicobacter pylori	CM, DiW, RW, BW, SW, A	T, NS	Mai et al. (83); Shahamat et al. (123); Cellini et al. (18)
Klebsiella pneumoniae	SW, BW	NS	Lopez-Torres et al. (81); Byrd et al. (14)
Klebsiella planticola	A	T, NS, DE	Heidelberg et al. (56)

Continued next page

Table 1. *Continued*

Species or group of species	Environment[a]	Parameters associated with the VBNC state[b] (when known)	Source or reference
Legionella pneumophila	TW, RW, SW	T, BA, BF	Hussong et al. (62); Paszko-Kolva et al. (107, 108); Bej et al. (9); Steinert et al. (130)
Micrococcus luteus	CM, FW	NS	Kaprelyants and Kell (71); Kaprelyants et al. (72)
Pasteurella piscicida	SW, SED	T, NS	Magarinos et al. (82)
Pseudomonas fluorescens	CM, EPW	T, NS	Amy et al. (4); Duncan et al. (37)
Rhizobium leguminosarum		BA	Alexander et al. (2)
Salmonella enteritidis	BFW, RW	T, NS	Roszak et al. (119); Chmielewski and Frank (23)
Salmonella typhi	FW	(NS)	Cho and Kim (25)
Salmonella typhimurium	FW, SW, S	T, NS, VL, S	Davies and Evison (33); Turpin et al. (133); Smith et al. (127)
Salmonella montevideo	FW, SW	T, VL, S(O)	Davies and Evison (33)
Salmonella oranienburg	FW, SW	T, VL, S(O)	Davies and Evison (33)
Salmonella spp.	WW	BF	
Serratia marcescens	A	T, NS, DE	Heidelberg et al. (56)
Shigella dysenteriae	WW, LW, RW, SW	(T), (NS)	Islam et al. (63); Rahman et al. (116)
Streptococcus pyogenes	CM	NS	Trainor et al. (132)
Vibrio anguillarum	SW	S (O)	Hoff (59)
Vibrio campbellii	SW	T	Wolf and Oliver (138)
Vibrio cholerae	BW, SW	T, NS, S (O), BS	Xu et al. (139); Linder and Oliver (79); Colwell et al. (28, 29); Xu and Colwell (140); Wolf and Oliver (138)
Vibrio harvey	CM	T, NS	Duncan et al. (37)
Vibrio mimicus	SW	T	Wolf and Oliver (138)
Vibrio natriegens	SW	T	Wolf and Oliver (138)
Vibrio parahaemolyticus	CM, SW	T, NS	Wolf and Oliver (138); Jiang and Chai (66)

Table 1. *Continued*

Species or group of species	Environment[a]	Parameters associated with the VBNC state[b] (when known)	Source or reference
Vibrio proteolyticus	SW	T	Wolf and Oliver (138)
Vibrio salmonicida	SW	S (O)	Hoff (59)
Vibrio vulnificus	CM, BW, SW, SF, A	T, NS, S	Linder and Oliver (79); Oliver and Wanucha (102); Oliver et al. (103); Nilsson et al. (100); Wolf and Oliver (138); Oliver et al. (104); Bryan et al. (12)
Vibrio sp. Ant-300 (marine)	SW	T	Preyer and Oliver (115)
Vibrio sp. S14 (marine)	MMM	NS	Holmquist and Kjelleberg (60)
Yersinia enterocolitica	SW	T, NS	Smith et al. (127)

[a]CM, culture medium; FW, fresh water; BFW, buffered fresh water; DiW, distilled water; TW, tap water; RW, river water; LW, lake water; DW, drinking water; WW, waste water; BW, brackish water; SW, sea water; EPW, endolithic pore water; S, soil; SED, sediment; A, air; LS, leaf surface; SF, shellfish; MMM, marine minimal medium.

[b]T, temperature; VL, visible light; UVL, ultraviolet light; O, osmolarity; A, aeration; NS, nutrient scarcity; S, salts; HA, humic acids; BA, biocidal agent; BF, biological factor; TO, toxic forms of oxygen; HOT, high oxygen tension; DE, desiccation. In parentheses, parameters for which the relation with the VNC state is not established.

simplified laboratory microcosms and almost exclusively pure bacterial strains, under well-controlled physicochemical conditions. A few in situ experiments bring additional information on how this process could develop under environmental conditions, both in monospecific populations enclosed in diffusion chambers or in natural complex bacterial communities of the open sea. Nevertheless, it can be assumed that the nonculturable response of bacteria is a universal process which can occur in any aquatic environment and in a variety of telluric biotopes such as sediments, soil, rhizosphere, and phyllosphere.

Only some of the studies dealing with entry of bacteria into the VBNC state provide informations on factors associated with this process. A number of studies simply evidenced VBNC forms of bacteria. Some of them described the resuscitation of unculturable cells through such physicochemical treatments as addition of nutrients or temperature upshift. Although not clearly establishing a connection between cause and effect, they strongly suggested relationships between cell culturability and environmental factors.

Data from Laboratory Experiments

Physical Parameters

Temperature. It has long been realized that the survival of enteric bacteria in natural waters is highly dependent on temperature (16, 40). Temperature is probably the most significant environmental physicochemical parameter that induces or favors the nonculturable response. However, it is worth noting that many experiments have been performed under conditions hindering the specific characterization of the effects due to temperature (for example, in cultures or microcosms where bacteria were also subjected to nutrient starvation). As far as we know, the effect of temperature as an inducer of the VBNC response differs with bacterial species. Vibrios are affected by low temperature (4 to 6°C), whereas all other bacteria tested to date enter more rapidly the VBNC state at high temperature (25 to 37°C) (37).

The inducing effect of low temperature was first established by Xu et al. (139) with *Vibrio cholerae*. The cells exhibited a rapid culturability decline in nutrient-free seawater when incubated at 4 to 6°C, whereas the direct viable count (DVC) method indicated a very low decrease in viability. In contrast, a slight decrease in culturability was observed at 10 and 25°C. Later on, it was demonstrated that a low temperature is able to induce solely the nonculturable response in *Vibrio vulnificus* (79, 101, 103, 138). This bacterium evolves towards the VBNC state in both nutrient-rich (102) and nutrient-poor (79) environments, and the transition to the VBNC state can be triggered by lowering temperature. Cells became nonculturable at 5°C but retained culturability for as long as 40 days in seawater when incubated at temperatures higher than 10°C (Fig. 1), where they underwent a normal starvation response (137, 138).

The unique causative role of temperature was confirmed by resuscitation of the whole nonculturable population at room temperature without addition of nutrient (100) (Fig. 2).

It should however be noted that *V. vulnificus* cells adapted to intermediate temperature (15°C), prior to change to 6°C, remain viable and culturable (12). There-

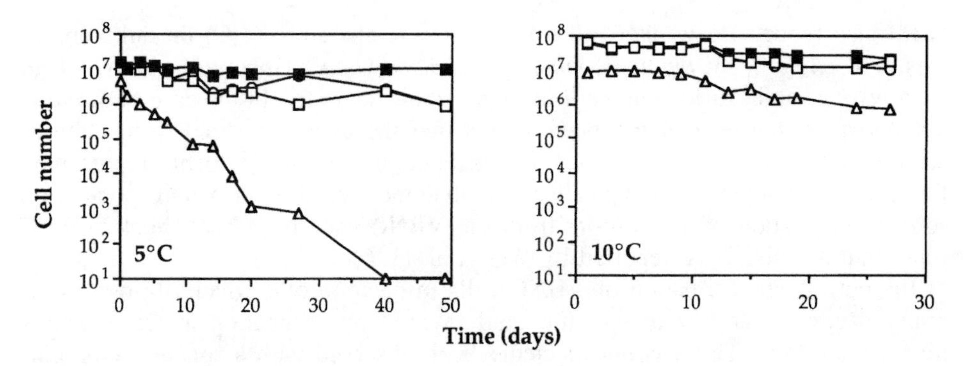

Figure 1. Effect of temperature on the nonculturable state of *V. vulnificus* in artificial seawater microcosms. Acridine orange total counts (■), plate counts (△), and direct viable counts determined by the INT (○) and Kogure et al. (75) (□) methods are shown. Reprinted with permission from Wolf and Oliver (138).

fore, *V. vulnificus* proves to be a very useful model for studies aiming to analyze the structural and physiological differences between growing and quiescent cells. In fact, various changes have been observed in nonculturable cells, including a reduction of cell division and transmembrane transport (100, 102), a reduction in the number of ribosomes, and an increase in short-chain fatty acid content of the cytoplasmic membrane (79). This underlines the pleiotropic influence of low temperature on *V. vulnificus* cells. In addition, it should be noted that cold shock induces considerable effects on vibrios, including a loss of proteins and carbohydrates and the alteration of membrane integrity (1), which could be at least partly responsible for initiating the VBNC response in these bacteria. Wolf and Oliver

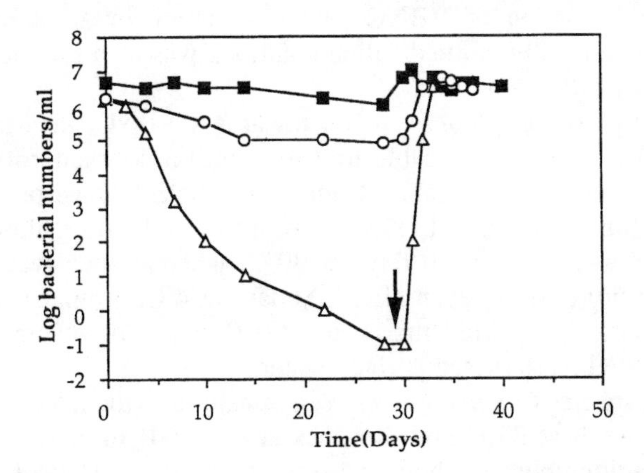

Figure 2. Effects of temperature down- and upshifts on stationary-phase cells of *V. vulnificus* in sterile salt solution. Acridine orange total counts (■), plate counts (△), and direct viable counts determined by the INT method (○) are shown. The arrow indicates when the 5°C microcosms were shifted to room temperature. Reprinted with permission from Nilsson et al. (100).

(138) also reported a weaker influence of low temperature (5°C) on culturability loss of *V. cholerae*, *V. mimicus*, *Vibrio parahaemolyticus*, *Vibrio natriegens*, *Vibrio proteolyticus* and *Vibrio campbelli* in oligotrophic seawater. In the case of *V. parahaemolyticus*, however, it has been shown that the cells recovered from cultures starved at low temperature (3.5°C) by a temperature upshift probably represented the regrowth of a few surviving cells rather than the revival of the VBNC population (66). Resuscitation of *V. cholerae* from the VBNC state by a heat shock at 45°C for 1 min has also been reported by Wai et al. (135).

In contrast, the formation of VBNC cells in bacteria other than vibrios is generally decreased at low temperature and favored by incubation at temperatures higher than 25°C. This has unexpectedly been observed with a species taxonomically close to vibrios, the fish pathogen *Aeromonas salmonicida.* As early as 1984, Allen-Austin et al. (3) showed that *A. salmonicida* lose culturability in river water within 15 days at 15°C and can be reactivated by the addition of nutrient more rapidly at 22°C than at 15°C. Whether this bacterium entered the VBNC state was not strictly stated, however, since its viability was not determined by a proper procedure. Later, Morgan et al. (90–92) reported that entry of *A. salmonicida* into the VBNC state may be increased at temperatures above 15°C and under low-nutrient conditions. Analogous results were obtained by Magarinos et al. (82) with the fish pathogen *Pasteurella piscicida* in seawater.

As for the *Enterobacteriaceae*, Xu et al. (139) reported that *E. coli* became more rapidly nonculturable (within 4 days) in brackish water at 25°C than at 10°C. A similar observation has been reported by Wang and Doyle (136) for the VBNC response of enterohemorrhagic *Escherichia coli* O157:H7 in municipal and lake waters, which was greater at 25°C and least at 8°C, regardless of the water source. Islam et al. (63) also reported *Shigella dysenteriae* to evolve towards the VBNC state after 15 to 21 days at 25°C in different surface waters (pond, lake, river, drain), although the authors did not describe explicitly the actual influence of temperature on the formation of VBNC cells. Formation of *Salmonella enteritidis* VBNC cells in chemically defined saline solutions was also reported as related to temperature downshift (7°C) (23).

The evolution of *Campylobacter jejuni* towards a VBNC stage in aquatic environments, together with its reversible transition into a dormant coccoid form, has also been investigated (11, 89, 117). Concerning the influence of temperature on this process, Rollins and Colwell (118) showed that cells lose culturability but not viability in river water within 10 days at 30°C, whereas they retain a significant level of culturability (0.01%) even after 120 days at 4°C. Similar results have been reported more recently by Medema et al. (88) (Fig. 3) for several strains of this bacterium incubated in sterilized surface water.

In the related species *Campylobacter coli*, transition to the NVC state was shown to occur within 48 h at 37°C and 2 weeks at 4°C (64) in sterile distilled water. Using an autoradiographic method to assess viability of *Helicobacter pylori* in sterile distilled water and river water, Mai et al. (83) and Shahamat et al. (123) reported several strains of this bacterium formed VBNC cells and showed that elevated temperatures (15, 22, and 37°C) resulted in a loss of culturability within 48 h and a loss of metabolic activity, whereas viable cells could be detected after

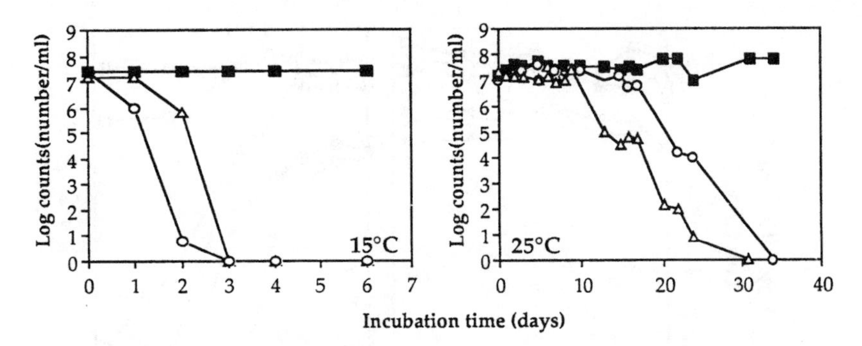

Figure 3. Survival of *C. jejuni* H354 in sterile surface water at 15 and 25°C. Acridine orange total counts (■), plate counts (○), and direct viable counts (△) are shown. Reprinted with permission from Medema et al. (88).

20 to 30 days at 4°C. Cellini et al. (18) also reported the entry of *H. pylori* into the VBNC state in cultures (Brucella-broth supplemented with calf serum) after 7 days at room temperature at 4°C, and after 15 days at 37°C, with evolution of the population toward the coccoid morphology.

Hussong et al. (62) reported that the human pathogen *Legionella pneumophila* enters the VBNC state more rapidly at 37°C (decimal rate of decline, 13 days) than at 4°C (decimal rate of decline, 29 days) in tap water. A similar finding has been reported more recently by Paszko-Kolva et al. (107, 108) for this bacterium.

Temperature may also have indirect effects on other factors inducing the VBNC response. Davies and Evison (33) analyzed the influence of temperature (5, 15, and 25°C) on the effect of visible light on entry of *E. coli*, *Salmonella oranienburg*, and *Salmonella montevideo* into the VBNC state in freshwater and in seawater. They found that, although a similar decrease in numbers of bacteria was observed at the three temperatures, numbers of VBNC cells were consistently higher at 25 than at 5°C in illuminated microcosms, whatever the medium. In the same way, it should be stressed that all the experiments performed in microcosms submitted to both stressing temperatures and nutrient starvation do not allow specific differentiation of the effects due to these environmental factors on the VBNC response.

Radiation. Both visible and UV solar radiation have long been considered the most drastic bactericidal agents in aquatic environments. On the basis of a careful comparative examination of previous studies on daily variations of numbers of fecal indicators in coastal seawaters, Chamberlin and Mitchell (19) concluded that solar irradiation is solely responsible for the decay of enteric bacteria in the sea. Recent results strongly suggest that light may significantly induce a VBNC response in bacteria, possibly depending on species. Whereas Hoff (59) found no effect of light on the culturability of either *Vibrio salmonicida* or *Vibrio anguillarum* in seawater, Barcina et al. (6) reported progressive dormancy of *E. coli* cells in river water microcosms under visible light irradiation. In illuminated systems, uptake and respiration of [^{14}C]glucose were drastically inhibited, and cells rapidly lost culturability without loss of viability (Fig. 4).

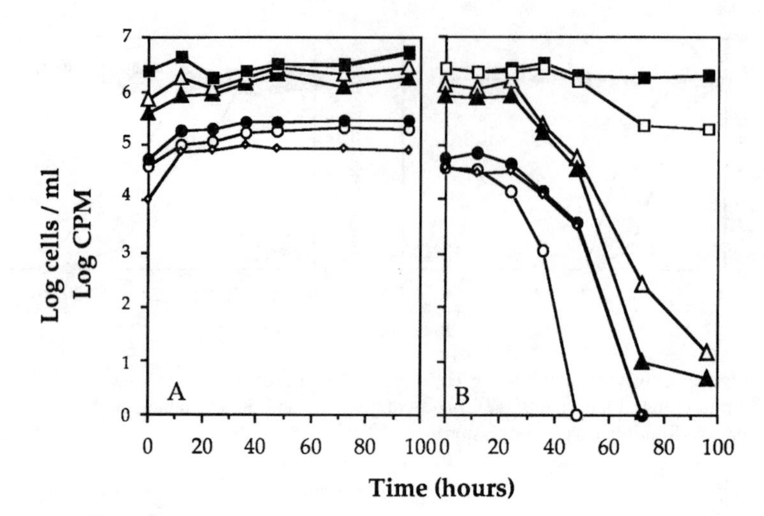

Figure 4. Survival of *E. coli* cells in nonilluminated (A) and illuminated (B) natural freshwater microcosms. Acridine orange total counts (■), metabolically active cells (□), plate counts on Trypticase soy agar (△) and eosin-methylene blue agar (▲), total uptake of [^{14}C]glucose (●), respired fraction of glucose taken up (◇), and assimilated fraction of glucose taken up (○) are shown. Reprinted with permission from Barcina et al. (6).

Further investigations by Barcina et al. (7) confirmed these observations for *E. coli* and *Enterococcus faecalis* both in fresh and marine systems. Davies and Evison (33) also reported the VBNC response of *E. coli*, *Salmonella typhimurium*, *Salmonella montevideo*, and *Salmonella oranienburg* in freshwater and seawater microcosms exposed to natural sunlight or artificial light of comparable intensity (with or without ultraviolet), and emphasized the crossed influence of temperature, salinity, and humic acids on the effects of light.

As suggested by Arana et al. (5), the effect of visible light on culturability loss of *E. coli* in freshwater could result from the photochemical generation of hydrogen peroxide in cells. Viability of cells was not determined in these experiments, however, and thus whether they became nonculturable or died cannot be stated. This hypothesis is supported by more recent data from Gourmelon et al. (52) showing that the drastic decrease of culturable *E. coli* cells in seawater microcosms submitted to visible light damage (about 40 klux) was lowered under anaerobic conditions and in the presence of scavengers of the reactive oxygen species (catalase and thiourea). These authors later on stressed the maintenance of virulence determinants of *E. coli* H10407 in seawater exposed to sunlight, even in the VBNC cells of this bacterium (113).

The study reported by Pedersen and Leser (109) is especially interesting in that it concerns the VBNC response of the enteric species *Enterobacter cloacae* in nonaquatic environments (soil and phylloplane). Cell suspensions were sprayed on bean (*Phaseolus vulgaris*) leaves in morning full sun and at dusk. In the evening, numbers of culturable, viable, and total cells were all of the same magnitude, indicating that most of the cells were viable, and culturable. In the morning, however, total counts were more than 10 times higher than culturable counts, whereas viable counts indicated that about 80% of the viable cells were in a VBNC form.

All these observations emphasize the effect of light as a powerful physical inducer of the nonculturable state in natural bacterial populations of biotopes subjected to solar irradiation. On the other hand, recent data from Pitonzo et al. (111, 112) strongly suggest that the induction of the VBNC response is not restricted to visible light irradiation. Exposure of the endolithic microflora from rocks of the Yucca Mountain (Nevada radioactive waste-deposit test site) to increasing doses of gamma radiations induced their entry in the VBNC state after a cumulative dose of 2.33 kGy. Injured heterotrophic bacteria could be resuscitated after storage for 2 months at 4°C.

Osmolarity. The influence of salinity on survival of telluric bacteria was soon evoked to explain, at least partly, their decay in natural or artificial saline waters (16, 105). During the last decade, several studies have indicated that the salt content of seawater might be responsible for the entry of enteric bacteria into the VBNC state. The pioneering study of Xu et al. (139) already reported a 40-fold higher number of VBNC forms of *E. coli* after 96 h of incubation in microcosms of 25‰ salinity in comparison to microcosms of 5‰ salinity. A later study by Roszak et al. (119) described a similar effect in *Salmonella enteritidis.* Roth et al. (122) found that 80 to 90% of *E. coli* cells lost culturability within 2 to 3 h but retained viability when exposed to a high salt concentration (0.8 M NaCl).

More recent studies reported by Gauthier et al. (43) and Munro et al. (95, 96), although not directly evidencing the VBNC response, strongly suggested that the loss of culturability of *Enterobacteriaceae* in seawater was due to high osmotic strength rather than to specific salts. This assumption was grounded on the observation that these bacteria acquire a high resistance to seawater (i.e., retain a high level of culturability in seawater) when previously grown at high osmolarity whatever the compound used to increase the osmotic strength of the growth medium (Fig. 5) (43, 95, 96).

The protective effect exerted by osmolytes such as betaine (44, 49, 122), potassium glutamate (45, 47) and trehalose (unpublished data) on *E. coli* cells in seawater strongly supports the hypothesis that osmotic strength is the actual physical stressing factor rather than salts. It also shows that osmoregulatory processes involving organic compatible osmolytes (31, 32) can help survival of enterobacteriaceae in saline aquatic environments. Whether osmolarity can induce the VBNC state remains, however, to be demonstrated. The resuscitation of *E. coli* from osmotic upshock by betaine (122) suggests, however, that osmotic stress may contribute to the physiology of the VBNC state. It should also be noted that osmotic upshift drastically inhibits transport of carbohydrates and amino acids (6, 102, 121) and thus may induce the VBNC response of enteric bacteria through nutrient starvation.

In contrast, salinity/osmolarity seems not to be a major inducer of the nonculturable response in autochtonous marine and estuarine bacteria naturally adapted to these environments. Xu et al. (139) found that the culturability of *V. cholerae* was not affected in microcosms of from 5 to 25‰ salinity. Furthermore, the ability

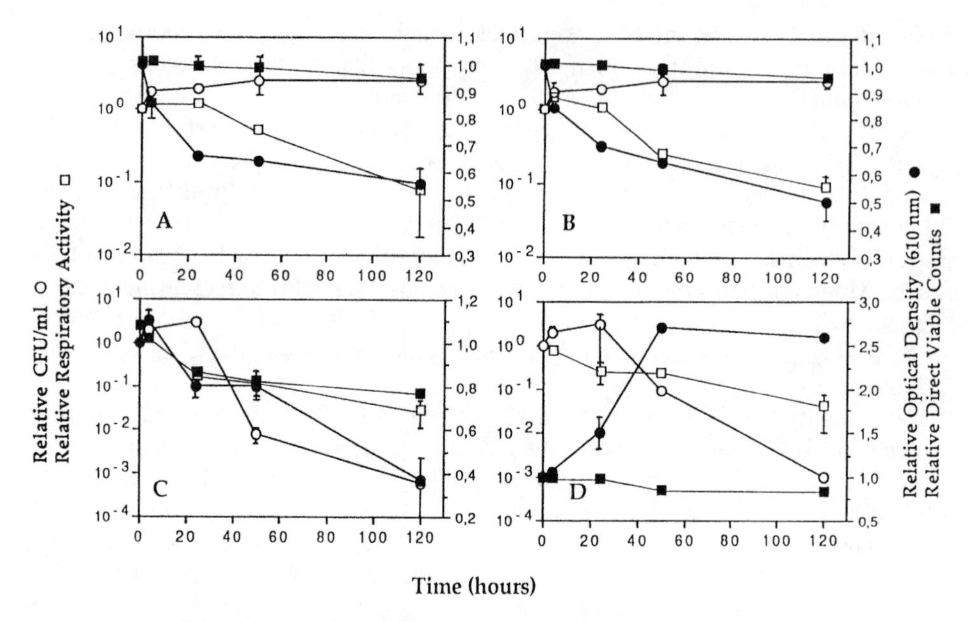

Figure 5. Modifications in viability and respiratory activity of *Vibrio* sp. S14 during starvation of carbon+nitrogen+phosphorus (A), carbon (B), nitrogen (C), and phosphorus (D). Plate counts (○), respiratory activity (□), optical density (610 nm) (●), and direct viable counts determined by the INT method (■) are shown. Reprinted with permission from Holmquist and Kjelleberg (60).

of *V. cholerae* to retain culturability in seawater microcosms is not modified by potassium glutamate and betaine (97), although cells actively transport both substrates in seawater. Hoff (59) also reported that the culturability of *V. salmonicida* and *V. anguillarum* did not significantly decline in microcosms of >20 and >10‰ salinity, respectively, without decrease in total cell counts for at least 4 weeks of incubation.

The higher sensitivity of *Enterobacteriaceae* to hypersalinity (hyperosmolarity) is usually considered as "not surprising," given the normal habitat of such bacteria (101). We would, however, point out that this ecological argument, if valid for most of the experimental studies, is questionable when regarding the natural processes. Most, if not all, experiments aiming to analyze the behavior of enteric bacteria in laboratory microcosms as well as in diffusion chambers were made using bacteria grown in media with osmolarity lower than the threshold of 250 mosM to which osmoregulatory processes are triggered (32). Now it is well documented that both the intestinal content (78, 134) and the urine (77) are media of high osmolarity in which enteric bacteria can grow and exhibit structural and physiological adaptations to hyperosmolarity (20–22, 77). Under natural conditions, however, cells are generally released into the environment and transported to the sea via wastewater or river water, two media with low osmolarity (1 to 100 mosM). The hypoosmotic shock experienced in wastewater is responsible for the loss of several cell compounds, including the osmolytes betaine, proline, and glutamate accumulated during growth at high osmolarity, followed by a dramatic loss of culturability (30, 45). A similar effect, discussed below, may also result from washing cells with fluids with low osmolarity before the test. However, cells grown in vivo or in vitro at high osmolarity retain adaptative structural elements such as OmpC porins (99) that are involved in the VBNC response of *E. coli* in seawater (50). Apart from specific osmotic effects, the transit in wastewater could also induce in enteric bacteria responses to several other stresses (low temperature, high oxygen content, light, perhaps low nutrient) that might be of high importance with regard to the VBNC response in various final receiving environments. This could explain the observed increased resistance of *E. coli* cells to seawater after a transient incubation in wastewater (30).

Davies and Evison (33) also reported the synergistic effect of salinity (osmolarity) on the induction of the VBNC response in *E. coli* and *Salmonella typhimurium* by natural sunlight in freshwater and seawater (Fig. 6).

More recently, Solic and Krstulovic (128) reported a similar effect of salinity on the detrimental action of sunlight on fecal coliforms in seawater, although viability of bacteria was not measured. This observation might be of high ecological and sanitary interest in natural surface and marine waters.

Chemical Parameters

Organic matter: nutritional effects. Over the last 20 years, a tremendous number of studies have aimed at elucidating the influence of nutrient starvation on the behavior of both autochthonous and allochthonous bacteria in natural environments. Most of them concerned the aquatic environment since surface and marine waters generally contain low amounts of organic matter (1 to 15 mg of carbon per liter) (120) and a lower quantity of bioavailable nutrients. To some extent, the fact that many bacterial species survive longer in a metabolically active state in marine sediments than in the water column (8, 34, 48, 61, 82) in spite of micropredation, evidences the protective influence of organic nutrients on the adaptation and the VBNC response of bacteria to environmental conditions.

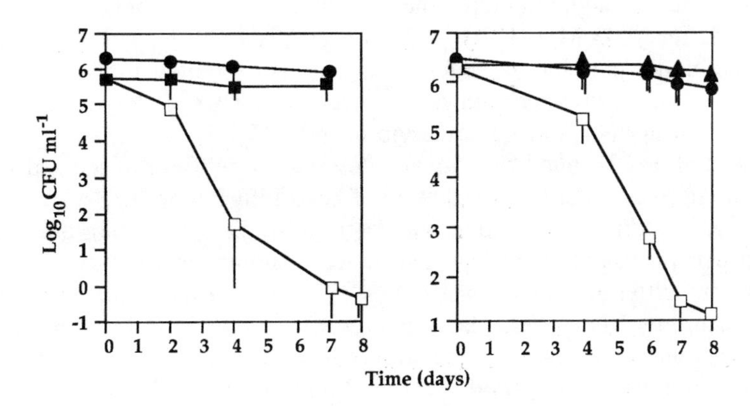

Figure 6. Survival in seawater of *E. coli* previously grown in nutrient broth of low osmotic strength (□) or supplemented with NaCl (0.5M) (●), LiCl (0.5M) (■), or saccharose (0.8M) (▲). Reprinted with permission from Munro et al. (96).

Nutritional strategies developed by bacteria under starvation conditions in oligotrophic environments have been extensively reviewed (51, 71, 74, 93, 94, 120). In this respect, two different bacterial groups have been individualized: oligotrophs and copiotrophs. Oligotrophic bacteria are naturally adapted to nutrient scarcity since they are able to grow on low-nutrient concentration and can exist over long starvation periods. In contrast, copiotrophs are bacteria requiring high-nutrient concentration for growth. It is noteworthy, however, that oligotrophs do not express the highest specific affinities for scarcely available substrates: facultative oligotrophs and even some copiotrophs may be equally effective in sequestering substrates under conditions of nutrient depletion (51).

The ability to remain viable through periods of starvation is obviously essential for the survival of bacteria in natural environments. The entry into a VBNC state may enable many bacterial species (and especially enteric copiotrophs) to persist in oligotrophic aquatic or telluric environments. Unlike other physicochemical putative or established inducing factors, nutrients only influence the entry of bacterial cells into the VBNC state when they are lacking. Therefore, most of the results suggesting that nutrient starvation may play a significant role in this way follow from two types of complementary investigations: an analysis of the viability of bacteria after different periods of starvation and the application of resuscitation procedures that evidenced the retrogression of the VBNC response in the presence of organic substrates. The influence of variations in nutrient concentrations on the VBNC response has not yet been determined. As recently reported by Trainor et al. (132), the starvation-VBNC response of *Streptococcus pyogenes* in stationary-phase cultures is induced by carbon and phosphorus limitations but not by nitrogen (as amino acids) limitation. Besides, Bryan et al. (12) reported that specific removal of iron from the growth medium prior to cold adaptation (6°C) decreased the viability of *V. vulnificus* by 2 log levels, suggesting that iron may play an important role in adaptation of this species to low temperatures.

Using comparative plate and microcolony counts, Ensign (39) reported that more than 65% of both spherical and rod-shaped cells of *Arthrobacter crystallopoietes* remained viable although not culturable after 60 days of incubation in nutrient-free phosphate buffer (0.03 M, pH 7.0) without significant cell lysis. Entrance into the VBNC state under nutrient starvation has been also reported for *V. cholerae* (28, 29), *Pseudomonas fluorescens* and several deep-subsurface endolytic isolates in an artificial pore water (4), and a marine vibrio (60) (Fig. 7).

Roszak et al. (119) found that *Salmonella enteritidis* was resuscitated from the VBNC state in river water by the addition of heart infusion broth. Such a protective effect of organic nutrients was confirmed later on for different copiotrophic species: *E. coli* in phosphate buffer by intestinal mucosal homogenate (125), *H. pylori* in freshwater by different organic substrates (83), *Micrococcus luteus* in minimal growth medium by lactate (70, 71), *Aeromonas salmonicida* and *Enterococcus faecalis* in lake water by tryptone and natural organic matter (80, 91), and *P. piscicida* in seawater by fresh culture broth (82). It should, however, be noted that the addition of growth substrates to the starvation medium does not necessarily result in an increase of cell viability or culturability. An acceleration of the loss of culturability has been reported to occur in starving cell suspensions of *Aerobacter aero-*

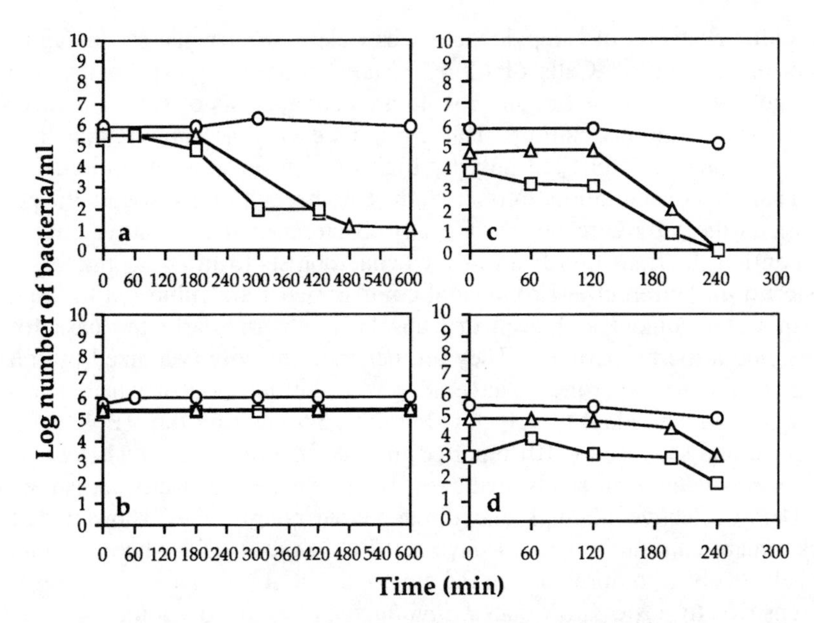

Figure 7. Survival of *E. coli* (a and b) and *S. typhimurium* (c and d) exposed to natural sunlight in seawater (a and c) and in freshwater (b and d). Plate counts on nonselective (△) and selective (□) media and direct viable counts determined by the method of Kogure et al. (75) (○) are shown. Reprinted with permission from Davies and Evison (33).

genes (114), *Streptococcus lactis* (131), and *Arthrobacter crystallopoietes* (39), but the phenomenon remains poorly understood.

Since *V. vulnificus* cells enter the VBNC state when incubated at 5°C even in nutrient-rich medium (102), it can be assumed that the nutrient level is not a parameter associated with the nonculturable response of this species. This unique example is interesting in that it suggests that VBNC response and nutrient starvation might be distinct processes corresponding to different survival strategies (101). This question is discussed below and debated in different chapters of this book.

Organic matter: nonnutritional effects. Davies and Evison (33) found that the presence of humic acids significantly decreased the entry of *E. coli* into the VBNC state in freshwater but not in seawater. They assumed that such protective effect could follow from the in situ absorption of some lethal solar radiation. This might be of ecological importance in soil and in soil-leaching waters. Pommepuy and coworkers (IFREMER, Brest, France, personal communication) have observed a reduction of the effect of sunlight on *E. coli* cells in brackish water and seawater, possibly due to the adsorption of lethal radiation by peptone. Compounds which eliminate reactive oxygen species (catalase, thioglycolate, pyruvate, thiourea) also exert a protective effect on *E. coli* cells (5, 52).

Aeration. Rollins and Colwell (118) reported a more rapid entry into the VBNC state of *C. jejuni* in microcosms incubated with shaking (ca. 3 days), as compared with stagnant microcosms (ca. 10 days). This effect could be related to the microaerophilic character of campylobacters, especially in suspensions containing low numbers of cells (126). Cells of *C. jejuni* are known to transform into dormant coccoid cells when exposed to air (73). In a recent analysis of *C. jejuni* coccal cell formation in continuous cultures, Harvey and Leach (54) showed that this morphological change was elicited only by high oxygen tension, in conjunction with reduced carbon concentration. Further studies are needed to investigate the possible influence on the VBNC response of enteric bacteria of the oxidative stress experienced during the transition from in vivo anaerobiosis to in situ aeration.

Desiccation. Different environmental compartments are subjected to large variations in water content and even dry up. This is particularly the case for soil, surfaces, and airborne particles. They are generally heavily colonized by microorganisms, and some experiments have addressed the question of whether and how these organisms can adapt to very low humidity by entering the VBNC state.

Pedersen and Jacobsen (110) reported the fate of *Enterobacter cloacae* and *Alcaligenes eutrophus* in a sandy loam soil with near-field-capacity moisture or air dried. Using selective platings and direct viable counts, they showed that both species remain culturable over 14 days at 17°C in moist soil, whereas about 99% of the cells evolved to dormancy in the air-dried soil. *Alcaligenes eutrophus* proved more sensitive to desiccation than *Enterobacter cloacae*. It should be noted that only a minor part of the test bacteria regained culturability after the dried soil was rewetted.

More recently, Heidelberg et al. (56) showed that aerosolization of *Serratia marcescens*, *Klebsiella planticola*, and *Cytophaga allerginae* in air of low humidity (20 to 25%) and mild temperature (20 to 22°C) over a 4-h incubation period led to the evolution of more than 90% of the bacteria toward the VBNC state, as determined by the INT dehydrogenase assay and the elongation test with nalidixic acid.

Although rather thin on the ground, such observations point out the powerful influence of desiccation as an inducer of the VBNC response of airborne bacteria and, more generally, of the microflora in environments subject to desiccation.

Biocidal agents. Some of the studies concerning the revival of bacterial cells injured by biocidal agents such as toxic metals, disinfectants, and biocides have produced interesting information on the possible role of these compounds as inducers of the VBNC response. For example, Palmer et al. (106) found that enterotoxigenic *E. coli* developed VBNC cells when exposed to natural freshwater containing toxic mineral and organic chemicals. According to Singh et al. (125), exposure of enterotoxigenic and invasive *E. coli* strains to copper (0.6–1 mg/liter) and chlorine (0.4–1.6 mg/liter) in buffered water at 4°C produced VBNC cells that retained enterotoxigenic activity when resuscitated. Alexander et al. (2) have recently determined that copper also induces nutrient-starved *Agrobacterium tumefaciens* and *Rhizobium leguminosarum* cells to become VBNC. Aluminum salt could also be able to induce a VBNC response in bacteria. Chowdhury et al. (26) showed that in suspensions of *V. cholerae* O1 in artificial seawater containing alum [$KAl(SO_4)_2$], a significant number of cells lost culturability but maintained viability as measured by direct viable counting.

VBNC cells of *L. pneumophila* were detected by Bej et al. (9) in growth medium treated with hypochlorite (100 ppm) for at least 10 min, whereas culturable cell counts fall to zero after 2 min of exposure. The study reported by Desmonts et al. (35) may be especially significant in that it examined the entry into the nonculturable state of salmonellas in raw and chlorinated wastewater. In some samples, 5 to 31.5% of the *Salmonella* spp. enumerated in chlorinated water by the direct viable count of Kogure et al. (75) coupled with immunofluorescence were substrate responsive. Dukan et al. (36) also reported the evolution to the VBNC state of about 10% of a population of *E. coli* suspended in phosphate buffer and treated with 10 μM HOCl.

On the other hand, an interesting parallel has been made by Mason et al. (84) between the activity of the quinolone antibiotic ciproflaxin and the concept of the VBNC state. Using the INT reduction technique together with traditional cultural counts and flow cytometric methods, the authors showed that over 90% of the *E. coli* cells exposed to 10 to 100 times the MIC of the antibiotic lost culturability, although they maintained their membrane potential and protein synthesis capacity. Further studies are obviously needed to confirm the role of antibiotics as inducers of the VBNC response of bacteria, both in vitro and in vivo.

Biological Parameters

Since the early 1970s, Colwell and coworkers have evidenced the importance of planktonic crustaceans (copepods) on the life cycle of vibrios in the estuarine and marine environments (27). In addition to the physicochemical parameters reviewed above, the living organisms may thus exert a protective effect on microorganisms in various environments. Steinert et al. (130) have recently reported the positive influence of a protozoan on the VBNC response of *L. pneumophila*. This bacterium enters a nonreplicative VBNC state under starved conditions; however, the addition of free-living *Acanthamoeba castellanii* amoebae to the dormant bacteria resulted in their resuscitation to a culturable state, without any change in their capacity to survive intracellularly in human monocytes. On the other hand, Chowdhury et al. (26) showed that the association of *V. cholerae* O1 and O139 cells to zooplankton (mainly copepods) in seawater lowered their entry into the VBNC state and protected them against alum and chlorine treatments.

These observations address the fundamental question of the possible protection afforded to VBNC bacteria by living organisms after ingestion or internalization. This question takes on multiple aspects and could have important consequences, especially when it concerns microorganisms pathogenic to humans.

Data from Field Studies

In addition to analytical laboratory experiments, a few studies have brought observational data that suggest the prominent influence of nutrient level on the VBNC response of bacteria in natural aquatic environments. Grimes and Colwell (53) reported the entry of *E. coli* and *V. cholerae* into the VBNC state in diffusion chambers immersed in semitropical ocean water but did not identify the inducing factor(s). Using the same technique, Carillo et al. (17) observed that *E. coli* could survive, remain physiologically active, and regrow at rates that were dependent on nutrient levels of the ambient waters collected from a tropical rain forest watershed (Mameyes River, Puerto Rico). Later on, Hazen et al. (55) and Lopez-Torres et al.

(81) examined the survival and respiratory activity (INT reduction) of *E. coli* and *K. pneumoniae* in different coastal areas of Puerto Rico. According to Lopez-Torres et al. (81), less than 10% of the total *K. pneumoniae* population was respiring after 96 h in water with low organic matter content, whereas 40% of the population was INT positive in water from a site receiving rum distillery effluents. Similar results were obtained with *E. coli*, although with higher overall proportions of active and respiring cells. The highly influential role of nutrient availability on the VBNC response of *V. cholerae* in natural estuarine and marine waters has also been suggested by Colwell et al. (28). In addition, the influence of low temperature on the entry of *V. vulnificus* into the VBNC state has clearly been evidenced in natural estuarine waters during winter months by means of diffusion chambers (104).

However, field studies sometimes deliver surprising results. For instance, Smith et al. (127) showed that survival, evolution toward dormancy, and recovery of enteric bacteria (*E. coli*, *Enterococcus faecalis*, *Salmonella typhimurium*, and *Yersinia enterocolitica*) in situ exposed to the antarctic polar marine environment apparently depended more tightly on nutrient starvation than on temperature in this very cold environment.

CONCLUSIONS: MULTIPLE-STRESS SITUATIONS AND THE NONCULTURABLE RESPONSE

It follows from data discussed above that the nonculturable response can be induced or favored by a variety of physicochemical environmental factors which have long been considered as "stressing factors." In most, if not all, natural environments, combined adverse conditions create multiple-stress situations that direct the adaptation of bacteria. The specific response of bacteria to specific stresses, such as heat shock, has been known for a long time. During the past decade, a number of studies have evidenced multiple-stress responses in bacteria and pointed out the prominent role played by nutrient starvation in this way, at least in the enteric models *E. coli* and *Salmonella typhimurium* (58, 85, 124). As previously outlined, the physiological and genetic analysis of survival of *E. coli* and *Salmonella typhimurium* under conditions prevailing during the stationary phase of cultures has led to a better understanding of the mechanisms allowing these bacteria to adapt to nutrient starvation. Nutritional stress induces a complex response, not strictly specific to nutrient starvation, that confers to cells a high level of resistance to other stresses such as oxidative, acid, thermal, osmotic, or radiative shock (38, 58, 65, 68). This starvation-induced multistress (SIMS) resistance depends on the induced synthesis of stress proteins, more or less specific to each stress and encoded by survival genes whose expression is controlled by sigma factors such as RpoS (KatF) (58, 85, 86).

The question now is whether a relationship exists between the SIMS resistance and the VBNC state. This point is debated elsewhere in this book. Several data reported in this chapter suggest mechanistic similarities in both responses. First is the fact that the same set of environmental parameters, known as stressing factors, induce (or favor) the VBNC response and the SIMS response. Second is the prominant role played by nutrient starvation in survival and maintenance of culturability of bacteria in hostile environments. For example, Preyer and Oliver (115) found that the marine psychrophilic *Vibrio* Ant-300 maintained a higher level of culturability in seawater when cells were carbon starved for 1 week prior to exposure to 17°C. They concluded that starvation conditions might be a significant factor in providing heat tolerance to marine psychrophiles. Third is the close dependence of both the VBNC response and the multistress adaptation on the age of the cells. It has clearly been established on the one hand, that the loss of culturability exhibited by copiotrophic bacteria in various environments is much lower in stationary-phase cells and, on the other hand, that the expression of genes involved in the SIMS response of *Enterobacteriaceae* is activated during the late exponential and stationary phases (58, 124). Even for *V. vulnificus*, in which the VBNC and starvation responses seem to be separate processes, Oliver et al. (103) reported that cells harvested during the stationary phase generally required about twice as many days to become nonculturable at 5°C than did logarithmically growing cells. Fourth is the dependence of both processes on de novo protein synthesis. We reported that the loss of culturability of *E. coli* in seawater is considerably reduced when cells undergo a thermal (48°C), oxidative (H_2O_2, 500 μM), acid (pH 4.5), osmotic (NaCl, 0.5 M), or nutritional (carbon and nitrogen) stress for 1 h prior to transfer to seawater (97). This protective effect depends partly on de novo protein synthesis and on the sigma factor RpoS (KatF) since it does not develop in a *katF* mutant strain. However, the protection afforded by osmotic stress proved independent of KatF, which suggests that influence of osmolarity on viability of enteric bacteria in saline environments is achieved through a more specific process. This might result from the fact that some genes do not depend on KatF (69). These are genes that have high importance in osmoregulation and have been shown to protect *E. coli* cells in seawater such as *proU* (betaine transport) and *bet* (betaine synthesis) (44, 96). It is noteworthy, however, that many findings in this field are specific to *Enterobacteriaceae* and related gram-negative copiotrophs. As reported by Ensign (39), cells of the gram-positive species *Arthrobacter crystallopoietes* lost viability more rapidly when taken from the stationary phase than did exponential-phase cells.

The existence of links between multistress and VBNC responses means that any stress undergone by bacteria during any phase of their life cycle, especially during growth, would influence their further adaptation in the environment. This is typically the case for enteric bacteria, since many physicochemical and biological factors influence their growth in the enteric environment and their behavior in natural waters (lake, river, estuary, sea). Enteric bacteria grow in intestinal and urinary tracts under highly specific conditions that adapt them to high temperature, hyperosmolarity, low oxygen pressure, high nutrient level, and (to some extent) tensioactive compounds (i.e., bile salts). Consequently, they undergo several stresses (hypothermal, hypoosmotic, hyperoxidative, hyperradiative, hypotrophic) in wastewater and natural surface freshwaters that modify their metabolic state. The release in seawater also generates a specific osmotic stress. The adaptative power of cells in a final receiving environment, and hence their possible entry into the VBNC state, will thereby depend on all these previous events.

Looking at it this way, one should underline the inadequacy of some experimental designs used to study the survival and the VBNC response of bacteria in laboratory microcosms with regard to natural processes. The most significant bias may have followed from the conditions under which cells were grown and then washed before inoculation of microcosms. Growth conditions (i.e., composition of culture media, aeration, temperature) largely differed between studies and were generally not representative of those prevailing in the enteric environment since bacteria were almost exclusively grown aerobically at low osmolarity. Moreover, cells were indiscriminately taken from the mid- or late-exponential phase or the stationary phase, frequently after a convenient overnight period of incubation. This yielded highly incomparable results with confused biological and ecological meaning. The washing step is also of high importance in this way, at least for enteric species, since it has been shown that hypoosmotic and hypothermal shocks generate structural or metabolic changes that could later on influence the survival of bacteria in laboratory microcosms (45, 103, 139).

Despite these limitations in the interpretation of previously published data, the present review finally suggests that a certain consistency in the adaptive response of microorgansims could lie under the apparent disparity of physicochemical parameters associated with the VBNC state of bacteria. Such a concept finds its most fundamental root in its compliance with the universal laws of simplicity and economy. However, the trees must not be allowed to mask the forest. Further extensive studies are needed to define more accurately the VBNC state and estimate its incidence within natural bacterial communities in order to confirm or invalidate, again, the existence of a functional unity under the phylogenetic diversity.

Our present knowledge of causative relationships between environmental parameters and the VBNC response remains very limited or even rudimentary. It is surprising to note the scarcity of studies in which the VBNC state has been clearly shown through a concomitant measurement of viable and culturable cells, with respect to the tremendous amount of work carried out by microbiologists to analyze survival of bacteria in the environment. The use of molecular genetic methods (see Chapter 6) will undoubtedly (it is hoped) allow us to fill the gap.

REFERENCES

1. **Adhikari, P. C.** 1975. Sensitivity of cholera and El Tor vibrios to cold shock. *J. Gen. Microbiol.* **87:**163–166.
2. **Alexander, E., D. Pham, and T. R. Steck.** 1999. The viable-but-nonculturable condition is induced by copper in *Agrobacterium tumefaciens* and *Rhizobium leguminosarum*. *Appl. Environ. Microbiol.* **65:**3754–3756.
3. **Allen-Austin, D., B. Austin, and R. R. Colwell.** 1984. Survival of *Aeromonas salmonicida* in river water. *FEMS Microbiol. Lett.* **21:**143–146.
4. **Amy, P. S., C. Durham, D. Hall, and D. L. Haldeman.** 1993. Starvation-survival of deep subsurface isolates. *Curr. Microbiol.* **26:**345–352.
5. **Arana, I., A. Muela, J. Iriberri, L. Egea, and I. Barcina.** 1992. Role of hydrogen peroxide in loss of culturability mediated by visible light in *Escherichia coli* in a freshwater ecosystem. *Appl. Environ. Microbiol.* **58:**3903–3907.
6. **Barcina, I., J. M. Gonzalez, J. Iriberri, and L. Egea.** 1989. Effect of visible light on progressive dormancy of *Escherichia coli* cells during the survival process in natural freshwater. *Appl. Environ. Microbiol.* **55:**246–251.
7. **Barcina, I., J. M. Gonzalez, J. Iriberri, and L. Egea.** 1990. Survival strategy of *Escherichia coli* and *Enterococcus faecalis* in illuminated fresh and marine systems. *J. Appl. Bacteriol.* **68:**189–198.

8. **Bauerfeind, S., G. G. Gerhardt, and G. Rheiheimer.** 1981. Investigations on the survival of fecal bacteria in experiments with or without sediments. *Zentbl. Bakt. Parasitikd. Infekt. Hyg.* **174:**364–374.
9. **Bej, A. K., M. H. Mahbubani, and R. M. Atlas.** 1991. Detection of viable *Legionella pneumophila* in water by polymerase chain reaction and gene probe methods. *Appl. Environ. Microbiol.* **57:**597–600.
10. **Beumer, R. R., J. de Vries, and F. M. Rombouts.** 1992. *Campylobacter jejuni* non-culturable coccoid cells. *Int. J. Food Microbiol.* **15:**153–163.
11. **Boucher, S. N., M. R. Adams, and A. H. L. Chamberlain.** 1993. Viability and culturability of *Campylobacter jejuni. In 62nd Annual Meeting of the Society for Applied Bacteriology, Nottingham, U.K.*
12. **Bryan, P. J., R. J. Steffan, A. DePaola, J. W. Foster, and A. K. Bej.** 1999. Adaptative response to cold temperature in *Vibrio vulnificus. Curr. Microbiol.* **38:**168–175.
13. **Byrd, J. J., and R. R. Colwell.** 1990. Maintenance of plasmids pBR322 and pUC8 in nonculturable *Escherichia coli* in the marine environment. *Appl. Environ. Microbiol.* **56:**2104–2107.
14. **Byrd, J. J., H.-S. Xu, and R. R. Colwell.** 1991. Viable but nonculturable bacteria in drinking water. *Appl. Environ. Microbiol.* **57:**875–878.
15. **Cappelier, J. M., B. Lazaro, A. Rossero, A. Fernandes-Astorga, and M. Federighi.** 1997. Double staining (CTC-DAPI) for detection and enumeration of viable but non-culturable *Campylobacter jejuni* cells. *Vet. Res.* **28:**547–555.
16. **Carlucci, A. F., and D. Pramer.** 1960. An evaluation of factors affecting the survival of *Escherichia coli* in seawater. II. Salinity, pH and nutrients. *Appl. Microbiol.* **8:**243–247.
17. **Carrillo, M., E. Estrada, and T. C. Hazen.** 1985. Survival and enumeration of the fecal indicators *Bifidobacterium adolescentis* and *Escherichia coli* in a tropical rain forest watershed. *Appl. Environ. Microbiol.* **50:**468–476.
18. **Cellini, L., I. Robuffo, E. Di Campli, S. Di Bartolomeo, T. Taraborelli, and B. Dainelli.** 1998. Recovery of *Helicobacter pylori* ATCC43504 from a viable but not culturable state: regrowth or resuscitation? *APMIS* **106:**571–579.
19. **Chamberlin, C. E., and R. Mitchell.** 1978. A decay model for enteric bacteria in natural waters, p. 325–348. *In* R. Mitchell (ed.), *Water Pollution Microbiology*, vol. 2. Wiley, New York, N.Y.
20. **Chambers, S. T., and C. M. Kunin.** 1985. The osmoprotective properties of urine for bacteria: the protective effect of betaine and human urine against low pH and high concentrations of electrolytes, sugars and urea. *J. Infect. Dis.* **152:**1308–1315.
21. **Chambers, S. T., and C. M. Kunin.** 1987. Isolation of glycine betaine and proline betaine form human urine. Assessment of their role as osmoprotective agents for bacteria and the kidney. *J. Clin. Invest.* **79:**731–737.
22. **Chambers, S. T., and C. M. Kunin.** 1987. Osmoprotective activity for *Escherichia coli* in mammalian renal inner medulla and urine: correlation of glycine and proline betaines and sorbitol with response to osmotic stress. *J. Clin. Invest.* **80:**1255–1260.
23. **Chmielewski, R. A., and J. F. Frank.** 1995. Formation of viable but nonculturable Salmonella during starvation in chemically defined solutions. *Lett. Appl. Microbiol.* **20:**380–384.
24. **Cho, J. C., and S. J. Kim.** 1999. Green fluorescent protein-based direct viable count to verify a viable but non-culturable state of *Salmonella typhi* in environmental samples. *J. Microbiol. Methods.* **36:**227–235.
25. **Cho, J. C., and S. J. Kim.** 1999. Viable, but non-culturable state of a green fluorescence protein-tagged environmental isolate of *Salmonella typhi* in ground water and pond water. *FEMS Microbiol. Lett.* **170:**257–264.
26. **Chowdhury, M. A., A. Huq, B. Xu, F. J. Madeira, and R. R. Colwell.** 1997. Effect of alum on free-living and copepod-associated *Vibio cholerae* O1 and O139. *Appl. Environ. Microbiol.* **63:**3323–3326.
27. **Colwell, R. R.** 1987. *Vibrios in the Environment.* John Wiley and Sons, New York, N.Y.
28. **Colwell, R. R., P. R. Brayton, D. J. Grimes, D. B. Roszak, S. A. Huq, and L. M. Palmer.** 1985. Viable but non-culturable *Vibrio cholerae* and related pathogens in the environment: implications for the release of genetically engineered microorganisms. *Bio/Technology* **3:**817–820.

29. **Colwell, R. R., M. L. Tamplin, P. R. Brayton, A. L. Gauzens, B. D. Tall, D. Herrington, M. M. Levine, S. Hall, A. Huq, and D. A. Sack.** 1990. Environmental aspects of *Vibrio cholerae* in transmission of cholera, p. 327–343. *In* R. B. Sack and Y. Zinnaka (ed.), *Advances on Cholera and Related Diarrheas*, KTK Scientific, Tokyo, Japan.
30. **Combarro, M.-P., M. J. Gauthier, G. N. Flatau, and R. L. Clément.** 1992. Effect of a transient incubation in wastewater on the sensitivity to seawater of *Escherichia coli* cells grown under intestinal-like conditions. *Biomed. Lett.* **47:**185–190.
31. **Csonka, L. N.** 1989. Physiological and genetic responses of bacteria to osmotic stress. *Microbiol. Rev.* **53:**212–247.
32. **Csonka, L. N., and A. D. Hanson.** 1991. Prokaryotic osmoregulation: genetics and physiology. *Annu. Rev. Microbiol.* **45:**569–606.
33. **Davies, C. M., and L. M. Evison.** 1991. Sunlight and the survival of enteric bacteria in natural waters. *J. Appl. Bacteriol.* **70:**265–274.
34. **Davies, C. M., J. A. Long, M. Donald, and N. J. Ashbolt.** 1995. Survival of fecal microorganisms in marine and freshwater sediments. *Appl. Environ. Microbiol.* **61:**1888–1896.
35. **Desmonts, C., J. Minet, R. R. Colwell, and M. Cormier.** 1990. Fluorescent-antibody method useful for detecting viable but nonculturable *Salmonella* spp. in chlorinated wastewater. *Appl. Environ. Microbiol.* **56:**1448–1452.
36. **Dukan, S., Y. Levi, and D. Touati.** 1997. Recovery of culturability of an HOCl-stressed population of *Escherichia coli* after incubation in phosphate buffer: resuscitation or regrowth? *Appl. Environ. Microbiol.* **63:**4204–4209.
37. **Duncan, S., L. A. Glover, K. Killham, and J. I. Prosser.** 1994. Luminescence-based detection of activity of starved and viable but nonculturable bacteria. *Appl. Environ. Microbiol.* **60:**1308–1316.
38. **Eisenstark, A., C. Miller, J. Jones, and S. Leven.** 1992. *Escherichia coli* genes involved in cell survival during dormancy: role of oxidative stress. *Biochem. Biophys. Res. Commun.* **188:**1054–1059.
39. **Ensign, J. C.** 1970. Long-term starvation survival of rod and spherical cells of *Arthrobacter crystallopoietes. J. Bacteriol.* **103:**569–577.
40. **Faust, M. A., A. E. Aotaky, and M. T. Hargadon.** 1975. Effect of physical parameters on the in situ survival of *Escherichia coli* MC-6 in an estuarine environment. *Appl. Microbiol.* **30:**800–806.
41. **Flint, K. P.** 1987. Long term survival of *Escherichia coli* in river water. *J. Appl. Bacteriol.* **63:**261–270.
42. **Fry, J. C., and T. Zia.** 1982. A method for estimating viability of aquatic bacteria by slide culture. *J. Appl. Bacteriol.* **53:**189–198.
43. **Gauthier, M. J., P. M. Munro, and S. Mohadjer.** 1987. Influence of salt and sodium chloride on the recovery of *Escherichia coli* from seawater. *Curr. Microbiol.* **15:**5–10.
44. **Gauthier, M. J., and D. LeRudulier.** 1990. Survival in seawater of *Escherichia coli* cells grown in marine sediments containing glycine betaine. *Appl. Environ. Microbiol.* **56:**2915–2918.
45. **Gauthier, M. J., G. N. Flatau, D. LeRudulier, R. L. Clément, and M. P. Combarro-Combarro.** 1991. Intracellular accumulation of potassium and glutamate specifically enhances survival of *Escherichia coli* in seawater. *Appl. Environ. Microbiol.* **57:**272–276.
46. **Gauthier, M. J., S. A. Benson, G. N. Flatau, R. L. Clément, V. A. Breittmayer, and P. M. Munro.** 1992. OmpC and OmpF porins influence viability and culturability of *Escherichia coli* cells incubated in seawater. *Microb. Releases,* **1:**47–50.
47. **Gauthier, M. J., G. N. Flatau, P. M. Munro, and R. L. Clément.** 1993. Glutamate uptake and synthesis by *Escherichia coli* cells in seawater : effects on culturability loss and glycine betaine transport. *Microb. Releases* **2:**53–59.
48. **Gerba, C. P., and J. S. McLeod.** 1976. Effect of sediments on the survival of *Escherichia coli* in marine waters. *Appl. Environ. Microbiol.* **32:**114–120.
49. **Ghoul, M., T. Bernard, and M. Cormier.** 1990. Evidence that *Escherichia coli* accumulates glycine betaine from marine sediments. *Appl. Environ. Microbiol.* **56:**551–554.
50. **Gonzalez, J. M., J. Iriberri, L. Egea, and I. Barcina.** 1992. Characterization of culturability, protistan grazing and death of enteric bacteria in aquatic ecosystems. *Appl. Environ. Microbiol.* **58:** 998–1004.

51. **Gottschal, J. C.** 1992. Substrate capturing and growth in various ecosystems. *J. Appl. Bacteriol.* **73:**39s–48s.
52. **Gourmelon, M., J. Cillard, and M. Pommepuy.** 1994. Visible light damage to *Escherichia coli* in seawater: oxidative stress hypothesis. *J. Appl. Bacteriol.* **77:**105–112.
53. **Grimes, D. J., and R. R. Colwell.** 1986. Viability and virulence of *Escherichia coli* suspended by membrane chamber in semitropical ocean water. *FEMS Microbiol. Lett.* **34:**161–165.
54. **Harvey, P., and S. Leach.** 1998. Analysis of coccal cell formation by *Campylobacter jejuni* using continuous culture techniques, and the importance of oxidative stress. *J. Appl. Microbiol.* **85:**398–404.
55. **Hazen, T. C., F. A. Fuentes, and J. W. Santo Domingo.** 1986. In situ survival and activity of pathogens and their indicators, p. 406–411. *In* F. Megusar and M. Gantar (ed.), *Perspectives in Microbial Ecology.* Slovene Society for Microbiology, Ljubljana, Slovenia.
56. **Heidelberg, J. F., M. Shahamat, M. Levin, I. Rahman, G. Stelma, C. Grim, and R. R. Colwell.** 1997. Effects of aerosolization on culturability and viability of gram-negative bacteria. *Appl. Environ. Microbiol.* **63:**3585–3588.
57. **Hengge-Aronis, R., R. Lange, N. Henneberg, and D. Fischer.** 1993. Osmotic regulation of *rpoS*-dependent genes in *Escherichia coli. J. Bacteriol.* **175:**259–265.
58. **Hengge-Aronis, R.** 1993. Survival of hunger and stress: the role of *rpoS* in early stationary phase gene regulation in *Escherichia coli. Cell* **72:**165–168.
59. **Hoff, K. A.** 1989. Survival of *Vibrio anguillarum* and *Vibrio salmonicida* at different salinities. *Appl. Environ. Microbiol.* **55:**1775–1786.
60. **Holmquist, L., and S. Kjelleberg.** 1993. Changes in viability, respiratory activity and morphology of the marine *Vibrio* sp. strain S14 during starvation of individual nutrients and subsequent recovery. *FEMS Microbiol. Ecol.* **12:**215–224.
61. **Hood, M. A., and G. E. Ness.** 1982. Survival of *Vibrio cholerae* and *Escherichia coli* in estuarine waters and sediments. *Appl. Environ. Microbiol.* **43:**478–584.
62. **Hussong, D., R. R. Colwell, M. O'Brien, E. Weiss, A. D. Pearson, R. M. Weiner, and W. D. Burge.** 1987. Viable *Legionella pneumophila* not detectable by culture on agar media. *Bio/Technology* **5:**947–950.
63. **Islam, M. S., M. K. Hasan, M. A. Miah, G. C. Sur, A. Felsenstein, M. Venkatesan, R. B. Sack, and M. J. Albert.** 1993. Use of the polymerase chain reaction and fluorescent-antibody methods for detecting viable but nonculturable *Shigella dysenteriae* type 1 in laboratory microcosms. *Appl. Environ. Microbiol.* **59:**536–540.
64. **Jacob, J., W. Martin, and C. Höller.** 1993. Characterization of viable but nonculturable stage of *Campylobacter coli*, characterized with respect to electron microscopic findings, whole cell protein and lipooligosaccharide LOS. patterns. *Zentralbl. Mikrobiol.* **148:**3–10.
65. **Jenkins, D. E., S. A. Chaisson, and A. Matin.** 1990. Starvation-induced cross protection against osmotic challenge in *Escherichia coli. J. Bacteriol.* **172:**2779–2781.
66. **Jiang, X., and T. J. Chai.** 1996. Survival of *Vibrio parahaemolyticus* at low temperature under starvation conditions and subsequent resuscitation of viable, nonculturable cells. *Appl. Environ. Microbiol.* **62:**1300–1305.
67. **Jones, D. M., E. M. Sutcliffe, and A. Curry.** 1991. Recovery of viable but non-culturable *Campylobacter jejuni. J. Gen. Microbiol.* **137:**2477–2482.
68. **Jouper–Jaan, A., A. E. Goodman, and S. Kjelleberg.** 1992. Bacteria starved for prolonged periods develop increased protection against lethal temperatures. *FEMS Microbiol. Ecol.* **101:**229–236.
69. **Kaasen, I., P. Falkenberg, O. B. Styrvold, and A. R. Strom.** 1992. Molecular cloning and physical mapping of the *otsA* genes, which encode the osmoregulatory trehalose pathway of *Escherichia coli*: evidence that transcription is activated by KatF (AppR). *J. Bacteriol.* **174:**889–898.
70. **Kaprelyants, A. S., and D. B. Kell.** 1992. Rapid assessment of bacterial viability and vitality by rhodamine 123 and flow cytometry. *J. Appl. Bacteriol.* **72:**410–422.
71. **Kaprelyants, A. S., and D. B. Kell.** 1993. Dormancy in stationary-phase cultures of *Micrococcus luteus*: flow cytometric analysis of starvation and resuscitation. *Appl. Environ. Microbiol.* **59:**3187–3196.
72. **Kaprelyants, A. S., J. C. Gottschal, and D. B. Kell.** 1994. Dormancy in non-sporulating bacteria. *FEMS Microbiol. Rev.* **104:**271–286.

73. **Karmali, M. A., and P. C. Fleming.** 1979. *Campylobacter* enteritis. *Can. J. Microbiol.* **120:**1525–1532.
74. **Kjelleberg, S. (ed.). 1993.** *Starvation in Bacteria.* Plenum Press, New York, N.Y.
75. **Kogure, K., U. Simidu, and N. Taga.** 1979. A tentative direct microscopic method for counting living marine bacteria. *Can. J. Microbiol.* **25:**415–420.
76. **Konishi, H., and Z. Yoshii.** 1986. Determination of the spiral conformation of *Aquaspirillum* spp. by scanning electron microscopy of elongated cells induced by cephalexin treatment. *J. Gen. Microbiol.* **132:**877–881.
77. **Kunin, C. M., T. H. Hua, L. Van Arsdale, and M. Villarejo.** 1992. Growth of *Escherichia coli* in human urine: role of salt tolerance and accumulation of glycine betaine. *J. Infect. Dis.* **166:**1311–1315.
78. **Leclerc, H., D. A. A. Mossel.** 1989. Le tube digestif, p. 141–162. *In* H. Leclerc and D. A. A. Mossel (ed.), *Microbiologie: le Tube Digestif, l'Eau et les Aliments.* Doin, Paris, France.
79. **Linder, K., and J. D. Oliver.** 1989. Membrane fatty acid and virulence changes in the viable but nonculturable state of *Vibrio vulnificus. Appl. Environ. Microbiol.* **55:**2837–2842.
80. **Lleo, M. D. M., M. C. Tafi, and P. Canepari.** 1998. Nonculturable *Enterobacter faecalis* cells are metabolically active and capable of resuming active growth. *Syst. Appl. Microbiol.* **21:**333–339.
81. **Lopez-Torres, A. J., L. Prieto, and T. C. Hazen.** 1988. Comparison of the in situ survival and activity of *Klebsiella pneumoniae* and *Escherichia coli* in tropical marine evironments. *Microb. Ecol.* **15:**41–57.
82. **Magarinos, B., J. L. Romalde, J. L. Barja, and A. E. Toranzo.** 1994. Evidence of a dormant but infective state of the fish pathogen *Pasteurella piscicida* in seawater and sediment. *Appl. Environ. Microbiol.* **60:**180–186.
83. **Mai, U. E. H., M. Shahamat, and R. R. Colwell.** 1990. Survival of *Helicobacter pylori* in the environment in a dormant but viable stage. *Rev. Esp. Enferm. Dig.* **78**(Suppl. 1)**:**17.
84. **Mason, D. J., E. G. Power, H. Talsania, I. Phillips, and V. A. Gant.** 1995. Antibacterial action of ciprofloxacin. *Antimicrob. Agents Chemother.* **39:**2752–2758.
85. **Matin, A.** 1991. The molecular basis of carbon-starvation-induced general resistance in *Escherichia coli. Mol. Microbiol.* **5:**3–10.
86. **McCann, M. P., J. P. Kidwell, and A. Matin.** 1991. The putative σ factor KatF has a central role in development of starvation-mediated general resistance in *Escherichia coli. J. Bacteriol.* **173:**4188–4194.
87. **McKay, A. M.** 1992. Viable but non-culturable forms of potentially pathogenic bacteria in water. *Lett. Appl. Microbiol.* **14:**129–135.
88. **Medema, G. J., F. M. Schets, A. W. van de Giessen, and A. H. Havelaar.** 1992. Lack of colonization of 1 day old chicks by viable, non-culturable *Campylobacter jejuni. J. Appl. Bacteriol.* **72:**512–516.
89. **Moran, A. P., and M. E. Upton.** 1986. A comparative study of the rod and coccoid forms of *Campylobacter jejuni* ATCC 29428. *J. Appl. Bacteriol.* **60:**103–110.
90. **Morgan, J. A. W., P. A. Cranwell, and R. W. Pickup.** 1991. Survival of *Aeromonas salmonicida* in lake water. *Appl. Environ. Microbiol.* **57:**1777–1782.
91. **Morgan, J. A. W., K. J. Clarke, G. Rhodes, and R. W. Pickup.** 1992. Non-culturable *Aeromonas salmonicida* in lake water. *Microb. Releases* **1:**71–78.
92. **Morgan, J. A. W., G. Rhodes, and R. W. Pickup.** 1993. Survival of nonculturable *Aeromonas salmonicida* in lake water. *Appl. Environ. Microbiol.* **59:**874–880.
93. **Moriarty, D. J. W., and R. T. Bell.** 1993. Bacterial growth and starvation in aquatic environments, p. 25–53. *In* S. Kjelleberg (ed.), *Starvation in Bacteria.* Plenum Press, New York, N.Y.
94. **Morita, R. Y.** 1993. Bioavailability of energy and the starvation state, p. 1–23. In S. Kjelleberg (ed.), *Starvation in Bacteria.* Plenum Press, New York, N.Y.
95. **Munro, P. M., F. Laumond, and M. J. Gauthier.** 1987. A previous growth of enteric bacteria on salted medium increases their survival in seawater. *Lett. Appl. Microbiol.* **4:**121–124.
96. **Munro, P. M., M. J. Gauthier, V. A. Breittmayer, and J. Bongiovanni.** 1989. Influence of osmoregulation processes on starvation survival of *Escherichia coli* in seawater. *Appl. Environ. Microbiol.* **55:**2017–2024.

97. **Munro, P. M., and M. J. Gauthier.** 1993. Uptake of glutamate by *Vibrio cholerae* in media of low and high osmolarity and in seawater. *Lett. Appl. Microbiol.* **18:**197–199.
98. **Munro, P. M., R. L. Clément, M. J. Gauthier, and G. N. Flatau.** 1993. Effect of thermal, oxidative, acidic, osmotic or nutritional stresses on subsequent culturability of *Escherichia coli* in seawater. *Microb. Ecol.* **27:**57–63.
99. **Nikaido, H., and M. Vaara.** 1987. Outer Membrane, p. 7–22. *In* F. C. Neidhardt, J. L. Ingraham, K. Brooks, B. Magasanik, M. Schaechter, and E. Umbarger (ed.), *Escherichia coli and Salmonella typhimurium: Cellular and Molecular Biology.* American Society for Microbiology, Washington, D.C.
100. **Nilsson, L., J. D. Oliver, and S. Kjelleberg.** 1991. Resuscitation of *Vibrio vulnificus* from the viable but nonculturable state. *J. Bacteriol.* **173:**5054–5059.
101. **Oliver, J. D.** 1993. Formation of viable but nonculturable cells, p. 239–272. *In* S. Kjelleberg (ed.), *Starvation in Bacteria.* Plenum Press, New York, N.Y.
102. **Oliver, J. D., and D. Wanucha.** 1989. Survival of *Vibrio vulnificus* at reduced temperatures and elevated nutrient. *J. Food Safety* **10:**79–86.
103. **Oliver, J. D., L. Nilsson, and S. Kjelleberg.** 1991. Formation of nonculturable *Vibrio vulnificus* cells and its relationship to the starvation state. *Appl. Environ. Microbiol.* **57:**2640–2644.
104. **Oliver, J. D., F. Hite, D. McDougald, N. L. Andon, and L. M. Simpson.** 1995. Entry into, and resuscitation from, the viable but nonculturable state by *Vibrio vulnificus* in an estuarine environment. *Appl. Environ. Microbiol.* **61:**2624–2630.
105. **Orlob, G. T.** 1956. Viability of sewage bacteria in seawater. *Sewage Ind. Wastes* **28:**1147–1167.
106. **Palmer, L. M., A. M. Baya, D. J. Grimes, and R. R. Colwell.** 1984. Molecular genetic and phenotypic alteration of *Escherichia coli* in natural water microcosms containing toxic chemicals. *FEMS Microbiol. Lett.* **21:**169–173.
107. **Paszko-Kolva, C., M. Shahamat, H. Yamamoto, T. Sawyer, J. Vives-Rego, and R. R. Colwell.** 1991. Survival of *Legionella pneumophila* in the aquatic environment. *Microbiol Ecol.* **22:**75–83.
108. **Paszko-Kolva, C., M. Shahamat, and R. R. Colwell.** 1993. Effect of temperature on survival of *Legionella pneumophila* in the aquatic environment. *Microb. Releases* **2:**73–79.
109. **Pedersen, J. C., and T. D. Leser.** 1992. Survival of *Enterobacter cloacae* on leaves. *Microb. Releases* **1:**95–102.
110. **Pedersen, J. C., and C. S. Jacobsen.** 1993. Fate of *Enterobacter cloacae* JP120 and *Alcaligenes eutrophus* AEO106 pRO101. in soil during water stress: effects on culturability and viability. *Appl. Environ. Microbiol.* **59:**1560–1564.
111. **Pitonzo, B. J., P. S. Amy, and M. Rudin.** 1999a. Effect of gamma radiation on native endolithic microorganisms from a radioactive waste deposit site. *Radiat. Res.* **152:**64–70.
112. **Pitonzo, B. J., P. S. Amy, and M. Rudin.** 1999b. Resuscitation of microorganisms after gamma irradiation. *Radiat. Res.* **152:**71–75.
113. **Pommepuy, M., M. Butin, A. Derrien, M. Gourmelon, R. R. Colwell, and M. Cormier.** 1996. Retention of enteropathogenicity by viable but nonculturable *Escherichia coli* exposed to seawater and sunlight. *Appl. Environ. Microbiol.* **62:**4621–4626.
114. **Postgate, J. R., and J. R. Hunter.** 1963. Acceleration of bacterial death by growth substrates. *Nature* **198:**273.
115. **Preyer, J. M., and J. D. Oliver.** 1993. Starvation-induced thermal tolerance as a survival mechanism in a psychrophilic marine bacterium. *Appl. Environ. Microbiol.* **59:**2653–2656.
116. **Rahman, I., M. Shahamat, P. A. Kirchman, E. Russek-Cohen, and R. R. Colwell.** 1994. Methionine uptake and cytopathogenicity of viable but nonculturable *Shigella dysenteriae* type 1. *Appl. Environ. Microbiol.* **60:**3573–3578.
117. **Rollins, D. M., D. Roszak, and R. R. Colwell.** 1985. Dormancy of *Campylobacter jejuni* in natural aquatic systems, p. 283–285. *In* A. D. Pearson and M. B. Skirrow (ed.), *Campylobacter III*, Proceedings of the Third International Symposium on *Campylobacter* Infections. Public Health Laboratory Service, London, United Kingdom.
118. **Rollins, D. M., and R. R. Colwell.** 1986. Viable but nonculturable stage of *Campylobacter jejuni* and its role in survival in the natural aquatic environment. *Appl. Environ. Microbiol.* **52:**531–538.
119. **Roszak, D. B., D. J. Grimes, and R. R. Colwell.** 1984. Viable but nonrecoverable stage of *Salmonella enteritidis* in aquatic systems. *Can. J. Microbiol.* **30:**334–338.

120. **Roszak, D. B., and R. R. Colwell.** 1987. Survival strategies of bacteria in the natural environment. *Microbiol. Rev.* **51:**365–379.
121. **Roth, W. G., M. P. Leckie, and D. N. Dietzler.** 1985. Osmotic stress drastically inhibits active transport of carbohydrates by *Escherichia coli. Biochem. Biophys. Res. Commun.* **126:**434–441.
122. **Roth, W. G., M. P. Leckie, and D. N. Dietzler.** 1988. Restoration of colony-forming activity in osmotically stressed *Escherichia coli. Appl. Environ. Microbiol.* **54:**3142–3146.
123. **Shahamat, M., U. Mai, C. Paszko-Kolva, M. Kessel, and R. R. Colwell.** 1993. Use of autoradiography to assess viability of *Helicobacter pylori* in water. *Appl. Environ. Microbiol.* **59:**1231–1235.
124. **Siegele, D. A., and R. Kolter.** 1992. Life after log. *J. Bacteriol.* **174:**345–348.
125. **Singh, A., R. Yeager, and G. A. McFeters.** 1986. Assessment of in vivo revival, growth, and pathogenicity of *Escherichia coli* strains after copper- and chlorine-induced injury. *Appl. Environ. Microbiol.* **52:**832–837.
126. **Smibert, R. M.** 1984. Genus *Campylobacter*, p. 111–118. *In* N. R. Krieg and J. G. Holt (ed.), *Bergey's Manual of Systematic Bacteriology*, vol. 1. Williams & Wilkins, Baltimore, Md.
127. **Smith, J. J., J. P. Howington, and G. A. McFeters.** 1994. Survival, physiological response and recovery of enteric bacteria exposed to polar marine environment. *Appl. Environ. Microbiol.* **60:** 2977–2984.
128. **Solic, M., and N. Krstulovic.** 1992. Separate and combined effects of solar radiation, temperature, salinity, and pH on the survival of faecal coliforms in seawater. *Mar. Pollut. Bul.* **24:**411–416.
129. **Stackebrandt, E.** 1992. Unifying phylogeny and phenotypic diversity, p. 19–47. *In* A. Balows, H. G. Trüper, M. Dworkin, W. Harder, and K.-H. Schleifer (ed.), *The Prokaryotes. A Handbook on the Biology of Bacteria: Ecophysiology, Isolation, Identification, Applications*, 2nd ed. Springer-Verlag, New York, N.Y.
130. **Steinert, M., L. Emody, R. Amann, and J. Hacker.** 1997. Resuscitation of viable but nonculturable *Legionella pneumophila* Philadelphia JR32 by *Acanthamoeba castellanii. Appl. Environ. Microbiol.* **63:**2047–2053.
131. **Thomas, T. D., and R. D. Batt.** 1968. Survival of *Streptococcus lactis* in starvation conditions. *J. Gen. Microbiol.* **50:**367–382.
132. **Trainor, V. C., R. K. Udy, P. J. Bremer, and G. M. Cook.** 1999. Survival of *Streptococcus pyogenes* under stress and starvation. *FEMS Microbiol. Lett.* **176:**421–428.
133. **Turpin, P. E., K. A. Maycroft, C. L. Rowlands, E. M. H. Wellington.** 1993. Viable but nonculturable salmonellas in soil. *J. Appl. Microbiol.* **74:**421–427.
134. **Wadolkowski, E. A., D. C. Laux, and P. S. Cohen.** 1988. Colonization of the streptomycin-treated mouse large intestine by a human fecal *Escherichia coli* strain: role of growth in mucus. *Infect. Immun.* **56:**1030–1035.
135. **Wai, S. N., T. Morita, K. Kondo, H. Misumi, and K. Amako.** 1996. Resuscitation of *Vibrio cholerae* O1 strain TSI-4 from a viable but nonculturable state by heat shock. *FEMS Microbiol. Lett.* **136:**187–191.
136. **Wang, G., and M. P. Doyle.** 1998. Survival of enterohemorrhagic *Escherichia coli* O157-H7 in water. *J. Food Prot.* **61:**662–667.
137. **Weichart, D., and S. Kjelleberg.** 1996. Stress resistance and recovery potential of culturable and viable but nonculturable cells of *Vibrio vulnificus. Microbiology* **142:**845–853.
138. **Wolf, P. W., and J. D. Oliver.** 1992. Temperature effects on the viable but nonculturable state of *Vibrio vulnificus. FEMS Microbiol. Ecol.* **101:**33–39.
139. **Xu, H.-S., N. Roberts, F. L. Singleton, R. W. Atwell, D. J. Grimes, and R. R. Colwell.** 1982. Survival and viability of nonculturable *Escherichia coli* and *Vibrio cholerae* in the estuarine and marine environment. *Microb. Ecol.* **8:**313–323.
140. **Xu, H.-S., and R. R. Colwell.** 1989. Overwintering of *Vibrio cholerae*—viable but non-culturable state and its determination. *J. Ocean. Univ. Qingdao* **19:**77–83. (In Chinese.)
141. **Zimmerman, R., E. Iturriaga, and J. Becker-Birck.** 1978. Simultaneous determination of the total number of aquatic bacteria and the number thereof involved in respiration. *Appl. Environ. Microbiol.* **36:**926–935.

Nonculturable Microorganisms in the Environment
Edited by R. R. Colwell and D. J. Grimes

Chapter 8

Starved and Nonculturable Microorganisms in Biofilms

Kevin C. Marshall

BIOFILMS AND THEIR SIGNIFICANCE

Casual observation of most solid surfaces immersed in aqueous environments reveals the presence of a slimy layer developing on the surfaces. These slimes, termed biofilms, form on exposed surfaces as a result of bacterial adhesion to, followed by growth and exopolymer production at, the solid-liquid interface (26, 28, 60, 67). The numbers and types of bacteria per unit volume of biofilm in different environments vary considerably, depending on factors such as the nature of the substratum, the trophic level of the aqueous phase, the flow rate, and the degree of turbulence (22). The complex communities of microorganisms found in biofilms ensure that these systems play a major role in microbially catalyzed reactions in natural environments, particularly in the degradation of organic molecules and in nitrification and other mineral transformation processes. Some of the most intensively studied biofilms, in terms of their structure, biology, and biogeochemistry, are the microbial mats found in shallow submerged or intermittently exposed littoral marine areas (25, 36, 97).

Development of Biofilms

Bacteria in nature exist in two distinct modes, the sessile and the planktonic states. Different species have developed various strategies to ensure that they take best advantage of either mode, depending on the environmental circumstances at any particular time (69). In nutrient-depleted habitats, bacteria benefit in the sessile state from organic molecules adsorbed at surfaces (67). Molecular fouling of immersed clean surfaces occurs very rapidly and precedes bacterial adhesion to and colonization of the surfaces. Adsorbed organic molecules, termed conditioning films (CF), alter the physical properties of the surface (78, 8) and, hence, the adhesion of bacteria (95). In addition, these CF molecules serve as a concentrated

Kevin C. Marshall • School of Microbiology and Immunology, The University of New South Wales, Sydney, New South Wales 2052, Australia.

energy substrate and stimulate active metabolism, cellular growth, and cellular reproduction of starved bacteria adhering to the surface (44, 46, 51, 55, 86, 92). Further colonization of surfaces and rapid reproduction of the various species of attached bacteria, combined with extensive production of heterogeneous extracellular polymers by the organisms, lead to the rapid development of biofilms (28, 59, 70).

Mechanisms of bacterial adhesion to "inert" substrata have been considered in detail in books by Marshall (67, 68), Berkeley et al. (13), Savage and Fletcher (93), ten Bosch (106), and Lappin-Scott and Costerton (60) and will not be dealt with in this review.

Fouling of Submerged Surfaces

The natural propensity for bacteria to adhere to surfaces and to form biofilms leads to problems on man-made structures associated with aquatic environments. Biofilms constitute the primary fouling on ship hulls, resulting in increased drag due to increased turbulence at the biofilm-water interface (22), and creating a more favorable surface for the settlement of larvae and spores of secondary fouling organisms, such as barnacles, mussels, and algae (65). Biofilm-induced turbulence (108) and drag (102) are also major problems in hydroelectric pipelines. Sloughing of biofilm components can cause health (regrowth of pathogens in biofilms [29]) and aesthetic (black iron or brown manganese deposits) problems in drinking water reticulation pipelines, whereas microbially induced corrosion of metals at the biofilm-substratum interface is a problem in pipelines and on offshore oil rigs (39). Biofilms can render heat exchangers almost useless as a result of an overall increase in heat transfer resistance across the metal surfaces of the exchanger. This resistance is the resultant of a small decrease in advective and a large increase in conductive heat transfer resistance within the heat exchange system (23).

Biofilms Associated with Humans

The significance of biofilms in human and animal health has only been recognized in recent times (29). The epithelial surfaces or mucus layers of different portions of the gastrointestinal tract (GIT) are colonized by specific groups of bacteria, termed the normal microbiota, forming biofilms and providing a reasonably effective barrier against invasion by intestinal pathogens (62). Plaque formation on teeth has long been recognized as biofilm formation, and intense research effort has gone into establishing the nature of the bacteria, their interactions, and their effects in terms of tooth decay (58). Of concern, in terms of human health, is the formation of biofilms on the surfaces of catheters and prosthetic devices implanted in the body (artificial hips and hearts, heart pacemakers, etc.) (29). In fact, a major impediment to long-term use of the total artificial heart is the sepsis resulting from sloughing of bacterial biofilms developed on the surface of the biomaterial employed in the manufacture of the artificial hearts (45).

Biofilms in Biotechnology

On a more positive note, useful applications of biofilms include controlled biodegradation in trickling filters in wastewater treatment plants and in biotechnological applications in fluidized-bed and other types of fixed-film reactors (22). Natural biofilms suffer from a lack of control over the rate and extent of biofilm formation and function, and consequently biotechnologists have introduced a variety of biomass support systems, which create very large surface areas that are favorable for microbial colonization, in order to optimize biofilm production (16). In order to achieve maximum control in some microbially catalyzed chemical transformations, a variety of techniques have been developed to artificially immobilize the microorganisms at suitable surfaces (16).

Control of Biofilms

Attempts to control biofilm development in medical, dental, or industrial situations have led to the realization that antibiotics and biocides in general are not very effective against bacteria immobilized in biofilms (15, 19, 21, 29, 80). Bacteria that are susceptible to antibiotics in the planktonic state are remarkably resistant to much higher concentrations of the same antibiotic in biofilms (6). The basis for this apparent resistance is not clear. Explanations range from growth-related phenomena (5, 15, 38, 42, 101) to diffusional limitations of the biocides in biofilms (79). Some success in biofilm control by simultaneous treatment with antibiotics and an electrical current has been reported by Blenkinsopp et al. (14) and by the selection of antibiotic treatments with known effectiveness against biofilms (5). For instance, Anwar et al. (7) suggest that *Staphylococcus aureus* biofilm infections may be eradicated at an early stage by combined treatment with tobramycin and cephalexin, yet this treatment is ineffective on well-established biofilms. Other possible approaches include silver coatings on (63) or antibiotic (gentamicin) addition to (20) the biomaterial employed for implants in humans.

Altered Gene Expression at Surfaces

Numerous observations have been recorded indicating changes in the physiology of microorganisms colonizing surfaces and growing in biofilms as compared to the same organisms in liquid culture (43). Recent studies using reporter gene technology have revealed altered gene expression in the target bacteria following transfer from liquid to agar (11, 12) and to solid (30, 32, 33, 49) surfaces. It is likely that many genes are either turned on or turned off when microorganisms colonize surfaces or grow in biofilms. Goodman and Marshall (43) have reviewed the range of physicochemical conditions at surfaces or in biofilms that could induce genetic responses as well as potential mechanisms of how these conditions might alter gene regulation.

The phenomenon of altered gene expression in microorganisms at surfaces or in biofilms is significant because it clearly shows that the physiological state of these immobilized microorganisms is different from that of the same organisms in the planktonic mode. Since most studies on the physiology of microorganisms are carried out on cells grown in liquid culture, one must question the relevance of such studies to cells immobilized within biofilms.

STRUCTURE OF BIOFILMS

The Changing Image of Biofilm Structure

Many biofilms, when viewed macroscopically, appear as uniform slime layers on the substrata. The opaque nature of biofilms makes study by conventional light microscopy difficult, and observations of ultrathin sections using transmission electron microscopy (TEM) gave the impression that biofilms consisted of relatively uniform arrays of bacteria embedded in a matrix of exopolymer of bacterial origin (22, 26). This impression was reinforced by examination of dehydrated biofilm material by scanning electron microscopy (SEM) and, more recently, of wet specimens by environmental scanning electron microscopy (ESEM) (64).

Measurements of distributions of thickness in a *Pseudomonas aeruginosa* biofilm revealed a mean steady-state thickness of approximately 33 mm, with individual measurements (13.3 to 60.0 mm) being indicative of a nonuniform, rough biofilm surface (99). This finding is in accord with the ripple pattern biofilm surface reported in naturally formed hydroelectric pipeline biofilms (103, 108). The conventional image of biofilm structure has been shattered by observations using the nondestructive scanning confocal laser microscopy (SCLM) technique, combined with fluorescent and pH-sensitive stains (18, 61). This study has revealed a more complex internal structure than previously envisaged, with biofilms found to consist of an extensive array of irregular columnar structures, each containing microorganisms embedded within the polymer matrix and surrounded by waterslide voids (Fig. 1). As will be seen below, this altered concept of biofilm structure has important ramifications in terms of mass transport to and from biofilms and the consequent nutritional status of microorganisms immobilized within these biofilms.

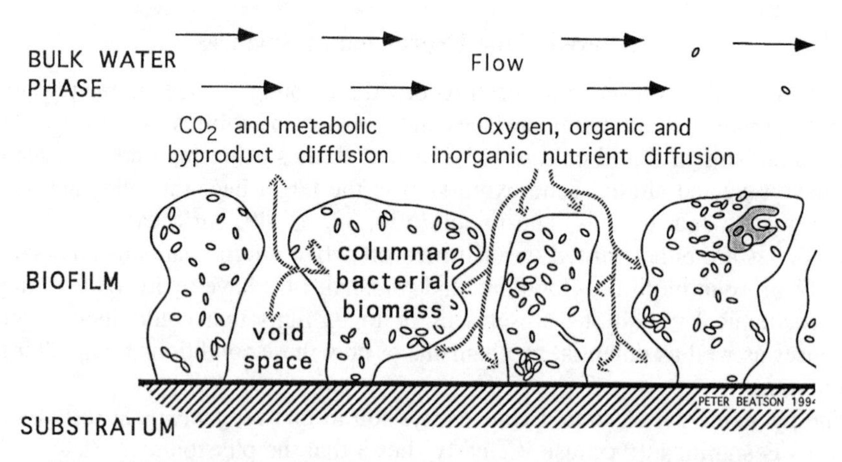

Figure 1. Diagrammatic representation of the current concept of biofilm structure, indicating the irregular columnar masses of bacteria embedded in an exopolymer matrix, the surrounding void spaces, and pathways of diffusion of oxygen, organic, and inorganic molecules in, and of carbon dioxide and metabolic by-products out of the biofilm.

Mass Transport in Biofilms

Diffusion of molecules is the major factor limiting microbial activity in ecosystems. The process of diffusion results from thermal energy causing random, chaotic, noncoherent motion of molecules, leading to the net movement of molecules from levels of high to low concentration (57). The flux, dq/dt, of material is given by Fick's law:

$$dq/dt = D \cdot A \cdot dC/dX$$

where q is the amount of material, t is the time, D is the diffusion constant, A is the surface area, and dC/dX is the concentration gradient. Obviously, any increase in transport via diffusion requires either a higher diffusion constant, an increased surface area, or an increased concentration gradient. The fundamentals of diffusion as related to bacterial behavior and to the movement of molecules through gels have been presented in detail by Koch (57).

Diffusion of organic substrates, inorganic nutrients, and oxygen from the water phase into the biofilm and of carbon dioxide and metabolic by-products from the biofilm to the water phase was thought to be controlled by the hydrodynamics of the water phase and the nature of the relatively homogeneous biofilm matrix and could be estimated by simple considerations of mass transport across the seemingly uniform water-biofilm interface (23).

Applications of microelectrode technology to diffusion of gases and inorganic molecules into and out of biofilms have had a marked impact on our understanding of the dynamics of such processes (31, 34, 87). The use of a combination of scanning confocal laser microscopy and oxygen microelectrodes in biofilms has revealed that the recently described columnar structural elements play a profound role in bacterial activity within the biofilm (35). Oxygen diffusion occurs freely in the void spaces, even at depth within the biofilm, but is limited in the polymer-bacterial columns (Fig. 1). Similarly, mass transport of nutrients into, and of carbon dioxide and metabolic by-products out of, biofilms is controlled to variable extents by the void-to-column ratio and tortuosity developed in different biofilms. To emphasize this point, de Beer et al. (35) have calculated that the observed exchange surface area in a columnar structured biofilm was twice that for a simple planar geometry biofilm.

Obviously, a radical rethink on the dynamics of mass transport of materials into and out of biofilms is necessary. The microstructure of each biofilm type will need to be ascertained and related to its specific mass transport characteristics before the nutritional status of microbes within the biofilms can be assessed with any degree of precision.

SPECIES DIVERSITY AND NUTRIENT AVAILABILITY IN BIOFILMS

Control of Population Distribution in Biofilms

Biofilms in nature are microbial communities comprising a range of populations with differing metabolic functions (27). The functioning of these populations within biofilms is restricted in part by limitations to mass transport and in part by direct interactions between the populations themselves. Strictly aerobic organisms obviously function best at the water-biofilm interface, including that found at the void-biomass boundaries (see Fig. 1). Oxygen penetration into the biofilm biomass is limited both by diffusion resistance in the polymer matrix and by rapid microbial utilization at or near the water-biofilm interface (87). Microaerophilic organisms are favored at sites where oxygen levels are reduced, whereas respiratory organisms capable of utilizing alternate electron acceptors (nitrate, sulfate, etc.) will function best in zones immediately where oxygen becomes depleted. Facultative anaerobes can exist throughout the biofilm structures, but strict anaerobes only function effectively in the central and lower zones of the columnar structures where oxygen is not likely to penetrate. All such zones within biofilms are not static, shifting positions with changes in water quality, temperature, etc. (24) and leading to changes in the physiological states of the organisms immobilized in the biofilm. Consequently, the presence of a particular metabolic group of microorganisms is not necessarily restricted to a particular zone, as evidenced by the demonstration of anaerobic sulfate-reducing bacteria in aerobic regions of biofilms (76). Cross-feeding between individuals and populations, resulting from excretion of metabolic by-products, will occur throughout biofilms, as will parasitism and predation by other microorganisms, particularly by the protozoans regularly associated with biofilms (40, 72).

In a relatively direct measurement of overall activity with depth in a biofilm, Kinniment and Wimpenny (53) generated a quasi-steady-state biofilm of *P. aeruginosa* and, following freezing and sectioning of the intact biofilm, assayed the adenylates in the sectioned material. Adenylate energy charge values throughout the biofilm were generally low, ranging from 0.2 at the base of the biofilm to a maximum of 0.6 units near the original water-biofilm interface. AMP (adenosine monophosphate) was the predominant nucleotide, particularly in the deeper portions of the biofilm, again indicating an overall low level of microbial activity in these regions.

Community Structure in Biofilms

Many principles in macroecology have evolved from detailed studies of population dynamics over space and time. Little progress has been possible in microbial population ecology because of our inability to isolate and describe even a minor proportion of microbial communities, including biofilm communities. Detailed analyses of biofilm microbial communities over space and time should allow the construction of accurate models to predict population responses to environmental perturbations, the establishment of gradients and microhabitats, energy input and its utilization, and biogeochemical cycling, particularly as to how these factors relate to the growth and/or starvation of microorganisms in the biofilms (71).

Broad physiological groups of microorganisms in biofilms can be defined, as discussed in the previous section, but no reliable estimates are available of complete population distributions or dynamics within biofilm communities. This state of affairs is true even for biofilms formed in extreme environments where species diversity in the biofilms is limited (110, 111, 114). In recent years, however, the application of a variety of molecular biological methods has provided the impetus for more detailed studies on species diversity, on the localization of specific populations, and on interactions between members of specific consortia in biofilms.

Pace et al. (84) introduced the concept of employing rRNA sequences to define and enumerate the components of mixed, natural microbial populations. These authors described two different approaches to this task: (i) isolating 5S rRNA from the mixed populations, sorting the species-specific molecules by high resolution gel electrophoresis, sequencing the individual 5S rRNA types, and comparing the sequences with those in a data bank (method only applicable to communities of limited complexity), and (ii) "shotgun cloning" of 16S rRNA genes using DNA purified from natural samples, with the rRNA genes clones selected as isolated recombinant bacteriophage, sorted, sequenced, and compared to complete and partial rRNA sequences in a data bank (a more widely applicable method). These approaches stimulated active research into defining the composition of natural microbial communities, and advances in methodology and interpretation of results obtained in a variety of ecosystems have been reviewed by Ward et al. (112).

These authors have concluded that rRNA methods are providing new, and even provocative, information on the dominant populations within the communities examined. For instance, hybridization probe studies (see next section) often show low probe activities, indicating that laboratory-cultivated species from a particular habitat may not be dominant members of the community but merely grow actively on isolation media. On the other hand, it is relatively simple to demonstrate with rRNA-targeted probes the dominance of certain species whose rRNA sequences are readily recovered from natural communities by cloning methods.

Methods based on rRNA sequencing of microbial populations should have a profound effect on the ultimate description of the individual species present in biofilm communities, their location within the biofilm, and their fluctuations over space and time.

Location of Specific Populations in Biofilms

A variety of techniques designed to pinpoint specific populations within biofilms have been reported recently. In particular, the use of enzyme- and fluorescence-labeled rRNA-targeted oligonucleotide probes promises to reveal the location and potential consortial interactions between populations within biofilms (1, 2, 3, 48, 50, 66, 85, 104, 109). Group-specific probes, such as those reacting with most members of the *Bacteria*, the *Archaea*, and the proteobacteria, have provided important information on these groups of bacteria in biofilms. Some difficulties have been experienced in the development of a wide range of species-specific probes, but improved knowledge of complications involved with certain rRNA oligonucleotide sequences should overcome these technical problems.

Bioluminescence has been employed as a sensitive marker for *Pseudomonas* cells in soils and rhizospheres, where either a naphthalene-inducible *luxCDABE* construct or a constitutive *luxAB* construct introduced into *Pseudomonas* cells emitted light following the addition of naphthalene or *n*-decyl aldehyde, respectively

(37, 52). Bioluminescence was detected by autophotography or optical fiber light measurements, and the limit of detection was 10^3 to 10^4 CFU/cm of plant root. Even greater sensitivity, the detection of single cells of a specific bacterium in soils, was reported by Silcock et al. (96), using charge coupled device-enhanced microscopy to detect light emission by *lux* gene constructs. *lux* gene probes have also been employed to detect and identify specific groups of luminescent bacteria (77). Rogers and Keevil (89) have employed episcopic differential contrast microscopy to specifically detect *Legionella pneumophila* in aquatic biofilms with the aid of immunogold- and fluorescein-labeled antibodies.

Ultimately, the ability to locate specific populations and to estimate the relative activities of these populations within biofilms (see a later section) may give a better understanding of the trophic status of individual cells and, possibly, an indication of the likelihood of encountering viable but nonculturable (VBNC) cells.

Starvation Conditions in Biofilms

Despite the fact that surfaces represent sites of nutrient enrichment for colonizing bacteria, most microorganisms within biofilms will be in a nutrient-depleted environment. Again, this depletion is the combined result of limited nutrient diffusion through the biofilm matrix combined with rapid utilization of nutrients by the biofilm microorganisms. Organisms in the inner regions are likely to be starving, unless they encounter a discontinuous source of nutrients originating from the release of metabolic by-products by or lysis of neighboring organisms.

The observations that AMP was the predominant nucleotide and that the adenylate energy charge was particularly low at depth in biofilms (53) are consistent with starvation conditions in such regions of the biofilms. Those organisms at the water-biofilm interface will gain most benefit from the oxygen and nutrients diffusing from the water phase (Fig. 1). Measurements of adenylate energy charge (53) suggest that, even at this advantageous site, microbial activity may not be at an optimal level. These results need to be considered with caution as they may be (i) an artifact created by the method employed to maintain the quasi-steady-state biofilm, (ii) a true reflection of partly nutrient-depleted conditions in the biofilm-water interfacial zone, or (iii) the net result of the combined activities of well-fed bacteria at the interface and essentially starving bacteria only micrometers into the main biofilm structure.

The starvation conditions prevailing in most areas within the biofilm matrix will have a profound effect on the culturability of many of the resident microorganisms (see next section). The molecular basis and ecological significance of starvation in bacteria have been reviewed recently (54).

VIABILITY AND NONCULTURABILITY IN BIOFILMS

The Question of Viability of Cells in Biofilms

As in other natural microbial communities, the numbers of viable cells recovered from mixed culture biofilms are only a small, but variable, fraction of the total numbers determined microscopically. Many of the microorganisms are known to have specialized nutritional or physiological requirements (e.g., the chemolithotrophic, methanogenic, and sulfate-reducing bacteria) and fail to grow on most standard laboratory plate count media. Biofilm samples could be plated on a wide range of media and incubated under a wide range of physical conditions to obtain an estimate of the numbers of these more fastidious microorganisms, but this is rarely undertaken because of the enormous effort required. In addition, there will still be large numbers that remain unaccounted for, because of our current inability to culture these organisms, because they are nonviable, or because they are in the VBNC state.

As-yet-unculturable microorganisms represent our ignorance of the specific conditions required for their growth and reproduction (41). Such organisms constitute the dominant populations in certain biofilm communities as, for example, revealed by directly probing microbial mats in hot springs (110, 111, 114) and in activated sludge (109). Polymerase chain reaction amplification of rRNA gene sequences from biofilms and other habitats has also been reported (10, 47). It is anticipated that knowledge gained from the identification of RNA sequences from these uncultured organisms in natural biofilms will stimulate more focused attempts to culture the organisms in the laboratory. This aim might be achieved by the application of steady-state diffusion gradients (17, 115) to mimic the subtle changes in environmental conditions found throughout a biofilm. The application of a two-dimensional, finite-element numerical transport model for advective diffusion transport to simulate concentration and flux profiles (115) showed good correlations with the observed profiles and microbial responses to the concentration gradients. It was concluded that these gradients provide the steady-state environments required to sustain the continuous interactions between populations that are characteristic of steady-state consortia. A combination of this steady-state diffusion system with the recently described optical trap system for manipulating and isolating bacteria from complex microbial communities (74) and with scanning confocal laser microscopy (18) may lead to increased isolations of as-yet-uncultured microorganisms.

Nonviable cells remain an enigma (see Chapter 17), with definitive criteria indicating death remaining elusive. One might expect that dead bacteria would lyse rapidly, at least in the case of gram-negative cells with their rather fragile cell walls. Is this true for other microorganisms? Those lacking a cell wall, such as the protozoa, lyse or are rapidly attacked by bacteria and degraded. Most fungi, on the other hand, have very thick cell walls that take time to degrade.

There is increasing emphasis on the use of a range of techniques to differentiate living and dead cells within biofilms, such as the reduction of artificial electron acceptor tetrazolium salts (indicative of cellular respiratory activity) to (i) insoluble formazans (105, 107, 117), (ii) water-soluble formazans (90), and (iii) a fluorescent product (88). Direct use of the fluorescent redox dye 5-cyano-2,3-ditoyl tetrazolium chloride (CTC) to examine bacterial activity in biofilms indicated that the number of CTC-reducing bacteria was greater than viable numbers on conventional media (116) and that the proportion of respiring bacteria to total cells in biofilms was higher (5 to 35%) than for the planktonic bacteria (1 to 10%) (94). The relative activity of individual cells in young and mature biofilms has been estimated by quantifying the cellular content of ribosomes using rRNA-targeted fluorescence hybridization probes in combination with digital microscopy (85). The technique has been employed to relate the cellular content of ribosomes to the growth rate of single cells of a specific population of sulfate-reducing bacteria in mixed anaerobic biofilms.

Further application of these techniques to biofilm situations should provide greater information on the activities of individual populations in the biofilms. Due regard must be given, however, to the possibility that some of the indicators of cellular activity may give misleading results (9).

The Question of Nonculturability of Cells in Biofilms

Oliver (82), Roszak and Colwell (91), and Colwell and Grimes (see chapter 1) have discussed in detail the concept of VBNC bacteria in natural habitats. The advantages and disadvantages of methods for determining the VBNC state, such as direct viable counts, evidence for respiratory activity, use of monoclonal antibodies, evidence for cellular lysis and loss of labeled internal products, ATP levels, and the application of flow cytometric techniques have been considered by Oliver (82). The VBNC state has been defined as a "cell which can be demonstrated to be metabolically active, while being incapable of undergoing the sustained cellular division required for growth in or on a medium normally supporting growth of that cell" (82). It is the definition of "metabolically active" that creates problems because, as pointed out by Barer et al. (9), gram-positive spores do not react to the histochemical tests applied to determine so-called viability and gram-negative minicells, which lack DNA and the ability to reproduce, do react to some of these tests, particularly the reduction of tetrazolium salts to give a microscopically observable product.

Is the nonculturable state of significance in biofilms? Biofilm polymer matrices provide something of a buffer against environmental changes in the associated aqueous phase, as exemplified by the apparent resistance of biofilm microbes to antibacterial agents (see Control of Biofilms, above, and Injury to Cells in Biofilms, below). This buffering effect, however, is not absolute. As indicated previously, microorganisms embedded in this polymer matrix are exposed to variable degrees of starvation stress and, as a result, a large proportion are prone to become nonviable or transform into the VBNC state (56). Few attempts have been made to determine the proportions of apparently viable, nonviable, and VBNC microorganisms in biofilms.

Nonculturability Versus Starvation

Bacteria in natural habitats possess variable abilities to survive starvation conditions, generally by becoming very much smaller, and these starvation forms rapidly respond to a nutrient input by growing back to normal size (75). This process (termed "starvation survival" by Morita) is accompanied by a marked change in the metabolic behavior of the bacteria, especially the synthesis of many stress proteins, including some that are specific to the starvation-induced stress (56). Some bacteria exhibit negligible loss in viability after prolonged storage in the absence of energy substrate, whereas others show either a slow or rapid decline in viability with time (4). Is this loss of viability on starvation associated with a transformation to the VBNC state? Generally, studies on loss of viability during starvation have not included an examination of cellular growth or activity of the stressed cells or the resuscitation of apparently VBNC cells following the cessation of the starvation-induced stress.

In a study of *Vibrio vulnificus* survival under starvation conditions at different temperatures, Oliver et al. (83) found that viability was retained at 20°C but the organism became nonculturable at 5°C. The VBNC cells appeared to resuscitate after a shift from 5 to 20°C (81), but later work indicated that growth of a few viable cells among the mainly VBNC cells may have been responsible for the apparent resuscitation (113). Nonculturable cells were obtained at 5°C whether the cells were held in a starvation or a nutrient-containing medium, indicating that the transformation of cells to the VBNC state was independent of starvation (81). This demonstration of separate starvation and VBNC functions in *V. vulnificus*, of course, does not rule out the possibility that starvation might be involved in inducing the VBNC state in other bacteria, particularly those experiencing starvation conditions in biofilms.

Injury to Cells in Biofilms

Biofilms in industrial and medical situations often are exposed to biocide and/or antibiotic treatments resulting in varying degrees of injury to a proportion of the cells, particularly those at or near the biofilm-water interface. Injury is defined as the sublethal physiological and structural consequences resulting from exposure to injurious factors within the biofilm environment (73). The practical manifestation of injury is an increased sensitivity of the stressed bacteria to selective media and in recovery of the ability to grow on selective media following incubation under suitable conditions (=repair) (73). The sublethal injury involved is thought to involve the loss of several functions mediated by the plasma membrane of the cells.

The question of injury is generally avoided in the context of VBNC (9, 82) but is of vital importance in biofilms that have been exposed to the effects of biocides or antibiotics. For instance, recovery of injured pathogens following chlorination and their regrowth in and subsequent release from biofilms formed in drinking water pipelines can create a potential health hazard. Repair of injured cells could be considered as equivalent to the resuscitation of VBNC, but does this suggestion withstand further scrutiny? If the cells are injured by a stressor such as chlorine, any regrowth requires repair of the damage to (presumably) the plasma membrane. When the stressor is starvation, the cells undergo major changes in size and metabolism (56), but there is no evidence of loss of major membrane functions. Short-term starvation of *Vibrio fluvialis*, for instance, resulted in a decrease in respiratory potential of about fivefold, but the cells maintained a proton motive force equivalent to that of growing cells (98). The rapid response of starved cells to the input of exogenous substrate also suggests that there is no damage to membrane integrity, as well as indicating that the cells do not require the resuscitation period expected if in the VBNC state. The possibility exists, however, that these effects may be the reaction of part or most of the population of starved cells, but some fraction of the cells may be injured or in the VBNC state.

SIGNIFICANCE OF THE NONCULTURABLE STATE IN BIOFILMS

Although the existence of the VBNC state in biofilm microorganisms has not been established beyond doubt, it is feasible that stresses within the biofilm, such as starvation, will induce this condition. More intensive study is required to account for the lack of culturability of many microorganisms in biofilms. This will entail the development of better techniques and more critical analysis of the results of a range of techniques to establish the differences between viable, nonviable, VBNC, and injured cells. Should these studies confirm the presence of the VBNC state in some biofilm cells, this phenomenon will assume considerable importance in public health and other functional attributes of biofilm organisms. Conversely, the future development of better methods for resuscitating VBNC cells may lead to more effective recoveries of these and other as-yet-uncultured microorganisms from biofilms.

Acknowledgments. I thank Bill Costerton for updating the reference list, Zbigniew Lewandowski and Gordon McFeters for providing access to their unpublished results, Dieter Weichart for help with some valuable references, and Peter Beatson for drawing the figure.

REFERENCES

1. **Amann, R., W. Ludwig, and K.-H. Schleifer.** 1992. Identification and in situ detection of individual bacterial cells. *FEMS Microbiol. Lett.* **100:**45–50.
2. **Amann, R. I., J. Stromley, R. Devereux, R. Key, and D. A. Stahl.** 1992. Molecular and microscopic identification of sulfate-reducing bacteria in multispecies biofilms. *Appl. Environ. Microbiol.* **58:**614–623.
3. **Amann, R. I., B. Zarda, D. A. Stahl, and K.-H. Schleifer.** 1992. Identification of individual prokaryotic cells by using enzyme-labeled, rRNA-targeted oligonucleotide probes. *Appl. Environ. Microbiol.* **58:**3007–3011.
4. **Amy, P. S., and R. Y. Morita.** 1983. Starvation-survival patterns of sixteen freshly isolated open-ocean bacteria. *Appl. Environ. Microbiol.* **45:**1109–1115.
5. **Anwar, H., and J. W. Costerton.** 1992. Effective use of antibiotics in the treatment of biofilm-associated infections. *ASM News* **58:**665–668.
6. **Anwar, H., M. K. Dasgupta, and J. W. Costerton.** 1990. Testing the susceptibility of bacteria in biofilms to antibacterial agents. *Antimicrob. Agents Chemother.* **34:**2043–2046.
7. **Anwar, H., J. L. Strap, and J. W. Costerton.** 1992. Establishment of aging biofilms: possible mechanism of bacterial resistance to antibiotic therapy. *Antimicrob. Agents Chemother.* **36:**1347–1351.
8. **Baier, R. E.** 1980. Substrata influences on adhesion of microorganisms and their resultant new surface properties, p. 59–104. *In* G. Bitton and K. C. Marshall (ed.), *Adsorption of Microorganisms to Surfaces*. Wiley-Interscience, New York, N.Y.
9. **Barer, M. R., L. T. Gribbon, C. R. Harwood, and C. E. Nwoguh.** 1993. The viable but nonculturable hypothesis and medical bacteriology. *Rev. Med. Microbiol.* **4:**183–191.
10. **Bej, A. K., M. H. Mahbubani, and R. M. Atlas.** 1991. Detection of viable *Legionella pneuophila* in water by polymerase chain reaction and gene probe methods. *Appl. Environ. Microbiol.* **57:**597–600.
11. **Belas, R., A. Mileham, M. Simon, and M. Silverman.** 1984. Transposon mutagenesis of marine *Vibrio* spp. *J. Bacteriol.* **158:**890–896.

12. **Belas, R., M. Simon, and M. Silverman.** 1986. Regulation of lateral flagella gene transcription in *Vibrio parahaemolyticus. J. Bacteriol.* **167:**210–218.
13. **Berkeley, R. C. W., J. M. Lynch, J. Melling, P. R. Rutter, and B. Vincent (ed.).** 1980. *Microbial Adhesion to Surfaces*. Ellis Horwood Publ., Chichester, U.K.
14. **Blenkinsopp, S. A., A. E. Khoury, and J. W. Costerton.** 1992. Electrical enhancement of biocide efficacy against *Pseudomonas aeruginosa* biofilms. *Appl. Environ. Microbiol.* **58:**3770–3773.
15. **Brown, M. R. W., D. G. Allison, and P. Gilbert.** 1988. Resistance of bacterial biofilms to antibiotics: a growth-rate related effect? *J. Antimicrob. Chemother.* **22:**777–783.
16. **Bryers, J. D.** 1990. Biofilms in biotechnology, p. 733–773. *In* W. G. Characklis and K. C. Marshall (ed.), *Biofilms*. Wiley-Interscience, New York, N.Y.
17. **Caldwell, D. E., S. H. Lai, and J. M. Tiedje.** 1973. A two-dimensional steady-state diffusion gradient for ecological studies. *Bull. Ecol. Res. Comm.-NFR* (Statens Naturvetensk, Forskningsrad, Sweden) **17:**151–158.
18. **Caldwell, D. E., D. R. Korber, and J. R. Lawrence.** 1992. Confocal laser microscopy and computer image analysis in microbial ecology. *Adv. Microb. Ecol.* **12:**1–67.
19. **Cargill, K. L., B. H. Pyle, R. L. Sauer, and G. A. McFeters.** 1992. Effects of culture conditions and biofilm formation on the iodine susceptibility of *Legionella pneumophila. Can. J. Microbiol.* **38:**423–429.
20. **Chang, C. C., and K. Merritt.** 1992. Microbial adherence on poly(methyl methacrylate) (PMMA) surfaces. *J. Biomed. Mater. Res.* **26:**197–207.
21. **Characklis, W. G.** 1990. Microbial biofouling control, p. 585–633. *In* W. G. Characklis and K. C. Marshall (ed.), *Biofilms*. Wiley-Interscience, New York, N.Y.
22. **Characklis, W. G., and K. C. Marshall (ed.).** 1990. *Biofilms*. Wiley-Interscience, New York, N.Y.
23. **Characklis, W. G., M. H. Turakia, and N. Zelver.** 1990. Transport and interfacial transfer phenomena, p. 265–340. *In* W. G. Characklis and K. C. Marshall (ed.), *Biofilms*. Wiley-Interscience, New York, N.Y.
24. **Characklis, W. G., G. A. McFeters, and K. C. Marshall.** 1990. Physiological ecology in biofilm systems, p. 341–394. *In* W. G. Characklis and K. C. Marshall (ed.), *Biofilms*. Wiley-Interscience, New York, N.Y.
25. **Cohen, Y., and E. Rosenberg (ed.).** 1989. *Microbial Mats: Physiological Ecology of Benthic Microbial Communities*. American Society for Microbiology, Washington, D.C.
26. **Costerton, J. W., K.-J. Cheng, G. G. Geesey, T. Ladd, J. C. Nickel, M. Dasgupta, and T. J. Marrie.** 1987. Bacterial biofilms in nature and disease. *Annu. Rev. Microbiol.* **41:**435–464.
27. **Costerton, J. W., Z. Lewandowski, D. DeBeer, D. Caldwell, D. Korber, and G. James.** 1994. Biofilms, the customized microniche. *J. Bacteriol.* **176:**2137–2142.
28. **Costerton, J. W., Z. Lewandowski, D. E. Caldwell, D. R. Korber, and H. M. Lappin-Scott.** 1995. Microbial biofilms. *Annu. Rev. Microbiol.* **49:**711–745.
29. **Costerton, J. W., P. S. Stewart, and E. P. Greenberg.** 1999. Bacterial biofilms: a common cause of persistent infections. *Science* **284:**1318–1322.
30. **Dagostino, L., A. E. Goodman, and K. C. Marshall.** 1991. Physiological responses induced in bacteria adhering to surfaces. *Biofouling* **4:**113–119.
31. **Dalsgaard, T., and N. P. Revsbech.** 1992. Regulating factors of dentrification in trickling filter biofilms as measured with the oxygen/nitrous oxide microsensor. *FEMS Microbial Ecol.* **101:**151–164.
32. **Davies, D. G., A. M. Chakrabarty, and G. G. Geesey.** 1993. Exopolysaccharide production in biofilms: substratum activation of alginate gene expression by *Pseudomonas aeruginosa. Appl. Environ. Microbiol.* **59:**1181–1186.
33. **Davies, D. G., M. R. Parsek, J. P. Pearson, B. H. Iglewski, J. W. Costerton, and E. P. Greenberg.** 1998. The involvement of cell-to-cell signals in the development of a bacterial biofilm. *Science* **280:**295–298.
34. **de Beer, D., J. C. van den Heuvel, and S. P. P. Ottengraf.** 1993. Microelectrode measurements of the activity distribution in nitrifying bacterial aggregates. *Appl. Environ. Microbiol.* **59:**573–579.
35. **de Beer, D., P. Stoodley, F. Roe, and Z. Lewandowski.** 1994. Effects of biofilm structures on oxygen distribution and mass transfer. *Biotechnol. Bioeng.* **43:**1131–1138.

36. **de Beer, D., P. Stoodley, and Z. Lewandowski.** 1997. Measurement of local diffusion coefficients in biofilms by microinjection and confocal microscopy. *Biotechnol. Bioeng.* **53:**151–158.
37. **de Weger, L. A., P. Dunbar, W. F. Mahafee, B. J. J. Lugtenberg, and G. S. Sayler.** 1991. Use of bioluminescence markers to detect *Pseudomonas* spp. in the rhizosphere. *Appl. Environ. Microbiol.* **57:**3641–3644.
38. **Evans, D. J., M. R. W. Brown, D. G. Allison, and P. Gilbert.** 1991. Susceptibility of *Pseudomonas aeruginosa* and *Escherichia coli* biofilms towards ciprofloxacin: effect of specific growth rate. *J. Antimicrob. Chemother.* **27:**177–184.
39. **Ford, T., and R. Mitchell.** 1990. The ecology of microbial corrosion. *Adv. Microb. Ecol.* **11:**231–262.
40. **Gerchakov, S. M., D. S. Marszalek, F. J. Roth, and L. R. Udey.** 1977. Succession of periphytic microorganisms on metal and glass surfaces, p. 203–211. *In* V. Romanovsky (ed.), *Proc. 4th Intern. Congr. Mar Corrosion Fouling*. Centre de Recherches et d'Etudes Oceanographiques, Boulogne, France.
41. **Gest, H.** 1993. Bacterial growth and reproduction in nature and in the laboratory. *ASM News.* **59:** 542–543.
42. **Gilbert, P., F. Collier, and M. R. W. Brown.** 1990. Influence of growth rate on susceptibility to antimicrobial agents: biofilms, cell cycle, dormancy, and stringent response. *Antimicrob. Agents Chemother.* **34:**1865–1868.
43. **Goodman, A. E., and K. C. Marshall.** 1995. Genetic responses of bacteria at surfaces, p. 80–98. *In* H. M. Lappin-Scott and J. W. Costerton (ed.), *Microbial Biofilms.* Cambridge University Press, Cambridge, U.K.
44. **Griffith, P. C., and M. Fletcher.** 1991. Hydrolysis of protein and model dipeptide substrates by attached and unattached marine *Pseudomonas* sp. strain NCMB 2021. *Appl. Environ. Microbiol.* **57:** 2186–2191.
45. **Gristina, A. G., J. J. Dobbins, M. S. Giammara, J. C. Lewis, and W. C. DeVries.** 1988. Biomaterial-centered sepsis and the total artificial heart. *JAMA* **259:**870–874.
46. **Hermansson, M., and K. C. Marshall.** 1985. Utilization of surface localized substrate by non-adhesive marine bacteria. *Microbiol. Ecol.* **11:**91–105.
47. **Herrick, J. B., E. L. Madsen, C. A. Batt, and W. C. Ghiorse.** 1993. Polymerase chain reaction amplification of naphthalenecatabolic and 16S rRNA gene sequences from indigenous sediment bacteria. *Appl. Environ. Microbiol.* **59:**687–694.
48. **Holben, W. E., B. M. Schroter, V. G. Calabrese, R. H. Olsen, J. K. Kukor, V. O. Biederbeck, A. E. Smith, and J. M. Tiedje.** 1992. Gene probe analysis of soil microbial populations selected by amendment with 2,4-dichlorophen-oxyacetic acid. *Appl. Environ. Microbiol.* **58:**3941–3948.
49. **Hoyle, B. D., L. J. Williams, and J. W. Costerton.** 1993. Production of mucoid exopolysaccharide during development of *Pseudomonas aeruginosa* biofilms. *Infect. Immun.* **61:**777–780.
50. **Kane, M. D., L. K. Poulsen, and D. A. Stahl.** 1993. Monitoring the enrichment and isolation of sulfate-reducing bacteria by using oligonucleotide hybridization probes designed from environmentally derived 16S rRNA sequences. *Appl. Environ. Microbiol.* **59:**682–686.
51. **Kefford, B., S. Kjelleberg, and K. C. Marshall.** 1982. Bacterial scavenging: Utilization of fatty acids localized at a solid-liquid interface. *Arch. Microbiol.* **133:**257–260.
52. **King, J. M. H., P. M. Digrazia, B. Applegate, R. Burlage, J. Sanseverino, P. Dunbar, F. Larimer, and G. S. Sayler.** 1990. Rapid, sensitive bioluminescent reporter technology for naphthalene exposure and biodegradation. *Science* **249:**778–781.
53. **Kinniment, S. L., and J. W. T. Wimpenny.** 1992. Measurements of the distribution of adenylate concentrations and adenylate energy charge across *Pseudomonas aeruginosa* biofilms. *Appl. Environ. Microbiol.* **58:**1629–1635.
54. **Kjelleberg, S.** 1993. *Starvation in Bacteria.* Plenum Press, New York, N.Y.
55. **Kjelleberg, S., B. A. Humphrey, B. A. Marshall, and K. C. Marshall.** 1982. The effect of interfaces on small starved marine bacteria. *Appl. Environ. Microbiol.* **43:**1166–1172.
56. **Kjelleberg, S., M. Hermansson, P. Mardén, and G. W. Jones.** 1987. The transient phase between growth and non-growth of heterotrophic bacteria, with emphasis on the marine environment. *Annu. Rev. Microbiol.* **41:**25–49.

57. **Koch, A. L.** 1990. Diffusion: The crucial process in many aspects of the biology of bacteria. *Adv. Microb. Ecol.* **11:**37–70.
58. **Kolenbrander, P. E., and J. London.** 1992. Ecological significance of coaggregation among oral bacteria. *Adv. Microb. Ecol.* **12:**183–217.
59. **Korber, D. R., J. R. Lawrence, H. M. Lappin-Scott, and J. W. Costerton.** 1995. Growth of microorganisms on surfaces, p. 15–45. *In* H. M. Lappin-Scott and J. W. Costerton (ed.), *Microbial Biofilms.* Cambridge University Press, Cambridge, U.K.
60. **Lappin-Scott, H. M., and J. W. Costerton (ed.).** 1995. *Microbial Biofilms.* Cambridge University Press, Cambridge, U.K.
61. **Lawrence, J. R., D. R. Korber, B. D. Hoyle, J. W. Costerton, and D. E. Caldwell.** 1991. Optical sectioning of microbial biofilms. *J. Bacteriol.* **173:**6558–6567.
62. **Lee, A.** 1985. Neglected niches: the microbial ecology of the gastrointestinal tract. *Adv. Microb. Ecol.,* **8:**115–162.
63. **Leung, J. W., G. T. Lau, J. J. Sung, and J. W. Costerton.** 1992. Decreased bacterial adherence to silver-coated stent material: an *in vitro* study. *Gastrointest. Endosc.* **38:**338–340.
64. **Little, B., P. Wagner, R. Ray, R. Pope, and R. Scheetz.** 1991. Biofilms: an ESEM evaluation of artifacts introduced during SEM preparation. *J. Indust. Microbiol.* **8:**213–222.
65. **Maki, J. S., D. Rittschof, D. Mitchell, and R. Mitchell.** 1992. Inhibition of larval barnacle attachment to bacterial films: an investigation of physical properties. *Microb. Ecol.* **23:**97–106.
66. **Manz, W., U. Szewzyk, P. Ericsson, R. Amann, K.-H. Schleifer, and T.-A. Stenstrom.** 1993. In situ identification of bacteria in drinking water and adjoining biofilms by hybridization with 16S and 23S rRNA-directed fluorescent oligonucleode probes. *Appl. Environ. Microbiol.* **59:**2293–2298.
67. **Marshall, K. C.** 1976. *Interfaces in Microbial Ecology*, Harvard University Press, Cambridge, Mass.
68. **Marshall, K. C.** 1984. *Microbial Adhesion and Aggregation.* Springer-Verlag, Berlin, Germany.
69. **Marshall, K. C.** 1992. Planktonic versus sessile life of prokaryotes, p. 262–275. *In* A. Balows, H. G. Trüper, M. Dworkin, W. Harder, and K. H. Schliefer (ed.), *The Prokaryotes, a Handbook on the Biology of Bacteria, Ecophysiology, Isolation, Identification, Applications*, 2nd ed. Springer-Verlag, New York, N.Y.
70. **Marshall, K. C.** 1992. Biofilms: an overview of bacterial adhesion, activity, and control at surfaces. *ASM News* **58:**202–207.
71. **Marshall, K. C.** 1993. Microbial ecology: whither goest thou?, p. 5–8. *In* R. Guerrero and C. Pedrós-Alió (ed.), *Trends in Microbial Ecology.* Spanish Society for Microbiology, Barcelona, Spain.
72. **Marszalek, D. S., S. M. Gerchakov, and L. R. Udey.** 1979. Influence of substrate composition on marine microfouling. *Appl. Environ. Microbiol.* **38:**987–995.
73. **McFeters, G. A.** 1990. Enumeration, occurrence, and significance of injured indicator bacteria in drinking water, p. 478–492. *In* G. A. McFeters (ed.), *Drinking Water Microbiology, Progress and Recent Development.* Springer-Verlag, New York, N.Y.
74. **Mitchell, J. G., R. Weller, M. Beconi, J. Sell, J. and J. Holland.** 1993. A practical optical trap for manipulating and isolating bacteria from complex microbial communities. *Microb. Ecol.* **25:**113–119.
75. **Morita, R. Y.** 1982. Starvation-survival of heterotrophs in the marine environment. *Adv. Microb. Ecol.* **6:**171–198.
76. **Muyzer, G., E. C. de Waal, and A. G. Uitterlinden.** 1993. Profiling of complex microbial populations by denaturing gradient gel electrophoresis analysis of polymerase chain reaction-amplified genes coding for 16S rRNA. *Appl. Environ. Microbiol.* **59:**695–700.
77. **Nealson, K. H., B. Wimpee, and C. Wimpee.** 1993. Identification of *Vibrio splendidus* as a member of the planktonic luminous bacteria from the Persian Gulf and Kuwait region with *luxA* probes. *Appl. Environ. Microbiol.* **59:**2684–2689.
78. **Neihof, R., and G. Loeb.** 1974. Dissolved organic matter in seawater and the electric charge of immersed surfaces. *J. Mar. Res.* **32:**5–12.
79. **Nichols, W. W., M. J. Evans, M. P. E. Slack, and H. L. Walmsley.** 1989. The penetration of antibiotics into aggregates of mucoid and nonmucoid *Pseudomonas aeruginosa. J. Gen. Microbiol.* **135:**1291–1303.

80. **Nickel, J. C., I. Ruseska, J. B. Wright, and J. W. Costerton.** 1985. Tobramycin resistance of *Pseudomonas aeruginosa* cells growing as a biofilm on urinary catheter material. *Antimicrob. Agents Chemother.* **27:**619–624.
81. **Nilsson, L., J. D. Oliver, and S. Kjelleberg.** 1991. Resuscitation of *Vibrio vulnificus* from the viable but nonculturable state. *J. Bacteriol.* **173:**5054–5059.
82. **Oliver, J. D.** 1993. Formation of viable but nonculturable cells, p. 239–272. *In* S. Kjelleberg (ed.), *Starvation in Bacteria*. Plenum Press, New York, N.Y.
83. **Oliver, J. D., L. Nilsson, and S. Kjelleberg.** 1991. Formation of nonculturable *Vibrio vulnificus* cells and its relationship to the starvation state. *Appl. Environ. Microbiol.* **57:**2640–2644.
84. **Pace, N. R., D. A. Stahl, D. J. Lane, and G. J. Olsen.** 1986. The analysis of natural microbial populations by ribosomal RNA sequences. *Adv. Microb. Ecol.* **9:**1–55.
85. **Poulsen, L. K., G. Ballard, and D. A. Stahl.** 1993. Use of rRNA fluorescence in situ hybridization for measuring the activity of single cells in young and established biofilms. *Appl. Environ. Microbiol.* **59:**1354–1360.
86. **Power, K., and K. C. Marshall.** 1988. Cellular growth and reproduction of marine bacteria on surface bound substrate. *Biofouling* **1:**163–174.
87. **Revsbech, N. P. and B. B. Jorgensen.** 1986. Microelectrodes: their use in microbial ecology. *Adv. Microb. Ecol.* **9:**293–352.
88. **Rodriquez, G. G., D. Phipps, K. Ishiguro, and H. F. Ridgway.** 1992. Use of a fluorescent redox probe for direct visualization of actively respiring bacteria. *Appl. Environ, Microbiol.* **58:**1801–1808.
89. **Rogers, J., and C. W. Keevil.** 1992. Immunogold and fluorescein immunolabelling of *Legionella pneumophila* within an aquatic biofilm visualized by using episcopic differential contrast microscopy. *Appl. Environ. Microbiol.* **58:**2326–2330.
90. **Roslev, P., and G. M. King.** 1993. Application of a tetrazolium salt with a water-soluble formazan as an indicator of viability in respiring bacteria. *Appl. Environ. Microbiol.* **59:**2891–2896.
91. **Roszak, D. B., and R. R. Colwell.** 1987. Survival strategies of bacteria in the natural environment. *Microbiol. Rev.* **51:**365–379.
92. **Samuelsson, M.-O., and D. L. Kirchman.** 1991. Degradation of adsorbed protein by attached bacteria in relation to surface hydrophobicity. *Appl. Environ. Microbiol.* **56:**3643–3648.
93. **Savage, D. C., and M. Fletcher.** 1985. *Bacteria Adhesion: Mechanisms and Physiological Significance*. Plenum Press, New York, N.Y.
94. **Schaule, G., H.-C. Flemming, and H. F. Ridgway.** 1993. Use of 5-cyano-2,3-ditoyl tetrazolium chloride for quantifying planktonic and sessile respiring bacteria in drinking water. *Appl. Environ. Microbiol.* **59:**3850–3857.
95. **Schneider, R. P., and K. C. Marshall.** 1994. Retention of the Gram-negative marine bacterium SW8 on surfaces—effects of microbial physiology, substratum nature and conditioning films. *Colloids & Surfaces, B: Biointerfaces* **2:**387–396.
96. **Silcock, D. J., R. N. Waterhouse, L. A. Glover, J. I. Prosser, and K. Killham.** 1992. Detection of a single genetically modified bacterial cell in soil by using charge coupled device-enhanced microscopy. *Appl. Environ. Microbiol.* **58:**2444–2448.
97. **Skyring, G. W., and J. Bauld.** 1990. Microbial mats in Australian coastal environments. *Adv. Microb. Ecol.* **11:**461–498.
98. **Smigielski, A. J., B. J. Wallace, and K. C. Marshall.** 1989. Changes in membrane functions during short-term starvation of *Vibrio fluvialis* strain NCTC 11328. *Arch. Microbiol.* **151:**336–347.
99. **Stewart, P. S., B. M. Peyton, W. J. Drury, and R. Murga.** 1993. Quantitative observations of heterogeneities in *Pseudomonas aeruginosa* biofilms. *Appl. Environ. Microbiol.* **59:**327–329.
100. **Stewart, P. S., T. Griebe, R. Srinivasan, C.-I. Chen, F. P. Yu, D. deBeer, and G. A. McFeters.** 1994. Comparison of respiratory activity and culturability during monochloramine disinfection of binary population biofilms. *Appl. Environ. Microbiol.* **60:**1690–1692.
101. **Stewart, P. S.** 1996. Theoretical aspects of antibiotic diffusion into microbial biofilms. *Antimicrob. Agents Chemother.* **40:**2517–2522.
102. **Stoodley, P., Z. Lewandowski, J. D. Boyle, and H. M. Lappin-Scott.** 1998. Oscillation characteristics of biofilm streamers in turbulent flowing water as related to drag and pressure drop. *Biotechnol. Bioeng.* **57:**536–544.

103. **Stoodley, P., I. Dodds, Z. Lewandowski, A. B. Cunningham, J. D. Boyle, and H. M. Lappin-Scott.** 1999. Influence of hydrodynamics and nutrients on biofilm structure. *J. Appl. Microbiol.* **85:** 19S–28S.
104. **Szewzyk, U., W. Manz, R. Amann, K.-H. Schleifer, and T.-A. Stenstrom.** 1994. Growth and *in situ* detection of a pathogenic *Escherichia coli* in biofilms of a heterotrophic water bacterium by use of 16S- and 23S-rRNA-directed fluorescent oligonucleotide probes. *FEMS Microbiol. Ecol.* **13:** 169–176.
105. **Tabor, P. S., and R. A. Neihof.** 1982. Improved method for determination of respiring individual microorganisms in natural waters. *Appl. Environ. Microbiol.* **43:**1249–1255.
106. **ten Bosch, J. J.** 1991. Physico-chemical aspects of biological adhesion. *Biofouling* **4(1-3):**1–247.
107. **Terzieva, S., J. Donnelly, V. Ulevicius, S. A. Grinshpun, K. Willeke, G. N. Selma, and K. P. Brenner.** 1996. Comparison of methods for detection and enumeration of airborne microorganisms collected by liquid impingement. *Appl. Environ. Microbiol.* **62:**2264–2272.
108. **Tyler, P. A., and K. C. Marshall.** 1967. Microbial oxidation of manganese in hydro-electric pipelines. *Antonie van Leeuwenhoek* **33:**171–183.
109. **Wagner, M., R. Amann, H. Lemmer, and K.-H. Schleifer.** 1993. Probing activated sludge with oligonucleotides specific for proteobacteria: inadequacy of culture-dependent methods for describing microbial community structure. *Appl. Environ. Microbiol.* **59:**1520–1525.
110. **Ward, D. M., R. Weller, and M. M. Bateson.** 1990. 16S rRNA sequences reveal numerous uncultured microorganisms in a natural community. *Nature* **345:**63–65.
111. **Ward, D. M., R. Weller, and M. M. Bateson.** 1990. 16S rRNA sequences reveal uncultured inhabitants of a well-studied thermal community. *FEMS Microbiol. Rev.* **75:**105–116.
112. **Ward, D. M., M. M. Bateson, R. Weller, and A. L. Ruff-Roberts.** 1992. Ribosomal RNA analysis of microorganisms as they occur in nature. *Adv. Microb. Ecol.* **12:**219–286.
113. **Weichart, D., J. D. Oliver, and S. Kjelleberg.** 1992. Low temperature induced nonculturability and killing of *Vibrio vulnificus*. *FEMS Microbiol. Lett.* **100:**205–210.
114. **Weller, R., M. M. Bateson, B. K. Heimbuch, E. D. Kopczynski, and D. M. Ward.** 1992. Uncultivated cyanobacteria, *Chloroflexus*-like inhabitants, and spirochete-like inhabitants of a hot spring microbial mat. *Appl. Environ. Microbiol.* **58:**3964–3969.
115. **Wolfaardt, G. M., J. R. Lawrence, M. J. Hendry, R. D. Robarts, and D. E. Caldwell.** 1993. Development of steady-state diffusion gradients for the cultivation of degradative microbial consortia. *Appl. Environ. Microbiol.* **59:**2388–2396.
116. **Yu, F. P. and G. A. McFeters.** 1994. Rapid *in situ* assessment of physiological activities in biofilms using fluorescent probes. *J. Microbiol. Methods* **20:**1–10.
117. **Zimmerman, R., R. Iturriaga, and J. Becker-Birck.** 1978. Simultaneous determination of the total number of aquatic bacteria and the number thereof involved in respiration. *Appl. Environ. Microbiol.* **36:**926–935.

Nonculturable Microorganisms in the Environment
Edited by R. R. Colwell and D. J. Grimes

Chapter 9

Phenotypic Plasticity in Bacterial Biofilms as It Affects Issues of Viability and Culturability

J. William Costerton

Microbiology is approaching a critical intellectual challenge, which is brought into focus by the discovery of viable but nonculturable organisms in nature. In a continuous operational sequence, beginning with Louis Pasteur one and a half centuries ago, we have been preoccupied by the planktonic bacterial cells that grow so rapidly and readily in our cunningly formulated media in vitro. Until very recently a bacterial cell was not considered to exist, if it could not make the very rapid transition from its phenotype in its natural environment to this stylized entity in the test tube. Modern direct observations of natural populations have shown hundreds of morphotypes of bacterial cells that yield no corresponding planktonic cells on culture, and even more modern nucleotide analyses have shown the presence of many organisms that have never yielded culturable cells. Microbiology is seen to have concentrated on the minority of bacteria that can be cultured from nature, partly because of our understandable preoccupation with organisms that cause specific problems such as acute diseases of ourselves or of our domestic plants and animals. For these special organisms we have always developed suitable media and culture methods, and many of these media and methods actually discourage the growth of "environmental" species. This approach has produced the vaccines and antibiotics that still control many bacterial diseases, but the science of microbiology has committed a grave intellectual error, and we know very little of the neglected organisms that didn't happen to like to grow in our specialized media.

While each cell of a multicellular eukaryotic plant or animal is totipotent in that it contains all the genetic information of the species, we know that a leaf cell and a liver cell cannot readily replicate the entire organism in culture because each has assumed a specialized phenotype. Some of these individual cells of multicellular organisms can readily reverse their phenotypic specialization and grow in culture, but most cannot. We have heretofore assumed that bacterial cells exist in a single planktonic phenotype, but a collective sinking feeling steals over us as we realize

J. William Costerton • Center for Biofilm Engineering, Montana State University, Bozeman, MT 59717.

that even stationary-phase planktonic cells have a distinct phenotype and that starvation survival involves the assumption of a profoundly different phenotype. In recent months it has become evident that sessile bacterial cells growing in biofilms also express a phenotype that is equally different from planktonic cells, and this revelation is devastating because the majority of the bacteria in the biosphere exist in either the starved or the biofilm phenotype. We must therefore begin to come to terms with the growing realization that not only have we concentrated almost all our attention on only one of many bacterial phenotypes, but that we have chosen the rare planktonic phenotype that exists in very small numbers in nature. The blind spot in microbiology can be blamed largely on the anthrocentric mindset of our very practical and very useful science.

BACTERIAL BIOFILMS AS A DIFFERENTIATED MULTICELLULAR COMMUNITY

While microbiology had its beginnings in the direct microscopic observations of such pioneers as Antonie van Leuvenhoek and Claude Zobell, the ennui induced by watching planktonic cells swimming in culture fluids soon converted most microbiologists into molecular biologists. Bacterial biofilms were examined by scanning and transmission electron microscopy (SEM and TEM, respectively) of dehydrated specimens, when they were examined at all, and their complex architecture was largely lost in drying and embedding artifacts. Only the use of confocal scanning laser microscopy (CSLM), with its ability to resolve three-dimensional detail in living, fully hydrated biofilms, has revealed the complexity and sophistication of the biofilms that are the epitome of the plasticity of phenotypic expression in bacteria. Figure 1 is a diagrammatic representation of the structure of a wide range of in vitro-grown and natural multispecies biofilms, as seen by CSLM and compiled by image analysis, and this basic pattern of microcolonies and water channels has been confirmed in many ecosystems (1). The sessile bacterial cells actually grow in microcolonies of a wide variety of shapes and sizes, in which they are embedded in an exopolysaccharide matrix material that comprises about 85% of the volume of each of these structural units of the biofilm. The microcolonies are separated by water channels, even in very thick biofilms, and these channels carry water and nutrients throughout the sessile community by convective flow (12). These direct observations of living biofilms, many of which were made and reported by biofilm engineers, clearly show two features heretofore held to characterize multicellular eukaryotes. These features are the formation of primitive but structurally organized "tissues" of accumulated cells and the presence of a primitive but functional circulatory system.

Whenever cells are immobilized in a structured mass, like a matrix-enclosed microcolony, the laws of chemistry and physics dictate that the local microenvironments of individual cells will differ. Peripheral cells will see nutrients, including oxygen, more readily than deeply embedded cells, and the phenotype of each sessile cell will be dictated by the vagaries of its own microniche (2) within the structural whole. Biofilm engineers have made direct measurements of such parameters as dissolved oxygen and pH at many specific locations within biofilms, and huge

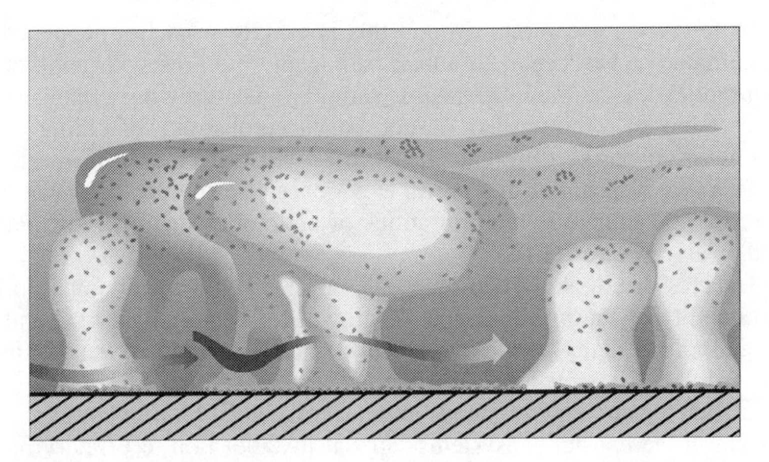

Figure 1. Diagrammatic representation of the structure of bacterial biofilms based on the systematic analysis of CSLM images of natural and laboratory biofilms. Note that the biofilm community is composed of microcolonies, in which sessile cells grow in matrix material, and that the microcolonies are separated by open water channels where convective flow occurs. Microcolonies are frequently seen to detach from these complex sessile communities.

variations have been found (6). The physiological diversity of sessile cells in a biofilm, in which some cells grow aerobically just tens of microns from where some other cells grow anaerobically (Fig. 2), can be interpreted as a form of cell specialization and recorded as another characteristic usually attributed to multicellular eukaryotes. If we add the diversity of sessile cells, some of which express the *rpo S* controlled genes of the stationary-growth phase (Pulcini, Stewart, and Camper, unpublished data) while others grow very rapidly, we begin to see the biofilm as a mosaic of differing phenotypes which are functionally integrated in a multicellular community. It is axiomatic that phenotypic diversity confers a measure

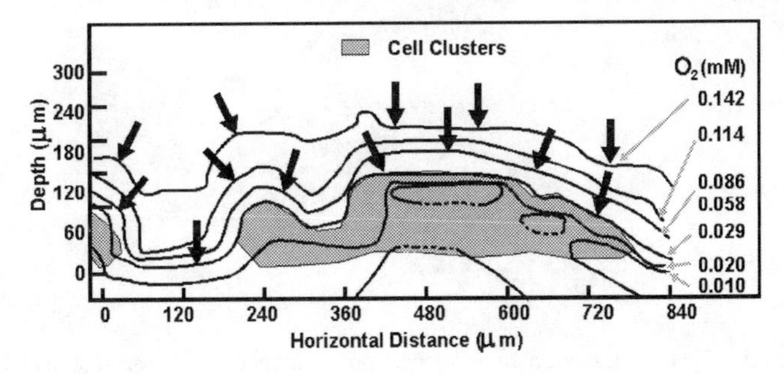

Figure 2. Diagrammatic representation of CSLM data locating a microcolony within a biofilm, on which dissolved oxygen data obtained by systematic analysis using a dissolved oxygen microelectrode has been superimposed to create a dissolved oxygen "map." Note that the center of this microcolony of matrix-enclosed cells of *P. aeruginosa*, growing in ambient air in a flow cell, is functionally anaerobic.

of survival fitness on a community, and this is exactly what has happened as the medical community has begun to attack pathogenic biofilms with antibiotics (3). Some antibiotics kill aerobic organisms, some kill fast-growing organisms, some work well at low pH while others do not, but the probability of killing all of the sessile cells in all of the different phenotypes with one or two agents is indeed remote. This fact is born out by clinical experience, which concludes that biofilms resist up to 1,000 times the concentrations of antibiotics that will kill planktonic cells of the same species (13).

The pleasures of teleology beckon to us, and we begin to ruminate on the exquisite suitability of the biofilm mode of growth for bacterial survival and proliferation, in all of the many periods in which they have constituted the dominant life form on earth. Sessile growth remains the predominant form of bacterial growth in all nutrient-rich aquatic environments, and bacterial/algal biofilm communities still predominate in the coastal ecosystems. Spatial juxtaposition to primary producers provides nutrients, as does location at a surface, and this situation is especially serendipitous if the surface that is colonized is also a nutrient, as in the case of cellulose (Fig. 3). The matrix of the biofilm concentrates nutrients, and sessile cells within the biofilm community cooperate in prodigious feats of metabolism (11) in which they digest complex substrates at speeds hundreds of times faster than those seen in pure planktonic cultures. Because the matrix material that comprises most of the volume of each microcolony is highly elastic, these structural units of the biofilms respond to shear stress by elongating and oscillating in the turbulent flow zone (14), which further enhances nutrient trapping. Thus the bacterial biofilm

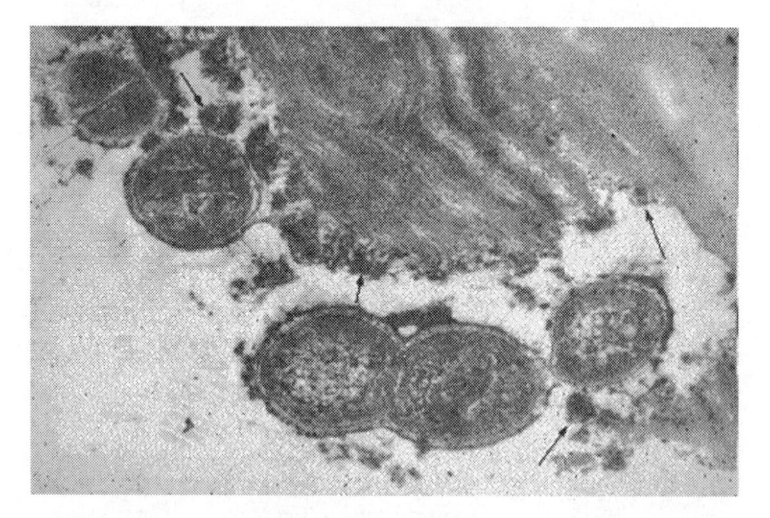

Figure 3. TEM of a section of a stained preparation of cellulose that had been incubated with cells of *Fibrobacter succinogenes*, showing the colonization of the insoluble cellulose by these cells, which have formed a biofilm on this nutrient surface. Some bacterial cells have produced a cavity in the surface of the cellulose, and many small fragments of the bacterial cell surface (arrows) have detached and are carrying out a focused enzymatic attack on this substrate.

seems to be the optimal mode of growth when these organisms had the earth pretty much to themselves, but this community structure also provided protection against two antibacterial challenges that surely occurred early in the evolutionary process. The matrix of the microcolonies excludes most bacteriophage and other viruses, and it makes sessile bacteria virtually completely resistant to uptake by amoeba and by modern day phagocytes (10).

THE BIOFILM PHENOTYPE

Direct observation of gene expression, using a reporter construct whose expression can be visualized by light microscopy, has allowed Center for Biofilm Engineering (CBE) scientists (4) to monitor the up regulation of a gene which must *a priori* be associated with biofilm formation. In *Pseudomonas aeruginosa*, the synthesis of the matrix polymer alginate is down regulated in planktonic cells, but direct observations of living cells show that it is up regulated shortly after the adhesion of these planktonic cells to a surface prior to biofilm formation (Fig. 4). The fusion of a *lac Z* reporter gene downstream from the promoter of the *alg C* gene, which codes for the phosphomannomutase enzyme of the alginate synthesis cascade, produces a reporter system that reveals up regulation of this gene. The use of this reporter strain, under a light microscope, shows that this gene is up regulated within 15 min of the adhesion of planktonic cells of this species to a surface (Fig. 5), and non-microscopic studies (Hongwei Yu, PhD thesis, University of Calgary, 1992) indicate that the *algD* gene is similarly regulated. Because both

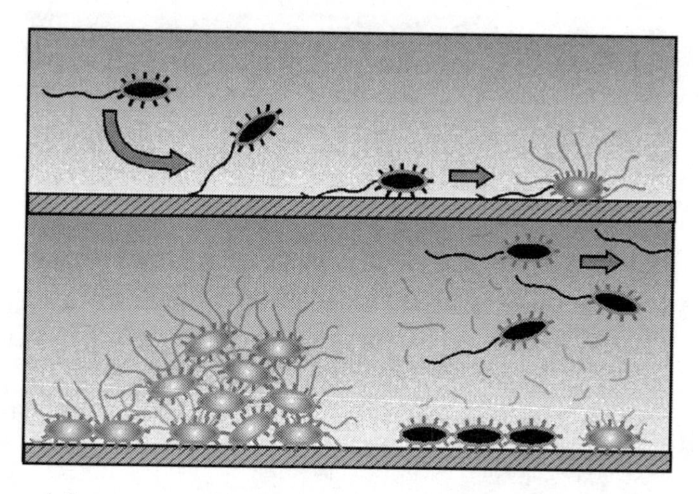

Figure 4. Cartoon summarizing some features of biofilm formation. Planktonic cells adhere to a surface, sometimes using their flagella as adhesins, and they adopt the biofilm phenotype and begin to produce exopolysaccharide matrix material soon after adhering. Sessile cells in the biofilm phenotype build highly structured biofilms (lower panel) composed of microcolonies within which some cells are programmed to return to the planktonic phenotype, digest the matrix material, and swim away to establish new biofilms.

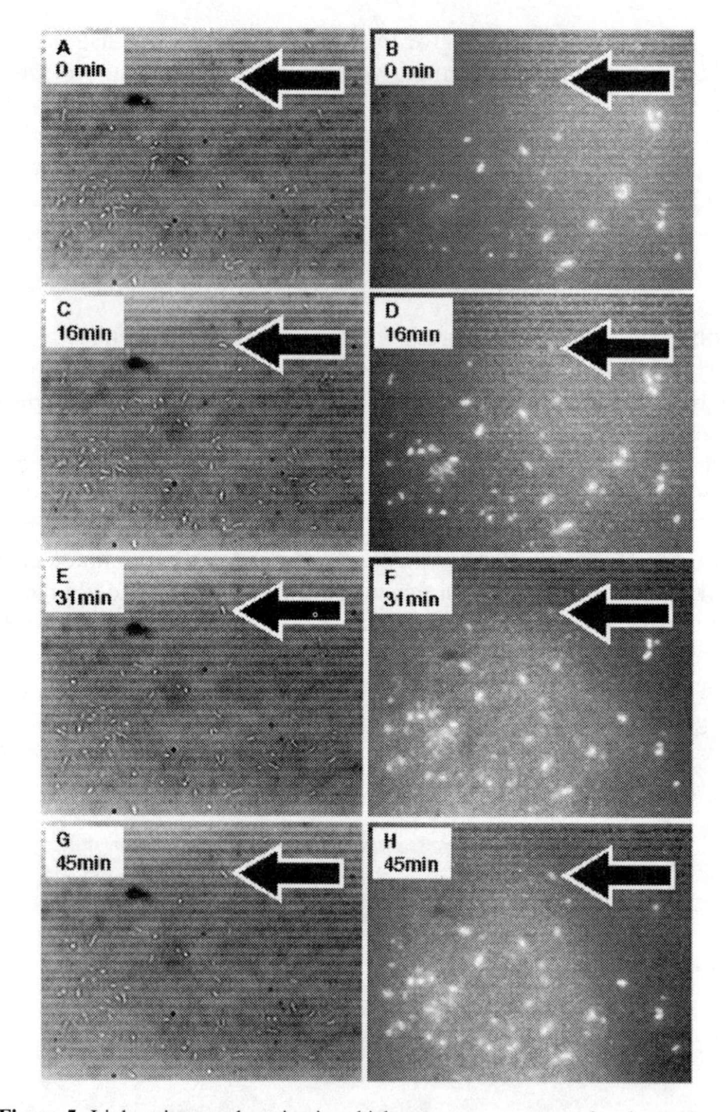

Figure 5. Light micrograph series in which a reporter construct has been cloned into cells of *P. aeruginosa*, by inserting a *lacZ* cassette downstream from the promoter controlling the expression of the *algC* gene of the alginate synthesis cascade. A planktonic cell adheres to the glass surface, as seen by phase contrast microscopy (left panel), and remains adherent. Within 45 min of its adhesion to this surface, the cell up regulates the *algC* gene, as indicated by the *lacZ*-mediated color reaction seen by epifluorescence microscopy (right panel).

algC and *algD* are regulated by a sigma factor coded by *algU* (7), we deduce that this regulatory gene must be activated by adhesion to a surface. The comparison of the gene products (proteins) present in both planktonic and sessile cells of *P. aeruginosa*, by two-dimensional electrophoresis, shows that the *rpoS* gene is up regulated in biofilms and in planktonic cells in the stationary phase of growth. These observations were expected, especially in the case of the alginate synthesis genes, but nothing prepared us for the revelation that even the outer membrane proteins (OMPs) of sessile cells of *P. aeruginosa* differ very profoundly from those of planktonic cells of the same species (Fig. 6). Because the OMPs of *P. aeruginosa* have been implicated in everything from adhesion to outer membrane permeability, these profound differences suggest that sessile cells in a biofilm comprise a phenotype that is as different from the planktonic phenotype as a seed is from a leaf.

To put the biofilm phenotype in context, we should crystallize our contention that this sessile phenotype is triggered soon after adhesion and as a precondition of biofilm formation since this process requires alginate synthesis. The sessile cells need alginate synthesis to form the matrices of the microcolonies in which they will live, and a very large number of secondary phenotypic changes occur thereafter in response to the conditions in each individual microniche occupied by a sessile cell. We should not assume that sessile bacterial cells respond to environmental stimuli, such as oxygen deprivation, in the same way as planktonic cells respond, because the basic phenotypes of the sessile cells are different. A muscle cell does not respond to oxygen deprivation in the same way as a lung cell. However, unlike most of the phenotypic differentiations of eukaryotic cells, the biofilm-planktonic variation in bacteria is reversible, and biofilms regularly undergo programmed detachment events (Fig. 4) in which cells growing in the biofilm phenotype revert to the planktonic phenotype and leave the sessile community.

THE REGULATION OF PHENOTYPIC VARIATION BY CHEMICAL SIGNALS

The very complex pattern of phenotypic variation in bacteria, in which both the planktonic and biofilm phenotypes can respond to environmental stimuli in sophis-

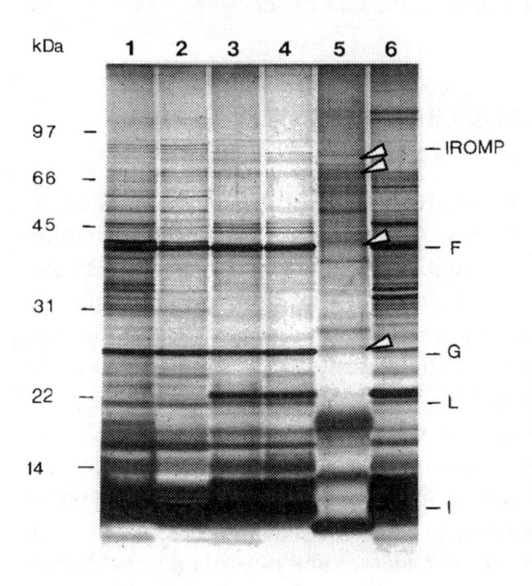

Figure 6. Polyacrylamide gel showing the OMPs extracted from the cell walls of planktonic and biofilm cells of *P. aeruginosa*. The OMPs seen in lanes 1 through 4 and 6 are those of planktonic cells grown in various iron concentrations, while those seen in lane 5 are those of biofilm cells of the same strain grown in the same medium as the planktonic cells whose OMPs are seen in lane 6. The OMPs of the planktonic phenotype are seen to differ very radically from those of the biofilm phenotype.

ticated ways, is under the control of simple chemical signals that resemble both hormones and pheromones in their functions. Biofilm formation in *P. aeruginosa* is controlled (5) by oxy-dodecanoyl homoserine lactone (ODdHL), coded for by the *lasI* gene, which may in turn be controlled by the polyphosphokinase (PPK) regulatory system (Kornberg, Rashid, and Davies, unpublished data). Reversion to the planktonic phenotype of this organism may be controlled by another quorum-sensing molecule called butryl homoserine lactone (BHL), and we expect that both biofilm formation and detachment will prove to be controlled by similar types of chemical signals in other gram-negative bacteria. Peptide signals seem to control the same activities in gram-positive organisms (8). These universal chemical signals were discovered because they control quorum sensing in planktonic cells (9), but we contend that their primary role in bacterial evolution may have been in their control of the formation and detachment of biofilms.

The search for biofilm control signals was initiated because we reasoned that the elaborate shapes of microcolonies and the open spaces that serve as water channels must be formed and maintained by a system of hormone-like signals that control cell replication and matrix production. Without such a system and in the absence of some form of control of cell growth and slime production, the biofilm would degenerate into a random and disorganized accretion of bacterial cells on a surface. This line of reasoning predicts, a priori, that many more phenotypes will be found in biofilms. An example would be a cell at the edge of a microcolony that had open access to nutrients and space to replicate, but that was constrained in replication by a signal molecule whose role it was to maintain open spaces in the biofilm. The uncontrolled replication of cells in a multicellular eukaryote constitutes cancer, and bacterial biofilms throughout the natural world resemble controlled tissue growth much more than they resemble cancer (Fig. 1). Perhaps the greatest damage done to microbiology by our unfortunate and exclusive emphasis on the planktonic phenotype has been the tragic delay in finding and describing these complex biofilm communities that clearly represent the apogee of prokaryotic development.

THE RECOVERY OF CULTURABLE BACTERIA FROM NATURAL ECOSYSTEMS

When bacteria are being enumerated by recovery and plating methods, enumeration is most accurate when planktonic cells are recovered from a fluid culture medium and spread on the surface of a chemically identical agar medium to determine how many colonies develop. Even in these idealized and totally unnatural systems, the cell count is often only one-tenth of the microscopic count, because planktonic cells clump, and clumps of cells yield only one colony. This is like counting plant seeds with a germination assay, because the planktonic cells in question are adapted to the liquid medium in which they are growing, and they grow well on the same medium with agar added. However, like plant seeds, some planktonic bacteria may fail to grow, because they have entered a dormant phase of growth or because local concentrations of a growth control signal may inhibit their replication. When we attempt to recover and enumerate planktonic bacterial cells from a mixed-species natural ecosystem in which they have entered the starvation-survival mode of growth, only a very small number of these cells may be able to grow on the agar chosen for the study. Some cells may be too far committed to the dormant mode of growth that results from starvation, and others may not find the nutrients offered in the medium suitable for rapid growth. We might liken this recovery operation to a germination assay of a very old sample of mixed plant seeds, some of which are too senescent to germinate, and some of which cannot germinate in the soil provided.

If we then consider the recovery and enumeration of sessile cells from well-established mixed-species natural biofilms, using the traditional "scrape and plate" methods, we see even more obstacles to success. Biofilms contain sessile cells in a very wide variety of different phenotypes that are modifications of the basic biofilm phenotype formed immediately after the adhesion of the initial planktonic cells to a surface (Fig. 1 and 4). In any biofilm some sessile cells will be found in the anaerobic phenotype, while others will have expressed the *rpoS* gene and entered senescence, and some will be in locations in which growth is inhibited by high concentrations of growth control signals. Some of these cells may fail to grow on the agar plate because they are in the wrong phenotype, and others may fail to grow because they require different nutrients. Because of their matrix-enclosed mode of growth (Fig. 1), many of the sessile cells will be recovered in coherent clumps (Fig. 7) that will yield only one colony, but some planktonic cells will also be recovered because biofilms shed these free-living cells continuously. Because they are in a phenotype that is well suited to growth on the recovery agar, the planktonic cells that are freshly shed from biofilms always dominate the recovered population, and these methods recover only a very small fraction of the sessile cells from the biofilm itself. In the analogy of the plant germination assay, this procedure is like cutting mature plants from a given area of mature growth and shaking a mixture of stems, leaves, and seeds onto a soil. The seeds will germinate if the soil is suitable, and some leaves may act as propagules, but the major part of the plant biomass will not be registered in the assay. The "scrape and plate" recovery and enumeration techniques yield little of value in ecological studies of biofilms, except some limited data on species composition and some indirect data on the rates at which planktonic cells of certain species are detaching from the biofilm.

Direct observations, during the past 9 years, have shown that biofilms are highly organized multicellular communities of bacteria, with many of the characteristics previously thought to be the exclusive properties of eukaryotic organisms. These communities have primitive circulatory systems, and their component cells show a measure of physiological diversity and functional specialization. The initiation and maintenance of these complex structures are controlled by a sophisticated system of chemical signals whose function is analogous to the hormones and pheromones of eukaryotic organisms, and we now know that biofilms are not simply random accretions of bacterial cells in an unstructured matrix. The distribution of biofilms is accomplished by the programmed detachment of planktonic cells whose function is similar to that of plant seeds, and we can regard these creatures with suitable

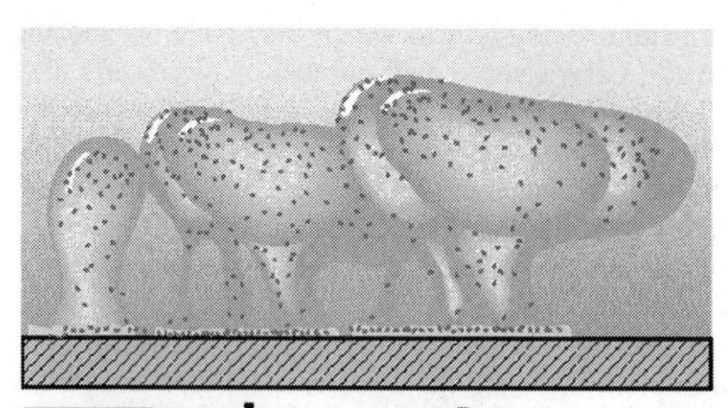

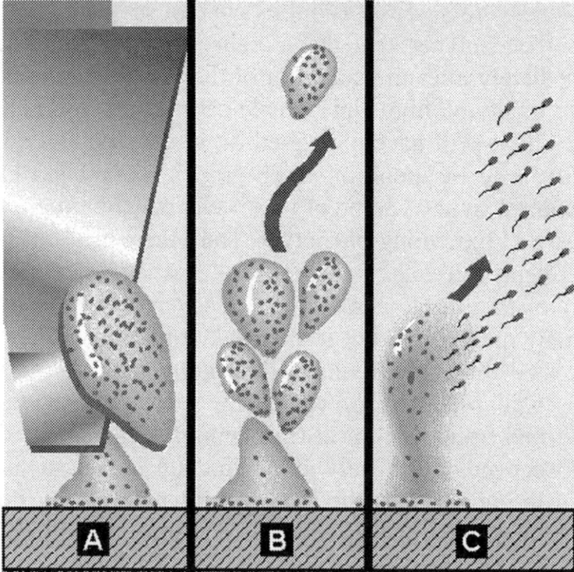

Figure 7. Cartoon showing the process of scraping to remove biofilms from surfaces for the enumeration of their component bacterial cells. The sterile scalpel blade removes some large microcolonies intact, fragments other microcolonies, and encounters some microcolonies that are in the process of shedding planktonic cells as a programmed activity. Each large fragment of a microcolony will yield a colony on plating, independent of its size or of the number of cells it contains, as will each planktonic cell harvested in this procedure. For this reason, shedding biofilms yield huge counts in scrape and plate assays, while nonshedding biofilms yield very low numbers, even though they contain millions of actively growing cells.

awe because this most ancient of all biological communities still dominates the biosphere.

A MODEST PROPOSAL FOR A BETTER SYSTEM FOR STUDYING MICROBIAL ECOLOGY

The excuse most commonly offered for our misplaced emphasis on planktonic cells in culture is that the subjects of our arcane scientific speciality are microscopic and that we could not see how they really grow in nature. This is simply not true. We can see bacterial biofilms on our teeth, on the surfaces of recovered medical devices, and on the coral reefs that delight us on tropical vacations. High school science projects have described the development of mixed bacterial/algal biofilms on the walls of tropical fish tanks, and the surfaces of the rocks in any neighborhood stream provide a biofilm that is both visible and slippery to the touch. Microscopes capable of resolving bacterial biofilms have been available since the mid-1700s, and the modern confocal scanning laser microscope has been used since the 1960s to image subcellular structures deep within plant and animal cells. Modern confocal microscopes are now virtually ubiquitous, entirely capable of providing a three-dimensional image of living microbial communities at a resolution of at least 0.6 microns and capable of imaging these communities on opaque surfaces.

Even organisms that cannot be cultured can be identified by PCR, and fluorescent in situ hybridization (FISH) oligonucleotide probes keyed to signature sequences in their genomes can be used to identify cells of that species in biofilms. These FISH probes cannot always be used to identify individual species in living biofilms because of penetration problems, but their use on cryosectioned biofilm samples has yielded excellent data (Fig. 8). Fluorescent antibodies can be used to identify cells of a particular species, or to locate antigenic cell products such as toxins and enzymes. The armamentarium of modern probes is virtually limitless, and some probes detect respiratory activity (Fig. 9), while others detect the viability of individual cells on the basis of membrane integrity (Fig. 10). Perhaps the most sophisticated in situ probes are reporter constructs in which reporter sequences (e.g. green fluorescent protein [GFP]) are inserted downstream from the promoter of interest, and the reporter system is turned on when the gene is up regulated (Fig. 5). Biofilms are complex communities in which the location of individual cells in any one of a vast spectrum of different microniches determines the phenotype that

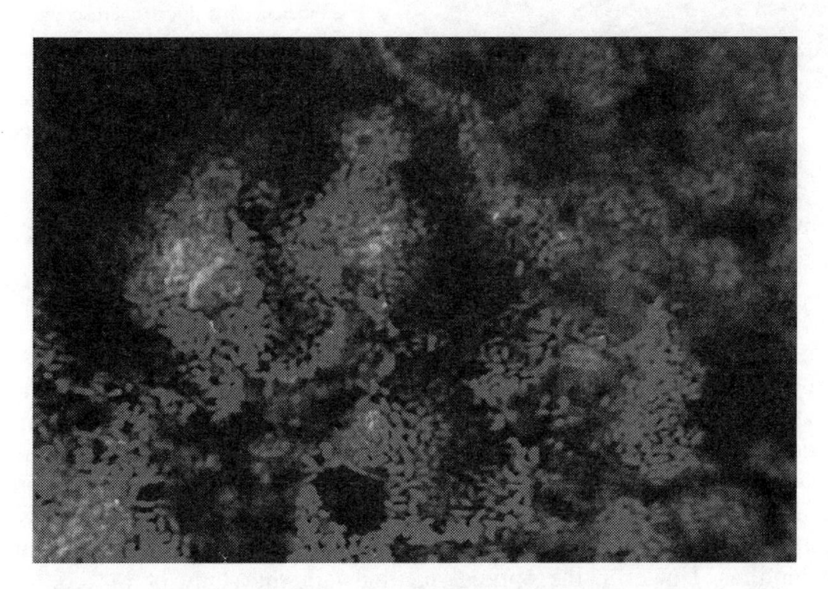

Figure 8. CSLM micrograph of a mixed biofilm containing cells of both *P. aeruginosa* and *Klebsiella pneumoniae*. In this x-z optical section, perpendicular to the colonized surface, the pseudomonas cells stain red and the klebsiella cells stain green with the probes used in this study. Note that the klebsiella cells tend to occupy the interior space of mushroom-like microcolonies formed by the pseudomonas cells, indicating some form of "tissue-like" organization.

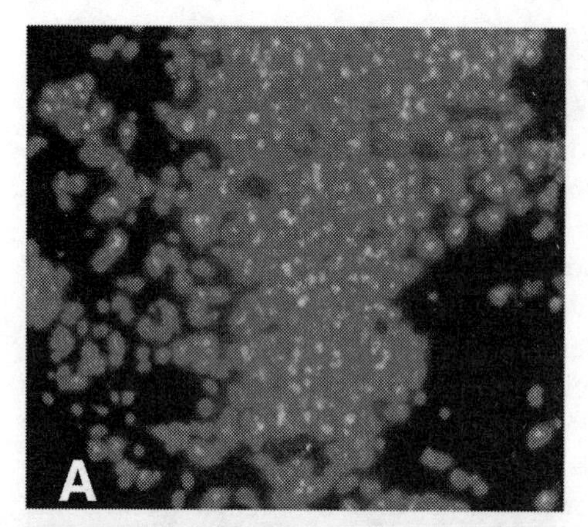

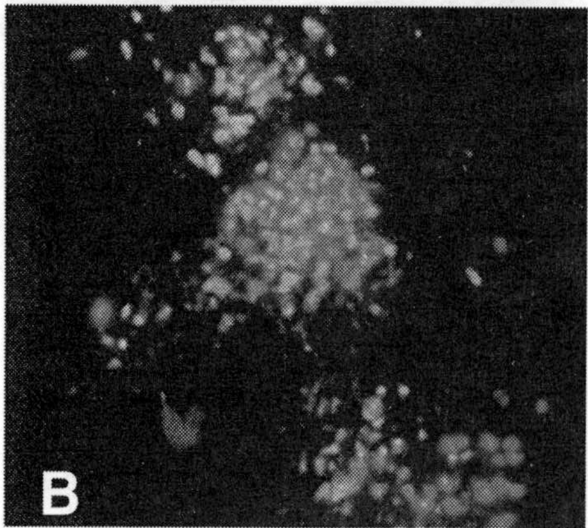

Figure 9. CSLM micrograph of frozen sections of biofilms of cells of *P. aeruginosa* that have been stained with CTC, which is a tetrazolium compound that is reduced to an orange-colored formazan by living bacteria capable of respiration. This x-z section, perpendicular to the colonized surface, shows a majority of living cells in the untreated biofilm (A), while only few living cells remain, in clumps within discrete microcolonies, in the biofilm (B) that has been treated with a biocide. This direct microscopic method, especially when combined with whole biofilm respirometry, gives very accurate data in the study of the efficacy of antibacterial agents in killing sessile bacteria.

the sessile cell will adopt, so morphological examinations of individual cells are really de rigeur in the study of these sessile populations.

The twin weaknesses of morphological methods are the fact that they are forced to concentrate on a limited area of the colonized surface and the fact that they can give only limited chemical information on integrated physiological processes. Biofilm engineers have developed a series of very accurate physical probes (microelectrodes) that can measure chemical parameters (e.g., dissolved oxygen concentration) with a resolution of <15 μm and can therefore map these parameters (Fig. 2). These maps of chemical parameters can be related to images showing the locations of microcolonies, and physiological conclusions such as the essential anaerobiasis of the centers of microcolonies can be supported by data generated in living biofilms. However, the optimal methods for the study of the cooperative physiology of biofilm populations are respirometry and other direct techniques that measure the chemical changes wrought by biofilms growing over large areas of colonized surfaces. If microscopy of selected areas of a colonized medical device showed only nonrespiring cells and cells whose membranes were no longer intact but respirometry of the whole device showed the consumption of oxygen and the

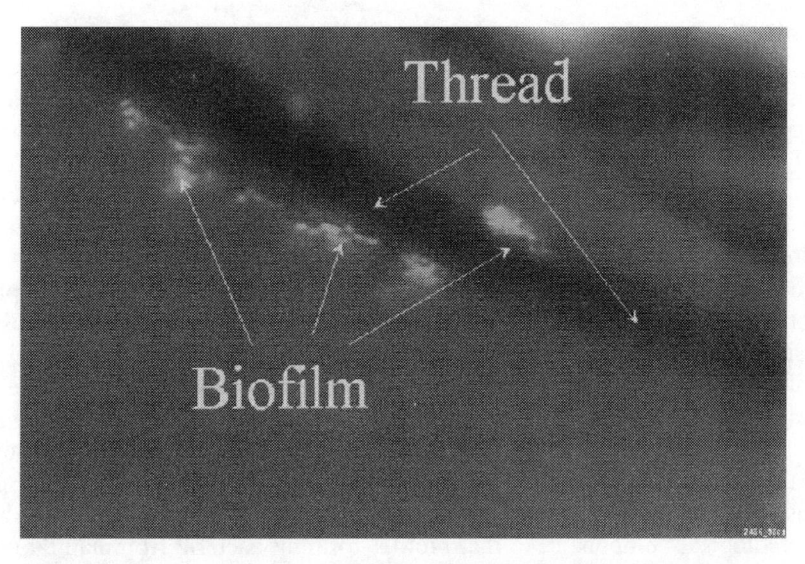

Figure 10. CSLM micrograph of a thread from the sewing cuff of a mechanical cardiac valve that had been incubated with cells of *Staphylococcus epidermidis* to assess its putative resistance to bacterial colonization and biofilm formation. The bacterial cells have adhered to the material and have formed a biofilm that is clearly visible in this living fully hydrated preparation. The viability of most of the sessile cells in this nascent biofilm is attested by the green color formed because the BacLite "live-dead" probe is excluded by their intact membranes. However, one cell is red because its damaged membrane failed to exclude this probe, and two more cells are orange or yellow because their membranes are compromised. Direct counts of living bacterial cells on any surface of interest can readily be made by this method.

generation of CO_2, we would conclude that some areas of the biofilm remained alive. The study of large biofilm populations by respirometry allows us to detect the final gaseous products of cooperative metabolic activities, such as respiration or methane generation, and the values obtained are the averages of hundreds or thousands of microcolonies over a very large area (e.g., 2 in^2).

VALETE

Never in the recent history of any biological science has one of these sciences been faced with such an urgent necessity of updating its concepts and methods, in order to correct a seriously mistaken emphasis and to accommodate radically new data. Microbiologists have discovered that the planktonic phenotype that they have studied so assiduously for >150 years is only one of many phenotypes adopted by bacteria in nature, and that this favored phenotype is neither the most prevalent nor the most interesting of these phenotypes. The starvation-survival phenotype is certainly the most prevalent form adopted by bacteria in the biosphere, and the biofilm phenotype is clearly the most complex mode of bacterial growth and the form that is more commonly encountered in the environments in which we live. The planktonic phenotype received disproportionate attention because it is most commonly found in acute human disease and in vitro cultures, but the starvation-survival phenotype and the biofilm phenotype are of much greater importance in microbial ecology.

This flood of new perceptions concerning bacterial phenotypes causes current practitioners of the important science of microbial ecology to question the traditional recovery and culture methods of studying bacteria and to rely increasingly on direct observations of living populations. These direct methods may be as simple-minded as confocal microscopy of living biofilms, as powerful as physiological studies of living mixed-species communities, or as sophisticated as PCR analysis of the nucleic acids extracted from mixed-species populations growing in natural environments. Their appeal, and their common thread, is that they are direct studies of whole populations without the intermediate selection and culture of the rare phenotypes and the rare species that happen to grow on selected agar surfaces when they are ripped from their natural habitats and deposited on recovery plates. Because the bacterial diseases that affect humans in the developed world have a major ecological component, in that biofilm-forming bacteria from natural ecosystems are invading our hospitals (3), the considerable resources of medical institutions (e.g., National Institutes of Health) will also now be thrown into this new activity. As in all systems in which a major error is corrected, we can now look forward to a renaissance in microbial ecology that will surpass even the phenomenal bursts of activity in bacterial physiology and in bacterial genetics that have excited microbiologists during the past century.

Acknowledgment. This work was supported by cooperative agreement ECD-8907039 from the National Science Foundation.

REFERENCES

1. **Costerton, J. W., Z. Lewandowski, D. E. Caldwell, D. R. Korber, and H. M. Lappin-Scott.** 1995. Microbial biofilms. *Annu. Rev. Micro.* **49:**711–745.
2. **Costerton, J. W., Z. Lewandowski, D. DeBeer, D. Caldwell, D. Korber, and G. James.** 1994. Biofilms, the customized microniche. *J. Bacteriol.* **176:**2137–2142.
3. **Costerton, J. W., P. S. Stewart, and E. P. Greenberg.** 1999. Bacterial biofilms: a common cause of persistent infections. *Science* **284:**1318–1322.
4. **Davies, D. G., and G. G. Geesey.** 1995. Regulation of the alginate biosynthesis gene *algC* in *Pseudomonas aeruginosa* during biofilm development in continuous culture. *Appl. Environ. Microbiol.* **61:**860–867.
5. **Davies, D. G., M. R. Parsek, J. P. Pearson, B. H. Iglewski, J. W. Costerton, and E. P. Greenberg.** 1998. The involvement of cell-to-cell signals in the development of a bacterial biofilm. *Science* **280:** 295–298.
6. **deBeer, D., P. Stoodley, F. Roe, and Z. Lewandowski.** 1994. Effects of biofilm structures on oxygen distribution and mass transport. *Biotechnol. Bioeng.* **43:**1131–1138.
7. **Deretic, V., M. J. Schurr, J. C. Boucher, and D. W. Martin.** 1994. Conversion of *Pseudomonas aeruginosa* to mucoidy in cystic fibrosis: environmental stress and regulation of bacterial virulence by alternative sigma factors. *J. Bacter.* **176:**2773–2780.
8. **Dunny, G. M., and B. A. Leonard.** 1997. Cell-cell communication in Gram-positive bacteria. *Ann. Rev. Microbiol.* **51:**527–564.
9. **Fuqua, W. C., E. P. Winans, and E. P. Greenberg.** 1994. Quorum sensing in bacteria: The Lux R–Lux I family of cell density-responsive transcriptional regulators. *J. Bacteriol.* **176:**269–275.

10. **Jensen, E. T., A. Kharazmi, K. Lam, and J. W. Costerton.** 1990. Human polymorphonuclear leukocyte response to *Pseudomonas aeruginosa* biofilms. *Infect. Immun.* **58:**2383–2385.
11. **Kudo, H., K.-J. Cheng, and J. W. Costerton.** 1987. Interactions between *Treponema bryantii* and cellulolytic bacteria in the *in vitro* digestion of straw cellulose. *Can. J. Microbiol.* **33:**244–248.
12. **Lewandowski, Z., S. A. Altobelli, and E. Fukushima.** 1993. NMR and microelectrode studies of hydrodynamics and kinetics in biofilms. *Biotechnol. Prog.* **9:**40–45.
13. **Nickel, J. C., I. Ruseska, J. B. Wright, and J. W. Costerton.** 1985. Tobramycin resistance of *Pseudomonas aeruginosa* cells growing as a biofilm on urinary catheter material. *Antimicrob. Agents Chemother.* **27:**619–624.
14. **Stoodley, P., I. Dodds, Z. Lewandowski, A. B. Cunningham, J. D. Boyle, and H. M. Lappin-Scott.** 1999. Influence of hydrodynamics and nutrients on biofilm structure. *J. Appl. Microbiol.* **85:** 19S–28S.

Nonculturable Microorganisms in the Environment
Edited by R. R. Colwell and D. J. Grimes

Chapter 10

Survival, Dormancy, and Nonculturable Cells in Extreme Deep-Sea Environments

Jody W. Deming and John A. Baross

An important conclusion from the initial application of 16S rRNA sequence analysis to problems in microbial ecology (263) is that most microorganisms (and frequently the dominant ones in an environment) do not exist in culture collections (6, 59, 124, 158). The basis for the incongruence between cultivated and naturally occurring strains has been attributed to microbial competition during the culturing process, to cell dormancy in nature due to starvation and other stresses, and (or) to insufficient understanding of the combined physical, chemical, and nutritional conditions required by the dominant strains, especially by symbionts and members of consortia. The ability to culture only a minor fraction of the microorganisms occurring in nature is particularly perplexing in specialized environments where limited diversity is anticipated and intensive culturing efforts have been pursued for decades, as in the famous hot springs of Yellowstone National Park (15, 336). On the other hand, the discovery that potentially key microbial players in pelagic marine environments do not yet exist in culture (70, 71, 75, 116, 118, 125, 237, 297) is less surprising. The oceans are spatially extensive, remote to sample, chemically and physically diverse on many scales, and frequently characterized by extreme conditions of temperature, pressure, or oligotrophy (300), promoting starvation and dormancy (230). Together these conditions make obtaining representative cultures, or even samples, of the major microbial players difficult and the results sometimes counterintuitive. For example, anaerobic microorganisms are now known to live in oxygenated sections of the water column. Biologically produced methane has been detected in such waters (208, 301), as have significant communities of uncultured (activity unknown) archaeal organisms, identified by rRNA sequence analysis as possibly related to known anaerobic genera (70, 116, 118, 237).

A basic premise of this chapter is that microbial diversity is greater in benthic and deep-sea environments than in upper layers of the ocean. Ranges in temperature, pressure, and sources (terrestrial and marine) and concentrations of organic

Jody W. Deming and John A. Baross • School of Oceanography, Box 357940, University of Washington, Seattle, WA 98195.

material are greater. Cold, deep waters rank among the most oligotrophic environments known, yet freshly delivered pulses of phytodetrital aggregates (213), significant concentrations of reactive organic colloidal material in abyssal seawater (341), and foraging abyssal animals (177) provide organic oases and disturbances to steady state that are expected to enhance diversity (105, 119). A wide variety of animal-associated habitats abound in the benthic environment, while a complete range of redox conditions persist within the seabed. With each year of exploration, the spatial and physicochemical bounds of submarine hot springs are adjusted upwards (181). In addition to providing electron acceptors and those elements required to support the growth of virtually all known microbial nutritional groups (160, 183, 208) (photoautotrophs being the apparent exception [but see reference 327]), vent environments provide a marked temperature contrast to the rest of the deep sea. Fluids, rocks, and sediments range from ambient seawater temperature (2°C) to >400°C (68), providing habitats for all known thermal groups of microorganisms from strict psychrophiles (229) to superthermophiles (84). The concept that ridge spreading centers may also provide vast subsurface microbial habitats characterized by expansive gradients in temperature, pressure, chemical and salt conditions draws increasing attention and evidence in support of it (17, 19, 20, 23, 69, 84, 128, 141, 150, 208, 311, 313, 342).

Unfortunately, direct measures of the true diversity of viable microorganisms in the deep ocean are few. Much early culturing work selected strongly for surface-derived bacteria by virtue of incubation conditions foreign to the cold deep sea (warm temperatures, atmospheric pressure). More recent culturing work, better simulating in situ conditions, has focused on the isolation and characterization of two groups of microorganisms: psychrophilic, barophilic copiotrophs (growing optimally at low temperature and deep-sea pressure in organic-rich media) from the cold deep sea; and hyperthermophiles (growing above 90°C) from hot vent sites. Obvious examples of a greater diversity revealed by methods other than culture include symbiotic or mutualistic associations between microorganisms and specific animals at vents and elsewhere in the deep sea (48, 51, 109, 223) and the extensive microbial mats observed at hydrothermal sites and cold seeps (163, 242) and at productive sediments in upwelling regions (111, 120). Only recently have molecular techniques been applied to whole samples of various free-living microbial communities from deep environments for the purpose of assessing phylogenetic diversity (34, 116, 139, 187, 234, 330). These results and the known diversity of deep-sea habitats make obvious the conclusion that only a minor fraction of the microorganisms resident in the deep sea, as in other environments, has been obtained in culture.

Our goal in this chapter is to examine potential reasons for the nonculturability of microorganisms persisting under the extreme physical and chemical conditions that typify deep-sea environments, first providing a perspective on the diversity of microorganisms (cultured or not) that are active in the deep sea. We define nonculturable microorganisms broadly as viable cells in the environment that resist culturing in the laboratory; they may or may not be metabolically active in situ (dormant) and may or may not be recoverable (culturable) with a change in conditions. Cells active in situ but recalcitrant to culturing are distinguished from cells that are dormant (inactive in situ due to starvation or other stresses) but recoverable. We consider active microorganisms as cells reproducing in situ or consuming (exogenous) nutrients from the environment. Dormant cells are either "frozen" metabolically, as in the case of spores or cells exposed to suboptimal temperatures, or have shifted from exogenous to endogenous metabolism to survive starvation conditions. However, the lines between active and dormant cells can be blurred as a result of time-dependent processes. Nonculturable cells without measurable metabolic activity or the ability to reproduce can be characterized as either dormant or dead (see chapter 18). Differentiating dead cells from dormant cells is difficult, requiring application of specific molecular or biochemical tests to individual cells. Genetically controlled survival mechanisms associated with stress responses figure strongly in dormancy. They will be addressed in some detail, given the working hypothesis that survival mechanisms (like reproductive processes) are essential to evolutionary fitness across the biosphere (9), but especially for microorganisms in the deep sea.

In a review of evidence for active and dormant nonculturable cells, we consider both hydrothermal vent (Fig. 1) and cold deep-sea (Fig. 2) habitats. We examine strategies for survival under conditions unfavorable to growth, including extremes in temperature and pressure and nutrient limitation, with brief reference to other stresses (oxygen, salt, metal toxicity). By considering nonculturable microorganisms of marine origin, we depart from the original concept of nonculturable cells as laboratory-reared pathogens that fail to form colonies after exposure to a natural environment or conditions mimicking one (as reviewed by Oliver [258]). Though each of us helped to identify this phenomenon in early research on the human health implications of nonculturable and (or) starving cells in natural milieu (25, 50, 91), here we hope to convince that the ecological and evolutionary implications of nonculturable cells in deep-sea and subsurface marine environments are more profound.

EVIDENCE FOR MICROORGANISMS ACTIVE BUT NONCULTURABLE IN THE DEEP SEA

The paradigm of bacterial abundance decreasing with increasing ocean depth holds today (85, 90) in spite of the occurrence of organic oases at great depth (92, 93, 185, 213, 363). However, the discovery that the bacterial fraction of total (infaunal) benthic biomass increases with greater ocean depth (292) may not be widely appreciated. A graph of data presented in tabular form by Rowe et al. (292) illustrates the point (Fig. 3). Biomass of the abyssal benthos is thus dominated by bacteria (and possibly protozoa, which were not measured [292]), which also appear to utilize available carbon sources more efficiently than their counterparts in shallower settings (283). Yet the relative fractions of active, dormant, and dead cells remain unquantified for any given habitat. In this section, we consider the evidence for a diversity of microorganisms active (cultured and nonculturable) in the deep sea.

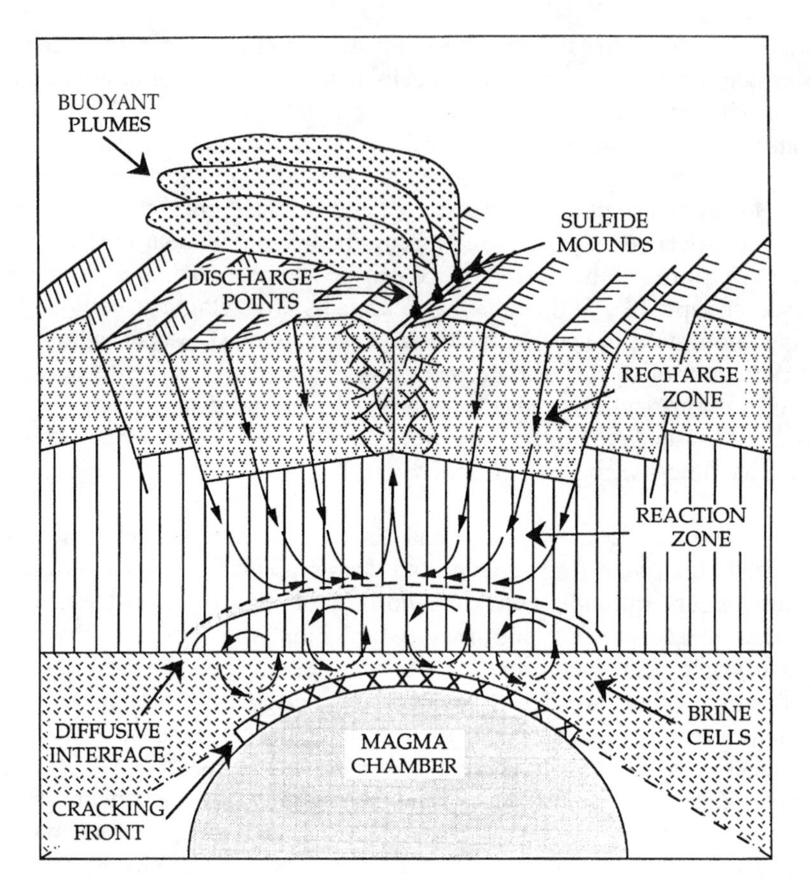

Figure 1. Examples of large-scale benthic and subsurface habitats for microorganisms at hydrothermal vents in the deep sea. Modified from Bischoff and Rosenbauer (36).

Diversity of Cultured Microorganisms Adapted to Deep-Sea Conditions

A limited number of strains clearly evolved for optimal activity under the extreme temperatures and pressures of the cold deep sea have been cultured and identified (Table 1; Fig. 4). The overwhelming influence of biotechnological and evolutionary interests in hyperthermophilic strains is apparent in the greater number of such isolates obtained from deep-sea hydrothermal vents (Table 1) (reviewed in references 19, 21, and 307). Increased interest in culturing microorganisms from deep polar oceans (80, 90, 343) and deep trenches (186, 218, 295, 314, 315), however, may soon shift the balance. The current list of microorganisms cultured from deep-sea samples, in general, would be more extensive if unidentified strains and isolates that also grow easily or preferentially at conditions foreign to the deep sea were included. The lengthy tables provided in reviews by Karl (183), Baross and Deming (19) and Baross and Holden (21) testify to the physiological diversity of culturable microorganisms from the more intensely studied (and variable) hydro-

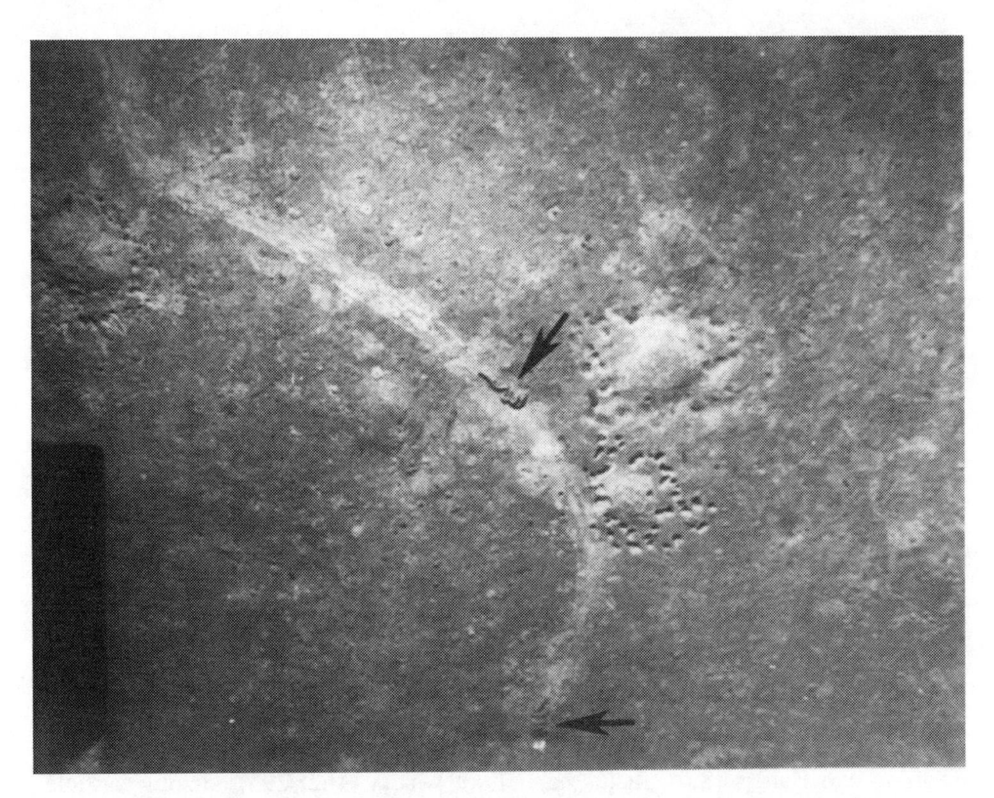

Figure 2. Examples of small-scale benthic habitats for microorganisms on the cold abyssal seafloor, from an in situ photograph obtained at 4715 m in the Bay of Biscay by D. Desbruyeres (79, 96): solid substrata protruding into the benthic boundary layer (worm tubes, seen as point shadows on the seafloor), echiurian mounds and associated pits of likely elevated organic content (351), freshly deposited holothurian fecal coils (arrows), and the undisturbed, aerobic sediment-seawater interface of recently arrived detrital material (darker areas). Additional subsurface layers characterized by increasing oxygen limitation and changing electron acceptors (85) or irrigated by worm tubes (3) not visible. Width of photo, 2 m.

thermal vent sites, although the majority can be categorized as either chemolithoautotrophic (hereinafter called chemoautotrophic) sulfur-oxidizers, chemoorganotrophs (hereinafter called heterotrophs), or methanogens. Reports of numerous culturable bacteria from cold deep-sea environments have also appeared regularly in the literature since the last century (49). These have focused almost exclusively on heterotrophic isolates (e.g., 278) and in particular (in recent decades) on those that are psychrophilic (e.g., 60, 144, 218, 252) and/or barophilic (for reviews see 85, 354, 355). Recent work by Takami et al. (315) and Vetriani et al. (330) reaffirms the considerable diversity of microorganisms that ultimately accumulate in, and can be recovered from, deep-sea sediments, even if they are incapable of activity in situ.

It is useful to realize that once appropriate protocols were developed (97), psychrophilic barophilic copiotrophs, elusive since the last century (49), were isolated

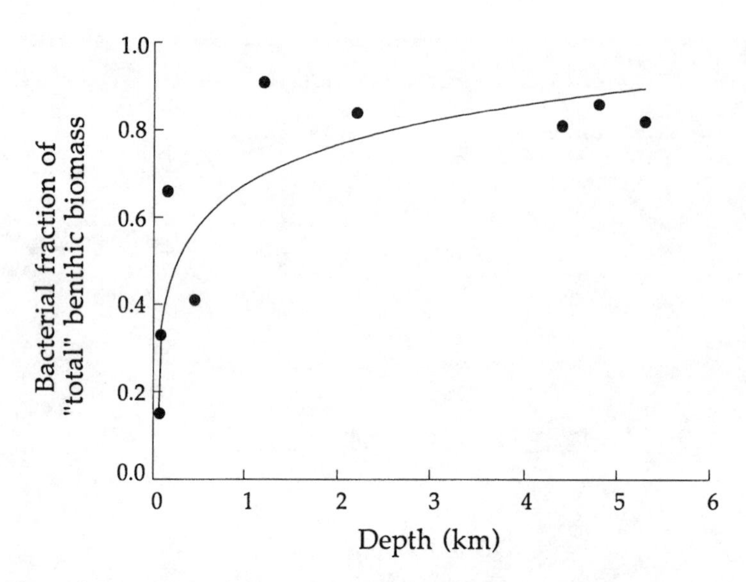

Figure 3. Bacterial fraction of "total" benthic biomass (excluding Protozoa which were not measured) as a function of depth in the ocean (data from Table 1 in Rowe et al. [292]).

readily from every cold deep-sea sample collected for that purpose (reviewed by Deming and Baross [85]), including bottom waters (162, 295), sinking particles and fecal pellets (78, 80, 95), sediments from both polar (80, 144) and low-latitude basins (78, 355), fresh benthic invertebrate guts (86, 92), and decaying amphipods (358). Evolving culturing protocols (81) continue to yield new isolates (186, 295). Although sample warming was shown fatal to psychrophilic barophilic bacteria during recovery in the laboratory, sample decompression was not (87, 356). In fact, the logarithmic growth phase of all known psychrophilic barophiles, even obligate strains (95, 186, 295, 359) and those obtained in the complete absence of decompression (165), is unaffected by repeated periods (if brief and at low temperature) of decompression and recompression.

Interestingly, no psychrophilic barophiles have yet been obtained in sustainable culture from shallow regions of the ocean (78). Nor have strict psychrophiles (unable to grow above 15°C, as defined by Morita [229]), regardless of their pressure requirements, been easy to isolate from shallow environments (sea ice being an exception; see 42, 131, 145, 179). Even in permanently cold (polar) water masses, psychrotolerant bacteria (previously known as psychrotrophs [229]; able to grow over a wider range of temperatures, from 0 to >20°C) appear to be the dominant culturable forms (22, 239). We return to this apparent linkage between psychrophily and barophily (80, 218), which may also pertain at the high end of the biological temperature scale (between hyperthermophily and barophily [84]), in later sections.

In terms of phylogenetic diversity, all of the known psychrophilic barophiles cultured and examined by 16S rRNA criteria fall within a closely related group of gamma Proteobacteria (Table 1) (76, 218). This measure of limited diversity is

Table 1. Strains of deep-sea microorganisms[a] obtained in pure culture and identified

Strain designation	Deep-sea origin	Reference(s)
Bacteria (psychrophilic, barophilic, heterotrophic[b])[c]		
Alteromonas strain F1A[b]	Seawater, Atlantic abyssal basin	348
Colwellia hadaliensis	Sinking particles, Puerto Rico Trench	95
Moraxella strain JT761	Seawater, Japan Trench	295
Moritella japonica	Sediment, Japan Trench	249
Moritella yayanosii	Sediment, Mariana Trench	248
Photobacterium profundum	Sediment, Ryukyu Trench	251
Shewanella benthica	Invertebrate gut, Atlantic abyssal basin; sediment, Pacific Ocean	94, 215, 250
Shewanella violaceae	Sediment, Pacific Ocean	250
Bacteria (hyperthermophilic, heterotrophic[b])		
Thermosipho melanesiensis	Mussel gills, Lau Basin	10
Archaea (hyperthermophilic, methanogenic)[d]		
Methanococcus fervens	Sediment, Guaymas Basin	167
Methanococcus jannaschi	Smoker wall, 21°N, East Pacific Rise	173
Methanopyrus kandleri	Sediment, Guaymas Basin	45, 154, 202
Methanococcus strains AG86, CS1	Sediment, Guaymas Basin	172, 361
Methanococcus vulcanius	Smoker wall, East Pacific Rise	167
Archaea (hyperthermophilic, mixotrophic)		
Archaeoglobus profundus	Sediment, Guaymas Basin	44
Archaeoglobus veneficus	Smoker wall, Snake Pit site, Mid-Atlantic Ridge	156
Archaea (hyperthermophilic, heterotrophic[b])[d]		
Desulfurococcus strains S, SY	Smoker wall, 11°N, East Pacific Rise	166
Pyrococcus strain GBD	Smoker wall, Guaymas Basin	166
Pyrococcus endeavouri ES4, AL2	Flange solids, Endeavour site, Juan de Fuca Ridge	271, 284, 285; J. A. Baross and R. J. Pledger (unpublished data)
Pyrodictium abyssi	Sediment, Guaymas Basin; smoker fluid, N. Fiji Basin	273, 307; 104, 217
Pyrolobus fumarii[e]	Smoker wall, TAG site, Mid-Atlantic Ridge	37
Staphylothermus marinus	Smoker fluid, 11°N, East Pacific Rise	362, 107
Thermococcus barossi	Flange solids, Endeavour site, Juan de Fuca Ridge	101
Thermococcus chitonophagous	Smoker wall, Guaymas Basin	157
Thermococcus fumicolans	Smoker wall, N. Fiji Basin	126
Thermococcus hydrothermalis	Smoker wall, 21°N, East Pacific Rise	127
Thermococcus paralvinelli ES1	Vent polychaete, Endeavour site, Juan de Fuca Ridge	270; J. A. Baross and R. J. Pledger (unpublished data)
Thermococcus peptonophilus	Vent fluid, Izu-Bonin forarc, Western Pacific	129

Continued next page

Table 1. *Continued*

Strain designation	Deep-sea origin	Reference(s)
Thermococcus profundus	100°C diffuse-flow fluid, Mid-Okinawa Trough	194
Thermococcus siculi	110–130°C diffuse-flow fluid, Mid-Okinawa Trough	134
Thermococcus strain AL1	Smoker fluid, Endeavour site, Juan de Fuca Ridge	284, 285
Thermococcus strain DT1331	Smoker wall, Minami-ensei Knoll	204

[a]Indigenous to the deep sea according to extreme temperature and (or) pressure requirements.
[b]Heterotrophic = chemoorganotrophic.
[c]Other strains appear closely related to *S. benthica* or *C. hadaliensis* by 16S rRNA criteria (76) or are under study (54, 186, 315).
[d]When tested, these have been shown to be barophilic or barotolerant (84).
[e]*P. fumari* grows to 113°C, the highest recorded temperature for a sustainable pure culture.

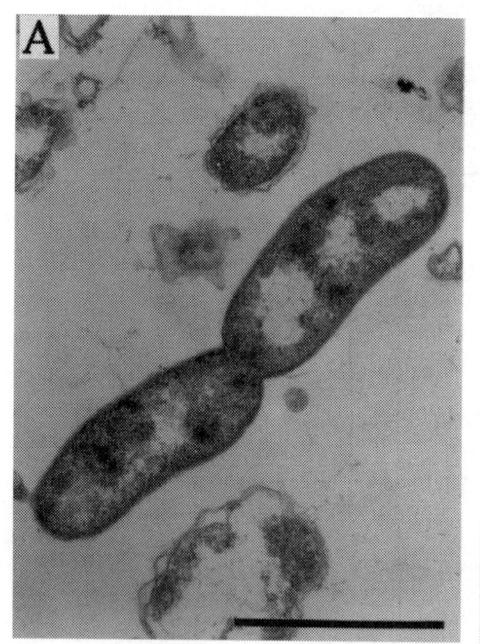

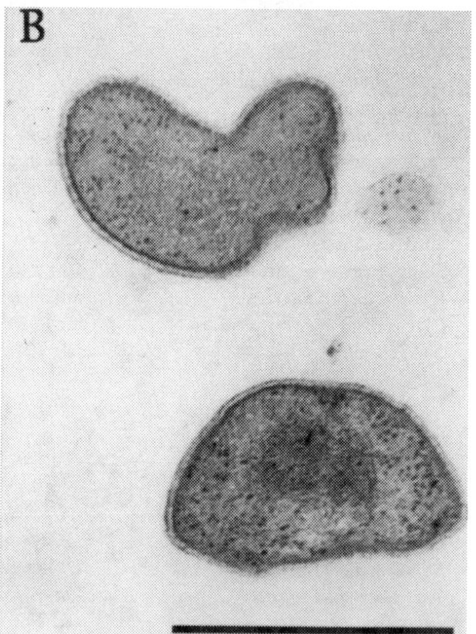

Figure 4. Examples of cultured and identified microorganisms indigenous to the deep sea (Table 1): (A) transmission electronic microscopy (TEM) of the psychrophilic and obligately barophilic bacterium, *Colwellia hadaliensis*, from the Puerto Rico Trench, cultured under simulated in situ conditions of 2°C and 740 atm (95); and (B) TEM of the hyperthermophilic (66 to 110°C; 271) and barophilic (272) archaeon, *Pyrococcus endeavouri* ES4 (271; J. A. Baross and R. J. Pledger, unpublished data), from the Endeavour Segment of the Juan de Fuca Ridge, cultured at 80°C and 3 atm. Bars, 1 μm.

likely an artifact of the singular use by all investigators of organic-rich heterotrophic media; a greater physiological diversity of psychrophilic barophiles will emerge as the nutritional menu provided in isolation media is diversified (85). Indeed, a greater diversity of deep-sea microorganisms (heterotrophs, chemoautootrophs, methanotrophs, methanogens, and halotolerant bacteria) has already been discovered in this way at a variety of temperatures at hydrothermal vent sites (Table 1; 19, 56, 63, 183, 188).

Diversity of Nonculturable Microorganisms Active in the Deep Sea

Evidence for greater phylogenetic, morphological, and physiological diversity among microorganisms active under extreme conditions of the deep sea comes from studies of microbial processes (e.g., colonization, endosymbiosis, chemical reductions, and oxidations) where culturing, if attempted, was either unsuccessful or accounted for only a minor fraction of the biomass and morphologies detected. The best known examples of cells active in the deep sea but resistant to culturing derive from hydrothermal vents (Table 2). There, and subsequently at hydrocarbon seeps and other reducing environments in the cold deep sea (52, 93, 108, 304), intensive research on symbiotic relationships between microorganisms and invertebrates, using a combination of enzyme, radiolabel, and molecular analyses, has revealed an extraordinary world of novel endosymbionts (Fig. 5) of varied physiological and biochemical capabilities, all actively providing nutritional support to their hosts and all unculturable (48, 51, 109). The extensive, thick mats of *Beggiatoa* and other filamentous microorganisms, clearly active in situ in the rapid colonization of recently formed vent sites (141, 243, 321), have also resisted culturing for the most part (163, 242, 316). Large volumes (flocs) of filamentous forms, streaming continuously into the ocean from subsurface habitats of unknown conditions and dimensions, have been observed from submersibles during exploration of new seafloor eruption sites (141, 321) but not obtained in culture. Animal and metalliferous sulfide surfaces at vents are densely colonized by a variety of morphological forms not known in culture (18, 47, 161, 180, 329; Fig. 6). Significant populations of particle-associated and suspended microcolonies that stain blue with the MnIV-specific dye benzidine hydrochloride (106) have been observed microscopically (Fig. 6) in regions of buoyant hydrothermal plumes (Fig. 1) known to support significant rates of microbial methane oxidation (65, 210), hydrogen oxidation (225), and manganese and iron scavenging (57, 210, 216, 346).

In general, the above measures of active but uncultured microorganisms derive from low-temperature vent environments. However, biochemical and activity measurements have also revealed hot habitats where hyperthermophilic archaea are present in great densities and (or) dominate the microbial community (reviewed by Baross and Deming [18, 19]). The most recent examples were obtained from flanges, outcroppings of sulfidic material (<1 to >10 m in width) that occur on large sulfide mounds (20 to >100 m in height [Fig. 1]) at the Endeavour Segment and other sites along the Juan de Fuca Ridge (68). Unlike the smokers on these mounds that release hot hydrothermal fluid rapidly and directly into the ocean (Fig. 1), flanges trap an inverted pool of hot fluids that diffuse upwards through a porous

Table 2. Examples of active but uncultured microorganisms at hydrothermal vent sites (Fig. 1)

Type of organism	Measure of activity	Deep-sea origin	Reference(s)
Endosymbiotic sulfur-oxidizing chemoautotrophs[a]	Microscopic data (Fig. 5), enzyme and tracer assays, stable isotope data, DNA analyses	Vent animals	48, 51, 241
Endosymbiotic aerobic methanotrophs	Microscopic data, enzyme and tracer assays, stable isotope data, DNA analyses	Vent animals	51, 52, 109, 241
Epibiotic aerobic methanotrophs	Microscopic data (Fig. 6), tracer assays	Vent animal surfaces	64
Epibiotic organisms, unknown trophic status	Microscopic data (Fig. 6), chemical profiles, DNA analyses	Vent animal and sulfide surfaces	18, 47, 161, 180
Free-living mixotrophic mats of *Beggiatoa*, *Thioploca*, S-oxidizers and unknowns	Visible mats and flocs, microscopic data, enzyme assays, DNA analyses, culture analogs[b]	Vent sediments, seafloor eruption sites	135, 163, 234, 242, 243, 316, 321
Free-living aerobic methanotrophs	Microscopic data, tracer assays, chemical profiles[b]	Buoyant vent plumes	65, 210; J. A. Baross (unpublished data)
Free-living Mn-oxidizers	Microscopic data (Fig. 6), chemical profiles[b]	Buoyant vent plumes	57, 216
Free-living heterotrophs and/or chemoautotrophs	Visible turbidity, microscopic data, tracer assays, chemical profiles[b]	Cool (diffuse) vent fluids	183, 185, 326
Free-living hyper-thermophilic *Archaea*	Lipid analyses (Fig. 7), DNA analyses[b]	Flange sulfides, smoker walls, deep hot sediment	139, 142, 313
Free-living methanogens and ammonia producers	Microscopic data, chemical profiles	Subsurface vent habitats	17, 23, 26
Free-living subsurface dwellers of largely unknown trophic status	Visible flocs, microscopic data, DNA analyses, lipid analyses[b]	Smoker fluids, subsurface vent habitats, deep hot sediments	19, 84, 141, 150, 310, 311, 313

[a]Chemoautotrophic = chemolithoautotrophic.
[b]Some pure culture evidence also available (see references cited).

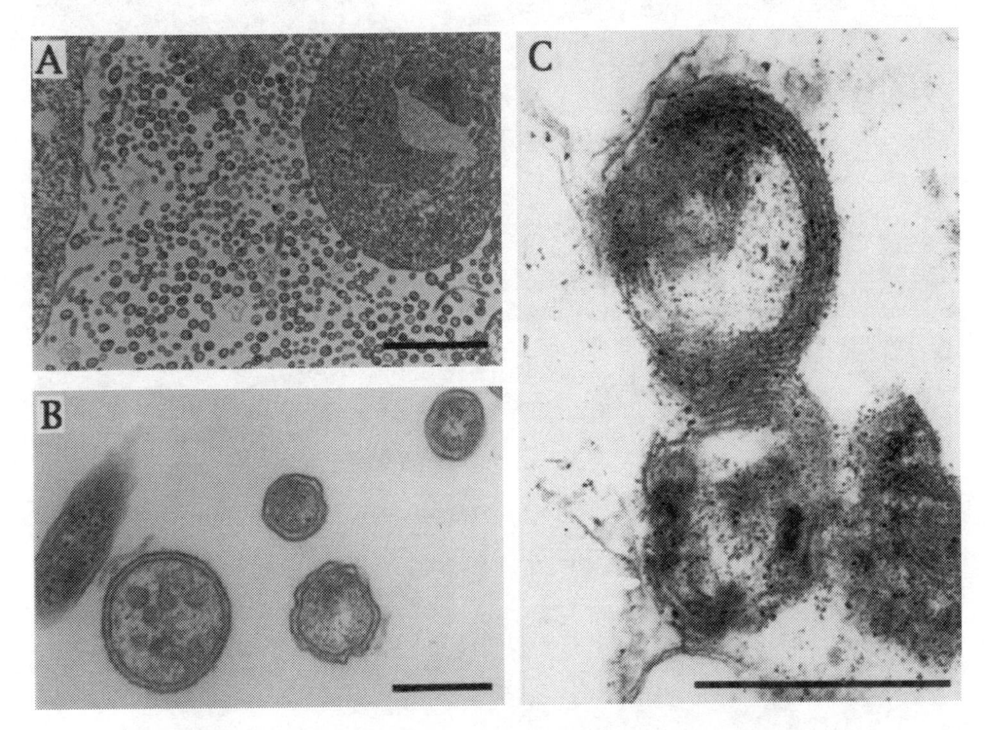

Figure 5. Examples of endosymbiotic hydrothermal vent microorganisms recalcitrant to culturing efforts but active in situ (as inferred from cell morphology and density): (A) TEM of high cell density in the gill tissue of a limpet from Endeavour Segment, Juan de Fuca Ridge (J. A. Baross, unpublished data); (B) enlargement of selected cells in (A) revealing unusual double membrane structure; and (C) TEM of putative (according to diagnostic internal membrane structure) dividing *Nitrosomonas* cells, nitrifying chemoautotrophic endosymbionts in gill tissue of the vent clam *Vesicomya gigas* found colonizing a whale fall on the seafloor (93). Bars, 5 μm (A) or 0.5 μm (B, C).

matrix creating a steep internal temperature gradient (Fig. 7). Hedrick et al. (142) extracted lipids from four zones of a subcore taken through a flange, including the upper silicate surface, colonized by tubeworms and bathed (in situ) by seawater at approximately 2°C, and the bottom chalcopyrite (Cu-Fe sulfides) layer that had been in contact with 350°C fluid. Only archaeal lipids were detected in zones hotter than the silica layer (Fig. 7; 142). The highest concentrations of archaeal lipids, detected in the second zone, could represent as many as 10^8 to 10^9 cells g^{-1} dry weight flange solids, assuming that lipids constitute about 9% of cell dry weight and that a (dry) cell weighs approximately 10^{-13} g. Substantial levels of archaeal lipids were also detected in the other zones of the flange believed to be at temperatures greater than 150°C. While some strains of hyperthermophilic archaea were cultured successfully from the second zone (271), none were isolated from the underlying hotter layers. Whether or not active but nonculturable dormant or dead cells exist in the hotter layers is unknown, but the stability of archaeal lipids

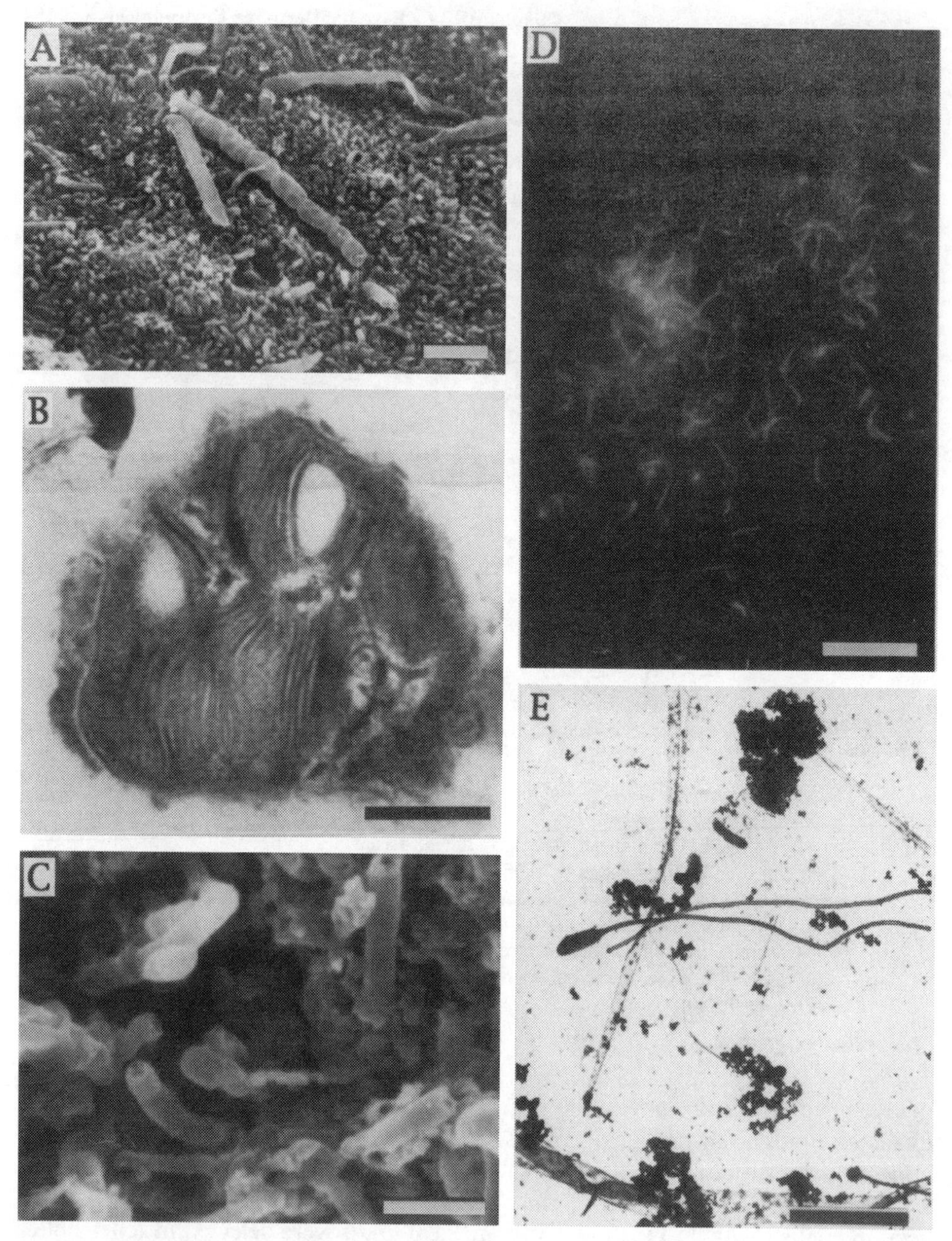

Figure 6. Examples of free-living hydrothermal vent microorganisms recalcitrant to culturing but active in situ (as inferred from cell morphology and density, microcolony formation and [or] activity measured in parallel samples): (A) TEM of dense epibiotic microorganisms on the surface of mussel shells from Galapagos vents (208); (B) putative (according to diagnostic internal membrane structure) methanotroph living epibiotically on limpet shell at Endeavour, Juan de Fuca Ridge (64); (C) scanning electron microscopy (SEM) of filamentous forms colonizing sulfides at 21°N (18); (D) epifluorescent micrograph of DAPI-stained microcolonies (duplicate samples of which were observed by light microscopy to stain with MnIV-specific benzidinium hydrochloride [106]), in the buoyant plume of Juan de Fuca vents at 17 km from the source, the plume zone where particulate Mn occurred preferentially (210; J. A. Baross, unpublished data); and (E) TEM of a prosthecate microorganism in the buoyant plume of Juan de Fuca vents (B. C. Crump and J. A. Baross, unpublished data). Bars, 2 μm (A, C–E) or 0.5 μm (B).

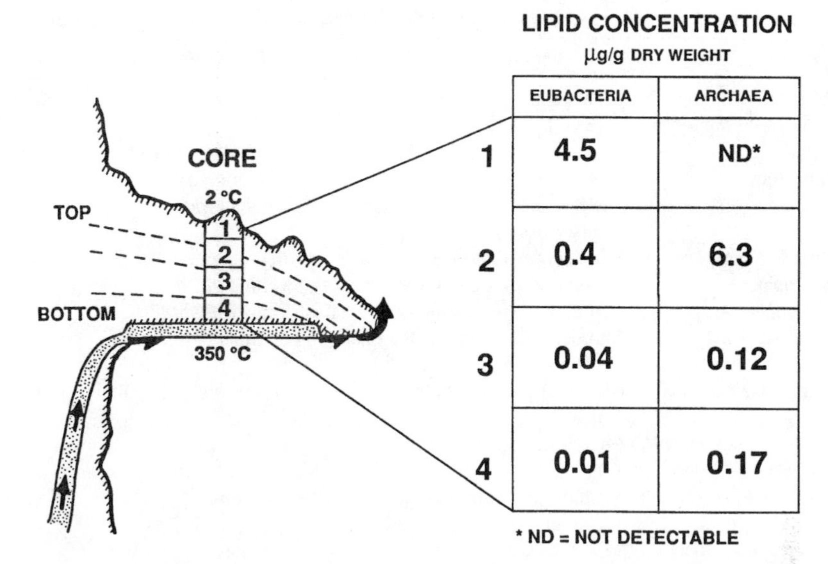

Figure 7. Bacterial and archaeal lipid concentrations extracted from sequential zones of a core taken through a flange formation, and therefore through a steep temperature gradient, retrieved from Endeavour Segment, Juan de Fuca Ridge (data from Hedrick et al. [142]). Note absence of bacterial lipids and preferential occurrence of archaeal lipids, characteristic of known hyperthermophiles, in zones of higher temperatures.

under extremely high temperatures is clear (and recently supported by the detection of similar lipids in a deep-sea drilling core of sediments exposed in situ to 130°C; 313). The possible role of extremely thermostable extracellular enzymes produced by active cells in the second zone of the flange for "foraging" purposes (333) in hotter zones is explored as a survival strategy in a later section.

The occurrence of active but nonculturable microorganisms at hydrothermal vents has captured much attention in recent years, but the cold deep sea is also home to a diversity of active organisms (albeit at constant temperature) not known in culture (Table 3). The bioluminescent lures of female deep-sea anglerfishes are packed with endosymbiotic, light-producing bacteria recalcitrant to culturing (140). Intestinal cells in the hindguts of deposit-feeding invertebrates, where animal defenses against bacteria are relaxed (269) and microbial activity is high even in abyssal animals (86, 92), harbor microorganisms not obtained in subsequent culturing efforts (Fig. 8; see also 223). High percentages of viable bacteria, measured by the Kogure et al. (195) method (the standard for detecting viable cells in the environment without culturing [291]), have been detected in abyssal invertebrate guts (86) and sinking aggregates (79). Some deep-sea invertebrates suffer high incidence of bacterial infection, as apparent from carapace lesions on the tanner crab (27). Other sites of organic enrichment in the cold deep sea, such as phytodetritus on the seafloor (213) and bits of chitin and cellulose buried in sediments at abyssal depths (85, 96), have been recovered heavily colonized by microbial

Table 3. Examples of active but uncultured microorganisms in the cold deep sea (Fig. 2)

Type of organism	Measure of activity	Deep-sea origin	Reference(s)
Endosymbiotic heterotrophs[a]	Light production, microscopic data, DNA analyses	Anglerfishes	140
Endosymbiotic chemoautotrophs[a]	Microscopic data (Fig. 5), enzyme and tracer assays, stable isotrope data, DNA analyses	Animals in reducing sediments, at whale falls	93, 108, 304
Mutualistic or symbiotic heterotrophs	Microscopic data (Fig. 8), tracer assays, DNA analyses[b]	Invertebrate hindguts	86, 92, 223
Pathogenic heterotrophs[a]	Visible lesions, microscopic data[b]	Tanner crabs	27
Free-living aerobic heterotrophs[a]	Microscopic data (Fig. 8), tracer assays, chemical profiles	Sediments, seawater	3, 85, 283
Free-living anaerobic methanotrophs	Chemical profiles, stable isotope data	Anaerobic sediments (Cariaco Trench)	4, 5, 280
Free-living magnetotactic anaerobes	Microscopic data, chemical profiles	Sediments	267
Free-living mixotrophic mats of *Thioploca, Beggiatoa,* and others	Visible mats, microscopic data, chemical profiles, stable isotope data	Peru upwelling sediments, whale bones	93, 111, 120
Consortia of aerobic and anaerobic heterotrophs[a] chemoautotrophs,[a] and methanogens	Electron acceptor and other chemical profiles[b]	Sediments, fecal material (Fig. 2)	35, 85, 238

[a]Heterotrophic = chemoorganotrophic; chemoautotrophic = chemolithoautotrophic.
[b]Some pure culture evidence also available (see references cited).

communities of complex morphology (Fig. 8). Even in the absence of an obvious organic bonanza, microorganisms morphologically akin to oligotrophic species from other environments (228, 274) have been observed to colonize solid substrates suspended in abyssal waters (Fig. 9). In deep benthic sediments, active microorganisms are apparent from chemical profiles indicating the disappearance of specific electron acceptors and sources of carbon and nitrogen (reviewed by Billen [35], Nealson [238], and Deming and Baross [85]), but also from simple physical assays, such as that for magnetotaxis (267). Though eukaryotic microorganisms are beyond the scope of this chapter, evidence exists for fungi (196, 279) and novel protozoa (325) active at great depth in the ocean (213; Fig. 9). We intentionally impart this view of a great diversity of active but nonculturable (or as yet uncultured) microorganisms in the deep sea to focus attention on the potential for discovery of novel genetic capabilities in this vast and varied environment.

EVIDENCE FOR MICROORGANISMS DORMANT IN THE DEEP SEA

Evidence for microbial dormancy in the deep sea derives from knowledge of cells that can be cultured in the laboratory and shown to enter a dormant state, either upon incubation under temperatures and pressures outside their cardinal growth limits or upon exposure to other stresses, particularly starvation. Dormancy

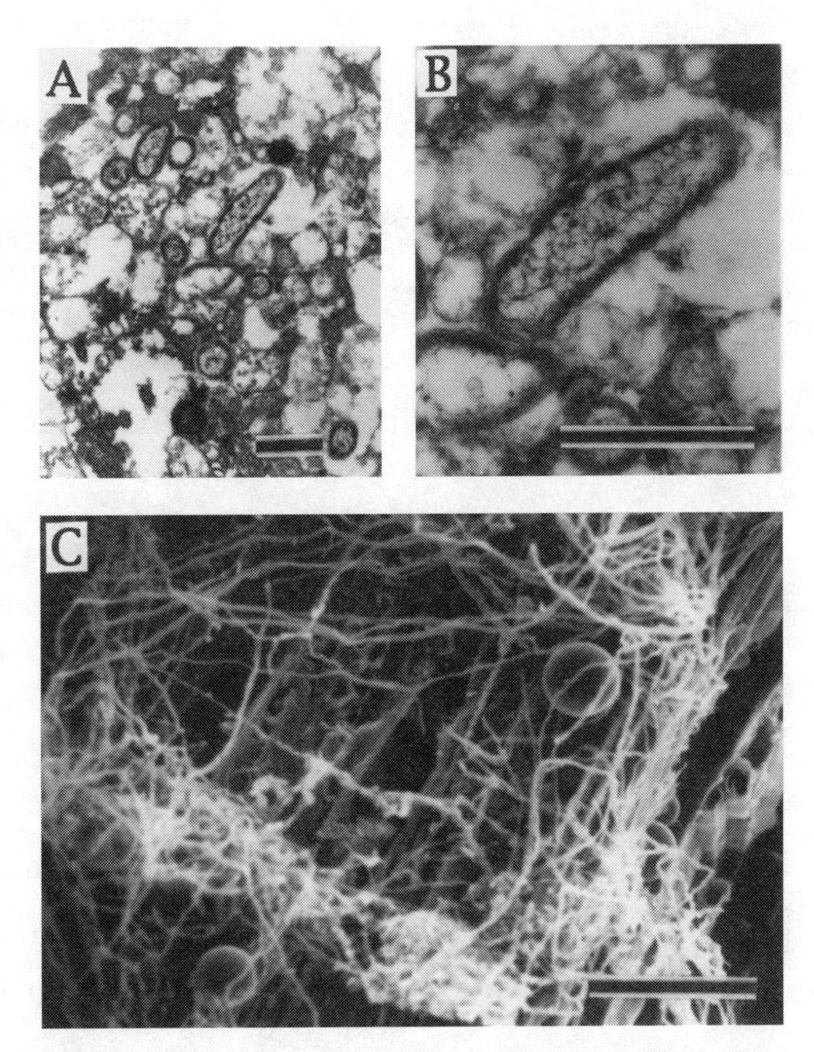

Figure 8. Examples of microorganisms recalcitrant to culturing efforts but active in situ (as inferred from cell density and activity measurements in parallel samples or from colonization rates) in sedimentary environments of the cold deep sea: (A, B) TEM of cells associated with the hindgut intestinal lining of an abyssal deposit-feeding holothurian (86); and (C) SEM of actinomycete-like forms colonizing a piece of cellulose buried in artificial sediments (glass beads, two of which are visible) placed on the seafloor at 4100 m in the Bay of Biscay for 6 months (85, 96). Bars, 1 μm (A, B) or 100 μm (C).

is also implicit in the general finding (we have already examined the exceptions) that the large number of intact microorganisms observable microscopically in deep-sea sediment (Fig. 3) does not translate into high (bulk mean) rate of microbial activity (85). One explanation for the discrepancy is that a negligible fraction of these organisms is adapted for activity under extreme temperatures and pressures

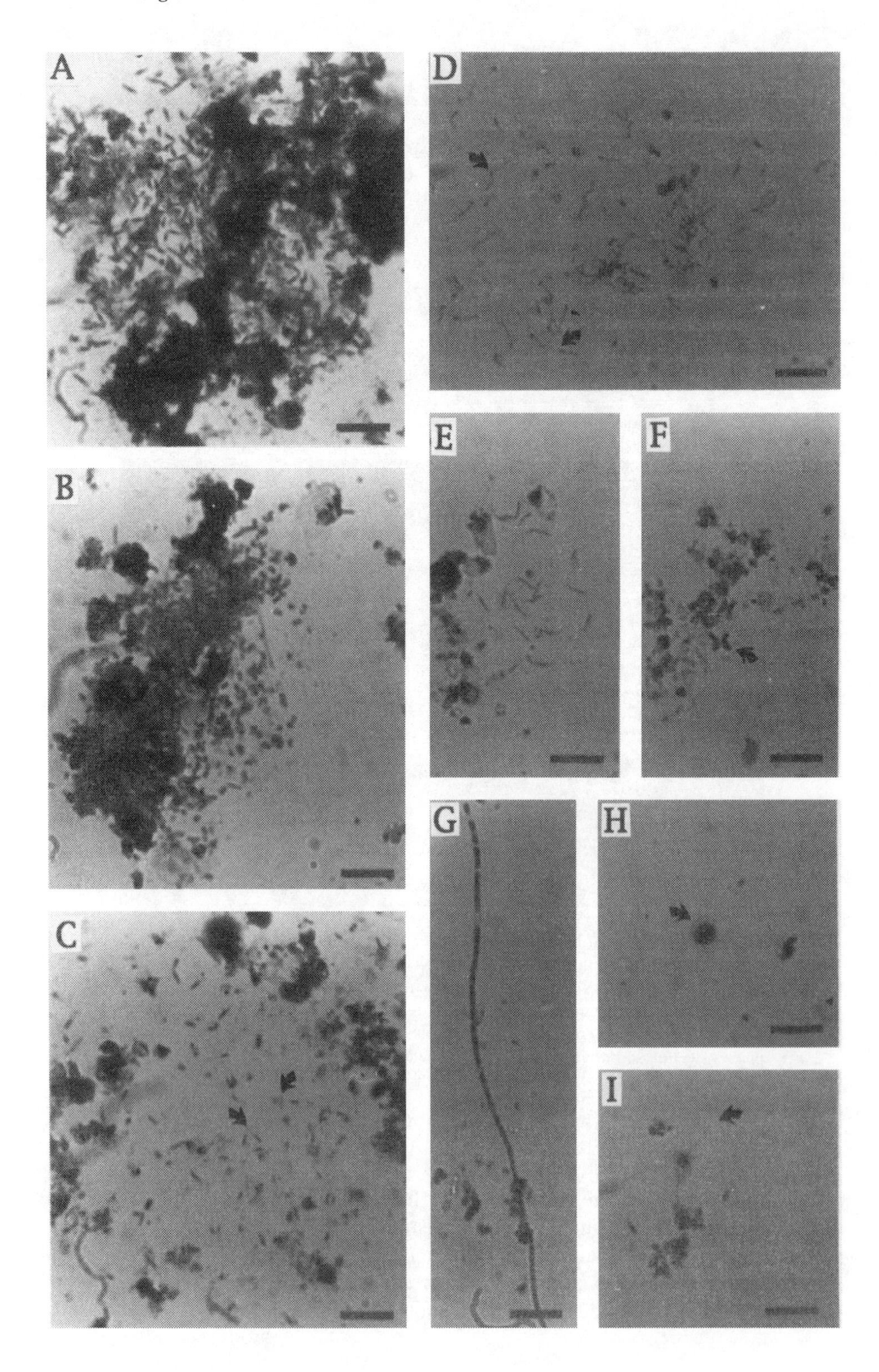
A
B
C
D
E
F
G
H
I

(159); the vast majority have derived from shallower environments and become inactivated upon entry into the ocean interior where they either remain dormant for short or long periods (25, 147, 324) or die (317). An alternative explanation is that many cells at abyssal depths are capable of significant activity under deep-sea temperatures and pressures but lack suitable nutrition (79, 164). Aspects of both explanations are true. We examine the phenomenon of starvation, including cellular strategies for surviving it and other stresses, and more explicitly define dormancy at the intracellular level in later sections, limiting this discussion to evidence that the physical trigger of a shift in temperature and (or) pressure can induce cell dormancy in the deep sea.

Dormancy of Shallow-Water Microorganisms in the Deep Sea

A review (258) of viable but nonculturable cells traces the recognition of dormancy as a common phenomenon in the marine environment to shallow-water research published in 1978 (61, 309). However, we trace the original concept to deep-sea research and the 19th-century voyages of two French scientists, Certes (49) and Regnard (282). Using shipboard incubation temperatures and pressure (atmospheric) milder than those encountered in situ, they cultured numerous bacterial strains from water and sediment samples retrieved from depths as great as 8,200 m. The question of whether or not these isolates were metabolically active under deep-sea conditions was first posed by Certes (49). Based on pressure experiments he conducted in Louis Pasteur's laboratory using a modified Cailletet press, Certes (49) concluded (as did Regnard [282]) that bacteria existed in the deep sea only in a state of suspended animation. For more than a century, microbiologists studying the deep sea have been preoccupied with this conclusion and tests of its validity (24, 92, 159, 213, 358, 363). Much enlightening evidence for active but nonculturable cells in the deep sea, as well as for culturable strains uniquely adapted for optimal growth under deep-sea conditions, is now available, as we have indicated (Tables 1–3; Fig. 4–9). Microbial activity uniquely restricted to the deep sea (whether attributable to culturable or nonculturable cells) makes the assertion that all sediment microorganisms are recent (vertical) descendants from surface waters (253) irrelevant to the deep sea. However, the notion of suspended animation still pertains to possibly vast numbers of microbial cells that did originate in other environments but settled to the deep seafloor via particles to become buried there. The sizable body of culturing work by Kriss (200) and others

Figure 9. Examples of microorganisms recalcitrant to culturing efforts but active in situ (as inferred from cell morphology, density, and colonization of glass slides suspended on a mooring for 6 months at 4,700 m in the East Pacific) in abyssopelagic environments of the cold deep sea. (A and B) dense microcolonies of rods; (C) microcolony of rods with putative attachment structures (arrows); (D) microcolony of thin budding rods (arrows); (E) microcolony of thin fusiform rods; (F) cells in unusual division mode; (G) dividing filamentous cells; (H) flagellated protozoan; (I) loricate protozoan. Slides were protected from surface waters during recovery, heat fixed shipboard, subsequently stained with methylene blue, and photographed by light microscopy (J. A. Baross, unpublished data). Bars, 5 μm.

(33, 278), detailing the types of heterotrophic bacteria that can be recovered from deep samples using incubation conditions foreign to the deep sea, amply demonstrates Certes' conclusion.

At least three types of shallow-water microorganisms are known from laboratory studies to enter a state of dormancy in the deep sea: spore-forming, moderately thermophilic soil bacteria that have washed into the sea by natural processes; mesophilic human enteric or sewage bacteria released in large numbers into the open ocean by major cities of the world (147); and mesophilic to psychrophilic marine bacteria indigenous to shallow waters. Spore-forming thermophiles inactive below temperatures of 36 to 46°C were recovered from cold (Arctic) marine sediments as early as 1938 by the Russian microbiologist Egorova (103). Bartholomew and Rittenberg (29) subsequently detected similar thermophilic bacilli in samples recovered from deep ocean basins. Baross et al. (25) studied mesophilic enteric and other sewage microorganisms, including the anaerobic spore-former *Clostridium perfringens*, under simulated deep-sea conditions. They coined the phrase "viable but inactive" to describe cell survival under and recovery from physical conditions more extreme than cardinal growth limits (25). Hill et al. (147) recently demonstrated recoverability of the *C. perfringens* spore as a useful indicator of the distribution of sewage sludge on the deep seafloor, thus confirming dormancy of *C. perfringens* in the deep ocean with field measurements.

For these first two cases, entry into the dormant state in the deep sea can be attributed readily to a temperature shift, specifically to a drop in temperature below the minimal growth limit (24, 25, 258; see ref. 258 for some exceptions to this analysis). Because increasing pressure theoretically has an effect similar to decreasing temperature (24, 25) and hydrostatic pressures on a cell sinking into the ocean interior do not become inhibitory (>200 atm [360]) until after subminimal temperatures have been reached, pressure effects on dormancy should be additive, promoting long-term dormancy in the deep ocean (22, 24, 25, 230). In support of this hypothesis is the finding that *Escherichia coli* (and other mesophilic species) is more resistant to pressure damage at temperatures near its minimal growth temperature than at its optimum (24).

It follows that marine bacteria indigenous to shallow seawater, the third type of shallow-water microorganisms studied under simulated deep-sea conditions, would suffer greater pressure damage during sinking, since their optimal growth temperatures are typically lower than those of most sewage microorganisms and certainly lower than those of thermophiles. This idea is supported most clearly by the marked barosensitivity of extremely psychrophilic bacteria isolated from shallow Antarctic waters (343) and of natural assemblages of marine bacteria found on sinking particles (324). Work with *Vibrio harveyi* (317) indicated cell death rather than dormancy under extreme deep-sea temperatures and pressures. It also underscored the fact that the nutritional history of cells can significantly affect the outcome of such experiments, as was first recognized by starvation experiments in Morita's laboratory a decade earlier (254–256; discussed below). However, a shift in temperature alone (without nutritional stress) can be sufficient to trigger (260, 338) or reverse (246) the nonculturable state for some shallow-water bacteria. Recent work with the mesophilic estuarine bacterium *Vibrio vulnificans* suggests a survival advantage to entering a nonculturable state as a result of a downshift in temperature: increased resistance to mechanical stress (sonication) and therefore possibly to predation pressure.

For microorganisms in general (123, 193, 258), but also for many in the deep sea that originated elsewhere, recoverability from the dormant state may also be a function of time spent under the suboptimal growth conditions, for at some point irreversible intracellular processes leading to death can occur. However, it is also clear that some microorganisms indigenous to deep sediments have persisted under extreme temperatures and pressures for extraordinary lengths of time. Results from the U.S. Department of Energy program on deep subsurface terrestrial environments have convincingly demonstrated the presence of viable (activity in situ as yet unknown) anaerobic thermophiles indigenous to Triassic sediments deposited millions of years ago (99; see also references 41, 112, 136, 201). A group of British microbiologists working under the Ocean Drilling Program have demonstrated the presence of numerous bacteria (10^7 g^{-1}) in marine sediments as deep as 500 m below the seafloor, deposited over 3 million years ago (264), including sediments heated in situ to 169°C (58). They also obtained microscopic and chemical evidence (dividing cells and methane production) for the activity of some of the cells (264) and established that DNA present in the sediments could be amplified by the PCR for future determination of phylogenetic diversity (288). Aspects of their work have been confirmed in other deep-sea drilling cores, including subsurface sediments heated in situ to 130°C (313), but the identification and in situ activities of the resident microorganisms remain largely unknown. Mechanisms to explain microbial persistence and activity under extreme conditions over geologic time scales are just beginning to be explored (342). In the case of persistence in a dormant state, cessation of intracellular or endogenous metabolism, as may occur in spores or cells "frozen" by temperatures (and further by elevated pressures) below minimal growth limits, seems a likely prerequisite.

Dormancy of Indigenous Microorganisms in the Deep Sea

Microorganisms indigenous to the deep sea (restricted to activity there by unique temperature and pressure requirements; Table 1) may also pass short or long periods in their natural environment in a dormant state. Starvation may be the primary factor leading to dormancy in the cold deep sea (as discussed below), but a downshift in temperature can trigger a similar process for microorganisms indigenous to warm or hot deep-sea vent habitats. Reductions in activity (though not specifically dormancy) due to low temperature have been inferred from a limited number of in situ measurements made at the 21°N East Pacific Rise vent site (326, 347). Virtually all hyperthermophiles isolated from deep-sea vent environments and tested in this way enter a dormant state when exposed to cold temperatures mimicking ambient bottom waters. Cultures can be stored under refrigeration (or even at room temperature) and successfully recovered months to years later without significant loss of viability; at low temperature, not even exposure of these obligate anaerobes to oxygen diminishes viability (155, 217, 285; J. A. Baross and R. J. Pledger, unpublished data). Dormancy of hyperthermophiles induced by suboptimal growth tem-

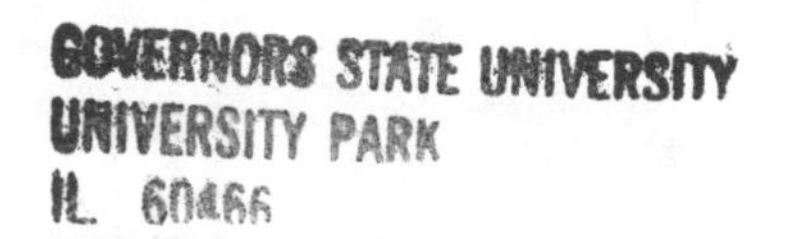

perature has also been demonstrated in the field. Huber et al. (155) detected anaerobic hyperthermophilic archaea by sampling oxygenated surface seawater after the eruption of a submarine volcano. Holden et al. (150) and Summit and Baross (311) cultured anaerobic hyperthermophiles, requiring temperatures >60°C for growth, from submersible samples of <20°C vent fluids at eruption sites on the Co-Axial Segment of the Juan de Fuca Ridge and on the North Gorda Ridge in the Pacific. These results have important implications for the concept of a hot subsurface biosphere (19, 84, 128), since no seafloor habitat hot enough for the strains was present. They may also bear upon the discoveries of archaeal DNA sequences, closely related to known thermophilic sequences, in cold seawater from the ocean interior (70, 118). Since hydrothermal plumes are known to occur globally (even at high latitudes) and stretch hundreds of kilometers into the ocean interior (181) and hyperthermophiles can survive for years at low temperature in an oxygenated milieu, perhaps some of the new sequences being detected in hydrocast samples from surface ships are dormant vent organisms rather than novel species of seawater archaea (but see 116, 237, 276, 306). Deep-sea psychrophiles rarely suffer in situ from a downshift in temperature, since the bulk of the deep sea is already cold (<4°C). Even at near-freezing (−1.7°C) temperatures in deep Arctic basins, significant microbial activity attributable to extreme psychrophiles continues (80, 90). However, deep-sea psychrophiles can experience a shift to warmer temperatures in the deep sea upon entrainment into vent fluids. The compensating effects of elevated hydrostatic pressure in the deep sea would allow some barophilic psychrophiles to continue activity at temperatures exceeding 20°C (94, 230, 353) rather than enter a dormant state. At higher temperatures (>30°C), all available information indicates that psychrophiles would succumb (lyse) rather than enter a dormant state (although the possibility for survival in a dormant, spore-like, or spore state remains). As a practical matter for in situ experiments at hydrothermal vents, sampler flushing by superheated effluents is assumed to eliminate all microorganisms other than hyperthermophiles or superthermophiles (84, 19, 310). We consider the heat-shock response (discussed below), which for short periods can protect all thermal classes of microorganisms from death upon exposure to supraoptimal growth temperatures, as distinct from cell dormancy, which is not limited to short duration. The significant discovery that prior starvation of a shallow-water psychrophile enhances its tolerance of supraoptimal growth temperatures (277) may also pertain to psychrophiles indigenous to the deep sea, but this possibility remains to be examined in deep-sea microorganisms, either psychrophiles or hyperthermophiles.

SURVIVAL STRATEGIES OF FREE-LIVING CELLS IN THE DEEP SEA

The issue of microbial survival strategies in nature has been the subject of intensive research in recent decades and of numerous reviews (14, 62, 193, 222, 231, 232, 240). Here, we highlight general findings of relevance to survival in the cold and (or) hot deep sea, a topic that has not received the same level of attention. In particular, we consider evidence for cellular and extracellular strategies that microorganisms may adopt to survive or overcome starvation under extreme temperatures and pressures, including the classic responses of reduction in cell size, attachment, and production of extracellular enzymes. Intracellular strategies such as modifications to membranes, shifts in metabolism and de novo synthesis of protective (heat-shock) proteins pertain not only to survival of starvation but also of other stresses in the deep sea. The potential links between stress conditions and genetic exchange in the deep sea need to be considered, but this subject remains largely unexplored.

In the deep sea, the primary stresses include exposure temperatures and pressures outside the bounds of growth, as discussed in earlier sections, and to oligotrophic or starvation conditions. In selected deep-sea environments, hypersalinity introduces an additional factor. Uniquely salt-adapted microorganisms are well known from shallower habitats (203), but information from the deep sea is scarce (188). In the anoxic brines of the Gulf of Mexico (322), salt conditions appear damaging to most microorganisms entering them, but in brine deep in the subsurface at hydrothermal vents (36; Fig. 1), microorganisms exposed to the combination of extreme temperature, pressure and salinity may be protected (19, 84, 188, 245). Some microorganisms in the deep sea can suffer additional stress according to specific types of metabolism, especially at hydrothermal vents and in symbiosis with invertebrates, e.g., oxygen limitations (for endosymbionts), oxygenation (for anaerobes in an active state), and harsh pH, sulfide, and metal toxicity.

The multiple and unique strategies for survival and maintenance of microbial endosymbionts in deep-sea invertebrate hosts are beyond the scope of this chapter, but the subject has been treated in numerous reviews (51, 108, 109, 140, 241). The survival strategies of many free-living (nonsymbiotic) cells in the deep sea may involve attachment to surfaces, including those of potential invertebrate hosts. Microorganisms attaching to invertebrate tissues (e.g., gut linings) in response to starvation or other stress may represent the first stage in the evolution or maintenance of some types of microbial-invertebrate symbioses (289) in the deep sea.

Strategies for Surviving Starvation

Microbial survival strategies have evolved in response to the "feast-or-famine" existence they must confront in every environment and perhaps especially in the deep sea. It is common practice in the microbiological literature to consider the deep sea a desert (159, 193, 231) and the microorganisms suspended in abyssal waters effectively stranded under oligotrophic conditions for hundreds or even thousands of years, according to water mass residence time. However, advances in chemical oceanography demand a reevaluation of this assumption. In the upper reaches of the ocean, the principal source of carbon and energy for microorganisms is believed to be the dissolved organic carbon (DOC) produced during phytoplankton blooms and grazing activities (53, 176, 184). The starvation state for these microorganisms may be short (hours) or long (months). That longer starvation is more likely in the deep sea, where DOC is thousands of years old and believed to be mostly unreactive (31), is true, but new information suggests that abyssopelagic bacteria need not be solely dependent on this refractory DOC for exogenous nutrition. Recent analyses of shallow and deep seawater samples have revealed high concentrations of small organic colloids (5 to 200 nm) exceeding 109 ml^{-1} at depths as great as 4,000 m (340, 341). Since they are nonsinking, their origin is presumed to be the larger particles that sink into the ocean interior and become disaggregated there, either by microbial hydrolysis or physical mechanisms (53, 184). Periodic delivery of phytodetrital aggregates, sometimes in high densities (213), to the deep sea has been known for some time. Chemical analyses of the newly discovered colloidal particles have revealed a significant fraction to be biologically reactive carbohydrates of high molecular weight (32, 197). Indigenous deep-sea bacteria may attach to particulate material larger than typical cell size or dense enough to sink, but in situ microbial assimilation of these submicrometer-sized colloids as nutritional DOC must be considered in the design and analysis of future experiments in the deep sea (197, 341).

Size Reduction

Starvation stress in shallow-water marine bacteria is first and most universally manifested by miniaturization of the cell (193, 254, 260). The reductive division that accompanies decrease in cell size in a population provides increased dispersal of the parent genome and chances of recovery in a less hostile environment, although not all of the miniaturized cells may retain viability (depending on the test isolate [8, 231]). Cell miniaturization improves uptake rate at low substrate concentration (178). In some cases, small starved cells also show an increased surface roughness and fimbrae-like structures, both of which can be associated with the ability to adhere to surfaces (193; see below) and resist detrimental pressure effects (256) and autolysis (257). In other cases, miniature cells are also accompanied by elongated, spiral-shaped cells after starvation (244). During the size-reduction starvation response of a marine psychrophile, the cell population remained constant in number (233), while many individuals appeared to enter a nonculturable state (recoverability unknown). Cell-size reduction and loss of culturability by vent microorganisms under starvation conditions has not been studied, but a psychrophilic barophile behaved in similar fashion to shallow-water bacteria when confronted with an absence of exogenous nutrition, adopting the cell-size reduction strategy (286).

At some point in the reduction process, a cell may lose elements essential to reproduction. Unless it can regain these elements from the environment (see below) or enter a spore-like stage, it may become permanently nonculturable or dead. The environmental paradox is that such evolutionary dead-ends, akin to laboratory-created minicells that lack genomic information, may continue to metabolize and affect their surroundings for unknown but potentially long periods under the "preserving" conditions of the cold deep sea. The fact that very small cells dominate in oligotrophic marine waters, including the deep sea, suggests that dormancy induced by starvation (or other factors) may be the rule. However, the nutritional and physiological status of small cells suspended in abyssal waters is largely unknown.

Attachment

In many cases, a monolayer or film of dissolved organic matter exists on surfaces in seawater, making attachment to that surface an improved circumstance for a starving microorganism (289), even in abyssopelagic waters (Fig. 9). Indeed, it is well known that copiotrophic marine bacteria under starvation conditions seek a surface rather than remain in suspension (193), while the tendency of oligotrophic microorganisms is to produce attachment appendages (228, 274; Fig. 9). However, even in the absence of nutrients on the target surface, attachment to a moving particle can increase the supply of nutrients to the cell by distorting (reducing) the diffusional shell that otherwise exists around a micron-sized cell (178). Note that although cold temperature has been considered in relevant diffusion models, the effects of hydrostatic pressure have not. Attachment can also promote eventual relief of starvation conditions by providing a vertical shuttle to the seabed, where nutrients in general are more available, or a lateral shuttle into a nutrient-rich animal gut. Abyssal invertebrates are known to feed selectively on recently arrived particles (130).

In shallow environments with strong spatial or seasonal temperature gradients, some mesophilic *Vibrio* spp. may use attachment as a mechanism to survive prolonged exposure to temperatures below their growth minima (182). This mechanism may not hold true for greater depths in the ocean, given the culture work of Smith and Oliver (302), also with a mesophilic *Vibrio* sp., showing that elevated pressures inhibit the attachment process. The deep-sea relevance of these results is weakened by use of room temperature in the experiments. However, if shallow-water microorganisms surviving the initial temperature and then pressure shifts upon entry into the ocean interior are indeed destined to suspension as small, unattached cells, then a simple physical explanation has been found for much of the earlier literature (159) showing only shallow-water bacteria in deep-water samples (that typically failed to capture particulate material). Studies of particles (78, 87, 213, 324) and other surfaces (invertebrate gut linings; 86) in the deep sea, including the pressure responses by the attached bacteria, have consistently documented enhanced barotolerance or barophily. Pure culture work by Rice and Oliver (286) with a psychrophilic barophile confirmed a greater tendency to attach under elevated hydrostatic pressure than at atmospheric pressure if first starved at 5°C. Furthermore, the extent of attachment under pressure increased with duration of starvation. Together these studies suggest that attachment may be a starvation strategy in the deep sea unique to indigenous deep-sea microorganisms and not shallow-water arrivals. This hypothesis could be tested by direct analysis of nucleic acid in deep-sea samples of sinking particles versus seawater, as fruitfully undertaken in shallower regions of the ocean (75), or to such samples as shown in Fig. 9, using 16S rRNA or other probes developed to distinguish barophiles from nonbarophiles. It is already partially supported by the recent chemical analyses of Druffel et al. (100), which provide some evidence that microbial incorporation of "old" DOC may be occurring preferentially on particulate matter.

The starvation-attachment strategy might also explain selective retention of barophilic gut flora by abyssal invertebrates during quiescent periods when the animal is not feeding, that is, during starvation conditions for the gut flora (89, 92, 98, 193, 303). The possibility of attachment favoring the culturability of cells in the deep sea has not been examined directly, but most of the psychrophilic barophiles available in culture collections have derived from surface-rich habitats such as abyssal animal guts, sediments, and sinking particles (85). Conversely, barophilic bacteria have been difficult to culture from abyssal seawater samples, although this pattern could also be explained by low frequency of occurrence in oligotrophic (particle-poor) abyssal water samples, as discussed above. New culturing efforts using deep-sea water samples may eventually alter the balance of surface-derived versus unattached barophiles in deep-sea culture collections (164, 295).

Many heterotrophic hyperthermophiles cultured from the deep sea are also known to attach to surfaces, specifically to elemental sulfur grains during cell growth in organic-rich culture media. Although research on starvation effects on hyperthermophiles has begun (217), relationships have not been reported between starvation and attachment by hyperthermophiles or other thermal classes of vent isolates under environmentally relevant temperatures and pressures. Such work could improve understanding of colonization (Fig. 6), establishment of endosymbioses with juvenile vent invertebrates (51, 241), and survival in the more extreme regions of flanges (Fig. 7), sulfide mounds, and the subsurface (Fig. 1).

Extracellular Enzymes

In the seabed and in sinking aggregates, where mineral surfaces and particulate organic material (POM) dominate the microbial environment, cell attachment appears to be the rule (66, 75, 333). To use POM as a food source, however, an immobilized microorganism must produce extracellular enzymes to "forage" and hydrolyze the source to membrane-transportable size (or depend on neighbors to do same; thus, the development of microbial consortia in sediment [85]) and then receive the hydrolysate in return. Under the starvation stress of limited dissolved organic matter, production of extracellular enzymes can be an alternative (332) or supplemental microbial survival strategy (2). In the confines of a sedimentary matrix, animal gut, fecal deposit, or sinking aggregate, return of hydrolysate may be enhanced (269, 333). Research on extracellular enzyme activity in the upper reaches of the ocean (reviewed by Hoppe [153]) and in shallow sediments (12, 219, 227, 331) is flourishing. Less is known about this phenomenon as a survival strategy in deep-sea habitats (38, 39, 275).

With regard to the cold deep sea, neither enzymes released into culture media by extreme psychrophiles and barophiles (144) nor extracellular activity assayed in slurries of polar sediments (227) have appeared to express any unique temperature or pressure characteristics that would enable rapid hydrolysis in situ. Because some tangential data suggest otherwise (12, 137, 158a, 191, 293; Huston et al., submitted for publication), more research on the potential psychrophilic or barophilic nature of extracellular enzymes produced in the deep sea is in order. On the other hand, some enzymes released into permanently cold environments may be long-lived under the stabilizing influence of low temperature (and elevated pressure in the deep sea) and provide hydrolysate at slow but useful rates to resident microorganisms (88, 331). Overproduction of enzymes (281) could be a deep-sea survival strategy that compensates for limitations on activity rates. Current methods for assaying enzyme activity in cold sediments measure the maximal potential rate of hydrolysis (226, 331), in effect, providing a crude estimate of amount of enzyme present. Results from an application of whole-core injection methods (226) to abyssal sediments from 3,700 m in the Greenland Basin (90) are noteworthy in this regard: rates of methylumbelliferyl (MUF)-glucosaminide hydrolysis in subcores incubated at −1°C and 370 atm were higher than those reported for shallower sediments incubated under milder in situ conditions (90); comparison to surface sediment controls at −1°C and atmospheric pressure indicated pressure-enhanced activity by a factor of five (Fig. 10); and significant hydrolysis continued in the subsurface sediment layers of the cores (Fig. 10). Production of extracellular enzymes may thus be an important life strategy for microorganisms in cold abyssal sediments and other confined POM-dominated habitats (as now appears certain for some polar sediments [40]). Whether or not the pressure-enhanced activity observed in the sediment surface layer reflects cell-free barophilic enzymes (191) or the activity of enzymes still under the control of barophilic cells (as argued for psychrophilic enzyme activity measured by Vetter and Deming [331]) remains to be determined.

The possible role of extracellular enzymes in the acquisition of nutrition for vent microorganisms in situ has not been considered. We infer a possible enzyme-related survival strategy for flange-inhabiting hyperthermophiles from studies of enzymes produced by *Pyrococcus endeavouri* ES4 (Table 1) in culture. The extracellular amylopullanase (hydrolyzes a1,6 glucosidic linkages) produced by this flange iso-

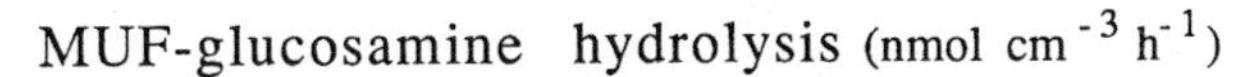

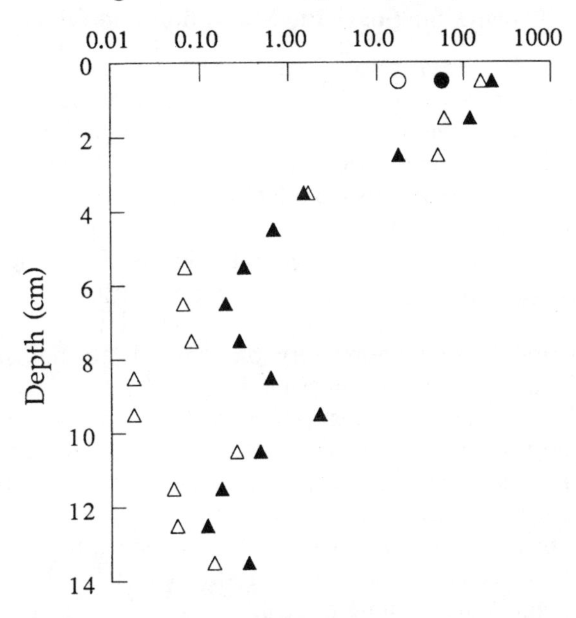

Figure 10. Depth profiles of extracellular enzyme activity, measured using the whole-core injection technique of Meyer-Reil (226) and fluorescently tagged MUF-glucosaminide as an analog for chitin, in sediment retrieved from 3700 m in the Greenland Basin (J. W. Deming, unpublished data). Replicate (open versus solid symbols) subcores of 1-cm length were incubated at −1°C and 1 atm for 4 h (circles); of 14-cm length, at −1°C and 370 atm for 4 h. Note log scale and five-fold increase in hydrolysis rate under simulated in situ conditions compared to 1-atm controls.

late is active to 143°C (298) or about 30°C higher than the maximal growth temperature of the organism (271). The half-life of the enzyme was extended from 6.5 to 20 h at 98°C and from 0.8 to 7 min at 122°C simply by adding Ca^{2+}. Furthermore, we and others have obtained evidence that hydrostatic pressure will greatly stabilize hyperthermophilic enzymes (28, 143, 312), though not related enzymes from mesophiles. It seems reasonable to predict significant lifetimes and activities for extracellular enzymes in zones of a flange structure too hot to support particular organisms. Such enzymes could thus provide a means to acquire dissolved organic matter (hydrolysis products from dead tubeworms, their chitinous tubes or microorganisms encased in the flange during its physical evolution) not otherwise obtainable by the microorganisms. The recent discovery of a hyperthermophile capable of degrading chitin at high temperatures, even if the chitinase itself has not yet been tested at supraoptimal growth temperatures (157), adds credence to the idea. If verified, the concept of enzyme foraging into temperature regimes too extreme for an organism would also have significant implications for microbial inhabitants of other environments subject to fluctuations or gradients in temperature, for example, hot sediments at Guaymas Basin (175), habitats in deep subsurface vent environments (17, 19, 23, 84; Fig. 1) and off-axis zones (313, 342), and wintertime sea ice (88).

Intracellular Strategies for Surviving Starvation and Other Stresses

All organisms that have been tested, regardless of environmental origin, demonstrate specific intracellular survival-enhancing reactions to various stresses, including extremes in temperature and pressure, starvation, chemical toxicity (2,4-dinitrophenol, ethanol, sodium azide, heavy metals), oxidants, pH, and ultraviolet light (232, 240). These reactions are referred to collectively as the heat-shock response. Cells exposed to one of these stresses typically respond by synthesizing new proteins (heat-shock proteins) and (or) elevated levels of existing proteins that degrade, stabilize, or repair other proteins denaturing as a result of the stress (211). In all cases, this response results in enhanced survival of the cell, although the extended survival period is usually temporary. Starvation is an anomalous stress in this respect, since both spore-forming microorganisms that enter the deep sea (as discussed earlier) and non-spore-forming marine bacteria have been shown to survive for extended periods without loss of viability (152, 193, 247, 255). While the starvation response is regulated by the same stress stimulons involved in heat shock, the net effects at both the biochemical and physiological levels are different. For many starved cells, maintaining some degree of endogenous metabolism appears essential to survival; for cells exposed to other stresses, stabilizing proteins subjected to denaturing conditions is more critical.

Shifts in Metabolism

The consensus is that under starvation conditions, microbial cells derive energy to meet their maintenance requirements by shifting from exogenous to endogenous metabolism (62). Although this idea is supported by observed decreases in the levels of protein, RNA, and DNA in some starved cells, such decreases do not always occur (8, 152, 193, 199, 233), nor is the evidence for consumption of storage polymers such as polyhydroxybutyrate and lipids during starvation consistent (193). The implication, however, is that the ability to recover a starved marine bacterium in culture may be dependent on organism retention of sufficient endogenous metabolites to allow conservation of essential rRNA and DNA (which may also be conserved or protected in other ways, e.g., by negative supercoiling [121]). In the final stages of dormancy for some organisms, cells may sacrifice (for maintenance energy) some nucleic acids and proteins that are essential to substrate transport or metabolism and thus become nonculturable. Unless such cells can obtain these essential molecules from the environment, they become nonculturable indefinitely or die.

Some conceptual models exist to explain the reversibility of this otherwise dead end. For example, in *E. coli*, starvation induces the accumulation of intracellular levels of cyclic AMP (cAMP) and the production of cAMP-dependent proteins, yet neither effect is required for survival while in the starvation stage (306). Since cAMP is known to be involved in cell attachment and substrate transport, two processes intimately linked to recovery from starvation (the return to exogenous metabolism), the starvation-induction of intracellular accumulation of cAMP and cAMP-dependent proteins may be advance preparation for return to the active state (299). A slightly altered version of this proposed *E. coli* strategy (299) is suggested for oceanic bacteria by the work of Ammerman and Azam (see ref. 7 and citations therein). They detected measurable levels of cAMP (about 10 pM) in oligotrophic seawater and high-affinity cAMP transport systems, controlled by concentrations and presumably quality of energy sources, in 2 to 7% of the naturally occurring microorganisms. An ability to acquire cAMP from the environment obviates the need to accumulate cAMP intracellularly in preparation for recovery from starvation. Similarly elegant and still different starvation recovery plans may have evolved in deep-sea microorganisms. How they manage to avoid the evolutionary dead end that can derive from prolonged starvation under their in situ temperatures and pressures remains to be seen.

Metabolic shifts by the cell to survive starvation may occur at another fundamental level: nutritional life mode. This suggestion comes from work with chemoautotrophic nitrifying bacteria (169, 171) and some chemoautotrophic denitrifiers (174). Such organisms enter dormancy without a measurable shift to endogenous metabolism. Instead, long-term survival (of the denitrifiers) in various anaerobic and nitrate-free environments (including marine and deep subsurface terrestrial sediments) is attributed to a possible shift to low-level fermentation to acquire maintenance energy (174). Little is known about the starvation survival strategies of other marine chemolithotrophic microorganisms (290) and whether they may be capable of shifting to alternate or esoteric metabolic survival processes.

Other types of metabolic shifts enhancing substrate transport under starvation conditions (and therefore survival) can be invoked for indigenous psychrophilic barophiles in the cold deep sea. Psychrophiles and barophiles are known to express both cold-adapted (11, 132) and pressure-adapted (30, 74) membrane-associated proteins and enzymes. Specific shifts in glucose transport proteins that favor substrate uptake under deep-sea pressure have been documented in a deep-sea psychrophilic barophile (74). If the pressure-adapted proteins were to express higher substrate affinity, then a strategic shift from copiotrophic to oligotrophic life-style, favoring survival in the deep sea, would be achieved. Some evidence for adoption of this strategy comes from starvation studies of a psychrophilic barophile under in situ temperature and pressure, as discussed in the next section.

Shifts in the structure of membrane lipids are also known to occur in response to environmental stress and to provide unique signatures of both psychrophiles and barophiles (138, 348, 352). Oliver and Stringer (259) observed significant changes in the membrane lipids of a psychrophilic *Vibrio* sp. during starvation: lipid phosphates decreased by as much as 65%, and longer carbon-backbone fatty acids were favored. Such shifts in membrane structure are believed to enhance membrane fluidity and maintain cell capacity to transport essential nutrients during periods of environmental stress (294). Significant pressure-induced changes in membrane lipids have also been documented for psychrophilic barophiles, including increases in the ratio of unsaturated to saturated fatty acids and the appearance of novel lipids (72, 73). The beauty of some stress-induced membrane adjustments is that they occur rapidly within the membrane of the affected cell, without the need for time- and energy-consuming protein induction to protect the stressed cell or for organism replication to leave adapted or protected progeny (294). Modifications in situ in the membrane without de novo synthesis therefore quickly facilitate higher affinity substrate transport as nutrient concentrations drop in the surrounding environment.

Less is known about specific metabolic reactions to the simultaneous stresses of starvation and sub- or supraoptimal temperatures (or pressures) on psychrophiles and barophiles; virtually nothing is known about the metabolic effects of starvation, with or without temperature or pressure stress, in deep-sea hyperthermophiles (but see reference 217). At temperatures below their growth optima, representatives of three major thermal groups (psychrotolerant, mesophilic, and moderately thermophilic microorganisms) have been shown to require higher-than-usual thresholds for uptake of exogenous organic energy supplies (239, 344, 345). Whether or not psychrophiles and hyperthermophiles (or barophiles) experience the same reduced affinity for substrate and thus "premature" entry into dormancy near their temperature growth minima (or some critical pressure) remains to be determined, as does the molecular basis for the general phenomenon. Some evidence for metabolic survival strategies that may partially compensate for reduced substrate-uptake affinity at cold temperature (350) is available: (i) psychrophilic microbial populations in Arctic settings have demonstrated higher incorporation efficiencies for amino acids than their counterparts in warmer low-latitude waters (90, 349), thus enhancing use of membrane-transportable levels of substrate; and (ii) bacteria in permanently cold seas tend to be larger than those in warmer habitats (287, 349), perhaps storing nutrients for leaner times.

De Novo Synthesis of Stress Proteins

Frequently, for microorganisms entering the dormant nonculturable state in response to some environmental stress, both the entry and recovery process are time dependent (136, 258, 300), implying a requirement for de novo synthesis of new molecular components. Rapid exposure to some stresses, allowing insufficient time for expression of protective proteins (assuming that the organism carries the appropriate genes), are clearly fatal: heat in the case of psychrophiles (356); decompression at supraoptimal temperature in the case of obligate barophiles (95, 357); and hypersalinity for most cells (322), except the extreme halophiles (188, 203). Rapid exposure to other stresses (such as suboptimal temperature, pressure, and subthreshold levels of nutrition) merely marks the onset of a decrease or cessation in metabolism or a shift from exogenous to endogenous (or other) metabolism (62), as discussed above.

Some proteins expressed when a cell is starved have been shown to provide increased tolerance to temperature (277) or pressure (256) as well. Of particular relevance to the strategy of de novo synthesis of protective proteins in the cold deep sea is early work, prior to the common application of protein analyses, by Novitsky and Morita (256). They showed that starvation of the shallow-water psychrophile ANT300 induced barotolerance, thus preparing the organism for survival in the ocean interior at higher pressures. However, for a psychrophilic barophile, already adapted to deep-sea pressure, starvation in the deep sea may simply shift the organism from a copiotrophic life-style to an oligotrophic one. In preliminary experiments addressing this question, the longer the starvation period, the more adapted the psychrophilic barophile *Shewanella benthica* became to growth under oligotrophic conditions (89; Fig. 11). Starvation of the barophile at in situ deep-sea temperature but atmospheric pressure caused an apparent loss of barophilic function; the longer the starvation period, the less likely the recovery of the barophilic growth trait, regardless of nutrient level provided in the recovery medium (Fig. 11). These observations can explain the failure to culture barophiles from shallow waters, even from permanently cold waters. They also strongly suggest the production and loss of protective stress proteins, depending on the pressure of incubation. Protein analyses were not made in this study, but others have detected the presence of unique, pressure-dependent membrane proteins (74) and genes (30) in other psychrophilic barophiles from the deep sea. Research on pressure-induced proteins in *E. coli* (339) may help characterize the phenomenon of pressure-shock response in microorganisms foreign to the ocean, but only protein and other molecular work with indigenous barophiles can elucidate their particular survival strategies in the deep sea, which are likely to differ from those of human enteric mesophiles.

The concept of heat-shock response in psychrophilic microorganisms from shallow environments has been examined at the molecular level in the absence of starvation (11, 220, 221), with the general finding that heat-shock proteins are produced at supraoptimal temperatures in a manner analogous to *E. coli* (132). Araki (11) documented differences in protein signatures of the shallow-water psychrophile ANT300 at 0°C compared to 13°C (11). However, the possibility that the proteins produced at 0°C represented a protective cold-shock response, as in *E. coli* (170), seems unlikely (132). Rather, they indicate cold adaptation and the enabling of growth at cold temperatures (11), a process quite distinct from survival under temperature stress (132).

Initial results of studies on how hyperthermophiles survive and thrive in the elevated temperatures of their environment indicate that the general heat-shock response applies to this group of organisms as to all others (21, 151). However,

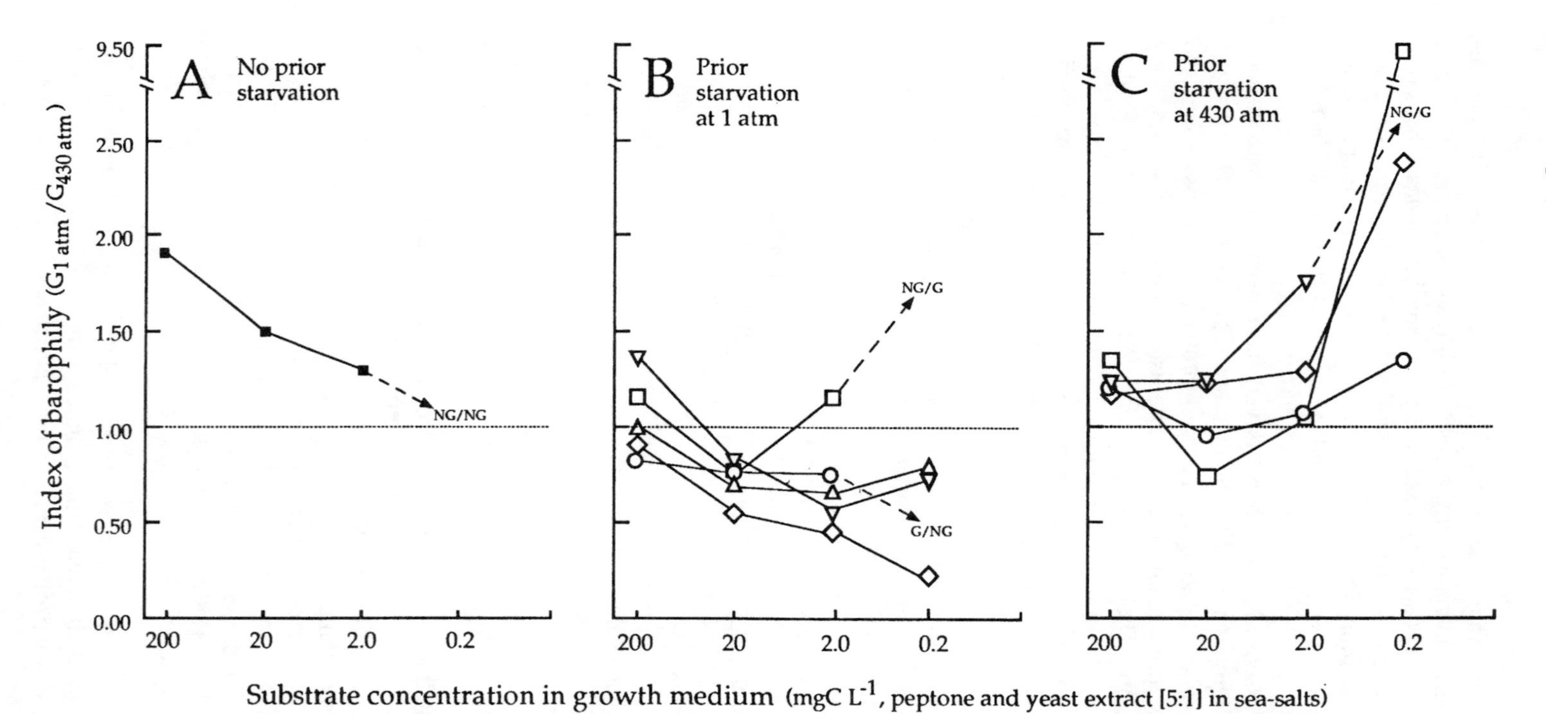

Figure 11. Results of starvation and recoverability experiments, using the psychrophilic barophile *Shewanella benthica* strain W145 (89 and J. W. Deming, unpublished data), in which an index of barophily (ratio of generation time at 1 atm [$G_{1\,atm}$] to generation time at in situ pressure [$G_{430\,atm}$]; an index >1 [dotted line] indicates barophily) was determined at 2°C and multiple substrate concentrations with and without prior starvation: (A) without prior starvation (solid squares)—barophilic at all substrate concentrations that enabled growth, but most barophilic in full-strength medium; (B) with prior starvation (open symbols) at 2°C and 1 atm for 23 (squares), 42 (triangles), 78 (circles), or 110 (diamonds) days—barophily reduced or lost with increasing starvation time, regardless of substrate concentration in recovery medium; (C) with prior starvation at 2°C and 430 atm for increasing periods (symbols as in B)—barophily retained in most cases and accentuated under oligotrophic conditions. G = growth detected in recovery medium; NG = no growth detected in recovery medium; NG/G = obligate barophily (NG at 1 atm, G at 430 atm); G/NG = marked barosensitivity (G at 1 atm, NG at 430 atm).

differences in the timing and molecular features of the expressed proteins represent stabilizing strategies that may be unique to the *Archaea* and further distinguished according to habitat depth. Unlike the *Bacteria* and *Eukarya*, which synthesize specific heat-shock proteins only upon exposure to stress, the first members of the domain *Archaea* to be examined, the shallow-water hyperthermophiles *Pyrodictium occultum* and *Sulfolobus shibatae*, were shown to produce a pair of identical heat-shock proteins providing a chaperone function over the entire range of their growth temperatures (268, 318, 319). When confronted with supraoptimal temperatures, all other protein levels dropped significantly. In contrast, levels of cellular protein in the first deep-sea archaeon examined, vent hyperthermophile *Pyrococcus endeavouri* ES4, remained constant upon hyperthermal stress (at >105°C after a milder stress at 102°C) while a 98-kDa protein, present only in low levels at suboptimal to optimal growth temperatures (75 to 95°C), was greatly overexpressed (147). Furthermore, hydrostatic pressure was found to increase the optimal and maximal growth temperatures of this deep-sea organism and therefore delay onset of the heat-shock response (149, 272). The kinetics of the heat-shock response and the emergence of the 98-kDa protein in ES4 are similar to the heat-shock response reported in the bacterium *E. coli* and the yeast *Saccharomyces cerevisiae* (296, 305), but the protein itself appears to be different. Two proteins responsible for enhanced thermotolerance in *E. coli* and *S. cerevisiae*, a 95-kDa protein ClpB and a 102-kDa protein Hsp104, respectively, are highly homologous in their amino acid sequences (265). Despite the apparent similarity in heat-shock response observed in *E. coli, S. cerevisiae*, and ES4, antibody for ClpB did not react with any proteins produced by ES4 (148), nor is the 98-kDa protein from ES4 a protease, the common function of heat-shock proteins in other organisms.

At the lower end of the temperature growth range for *P. endeavouri* ES4 (at 76°C), two proteins (90 kDa and 150 kDa) are produced in overabundance, suggesting that ES4 may also express a cold-shock response for stabilizing other proteins at suboptimal temperatures (148). The fact that an aldehyde ferredoxin oxidoreductase, a tungsten-containing iron-sulfur protein from *Pyrococcus furiosus* active in vivo at high temperatures, loses activity in vitro at low temperatures (4 or 23°C under either anaerobic or aerobic conditions [236]) suggests that stabilizing cold-shock proteins may also be produced by this archaeon. Studies on the combined effect of low temperature and elevated hydrostatic pressure are under way for *P. endeavouri* ES4. Potential relationships between starvation stress and the extremes of temperature or pressure remain to be examined. However, hydrostatic pressure as a stabilizing factor in the growth and survival of deep-sea vent hyperthermophiles at supraoptimal temperature and the general potential for discovering new molecular strategies for surviving stress are significant (19, 21, 46, 149, 217, 272).

Genetic Modification under Stress Conditions

Consensus is building that lateral transfer of genetic material, whether by transformation, transduction, or conjugation, may be widespread in natural environments (114, 115, 146, 206). Although the largest portion of the Earth's biosphere has not yet been included in this assessment, potential genetic events in the deep sea may be extrapolated or predicted from synthesis of available information from other environments. Significant concentrations of cell-free DNA (67, 266, 323) and high numbers (117) and diversity (189) of bacteriophage, the agents of transformation and transduction, respectively, have been detected in oligotrophic (shallow) marine environments. Natural populations of lysogenic bacteria have been detected in estuarine, coastal, and open-ocean environments (168), while predation pressure appears to enhance gene transfer among marine *Vibrio* strains (262). Although only a few specific genotypes transferred in nature or the laboratory have been determined (114), information on the required physiological state of a cell for uptake of DNA or susceptibility to lysis by bacteriophage is available, if largely from studies of nonmarine bacteria. The frequency of cell competence that an organism must develop for transformation can be increased in a population by modifying environmental parameters (214), for example, by growing cells in minimal media, denying key nutrients such as nitrogen, phosphorus, or essential cations, or holding the cells in stationary phase (214). The highest frequency of transformation in a *Vibrio* sp. occurred at suboptimal growth temperature (113). Such results clearly suggest a linkage between cell stress and transformation competence. Specific links between heat-shock response and competence have been demonstrated: the competence-regulating mecA and mecB genes in *Bacillus subtilis* also control ability to grow at high temperature (198, 235); and the mecB gene is closely related to the ClpC ATPase family of heat-shock proteins found in *E. coli* and other organisms (235).

However, the fate of DNA entering a cell may not always be transformation of the genome. It can aid in DNA repair, be used as a nutrient during times of starvation, or offer protection from lysis by phages (214). The fate of DNA introduced into the cell by phage can also be questioned. In the marine environment, many consider that phages are primarily lytic agents, yet little is known of the frequency of lysogeny (incorporation of the phage DNA into the microbial genome for a quiescent period) or potential environmental cues for induction to lyse (168), especially for deep-sea inhabitants not exposed to ultraviolet light or other inducing agents.

Whether or not linkage between environmental stress and genetic modification by transformation or transduction exists for deep-sea microorganisms remains to be seen; however, the variety and frequently unique combinations of stresses experienced by microorganisms living at great depths in the sea suggest at least an equal variety of molecular mechanisms for genetic exchange, some of which may well prove to be novel. If environmental stress is key to genetic exchange, then the concept of deep-sea habitats, especially those that are confined and surface rich (aggregates, sediments, animal guts, flanges), as environments where genes are readily recaptured is a tempting hypothesis. Further research on stress, genetic competency, and exchange in microorganisms indigenous to the deep sea will eventually elucidate not only the frequency of genetic modification in situ but also the genetic basis of, and microbial flexibility in acquiring, the unique lifestyles of the deep sea—psychrophily, barophily, and hyperthermophily. Some of the early culture work has shown that bacteriophage can modulate maximal and minimal growth temperatures in *Pseudomonas* spp., thereby providing rapid adaptation to seasonal or deep-sea conditions (261). Although these experimental results have not been confirmed (to our knowledge), other reports on mutations in *Pseudomonas aeruginosa*, extending the minimal growth temperature by 12°C, indicated that only one or a few genes may be involved in establishing psychrophily (13, 261). The possibility that thermal characteristics may be transferable in procaryotes is receiving new attention (212). The genetic basis of hyperthermophily, and most likely barophily, both evolving before psychrophily (hot deep marine subsurface environments having existed before the Earth and its ocean cooled [16, 20, 84, 128]), can be expected to be more complex and perhaps more rewarding in the unravelling.

CONCLUDING REMARKS

One of the most fundamental and as yet unanswered questions in marine microbial ecology relates to the physiological state of microorganisms in situ: what portion of a given microbial population is actively metabolizing and growing or capable of growth and what portion has entered a dormant (perhaps nonrecoverable) state due to starvation or other stresses? Assessing activities of indigenous microorganisms under conditions too extreme for shallow-water intruders appears to be synonymous with quantifying rates of mineralization of organic material entering the ocean interior and forming deposits on the seafloor. It also seems a prerequisite to understanding the potential of microbial consortia and symbioses uniquely established in the deep sea. Better assessment of the many physical, physiological, and molecular mechanisms involved in the survival strategies of active versus dormant, culturable versus nonculturable microorganisms (both indigenous forms and intruders), under the extreme conditions of the deep sea promises an improved basis for predicting microbial behavior in unexplored realms of the Earth and perhaps elsewhere in the Universe. With new efforts by the Ocean Drilling Program to plumb the subsurface depths of the ocean (264, 313, 342), and by NASA to explore planetary bodies that may support (or once supported) an ocean and, by implication, life (88, 110, 190, 224), the need to advance understanding of these microbial phenomena is timely.

With increasing concerns about deteriorating quality of the environment, the possible links between microbial responses to naturally imposed stresses and anthropogenic inputs of toxic organic compounds, heavy metals, sewage, and pharmaceutical wastes await exploitation. A recent treatment of the subject of microbial dormancy in the deep sea (207) suggested that dormant microorganisms may be more susceptible to pollution than their counterparts in shallow waters or soils. Our analysis of the heat-shock response and links between starvation, temperature, and pressure stress suggest the opposite: microorganisms stressed by natural deep-sea conditions may be better prepared to tolerate toxic compounds. Recent work has revealed close phylogenetic and physiological relationships between the dominant bacteria in cold (shallow) oligotrophic seawater (335) and those involved in the active degradation of organic pollutants in marine sediments (122). Directed studies of nonculturable bacteria in the deep sea may yield novel strains, metabolic mechanisms, or biosensors exploitable for bioremediation and biotechnology (82, 83,

132, 334). The precedent for success has already been set by studies of stressed microorganisms (102, 192, 328) in deep terrestrial environments (43). In fact, a burgeoning biotechnology industry based on thermostable enzymes from deep-sea thermophiles has been established. The recent development of a rich selection of molecular methods for addressing questions in microbial ecology, including fluorescently labelled rRNA probes, mRNA probes, and bioluminescence gene fusions (6, 71, 77, 205, 263, 328, 337), suggests that the study of active, dormant, and stressed cells, even in the relatively inaccessible depths of the ocean, can now be accelerated. Specific probes to heat shock genes or transcripts (151) should be particularly revealing.

Throughout this chapter we have counterposed knowledge and status of research on microorganisms from the cold deep sea and from hydrothermal vents. In the eukaryotic world of the deep sea, a clear distinction between these two environments is defined at the taxonomic level. High diversity exists in the cold deep sea at the lowest levels of genus and species, where an extraordinary number of 5 to 10 million species has been estimated (119, 133), yet diversity at higher phylogenetic levels is quite limited. Conversely, species diversity at vent sites appears limited to fewer than 300 species (320), while a rich array of higher taxa, including eight phyla, have been discovered there (320). These distinctions can be attributed to the relative constancy of conditions in the cold deep sea, where only evolutionary fine-tuning at the species level has been required for niche adaptation, and the more extreme and varied conditions that exist at hydrothermal vents sites. We have made a case for high microbial diversity throughout the deep sea, based on the range of activities and habitats encountered at the microbial scale. However, the range of microbial habitats at vent sites is considerably broader than that of vent Eukarya, and for all biota in the cold deep sea, primarily because of extremes in temperature and, in the subsurface realm, pressure. The fossil record and other considerations trace the earliest inhabitants on Earth to microbial ecosystems >3.8 billion years ago at hydrothermal vents (1, 20). In contrast, the cold deep sea as we know it today is less than 60 million years old (181). By inference, all life on Earth evolved from vent microorganisms (16). Considering the wide variations and instabilities in physical conditions at vents, it is likely that stress responses also evolved early in vent microorganisms (21). We believe that further studies of the diversity of vent microorganisms and their stress responses will illuminate the origin and evolution of molecular mechanisms for surviving extreme conditions outside the normal growth ranges of extant organisms. A better understanding of mechanisms for surviving starvation, the nonculturable state and long-term dormancy may come from studies of microorganisms in the cold deep sea.

Acknowledgments. Support for preparation of the manuscript was provided by grants from the National Science Foundation (OCE-9144237 to JWD and BCS-9320070 to JAB) and the Office of Naval Research (N00014-92-J-1578 and N00014-94-1-0264 to JWD). Some of the unpublished data were obtained under earlier NSF (DPP-8800401 to JWD) and ONR grants (N00014-89-J-1048 to JWD and N00014-91-J-1678 to JAB). We thank Shelly Carpenter for assistance with graphics.

REFERENCES

1. **Achenbach-Richter, L., R. Gupta, K. O. Stetter, and C. R. Woese.** 1987. Were the original eubacteria thermophiles? *Syst. Appl. Microbiol.* **9:**34.

2. **Albertson, N. H., T. Nystrom, and S. Kjelleberg.** 1990. Exoprotease activity of two marine bacteria during starvation. *Appl. Environ. Microbiol.* **56:**218–223.
3. **Aller, R. C., and J. Y. Aller.** 1986. Evidence for localized enhancement of biological activity associated with tube and burrow structures in deep-sea sediments at the HEBBLE site, western North Atlantic. *Deep-Sea Res.* **33:**755–790.
4. **Alperin, M. J., and W. S. Reeburgh.** 1984. Geochemical observations supporting anaerobic methane oxidation, p. 282–289. *In* R. L. Crawford and R. S. Hanson (ed.), *Microbial Growth on C-1 Compounds*. American Society for Microbiology, Washington, D.C.
5. **Alperin, M. J., W. S. Reeburgh, and M. J. Whiticar.** 1988. Carbon and hydrogen isotope fractionation resulting from anaerobic methane oxidation. *Glob. Biogeochem. Cycles* **2:**279–288.
6. **Amann, R. I., W. Ludwig, and K.-H. Schleifer.** 1995. Phylogenetic identification and in situ detection of individual microbial cells without cultivation. *Microbiol. Rev.* **59:**143–169.
7. **Ammerman, J. W., and F. Azam.** 1987. Characteristics of cyclic AMP transport by marine bacteria. *Appl. Environ. Microbiol.* **53:**2963–2966.
8. **Amy, P. S., and R. Y. Morita.** 1983. Starvation-survival of sixteen freshly isolated open ocean bacteria. *Appl. Environ. Microbiol.* **55:**788–793.
9. **Andrews, J. H.** 1991. *Comparative Ecology of Microorganisms and Macroorganisms.* Springer-Verlag, New York, N.Y.
10. **Antoine, E., C. Cilia, J. R. Meunier, J. Guezennec, F. Lesongeur, and G. Barbier.** 1997. *Thermosipho melanesiensis* sp. nov., a new thermophilic anaerobic bacterium belonging to the order Thermotogales, isolated from deep-sea hydrothermal vents in the Southwestern Pacific Ocean. *Intl. J. Syst. Bacteriol.* **47:**1118–1123.
11. **Araki, T.** 1991. The effect of temperature shifts on protein synthesis by the psychrophilic bacterium *Vibrio* sp. strain ANT-300. *J. Gen. Microbiol.* **137:**817–826.
12. **Arnosti, C.** 1998. Rapid potential rates of extracellular enzymatic hydrolysis in Arctic sediments. *Limnol. Oceanogr.* **43:**315–324.
13. **Azuma, Y., S. B. Newton, and L. D. Witter.** 1962. Production of psychrophilic mutants from mesophilic bacteria by ultraviolet irradiation. *J. Dairy Sci.* **45:**1529–1530.
14. **Barcina, I., P. Lebaron, and J. Vives-Rego.** 1997. Survival of allochthonous bacteria in aquatic systems: a biological approach. *FEMS Microbiol. Ecol.* **23:**1–9.
15. **Barns, S. M., R. E. Fundyga, M. W. Jeffries, and N. R. Pace.** 1994. Remarkable archaeal diversity detected in a Yellowstone National Park hot spring environment. *Proc. Natl. Acad. Sci. USA* **91:** 1609–1613.
16. **Baross, J. A.** 1998. Do the geological and geochemical records of the early earth support the prediction from global phylogenetic models of a thermophilic cenancestor? p. 3–18. *In* J. Weigel and M. Adams (ed.), *Thermophiles: the Keys to Molecular Evolution and the Origin of Life*. Taylor and Francis, London, UK.
17. **Baross, J. A., and J. W. Deming.** 1983. Growth of "black smoker" bacteria at temperatures of at least 250°C. *Nature* **303:**423–426.
18. **Baross, J. A., and J. W. Deming.** 1985. The role of bacteria in the ecology of black smoker environments. *In* M. Jones (ed.), *The Hydrothermal Vents of the Eastern Pacific: an Overview, Biol. Soc. of Washington Bull.* **6:**355–371.
19. **Baross, J. A., and J. W. Deming.** 1995. Bacterial growth at high temperatures: isolation and taxonomy, physiology and ecology, p. 169–217. *In* D. M. Karl (ed.), *Microbiology of Deep-Sea Hydrothermal Vents.* CRC Press, New York, N.Y.
20. **Baross, J. A., and S. E. Hoffman.** 1985. Submarine hydrothermal vents and associated gradient environments as sites for the origin and evolution of life. *Origins Life* **15:**327–345.
21. **Baross, J. A., and J. F. Holden.** 1996. Overview of hyperthermophiles and their heat-shock proteins. *Adv. Protein Chem.* **48:**1–34.
22. **Baross, J. A., and R. Y. Morita.** 1978. Microbial life at low temperatures: ecological aspects, p. 9–71. *In* D. J. Kushner (ed.), *Microbial Life in Extreme Environments.* Academic Press, New York, N.Y.
23. **Baross, J. A., J. W. Deming, and R. R. Becker.** 1984. Evidence for microbial growth in high pressure, high temperature environments, p. 186–195. *In* M. J. Klug and C. A. Reddy (ed.), *Current*

Perspectives in Microbial Ecology: Third International Symposium on Microbial Ecology. American Society for Microbiology, Washington, D.C.

24. **Baross, J. A., F. J. Hanus, and R. Y. Morita.** 1974. Effects of hydrostatic pressure on uracil uptake, ribonucleic acid synthesis, and growth of three obligately psychrophilic marine vibrios, *Vibrio alginolyticus*, and *Escherichia coli*, p. 180–202. *In* R. R. Colwell and R. Y. Morita (ed.), *Effect of the Ocean Environment on Microbial Activities*. University Park Press, Baltimore, Md.
25. **Baross, J. A., F. J. Hanus, and R. Y. Morita.** 1975. Survival of human enteric and other sewage microorganisms under simulated deep-sea conditions. *Appl. Microbiol.* **30:**309–318.
26. **Baross, J. A., M. D. Lilley, and L. I. Gordon.** 1982. Is the CH_4, H_2, and CO venting from submarine hydrothermal systems produced by thermophilic bacteria? *Nature* **298:**366–368.
27. **Baross, J. A., P. A. Tester, and R. Y. Morita.** 1978. Incidence, microscopy, and etiology of exoskeleton lesions in the tanner crab, *Chionoecete tanneri*. *J. Fish. Res. Board Can.* **35:**1141–1149.
28. **Baross, J. A., J. F. Holden, B. C. Crump, M. Summit, and E. J. Mathur.** 1994. Pressure and temperature effects on growth, survival and enzyme stability in deep-sea hyperthermophiles. Abstract, I&EC 0121. American Chemical Society, San Diego, Calif.
29. **Bartholomew, J. W., and S. C. Rittenberg.** 1949. Thermophilic bacteria from deep ocean bottom cores. *J. Bacteriol.* **57:**658.
30. **Bartlett, D. H., C. Kato, and K. Horikoshi.** 1995. High pressure influences on gene and protein expression. *Res. Microbiol.* **146:**697–706.
31. **Bauer, J. E., P. M. Williams, and E. R. M. Druffel.** 1992. ^{14}C activity of dissolved organic carbon fractions in the north-central Pacific and Sargasso Sea. *Nature* **357:**667–668.
32. **Benner, R., J. D. Pakulski, M. McCarthy, J. I. Hedges, and P. G. Hatcher.** 1992. Bulk chemical characteristics of dissolved organic matter in the ocean. *Science* **255:**1561–1564.
33. **Bensoussan, M. G., P.-M. Scoditti, and A. J. M. Bianchi.** 1984. Bacterial flora from echinoderm guts and associated sediment in the abyssal Veam Fault. *Mar. Biol.* **79:**1–10.
34. **Bidle, K. A., M. Kastner, and D. H. Bartlett.** 1999. A phylogenetic analysis of microbial communities associated with methane hydrate containing marine fluids and sediments in the Cascadia margin (ODP site 892B). *FEMS Microbiol. Lett.* **177:**101–108.
35. **Billen, G.** 1982. Modelling the processes of organic matter degradation and nutrient in sedimentary systems, p. 15–52. *In* D. B. Nedwell and C. M. Brown (ed.), *Sediment Microbiology*. Academic Press, New York, N.Y.
36. **Bischoff, J. L., and R. J. Rosenbauer.** 1989. Salinity variations in submarine hydrothermal systems by layered double-diffusive convection. *J. Geol.* **97:**613–623.
37. **Blochl, E., R. Rachel, S. Burggraf, D. Hafenbradl, H. W. Jannasch, and K. O. Stetter.** 1997. *Pyrolobus fumarii*, gen. and sp. nov., represents a novel group of archaea, extending the upper temperature limit for life to 113°C. *Extremophiles* **1:**14–21.
38. **Boetius, A., and E. Damm.** 1998. Benthic oxygen uptake, hydrolytic potentials and microbial biomass at the Arctic continental slope. *Deep-Sea Res. I* **45:**239–275.
39. **Boetius, A., and K. Lochte.** 1994. Regulation of microbial enzymatic degradation of organic matter in deep-sea sediments. *Mar. Ecol. Prog. Ser.* **104:**299–307.
40. **Boetius, A., and K. Lochte.** 1996. High proteolytic activities of deep-sea bacteria from oligotrophic polar sediments. *Aquat. Microb. Ecol.* **48:**269–276.
41. **Boivin-Jahns, V., R. Ruimy, A. Bianchi, S. Daumas, and R. Christen.** 1996. Bacterial diversity in a deep-subsurface clay environment. *Appl. Environ. Microbiol.* **62:**3405–3412.
42. **Bowman, J. P., S. A. McCammon, M. V. Brown, D. S. Nichols, and T. A. McMeekin.** 1997. Diversity and association of psychrophilic bacteria in Antarctic sea ice. *Appl. Environ. Microbiol.* **63:**3068–3078.
43. **Brockman, F. J., B. A. Denovan, R. J. Hicks, and J. K. Fredrickson.** 1989. Isolation and characterization of quinoline-degrading bacteria from subsurface sediments. *Appl. Environ. Microbiol.* **55:**1029–1032.
44. **Burggraf, S., H. W. Jannasch, B. Nicolaus, and K. O. Stetter.** 1990. *Archaeoglobus profundus* sp. nov., represents a new species within the sulfate-reducing archaebacteria. *Syst. Appl. Microbiol.* **13:**24–28.
45. **Burggraf, S., K. O. Stetter, P. Rouviere, and C. R. Woese.** 1991. *Methanopyrus kandleri*: an archaeal methanogen unrelated to all other known methanogens. *Syst. Appl. Microbiol.* **14:**346–349.

46. **Canganella, F., J. M. Gonzalez, M. Yanagibayashi, C. Kato, and K. Horikoshi.** 1997. Pressure and temperature effects on growth and viability of the hyperthermophilic archaeon *Thermococcus peptonophilus*. *Arch. Microbiol.* **168:**1–7.
47. **Cary, S. C., M. T. Cottrell, J. L. Stein, F. Camacho, and D. Desbruyeres.** 1997. Molecular identification and localization of filamentous symbiotic bacteria associated with the hydrothermal vent annelid *Alvinella pompejana*. *Appl. Environ. Microbiol.* **63:**1124–1130.
48. **Cavanaugh, C. M.** 1994. Microbial symbiosis patterns of diversity in the marine environment. *Am. Zool.* **34:**79–89.
49. **Certes, A.** 1884. Sur la culture, a l'abri des germes atmospheriques, des eaux et des sediments rapportes par les expeditions du Travailleur et du Talisman, 1882–1883. *C. R. Acad. Sci.* **98:**690–693.
50. **Chappelle, E. W., G. L. Picciolo, and J. W. Deming.** 1978. Determination of bacterial content in fluids. *Methods Enzymol.* **57:**65–72.
51. **Childress, J. J., and C. R. Fisher.** 1992. The biology of hydrothermal vent animals: physiology, biochemistry, and autotrophic symbioses. *Oceanogr. Mar. Biol. Annu. Rev.* **30:**337–441.
52. **Childress, J. J., C. R. Fisher, J. M. Brooks, M. C. Kennicutt II, R. Bidigare, and A. E. Anderson.** 1986. A methanotrophic marine moluscan (Bivalvia, Mytilidae) symbiosis: mussels fueled by gas. *Science* **233:**1306–1308.
53. **Cho, B. C., and F. Azam.** 1988. Major role of bacteria in biogeochemical fluxes in the ocean's interior. *Nature* **332:**441–442.
54. **Colquhoun, J. A., J. Mexson, M. Goodfellow, A. C. Ward, K. Horikoshi, and A. T. Bull.** 1998. Novel rhodococci and other mycolate actinomycetes from the deep sea. *Antonie van Leeuwenhoek* **74:**27–40.
56. **Cowen, J. P.** 1989. Positive pressure effect on manganese binding by bacteria in deep-sea hydrothermal plumes. *Appl. Environ. Microbiol.* **55:**764–766.
57. **Cowen, J. P., G. J. Massoth, and R. A. Feely.** 1990. Scavenging rates of dissolved manganese in a hydrothermal vent plume. *Deep-Sea Res.* **37:**1619–1637.
58. **Cragg, B. A., and R. J. Parkes.** 1994. Bacterial profiles in hydrothermally active deep sediment layers from middle valley (NE Pacific), sites 857 and 858. *Proc. Ocean Drilling Program Sci. Results* **139:**509–516.
59. **Crump, B. C., E. V. Armbrust, and J. A. Baross.** 1999. Phylogenetic analysis of particle-attached and free-living bacterial communities in the Columbia River, estuary and adjacent coastal ocean. *Appl. Environ. Microbiol.* **65:**3192–3204.
60. **Dahlback, B., L. A. H. Gunnarsson, M. Hermansson, and S. Kjelleberg.** 1982. Microbial investigations of surface microlayers, water column, ice and sediment in the Arctic Ocean. *Mar. Ecol. Prog. Ser.* **9:**101–109.
61. **Dawe, L. L., and W. R. Penrose.** 1978. "Bactericidal" property of seawater: death or debilitation? *Appl. Environ. Microbiol.* **35:**829–833.
62. **Dawes, E. A.** 1985. Starvation, survival and energy reserves, p. 43–79. *In* M. Fletcher and G. D. Floodgate (ed.), *Bacteria in Their Natural Environment.* Special Publications of the Society for General Microbiology, no. 16. Academic Press, New York, N.Y.
63. **de Angelis, M. A., J. A. Baross, and M. D. Lilley.** 1991. Enhanced microbial methane oxidation in water from a deep-sea hydrothermal vent field at simulated *in situ* hydrostatic pressures. *Limnol. Oceanogr.* **36:**565–569.
64. **de Angelis, M. A., A.-L. Reysenbach, and J. A. Baross.** 1991. Surfaces of hydrothermal vent invertebrates: Sites of elevated microbial CH_4 oxidation activity. *Limnol. Oceanogr.* **36:**570–577.
65. **de Angelis, M., M. D. Lilley, E. Olson, and J. A. Baross.** 1993. Microbial methane oxidation in deep-sea hydrothermal plumes of the Endeavour Segment of Juan de Fuca Ridge. *Deep-Sea Res.* **40:**1169–1186.
66. **DeFlaun, M. F., and L. M. Mayer.** 1983. Relationships between bacteria and grain surfaces in intertidal sediments. *Limnol. Oceanogr.* **28:**873–881.
67. **DeFlaun, M. F., J. F. Paul, and W. H. Jeffrey.** 1987. Distribution and molecular weight of dissolved DNA in subtropical estuarine and oceanic environments. *Mar. Ecol. Prog. Ser.* **38:**65–73.

68. **Delaney, J. R., V. Robigou, R. E. McDuff, and M. K. Tivey.** 1992. Detailed geologic relationships of a vigorous hydrothermal system: the Endeavour Vent Field, Northern Juan de Fuca Ridge. *J. Geophys. Res.* **97:**19,663–19,682.
69. **Delaney, J. R., D. S. Kelley, M. D. Lilley, D. A. Butterfield, J. A. Baross, W. S. D. Wilcock, R. W. Embley, and M. Summit.** 1998. The quantum event of oceanic crustal accretion: impacts of diking at mid-ocean ridges. *Science* **281:**222–230.
70. **DeLong, E. F.** 1992. Archaea in coastal marine environments. *Proc. Natl. Acad. Sci. USA* **89:**5685–5689.
71. **DeLong, E. F.** 1997. Marine microbial diversity: the tip of the iceberg. *Trends Microbiol.* **15:**203–207.
72. **DeLong, E. F., and A. A. Yayanos.** 1985. Adaptation of the membrane lipids of a deep-sea bacterium to changes in hydrostatic pressure. *Science* **228:**1101–1102.
73. **DeLong, E. F., and A. A. Yayanos.** 1986. Biochemical function and ecological significance of novel bacterial lipids in deep-sea procaryotes. *Appl. Environ. Microbiol.* **51:** 730–737.
74. **DeLong, E. F., and A. A. Yayanos.** 1987. Properties of the glucose transport system in deep-sea bacteria. *Appl. Environ. Microbiol.* **53:**527–532.
75. **DeLong, E. F., D. G. Franks, and A. L. Alldredge.** 1993. Phylogenetic diversity of aggregate-attached vs. free-living marine bacterial assemblages. *Limnol. Oceanogr.* **38:**924–934.
76. **DeLong, E. F., D. G. Franks, and A. Yayanos.** 1997. Evolutionary relationships of cultivated psychrophilic and barophilic deep-sea bacteria. *Appl. Environ. Microbiol.* **63:**2105–2108.
77. **DeLong, E. F., G. S. Wickham, and N. R. Pace.** 1989. Phylogenetic stains: Ribosomal RNA-based probes for the identification of single cells. *Science* **243:**1360–1363.
78. **Deming, J. W.** 1985. Bacterial growth in deep-sea sediment trap and boxcore samples. *Mar. Ecol. Prog. Ser.* **25:**305–312.
79. **Deming, J. W.** 1986. Ecological strategies of barophilic bacteria in the deep ocean. *Microbiol. Sci.* **3:**205–211.
80. **Deming, J. W.** 1993. Psychrophily in the deep sea, p. 33–36. *In* R. Guerrero and C. Pedros-Alio (ed.), *Trends in Microbial Ecology.* Spanish Society for Microbiology, Barcelona, Spain.
81. **Deming, J. W.** 1997. Unusual or extreme high-pressure marine environments, p. 366–376. *In* C. J. Hurst (ed.), *Manual of Environmental Microbiology.* ASM Press, Washington, D.C.
82. **Deming, J. W.** 1998. Marine bioremediation, p. 19–29. *In* Proceedings of the US:EU Workshop on Marien Microorganisms—Research Issues for Biotechnology. European Commission on Marine Science and Technology, Brussels, Belgium.
83. **Deming, J. W.** 1998. Deep ocean environmental biotechnology. *Curr. Opin. Biotechnol.* **9:**283–287.
84. **Deming, J. W., and J. A. Baross.** 1993. Deep-sea smokers: windows to a subsurface biosphere? *Geochim. Cosmochim. Acta* **57:**3219–3230.
85. **Deming, J. W., and J. A. Baross.** 1993. The early diagenesis of organic matter: bacterial activity, p. 119–144. *In* M. H. Engel and S. A. Macko (ed.), *Organic Geochemistry.* Plenum Press, New York, N.Y.
86. **Deming, J. W., and R. R. Colwell.** 1982. Barophilic bacteria associated with the digestive tracts of abyssal holothurians. *Appl. Environ. Microbiol.* **44:**1222–1230.
87. **Deming, J. W., and R. R. Colwell.** 1985. Observations of barophilic microbial activity in samples of sediment and intercepted particulates from the Demerara Abyssal Plain. *Appl. Environ. Microbiol.* **50:**1002–1006.
88. **Deming, J. W., and A. L. Huston.** An oceanographic perspective on microbial life at low temperatures with implications for polar ecology, biotechnology and astrobiology. *In* J. Seckbach (ed.), *Cellular Origins and Life in Extreme Habitats.* Kluwer Academic Publishers, Dordrecht, The Netherlands, in press.
89. **Deming, J. W., and J. D. Wilkins.** 1987. Effects of starvation on growth of *Shewanella benthica*, a barophilic bacterium from the deep ocean, abstr. N-82. *In Abstr. Annu. Meet. Am. Soc. Microbiol. 1987.* American Society for Microbiology, Washington, D.C.
90. **Deming, J. W., and P. L. Yager.** 1992. Natural bacterial assemblages in deep-sea sediments: towards a global view, p. 11–27. *In* G. T. Rowe and V. Pariente (ed.), *Deep-Sea Food Chains and the Global Carbon Cycle.* Kluwer Academic Publishers, Dordrecht, The Netherlands.

91. **Deming, J. W., G. L. Picciolo, and E. W. Chappelle.** 1979. Important factors in ATP determinations using firefly luciferase: applicability of the assay to studies of native aquatic bacteria, p. 89–98. *In* J. W. Costerton and R. R. Colwell (ed.), *Native Aquatic Bacteria, Enumeration, Activity, and Ecology.* ASTM Special Technical Publication 695. ASTM, Philadelphia, Pa.
92. **Deming, J. W., P. S. Tabor, and R. R. Colwell.** 1981. Barophilic growth of bacteria from intestinal tracts of deep-sea invertebrates. *Microb. Ecol.* **7:**85–94.
93. **Deming, J. W., A. L. Reysenbach, S. A. Macko, and C. R. Smith.** 1997. Evidence for the microbial basis of a chemoautotrophic invertebrate community at a whale fall on the deep seafloor: Bone-colonizing bacteria and invertebrate endosymbionts. *Microsc. Res. Tech.* **37:**162–170.
94. **Deming, J. W., H. Hada, R. R. Colwell, K. R. Luehrsen, and G. E. Fox.** 1984. The ribonucleotide sequence of 5S rRNA from two strains of deep-sea barophilic bacteria. *J. Gen. Microbiol.* **130:** 1911–1920.
95. **Deming, J. W., L. K. Somers, W. L. Straube, D. G. Swartz, and M. T. MacDonell.** 1988. Isolation of an obligately barophilic bacterium and description of a new genus, *Colwellia* gen. nov. *Syst. Appl. Microbiol.* **10:**152–160.
96. **Desbruyeres, D., J. W. Deming, A. Dinet, and A. Khripounoff.** 1985. Reactions de l'ecosysteme benthique profond aux perturbations: nouveaux resultats experimentaux, p. 193–208. *In* L. Laubier and C. Monniot (ed.), *Peuplements Profonds du Golfe de Gascogne.* IFREMER (Institute Francais de Recherche pour l'Exploitation de la Mer), Brest, France.
97. **Dietz, A. S., and A. A. Yayanos.** 1978. Silica gel media for isolating and studying bacteria under hydrostatic pressure. *Appl. Environ. Microbiol.* **36:**966–968.
98. **Dilmore, L. A., and M. A. Hood.** 1986. Vibrios of some deep-water invertebrates. *FEMS Microbiol. Lett.* **35:**221–224.
99. **DOE Subsurface Science Program's Taylorsville Basin Working Group.** 1994. D.O.E. seeks origins of deep subsurface bacteria. *EOS Trans. Am. Geophys. Union* **75:**385,395–396.
100. **Druffel, E. R. M., J. E. Bauer, P. M. Williams, S. Griffin, and D. Wolgast.** 1996. Seasonal variability of particulate organic radiocarbon in the northeast Pacific Ocean. *J. Geophys. Res.* **101:** 20,543–20,552.
101. **Duffaud, G. D., O. B. d'Hennezel, A. S. Peek, A.-L. Reysenbach, and R. M. Kelly.** 1998. Isolation and characterization of *Thermococcus barossi*, sp. nov., a hyperthermophilic Archaeon isolated from a hydrothermal vent flange formation. *Syst. Appl. Microbiol.* **21:**40–49.
102. **Duncan, S., L. A. Glover, K. Killham, and J. I. Prosser.** 1994. Luminescence-based detection of activity of starved and viable but nonculturable bacteria. *Appl. Environ. Microbiol.* **60:**1308–1316.
103. **Egorova, A. A.** 1938. Thermophile bacteria in Arctic. *C. R. (Doklady) Acad. Sci. URSS* **19:**649–651.
104. **Erauso, G., A.-L. Reysenbach, A. Godfroy, J.-R. Meunier, B. Crump, F. Partensky, J. A. Baross, V. T. Marteinsson, N. R. Pace, G. Barbier, and D. Prieur.** 1993. *Pyrococcus abyssi* sp. nov., a new hyperthermophilic archaeon isolated from a deep-sea hydrothermal vent. *Arch. Microbiol.* **160:**338–349.
105. **Etter, R. J., and F. Grassle.** 1992. Patterns of species diversity in the deep sea as a function of sediment particle size diversity. *Nature* **360:**576–578.
106. **Feigl, F.** 1958. Spot Tests in Inorganic Analyses, p. 600. D. Van Nostrand Co., Inc., Princeton, N.J.
107. **Fiala, G., K. O. Stetter, H. W. Jannasch, T. A. Langworthy, and J. Madon.** 1986. *Staphylothermus marinus* sp. nov. represents a novel genus of extremely thermophilic submarine heterotrophic archaebacteria growing up to 98°C. *Syst. Appl. Microbiol.* **8:**106–113.
108. **Fiala-Medioni, A., and H. Felbeck.** 1990. Autotrophic processes in invertebrate nutrition: bacterial symbioses in bivalve molluscs. *Comp. Physiol.* **5:**49–69.
109. **Fisher, C. R.** 1990. Chemoautotrophic and methanotrophic symbioses in marine invertebrates. *Rev. Aquat. Sci.* **2:**399–436.
110. **Fisk, M. R., and S. J. Giovannoni.** 1999. Sources of nutrients and energy for a deep biosphere on Mars. *J. Geophys. Res. Planets* **104:**11805–11815.
111. **Fossing, H., V. A. Gallardo, B. B. Jorgensen, M. Huttel, L. P. Nielsen, H. Schulz, D. E. Canfield, S. Forster, R. N. Glud, J. K. Gunderson, J. Kuver, N. B. Ramsing, A. Teske, B. Thamdrup,**

and O. Ulloa. 1995. Concentration and transport of nitrate by the mat-forming sulphur bacterium *Thioploca. Nature* **374:**713–715.

112. **Fredrickson, J. K., and T. C. Onstott.** 1996. Microbiology of deep subsurface environments. *Sci. Am.* **274:**42–47.
113. **Frischer, M. E., J. M. Thurmond, and J. H. Paul.** 1993. Factors affecting competence in a high frequency of transformation *Vibrio. J. Gen. Microbiol.* **139:**753–761.
114. **Frischer, M. E., G. J. Stewart, and J. H. Paul.** 1994. Plasmid transfer to indigenous marine bacterial populations by natural transformation. *FEMS Microb. Ecol.* **15:**127–136.
115. **Fry, J. C., and M. J. Day (ed.).** 1990. *Bacterial Genetics in Natural Environments*. Chapman and Hall, Ltd., London, U.K.
116. **Fuhrman, J. A., and A. A. Davis.** 1997. Widespread Archaea and novel Bacteria from the deep sea as shown by 16S rRNA gene sequences. *Mar. Ecol. Prog. Ser.* **150:**275–285.
117. **Fuhrman, J. A., and C. A. Suttle.** 1993. Viruses in marine planktonic systems. *Oceanogr.* **6:**51–63.
118. **Fuhrman, J. A., K. McCallum, and A. A. Davis.** 1992. Novel major archaebacterial group from marine plankton. *Nature* **356:**148–149.
119. **Gage, J. D.** 1996. Why are there so many species in deep-sea sediments? *J. Exp. Mar. Biol. Ecol.* **200:**257–286.
120. **Gallardo, V. A.** 1977. Large benthic microbial communities in sulphide biota under Peru-Chile subsurface countercurrent. *Nature* **268:**331–332.
121. **Gauthier, M. J., B. Labedan, and V. A. Breittmayer.** 1992. Influence of DNA supercoiling on the loss of culturability of *Escherichia coli* cells incubated in seawater. *Mol. Ecol.* **1:**183–190.
122. **Geiselbrecht, A. D., R. P. Herwig, J. W. Deming, and J. T. Staley.** 1996. Enumeration and phylogenetic analysis of polycyclic aromatic hydrocarbon-degrading marine bacteria from Puget Sound sediments. *Appl. Environ. Microbiol.* **62:**3344–3349.
123. **Gest, H., and J. Mandelstam.** 1987. Longevity of microorganisms in natural environments. *Microbiol. Sci.* **4**(3)**:**69–71.
124. **Giovannoni, S. J., and S. C. Cary.** 1993. Probing marine systems with ribosomal RNAs. *Oceanography* **6:**95–104.
125. **Giovannoni, S. J., T. B. Britschgi, C. L. Moyer, and K. G. Field.** 1990. Genetic diversity in Sargasso Sea bacterioplankton. *Nature* **345:**60–63.
126. **Godfroy, A., J.-R. Meunier, J. Guezennec, F. Lesongeur, G. Raguenes, A Rimbault, and G. Barbier.** 1996. *Thermococcus fumicolans* sp. nov., a new hyperthermophilic archaeon isolated from a deep-sea hydrothermal vent in the North Fiji Basin. *Intl. J. Syst. Bacteriol.* **46:**1113–1119.
127. **Godfroy, A., F. Lesongeur, G. Raguenes, J. Querellou, E. Antoine, J.-R. Meunier, J. Guezennec, and G. Barbier.** 1997. *Thermococcus hydrothermalis* sp. nov., a new hyperthermophilic archaeon isolated from a deep-sea hydrothermal vent. *Int. J. Syst. Bacteriol.* **47:**622–626.
128. **Gold, T.** 1992. The deep, hot biosphere. *Proc. Natl. Acad. Sci. USA* **89:**6045–6049.
129. **Gonzalez, J. M., C. Kato, and K. Horikoshi.** 1995. *Thermococcus peptonophilus* sp. nov., a fast growing, extremely thermophilic archaebacterium isolated from deep-sea hydrothermal vents. *Arch. Microbiol.* **164:**159–164.
130. **Gooday, A. J., and C. M. Turley.** 1990. Responses by benthic organisms to inputs of organic material to the ocean floor: A review. *Philos. Trans. R. Soc. London Ser.* A **331:**119–138.
131. **Gosink, J. J., and J. T. Staley.** 1995. Biodiversity of gas vacuolate bacteria from antarctic sea ice and water. *Appl. Environ. Microbiol.* **61:**3486–3489.
132. **Gounot, A.-M.** 1991. Bacterial life at low temperature: physiological aspects and biotechnological implications. *J. Appl. Bacteriol.* **71:**386–397.
133. **Grassle, J. F., and N. J. Maciolek.** 1992. Deep-sea species richness: regional and local diversity estimates from quantitative bottom samples. *Am. Nat.* **139:**313–341.
134. **Grote, R., L. Li, J. Tamaoka, C. Kato, K. Horikoshi, and G. Antranikian.** 1999. *Thermococcus siculi* sp. nov., a novel hyperthermophilic archaeon isolated from a deep-sea hydrothermal vent at the Mid-Okinawa Trough. *Extremophiles* **3:**55–62.
135. **Gundersen, J. K., B. B. Jorgensen, E. Larsen, and H. W. Jannasch.** 1992. Mats of giant sulphur bacteria on deep-sea sediments due to fluctuating hydrothermal flow. *Nature* **360:**454–456.

136. **Haldeman, D. L., P. S. Amy, D. C. White, and D. B. Ringelberg.** 1994. Changes in bacteria recoverable from subsurface volcanic rock samples during storage at 4°C. *Appl. Environ. Microbiol.* **60:**2697–2703.
137. **Hamamoto, T., and K. Horikoshi.** 1991. Characterization of an amylase from a psychrotrophic *Vibrio* isolated from a deep-sea mud sample. *FEMS Microbiol. Lett.* **84:**79–84.
138. **Hamamoto, T., N. Takata, T. Kudo, and K. Horikoshi.** 1995. Characteristic presence of polyunsaturated fatty acids in marine psychrophilic vibrios. *FEMS Microbiol. Lett.* **129:**51–56.
139. **Harmsen, H. J. M., D. Prieur, and C. Jeanthon.** 1997. Distribution of microorganisms in deep-sea hydrothermal vent chimneys investigated by whole-cell hybridization and enrichment culture of thermophilic subpopulations. *Appl. Environ. Microbiol.* **63:**2876–2883.
140. **Haygood, M., and D. L. Distel.** 1993. Bioluminescent symbionts of flashlight fishes and deep-sea anglerfishes form unique lineages related to the genus *Vibrio*. *Nature* **363:**154–156.
141. **Haymon, R. M., D. J. Fornari, K. L. Von Damm, M. D. Lilley, M. R. Perfit, J. M. Edmond, W. C. Shanks III, R. A. Lutz, J. M. Grebmeier, S. Carbotte, D. Wright, E. McLaughlin, M. Smith, N. Beedle, and E. Olson.** 1993. Volcanic eruption of the mid-ocean ridge along the East Pacific Rise Crest at 9°45–52′N. I. Direct submersible observations of seafloor phenomena associated with an eruption event in April, 1991. *Earth Planet. Sci. Lett.* **119:**85–119.
142. **Hedrick, D. B., R. J. Pledger, D. C. White, and J. A. Baross.** 1992. In situ microbial ecology of hydrothermal vent sediments. *FEMS Microb. Ecol.* **101:**1–10.
143. **Hei, D. J., and D. S. Clark.** 1994. Pressure stabilization of proteins from extreme thermophiles. *Appl. Environ. Microbiol.* **60:**932–939.
144. **Helmke, E., and H. Weyland.** 1986. Effect of hydrostatic pressure and temperature on the activity and synthesis of chitinases of Antarctic Ocean bacteria. *Mar. Biol.* **91:**1–7.
145. **Helmke, E., and H. Weyland.** 1995. Bacteria in sea ice and underlying water of the eastern Weddell Sea in midwinter. *Mar. Ecol. Prog. Ser.* **117:**269–287.
146. **Hermansson, M., and C. Linberg.** 1994. Gene transfer in the marine environment. *FEMS Microb. Ecol.* **15:**47–54.
147. **Hill, R. T., I. T. Knight, M. S. Anikis, and R. R. Colwell.** 1993. Benthic distribution of sewage sludge indicated by *Clostridium perfringens* at a deep-ocean dump site. *Appl. Environ. Microbiol.* **59:**47–51.
148. **Holden, J. F., and J. A. Baross.** 1993. Enhanced thermotolerance and temperature-induced changes in protein composition in the hyperthermophilic archaeon ES4. *J. Bacteriol.* **175:**2839–2843.
149. **Holden, J. F., and J. A. Baross.** 1995. Enhanced thermotolerance by hydrostatic pressure in the deep-sea hyperthermophile *Pyrococcus* strain ES4. *FEMS Microb. Ecol.* **18:**27–33.
150. **Holden, J. F., M. Summit, and J. A. Baross.** 1998. Thermophilic and hyperthermophilic microorganisms in 3–30°C hydrothermal fluids following a deep-sea volcanic eruption. *FEMS Microbiol. Ecol.* **25:**33–41.
151. **Holden, J. F., M. W. W. Adams, and J. A. Baross.** Heat-shock response in hyperthermophilic microorganisms. *In Stress Genes*: *Role in Physiological Ecology*, *Progress in Microbial Ecology*. Proceedings of the 8th International Symposium on Microbial Ecology, Halifax, Canada, in press.
152. **Hood, M. A., J. B. Guckert, D. C. White, and F. Deck.** 1986. Effect of nutrient deprivation on lipid, carbohydrate, DNA, RNA and protein levels in *Vibrio cholerae*. *Appl. Environ. Microbiol.* **52:**788–793.
153. **Hoppe, H.-G.** 1991. Microbial extracellular enzyme activity: a new key parameter in aquatic ecology, p. 60–83. *In* R. Chrost (ed.), *Microbial Enzymes in Aquatic Environments*. Springer-Verlag, New York, N.Y.
154. **Huber, R., M. Kurr, H. W. Jannasch, and K. O. Stetter.** 1989. A novel group of abyssal methanogenic archaebacteria (*Methanopyrus*) growing at 110°C. *Nature* **342:**833–834.
155. **Huber, R., P. Stoffers, J. L. Cheminee, H. H. Richnow, and K. O. Stetter.** 1990. Hyperthermophilic archaebacteria within the crater and open-sea plume of erupting Macdonald Seamount. *Nature* **345:**179–181.
156. **Huber, R., H. Jannasch, R. Rachel, T. Fuchs, and K. O. Stetter.** 1997. *Archaeoglobus veneficus* sp. nov., a novel facultative chemolithoautotrophic hyperthermophilic sulfite reducer, isolated from abyssal black smokers. *Syst. Appl. Microbiol.* **20:**374–380.

157. **Huber, R., J. Stoehr, S. Hohenhaus, R. Rachel, S. Burggraf, H. W. Jannasch, and K. O. Stetter.** 1995. *Thermococcus chitonophagus* sp. nov., a novel, chitin-degrading, hyperthermophilic archaeum from a deep-sea hydrothermal vent environment. *Arch. Microbiol.* **164:**255–264.
158. **Hugenholtz, P., B. M. Goebel, and N. R. Pace.** 1998. Impact of culture-independent studies on the emerging phylogenetic view of bacterial diversity. *J. Bacteriol.* **180:**4765–4774.
158a. **Huston, A. L., B. B. Krieger-Brockett, and J. W. Deming.** Remarkably low temperature optima for extracellular enzyme activity from arctic bacteria and sea ice. *Environ. Microbiol.*, in press.
159. **Jannasch, H. W.** 1979. Microbial turnover of organic matter in the deep sea. *BioSci.* **29:**228–232.
160. **Jannasch, H. W., and M. Mottl.** 1985. Geomicrobiology of deep-sea hydrothermal vents. *Science* **229:**717–725.
161. **Jannasch, H. W., and C. O. Wirsen.** 1981. Morphological survey of microbial mats near deep-sea thermal vents. *Appl. Environ. Microbiol.* **41:**528–538.
162. **Jannasch, H. W., and C. O. Wirsen.** 1984. Variability of pressure adaptations in deep sea bacteria. *Arch. Microbiol.* **139:**281–288.
163. **Jannasch, H. W., D. C. Nelson, and C. O. Wirsen.** 1989. Massive natural occurrence of unusually large bacteria (*Beggiatoa* sp.) at a hydrothermal deep-sea vent site. *Nature* **342:**834–836.
164. **Jannasch, H. W., C. O. Wirsen, and K. W. Doherty.** 1996. A pressurized chemostat for the study of marine barophilic and oligotrophic bacteria. *Appl. Environ. Microbiol.* **62:**1593–1596.
165. **Jannasch, H. W., C. O. Wirsen, and C. D. Taylor.** 1982. Deep-sea bacteria: isolation in the absence of decompression. *Science* **216:**1315–1317.
166. **Jannasch, H. W., C. O. Wirsen, S. J. Molyneaux, and T. A. Langworthy.** 1988. Extremely thermophilic fermentative archaebacteria of the genus *Desulfurococcus* from deep-sea hydrothermal vents. *Appl. Environ. Microbiol.* **54:**1203–1209.
167. **Jeanthon, C., S. L'Haridon, A.-L. Reysenbach, E. Corre, M. Vernet, P. Messner, U. B. Sleytr, and D. Prieur.** 1999. *Methanococcus vulcanius* sp. nov., a novel hyperthermophilic methanogen isolated from East Pacific Rise, and identification of *Methanococcus* sp. DSM 4213T as *Methanococcus fervens* sp. nov. *Intl. J. Syst. Bacteriol.* **49:**583–589.
168. **Jiang, S. C., and J. H. Paul.** 1997. Occurrence of lysogenic bacteria in marine microbial communities as determined by prophage induction. *Mar. Ecol. Prog. Ser.* **142:**27–38.
169. **Johnstone, B. H., and R. D. Jones.** 1988. Effects of light and CO on the survival of a marine ammonium-oxidizing bacterium during energy source deprivation. *Appl. Environ. Microbiol.* **54:** 2890–2893.
170. **Jones, P. G., R. A. VanBogelen, and F. C. Neidhardt.** 1987. Induction of proteins in response to low temperature in *Escherichia coli*. *J. Bacteriol.* **169:**2092–2095.
171. **Jones, R. D., and R. Y. Morita.** 1985. Survival of a marine ammonium oxidizer under energy-source deprivation. *Mar. Ecol. Prog. Ser.* **26:**175–179.
172. **Jones, W. J., C. E. Stugard, and H. W. Jannasch.** 1989. Comparison of thermophilic methanogens from submarine hydrothermal vents. *Arch. Microbiol.* **151:**314–318.
173. **Jones, W. J., J. A. Leigh, F. Mayer, C. R. Woese, and R. S. Wolfe.** 1983. *Methanococcus jannaschii* sp. nov., an extremely thermophilic methanogen from a submarine hydrothermal vent. *Arch. Microbiol.* **136:**254–261.
174. **Jorgensen, K. S., and J. M. Tiedje.** 1993. Survival of denitrifiers in nitrate-free, anaerobic environments. *Appl. Environ. Microbiol.* **59:**3297–3305.
175. **Jorgensen, B. B., M. F. Isaksen, and H. W. Jannasch.** 1992. Bacterial sulfate reduction above 100°C in deep-sea hydrothermal vent sediments. *Science* **258:**1756–1757.
176. **Jumars, P. A., D. L. Penry, J. A. Baross, M. J. Perry, and B. W. Frost.** 1989. Closing the microbial loop: dissolved carbon pathway to heterotrophic bacteria from incomplete ingestion, digestion, and absorption in animals. *Deep-Sea Res.* **36:**483–495.
177. **Jumars, P. A., L. M. Mayer, J. W. Deming, J. A. Baross, and R. A. Wheatcroft.** 1990. Deep-sea deposit-feeding strategies suggested by environmental and feeding constraints. *Philos. Trans. R. Soc. London Ser. A* **331:**85–101.
178. **Jumars, P. A., J. W. Deming, P. S. Hill, P. L. Yager, and L. Karp-Boss.** 1993. Physical determinants of diffusional advanatge in free-living, planktonic bacteria. *Mar. Microb. Food Webs* **7:** 121–159.

179. **Junge, K., J. T. Staley, and J. W. Deming.** 1999. Phylogenetic diversity of numerically important bacteria cultured at subzero temperature from Arctic sea ice, abstr. N-159. *In Gen. Meet. Soc. Microbiol. 1999.* American Society for Microbiology, Washington, D.C.
180. **Juniper, S. K., and B. M. Tebo.** 1995. Microbe-metal interactions and mineral deposition at hydrothermal vents, p. 219–253. *In* D. M. Karl (ed.), *Microbiology of Deep-Sea Hydrothermal Vents.* CRC Press, New York, N.Y.
181. **Kadko, D., J. A. Baross, and J. Alt.** 1995. The magnitude and global implications of hydrothermal flux, p. 446–466. *In* S. Humphris, R. Zierenberg, L. Mullineaux, and R. Thomson (ed.), *Seafloor Hydrothermal Systems: Physical, Chemical, Biological and Geological Interactions, Geophysical Monograph 91.* American Geophysical Union, Washington, D.C.
182. **Kaneko, T., and R. R. Colwell.** 1975. Adsorption of *Vibrio parahaemolyticus* onto chitin and copepods. *Appl. Environ. Microbiol.* **29:**269–274.
183. **Karl, D. M.** 1995. Ecology of free-living, hydrothermal vent microbial communities, p. 35–124. *In* D. M. Karl (ed.), *Microbiology of Deep-Sea Hydrothermal Vents.* CRC Press, New York, N.Y.
184. **Karl, D. M., G. A. Knauer, and J. H. Martin.** 1988. Downward flux of particulate organic matter in the ocean: a particle decomposition paradox. *Nature* **332:**438–441.
185. **Karl, D. M., C. O. Wirsen, and H. W. Jannasch.** 1980. Deep-sea primary production at the Galapagos hydrothermal vents. *Science* **207:**1345–1347.
186. **Kato, C., N. Masui, and K. Horikoshi.** 1996. Properties of obligately barophilic bacteria isolated from a sample of deep-sea sediment from the Izu-Bonin trench. *J. Mar. Biotechnol.* **4:**96–99.
187. **Kato, C., L. Li, J. Tamaoka, and K. Horikoshi.** 1997. Molecular analyses of the sediment of the 11,000-m deep Mariana Trench. *Extremophiles* **1:**117–123.
188. **Kaye, J. Z., and J. A. Baross.** 1998. Salt-tolerant microbes isolated from hydrothermal-vent environments. *EOS Transactions* **79(45):**F59.
189. **Kellogg, C. A., J. B. Rose, S. C. Jiang, J. M. Thurmond, and J. H. Paul.** 1996. Genetic diversity of related vibriophages isolated from marine environments around Florida and Hawaii, USA. *Mar. Ecol. Prog. Ser.* **120:**89–98.
190. **Kerr, R. A.** 1997. Once, maybe still, an ocean on Europa. *Science* **277:**764–765.
191. **Kim, J., and C. E. ZoBell.** 1972. Agarase, amylase, cellulase, and chitinase activity at deep-sea pressures. *J. Oceanogr. Soc. Jpn.* **28:**1–7.
192. **King, J. M. H., P. M. DiGrazia, B. Applegate, R. Burlage, J. Sanseverino, P. Dunbar, F. Larimer, and G. S. Sayler.** 1990. Rapid, sensitive bioluminescent reporter technology for naphthalene exposure and biodegradation. *Science* **249:**778–781.
193. **Kjelleberg, S., K. B. G. Flardh, T. Nystrom, and D. J. W. Moriarty.** 1993. Growth limitation and starvation of bacteria, p. 289–320. *In* T. E. Ford (ed.), *Aquatic Microbiology—an Ecological Approach.* Blackwell Scientific Publ., Inc., Cambridge, Mass.
194. **Kobayashi, T., Y. S. Kwak, T. Akiba, T. Kudo, and K. Horikoshi.** 1994. *Thermococcus profundus* sp. nov., a new hyperthermophilic archaeon isolated from a deep-sea hydrothermal vent. *Syst. Appl. Microbiol.* **17:**232–236.
195. **Kogure, K., U. Simidu, and N. Taga.** 1979. A tentative direct microscopic method for counting living marine bacteria. *Can. J. Microbiol.* **25:**415–420.
196. **Kohlmeyer, J., and E. Kohlmeyer.** 1979. *Marine Mycology: the Higher Fungi.* Academic Press, New York, N.Y.
197. **Koike, I., S. Hara, K. Terauchi, and K. Kogure.** 1990. Role of sub-micron particles in the ocean. *Nature* **345:**242–244.
198. **Kong, L., and D. Dubnau.** 1994. Regulation of competence-specific gene expression by mec-mediated protein-protein interaction in *Bacillus subtilis. Proc. Natl. Acad. Sci. USA* **91:**5793–5797.
199. **Kramer, J. G., and F. L. Singleton.** 1992. Variations in rRNA content of marine *Vibrio* spp. during starvation-survival and recovery. *Appl. Environ. Microbiol.* **58:**201–207.
200. **Kriss, A. E.** 1963. *Marine Microbiology (Deep Sea).* Oliver and Boyd, Edinburgh, Scotland.
201. **Krumholz, L. R.** 1997. Confined subsurface microbial communities in Cretaceous rock. *Nature* **386:**64–66.
202. **Kurr, M., R. Huber, H. K. Konig, H. W. Jannasch, H. Fricke, A. Trincome, J. K. Kristjansson, and K. O. Stetter.** 1991. *Methanopyrus kandleri,* gen. and sp. nov. represents a novel group of hyperthermophilic methanogens growing at 100°C. *Arch. Microbiol.* **156:**239–247.

203. **Kushner, D. J.** 1985. The Halobacteriaceae, p. 171–214. *In* C. R. Woese and R. S. Wolfe (ed.), *The Bacteria: a Treatise on Structure and Function, Vol. VIII. Archaebacteria.* Academic Press, Inc., Orlando, Fla.
204. **Kwak, Y. S., T. Kobayashi, T. Akiba, K. Horikoshi, and Y. B. Kim.** 1995. A hyperthermophilic sulfur-reducing archaebacterium *Thermococcus* sp. DT 1331 isolated from a deep-sea hydrothermal vent. *Biosci. Biotechnol. Biochem.* **59:**1666–1669.
205. **Lee, S., and P. F. Kemp.** 1994. Single-cell RNA content of natural marine planktonic bacteria measured by hybridization with multiple 16S rRNA-targeted fluorescent probes. *Limnol. Oceanogr.* **39(4):**869–879.
206. **Levy, S. B., and R. V. Miller.** 1989. *Gene Transfer in the Environment.* McGraw-Hill Book Co., New York, N.Y.
207. **Lewis, D. L., and D. K. Gattie.** 1991. The ecology of quiescent microbes. *ASM News* **57:**27–32.
208. **Lilley, M. D., J. A. Baross, and L. I. Gordon.** 1983. Reduced gases and bacteria in hydrothermal fluids: the Galapagos spreading center and 21°N East Pacific Rise, p. 411–449. *In* P. Rona, K. Bostrom, L. Laubier, and K. Smith (ed.), *Hydrothermal Processes at Seafloor Spreading Centers.* Plenum Press, New York, N.Y.
209. **Lilley, M. D., D. A. Butterfield, J. E. Lupton, S. A. Macko, and R. E. McDuff.** 1993. Anomalous CH_4 and NH^{4+} concentrations at a mid-ocean ridge hydrothermal system: combined effects of sediments and phase separation. *Nature* **364:**45–47.
210. **Lilley, M. D., R. A. Feely, and J. H. Trefry.** 1995. Chemical and biological transformations in hydrothermal plumes, p. 369–391. *In* S. Humphris, R. Zierenberg, L. Mullineaux, and R. Thomson (ed.), *Seafloor Hydrothermal Systems: Physical, Chemical, Biological and Geological Interactions, Geophysical Monograph 91.* American Geophysical Union, Washington, D.C.
211. **Lindquist, S.** 1992. Heat-shock proteins and stress tolerance in microorganisms. *Curr. Opin. Genet. Dev.* **2:**748–755.
212. **Lindsay, J. A.** 1995. Is thermophily a transferrable property in bacteria? *Crit. Rev. Microbiol.* **21:** 165–174.
213. **Lochte, K., and C. M. Turley.** 1988. Bacteria and cyanobacteria associated with phytodetritus in the deep sea. *Nature* **333:**67–69.
214. **Lorenz, M. G., and W. Wackernagel.** 1994. Bacterial gene transfer by natural genetic transformation in the environment. *Microbiol. Rev.* **58:**563–602.
215. **MacDonell, M. T., and R. R. Colwell.** 1985. Phylogeny of the Vibrionaceae, and recommendation for two new genera, *Listonella* and *Shewanella. Syst. Appl. Microbiol.* **6:**171–182.
216. **Mandranack, K. W., and B. M. Tebo.** 1993. Manganese scavenging and oxidation at hydrothermal vents and in vent plumes. *Geochim. Cosmochim. Acta* **57:**3907–3923.
217. **Marteinsson, V. T., P. Moulin, J. L. Birrien, A. Gambacorta, M. Vernet, and D. Prieur.** 1997. Physiological responses to stress conditions and barophilic behavior of the hyperthermophilic vent archaeon *Pyrococcus abyssi. Appl. Environ. Microbiol.* **63:**1230–1236.
218. **Maruyama, A., R. Taniguchi, H. Tanaka, H. Ishiwata, and T. Higashihara.** 1997. Low-temperature adaptation of deep-sea bacteria isolated from the Japan Trench. *Mar. Biol.* **128:**705–711.
219. **Mayer, L. M.** 1989. Extracellular proteolytic activity in sediments of an intertidal mudflat. *Limnol. Oceanogr.* **34(6):**973–981.
220. **McCallum, K. L., and W. E. Inniss.** 1990. Thermotolerance, cell filamentation and induced protein synthesis in psychrophilic and psychrotrophic bacteria. *Arch. Microbiol.* **153:**585–590.
221. **McCallum, K. L., J. J. Heikkila, and W. E. Inniss.** 1986. Temperature-dependent pattern of heat-shock protein synthesis in psychrophilic and psychrotrophic micro-organisms. *Can. J. Microbiol.* **32:**516–521.
222. **McDougald, D., S. A. Rice, D. Weichart, and S. Kjelleberg.** 1998. Nonculturability: adaptation or debilitation? *FEMS Microbiol. Ecol.* **25:**1–9.
223. **McInerney, J. O., M. Wilkinson, J. W. Patching, T. M. Embley, and R. Powell.** 1995. Recovery and phylogenetic analysis of novel archaeal rRNA sequences from a deep-sea deposit-feeder. *Appl. Environ. Microbiol.* **61:**1646-1648.

224. **McKay, D. S., E. K. Gibson, K. L. Thomas-Keptra, H. Vali, C. S. Romanek, S. J. Clemett, X. D. F. Bhillier, C. R. Maechling, and R. N. Zare.** 1996. Search for past life on Mars: possible relic biogenic activity in Martian meteorite ALH84001. *Science* **273:**924–930.
225. **McLaughlin, B.** 1997. Hydrogen oxidation in hydrothermal plumes. Ph.D. Thesis. University of Washington, Seattle.
226. **Meyer-Reil, L.-A.** 1986. Measurement of hydrolytic activity and incorporation of dissolved organic substrates by microorganisms in marine sediments. *Mar. Ecol. Prog. Ser.* **31:**143–149.
227. **Meyer-Reil, L.-A., and M. Koster.** 1992. Microbial life in pelagic sediments: the impact of environmental parameters on enzymatic degradation of organic material. *Mar. Ecol. Prog. Ser.* **81:** 65–72.
228. **Morgan, P., and C. S. Dow.** 1986. Bacterial adaptations for growth in low nutrient environments, p. 187–214. *In* R. A. Herbert and B. A. Codd (ed.), *Microbes in Extreme Environments.* Academic Press, London, U.K.
229. **Morita, R. Y.** 1975. Psychrophilic bacteria. *Bacteriol. Rev.* **39:**144–167.
230. **Morita, R. Y.** 1976. Survival of bacteria in cold and moderate hydrostatic pressure environments with special reference to psychrophilic and barophilic bacteria, p. 279–298. *In* T. G. R. Gray and J. R. Postgate (ed.), *The Survival of Vegetative Microbes.* Cambridge University Press, Cambridge, U.K.
231. **Morita, R. Y.** 1997. *Bacteria in Oligotrophic Environments: Starvation-Survival Lifestyle.* Chapman and Hall Microbiology Series, Chapman and Hall, New York, N.Y.
232. **Mortimoto, R. I., A. Tissieres, and C. Georgopoulos.** 1990. The stress response, function of the proteins, and perspectives, p. 1–36. *In* R. I. Mortimoto, A. Tissieres, and C. Georgopoulos (ed.), *Stress Proteins in Biology and Medicine.* Cold Spring Harbor Laboratory Press, Cold Spring Harbor, New York, N.Y.
233. **Moyer, C. L., and R. Y. Morita.** 1989. Effect of growth rate and starvation-survival on the viability and stability of a psychrophilic marine bacterium. *Appl. Environ. Microbiol.* **55:**1122–1127.
234. **Moyer, C. L., F. C. Dobbs, and D. M. Karl.** 1995. Phylogenetic diversity of the bacterial community from a microbial mat at an active, hydrothermal vent system, Loihi Seamount, Hawaii. *Appl. Environ. Microbiol.* **61:**1555–1562.
235. **Msadek, T., F. Kunst, and G. Rapoport.** 1994. MecB *Bacillus subtilis*, a member of the ClpC ATPase family, is a pleotropic regulator controlling competence, gene expression and growth at high temperature. *Proc. Natl. Acad. Sci. USA* **91:**5788–5792.
236. **Mukund, S., and M. W. W. Adams.** 1991. The novel tungsten-iron-sulfur protein of the hyperthermophilic archaebacterium *Pyrococcus furiosus* is an aldehyde ferredoxin oxidoreductase. *J. Biol. Chem.* **266:**14208–14216.
237. **Murray, A. E., K. Y. Wu, C. L. Moyer, D. M. Karl, and E. F. DeLong.** 1999. Evidence for circumpolar distribution of planktonic Archaea in the Southern Ocean. *Aquat. Microb. Ecol.* **18:** 263–273.
238. **Nealson, K. H.** 1982. Bacterial ecology of the deep sea, p. 179–200. *In* W. G. Ernst and J. G. Morin (ed.), *The Environment of the Deep Sea, Rubey Vol. II.* Prentice-Hall, Inc., Englewood Cliffs, N.J.
239. **Nedwell, D. B., and M. Rutter.** 1994. Influence of temperature on growth rate and competition between two psychrotolerant Antarctic bacteria: low temperature diminishes affinity for substrate uptake. *Appl. Environ. Microbiol.* **60:**1984–1992.
240. **Neidhardt, F. C., and R. A. VanBogelen.** 1987. Heat shock response, p. 1334–1345. *In* F. C. Neidhardt, J. L. Ingraham, K. B. Low, B. Magasanik, M. Schaechter, and H. E. Umbarger (ed.), *Escherichia coli and Salmonella typhimurium: Cellular and Molecular Biology.* American Society for Microbiology, Washington, D.C.
241. **Nelson, D. C., and C. R. Fisher.** 1995. Chemoautotrophic and methanotrophic endosymbiotic bacteria at deep-sea vents and seeps, p. 125–167. *In* D. M. Karl (ed.), *The Microbiology of Deep-Sea Hydrothermal Vents.* CRC Press, New York, N.Y.
242. **Nelson, D. C., C. O. Wirsen, and H. W. Jannasch.** 1989. Characterization of large, autotrophic *Beggiatoa* spp. abundant at hydrothermal vents of the Guaymas Basin. *Appl. Environ. Microbiol.* **55:**2909–2917.

243. **Nelson, D., R. M. Haymon, M. Lilley, and R. Lutz.** 1991. Rapid growth of unusual hydrothermal bacteria observed at new vents during ADVENTURE dive program to the EPR crest at 9°45′–52′N. *EOS, Trans. Amer. Geophys. Un.* **72**(Suppl.)**:**481.
244. **Nelson, D. R., Y. Sadlowski, M. Eguchi, and S. Kjelleberg.** 1997. The starvation-stress response of *Vibrio* (*Listonella*) *anguillarum. Microbiology* **143:**2305–2312.
245. **Nickerson, K. W.** 1984. An hypothesis on the role of pressure in the origin of life. *Theoret. Biol.* **110:**487–499.
246. **Nilsson, L., J. D. Oliver, and S. Kjelleberg.** 1991. Resuscitation of *Vibrio vulnificus* cells from the viable but nonculturable state. *J. Bacteriol.* **173:**5054–5059.
247. **Nissen, H.** 1987. Long term starvation of a marine bacterium, *Alteromonas denitrificans*, isolated from a Norwegian fjord. *FEMS Microbiol. Ecol.* **45:**173–184.
248. **Nogi, Y., and C. Kato.** 1999. Taxonomic studies of extremely barophilic bacteria isolated from the Mariana Trench and description of *Moritella yayanosii* sp. nov., a new barophilic bacterial isolate. *Extremophiles* **3:**71–77.
249. **Nogi, Y., C. Kato, and K. Horikoshi.** 1998. *Moritella japonica* sp. nov., a novel barophilic bacterium isolated from a Japan Trench sediment. *J. Gen. Appl. Microbiol.* **44:**289–295.
250. **Nogi, Y., C. Kato, and K. Horikoshi.** 1998. Taxonomic studies of deep-sea barophilic *Shewanella* strains and description of *Shewanella violacea* sp. nov. *Arch. Microbiol.* **170:**331–338.
251. **Nogi, Y., N. Masui, and C. Kato.** 1998. *Photobacterium profundum* sp. nov., a new, moderately barophilic bacterial species isolated from a deep-sea sediment. *Extremophiles* **2:**1–7.
252. **Norkrans, B., and B. O. Stehn.** 1978. Sediment bacteria in the deep Norwegian Sea. *Mar. Biol.* **47:**201–209.
253. **Novitsky, J. A.** 1990. Evidence for sedimenting particles as the origin of the microbial community in a coastal marine sediment. *Mar. Ecol. Prog. Ser.* **60:**161–167.
254. **Novitsky, J. A., and R. Y. Morita.** 1976. Morphological characteristics of small cells resulting from nutrient starvation of a psychrophilic marine vibrio. *Appl. Environ. Microbiol.* **32:**617–622.
255. **Novitsky, J. A., and R. Y. Morita.** 1977. Survival of a psychrophilic marine vibrio under long-term nutrient starvation. *Appl. Environ. Microbiol.* **33:**635–641.
256. **Novitsky, J. A., and R. Y. Morita.** 1978. Starvation-induced barotolerance as a survival mechanism of a psychrophilic marine vibrio in the waters of the Antarctic Convergence. *Mar. Biol.* **49:**7–10.
257. **Nystrom, T., and S. Kjelleberg.** 1989. Role of protein synthesis in cell division and starvation induced resistance to autolysis of a marine *Vibrio* during the initial phase of starvation. *J. Gen. Microbiol.* **135:**1599–1606.
258. **Oliver, J. D.** 1993. Formation of viable but nonculturable cells, p. 239–272. *In* S. Kjelleberg (ed.), *Starvation in Bacteria.* Plenum Press, New York, N.Y.
259. **Oliver, J. D., and W. F. Stringer.** 1984. Lipid composition of a psychrophilic marine *Vibrio* sp. during starvation-induced morphogenesis. *Appl. Environ. Microbiol.* **47:**461–466.
260. **Oliver, J. D., L. Nilsson, and S. Kjelleberg.** 1991. Formation of nonculturable *Vibrio vulnificus* cells and its relationship to the starvation state. *Appl. Environ. Microbiol.* **57:**2640–2644.
261. **Olsen, R. H., and E. S. Metcalf.** 1968. Conversion of mesophiles to psychrophilic bacteria. *Science* **162:**1288–1289.
262. **Otto, K., D. Weichart, and S. Kjelleberg.** 1997. Plasmid transfer between marine *Vibrio* strains during predation by the heterotrophic microflagellate *Cafeteria roenbergensis. Appl. Environ. Microbiol.* **63:**749–752.
263. **Pace, N. R., D. A. Stahl, D. L. Lane, and G. J. Olsen.** 1986. The analysis of natural microbial populations by ribosomal RNA sequences. *Adv. Microb. Ecol.* **9:**1–55.
264. **Parkes, R. J., B. A. Cragg, J. C. Fry, R. A. Herbert, and J. W. T. Wimpenny.** 1989. Bacterial biomass and activity in deep sediment layers from the Peru margin. *In* H. Charnook, J. M. Edmond, I. N. McCave, A. L. Rice, and T. R. S. Wilson (ed.), *The Deep Sea Bed: Its Physics, Chemistry and Biology. Philos. Trans. R. Soc. London Ser. A* **331:**139–153.
265. **Parsell, D. A., Y. Sanchez, J. D. Stitzel, and S. Lindquist.** 1991. Hsp104 is a highly conserved protein with two essential nucleotide-binding sites. *Nature* **353:**270–273.
266. **Paul, J. H., W. H. Jeffrey, and M. F. DeFlaun.** 1987. Dynamics of extracellular DNA in the marine environment. *Appl. Environ. Microbiol.* **53:**170–179.

267. **Petermann, H., and U. Bleil.** 1993. Detection of live magnetotactic bacteria in South Atlantic deep-sea sediments. *Earth Planet Sci. Lett.* **117:**223–228.
268. **Phipps, B. M., A. Hoffman, K. O. Stetter, and W. Baumeister.** 1991. A novel ATPase complex selectively accumulated upon heat shock is a major cellular component of thermophilic archaebacteria. *EMBO J.* **10:**1711–1722.
269. **Plante, C. J., P. A. Jumars, and J. A. Baross.** 1990. Digestive associations between marine detritivores and bacteria. *Annu. Rev. Ecol. Syst.* **21:**93–127.
270. **Pledger, R. J., and J. A. Baross.** 1989. Characterization of an extremely thermophilic archaebacterium isolated from a black smoker polychaete (*Paralvinella* sp.) at the Juan de Fuca Ridge. *Syst. Appl. Microbiol.* **12:**249–256.
271. **Pledger, R. J., and J. A. Baross.** 1991. Preliminary description and nutritional characterization of a heterotrophic archaebacterium growing at temperatures of up to 110°C isolated from a submarine hydrothermal vent environment. *J. Gen. Microbiol.* **137:**203–211.
272. **Pledger, R. J., B. C. Crump, and J. A. Baross.** 1994. A barophilic response by two hyperthermophilic, hydrothermal vent Archaea: an upward shift in the optimal temperature and acceleration of growth rate at supra-optimal temperatures by elevated pressure. *FEMS Microbiol. Ecol.* **14:**233–242.
273. **Pley, U., J. Schipka, A. Gambacorta, H. W. Jannasch, H. Fricke, R. Rachel, and K. O. Stetter.** 1991. *Pyrodictium abyssi* sp. nov. represents a novel heterotrophic marine Archaeal hyperthermophile growing at 110°C. *Syst. Appl. Microbiol.* **14:**245–253.
274. **Poindexter, J. S.** 1987. Bacterial responses to nutrient limitation, p. 283–317. *In* M. Fletcher, T. R. G. Gray and J. G. Jones (ed.), *Ecology of Microbial Communities.* Cambridge Univ. Press, London, U.K.
275. **Poremba, K.** 1995. Hydrolytic enzymatic activity in deep-sea sediments. *FEMS Microb. Ecol.* **16:** 213–222.
276. **Preston, C. M., K. Y. Wu, T. F. Molinski, and E. F. DeLong.** 1996. A psychrophilic crenarchaeon inhabits a marine sponge: *Cenarchaeum symbiosum* gen. nov., sp. nov. *Proc. Natl. Acad. Sci. USA* **93:**6241–6246.
277. **Preyer, J. M., and J. D. Oliver.** 1993. Starvation-induced thermal tolerance as a survival mechanism in a psychrophilic marine bacterium. *Appl. Environ. Microbiol.* **59:**2653–2656.
278. **Quigley, M. M., and R. R. Colwell.** 1968. Properties of bacteria isolated from deep-sea sediments. *J. Bacteriol.* **95:**211–220.
279. **Raghukumar, C., and S. Raghukumar.** 1998. Barotolerance of fungi isolated from deep-sea sediments of the Indian Ocean. *Aquat. Microb. Ecol.* **15:**153–163.
280. **Reeburgh, W. S.** 1976. Methane consumption in the Cariaco Trench waters and sediments. *Earth Planet. Sci. Lett.* **28:**337–344.
281. **Reichardt, W.** 1988. Impact of the Antarctic benthic fauna on the enrichment of biopolymer-degrading psychrophilic bacteria. *Microb. Ecol.* **15:**311–321.
282. **Regnard, P.** 1884. Note sur les conditions de la vie dans les profondeurs de la mer. *C. R. Soc. Biol.* **36:**164–168.
283. **Relexans, J.-C., J. Deming, A. Dinet, J. F. Gaillard, and M. Sibuet.** 1996. Sedimentary organic matter and micro-meiobenthos with relation to trophic conditions in the tropical northeast Atlantic. *Deep-Sea Res.* **43:**1343–1368.
284. **Reysenbach, A-L., and J. W. Deming.** 1991. Effects of hydrostatic pressure on growth of hyperthermophilic archaebacteria from the Juan de Fuca Ridge. *App. Environ. Microbiol.* **57:**1271–1274.
285. **Reysenbach, A-L., R. Pledger, G. Erauso, D. Prieur, J. Baross, J. W. Deming, and N. R. Pace.** 1993. Phylogenetic characterization of sulfur-reducing archaebacteria from two geographically distinct deep-sea hydrothermal vent sites. Abstract, Sixth International Symposium on Microbial Ecology, 6–11 Sept, Barcelona, Spain.
286. **Rice, S. A., and J. D. Oliver.** 1992. Starvation response of the marine barophile CNPT-3. *Appl. Environ. Microbiol.* **58:**2432–2437.
287. **Ritzrau, W.** 1997. Pelagic microbial activity in the Northeast Water Polynya, summer 1992. *Polar Biol.* **17:**259–268.

288. **Rochelle, P. A., J. C. Fry, R. J. Parkes, and A. J. Weightman.** 1992. DNA extraction for 16S rRNA gene analysis to determine genetic diversity in deep sediment communities. *FEMS Microbiol. Lett.* **100:**59–66.
289. **Rosenberg, M., and S. Kjelleberg.** 1986. Hydrophobic interactions: role in bacterial adhesion. *Adv. Microb. Ecol.* **9:**353–393.
290. **Roslev, P., and G. M. King.** 1994. Survival and recovery of methanotrophic bacteria starved under oxic and anoxic conditions. *Appl. Environ. Microbiol.* **60:**2602–2608.
291. **Roszak, D., and R. R. Colwell.** 1987. Survival strategies of bacteria in the natural environment. *Microb. Rev.* **51:**365–379.
292. **Rowe, G. T., M. Sibuet, J. W. Deming, A. Khripounoff, J. Tietjen, S. Macko, and R. Theroux.** 1991. "Total" sediment biomass and preliminary estimates of organic carbon residence time in deep-sea benthos. *Mar. Ecol. Prog. Ser.* **79:**99–114.
293. **Ruger, H.-J.** 1988. Substrate-dependent cold adaptations in some deep-sea sediment bacteria. *Syst. Appl. Microbiol.* **11:**90–93.
294. **Russell, N. J.** 1984. Mechanisms of thermal adaptation in bacteria: blueprints for survival. *TIBS* **9(3):**108–112.
295. **Sakiyama, T., and K. Ohwada.** 1997. Isolation and growth characteristics of deep-sea barophilic bacteria from the Japan trench. *Fish. Sci.* (Tokyo) **63:**228–232.
296. **Sanchez, Y., and S. L. Lindquist.** 1990. HSP104 required for induced thermotolerance. *Science* **148:**1112–1115.
297. **Schmidt, T. M., E. F. DeLong, and N. R. Pace.** 1991. Analysis of a marine picoplankton community by 16S rRNA gene cloning and sequencing. *J. Bacteriol.* **173:**4371–4378.
298. **Schuliger, J. W., S. H. Brown, J. A. Baross, and R. M. Kelly.** 1993. Purification and characterization of a novel amylolytic enzyme from ES4, a marine hyperthermophilic archaebacterium. *Mol. Mar. Biol. Biotechnol.* **2:**76–87.
299. **Schultz, J. E., G. I. Latter, and A. Matin.** 1988. Differential regulation by cyclic AMP of starvation protein synthesis in *Escherichia coli. J. Bacteriol.* **170:**3903–3909.
300. **Shut, F., R. A. Prins, and J. C. Gottschal.** 1997. Oligotrophy and pelagic marine bacteria: facts and fiction. *Aquat. Microb. Ecol.* **12:**177–202.
301. **Sieburth, J. McN.** 1987. Contrary habitats for redox-specific processes: methanogenesis in oxic waters and oxidation in anoxic waters, p. 11–38. *In* M. A. Sleigh (ed.), *Microbes in the Sea.* Ellis Horwood Limited, West Sussex, U.K.
302. **Smith, J. E., and J. D. Oliver.** 1991. The effects of hydrostatic pressure on bacterial attachment. *Biofouling* **3:**305–310.
303. **Sochard, M. R., D. F. Wilson, B. Austin, and R. R. Colwell.** 1979. Bacteria associated with the surface and gut of marine copepods. *Appl. Environ. Microbiol.* **37:**750–759.
304. **Southward, A. J., E. C. Southward, P. R. Dando, G. H. Rau, H. Felbeck, and H. Flugel.** 1981. Bacterial symbionts and low $^{13}C/^{12}C$ ratios in tissues of Pogonophora indicate unusual nutrition and metabolism. *Nature* **293:**616–620.
305. **Squires, C. L., S. Pedersen, B. M. Ross, and C. Squires.** 1991. ClpB is the *Escherichia coli* heat shock protein F84.1. *J. Bacteriol.* **173:**4254–4262.
306. **Stein, J. L., T. L. Marsh, K. Y. Wu, H. Shizuya, and E. F. DeLong.** 1996. Characterization of uncultivated prokaryotes: isolation and analysis of a 40-kilobase-pair genome fragment from a planktonic marine archaeon. *J. Bacteriol.* **178:**591–599.
307. **Stetter, K. O., G. Fiala, G. Huber, R. Huber, and A. Segerer.** 1990. Hyperthermophilic microorganisms. *FEMS Microbiol. Rev.* **75:**117–124.
308. **Stetter, K. O.** 1996. Hyperthermophilic procaryotes. *FEMS Microbiol. Rev.* **18:**149–158.
309. **Stevenson, H. L.** 1978. A case for bacterial dormancy in aquatic systems. *Microb. Ecol.* **4:**127–133.
310. **Straube, W. L., J. W. Deming, C. C. Somerville, R. R. Colwell, and J. A. Baross.** 1990. Particulate DNA in smoker fluids: evidence for existence of microbial populations in hot hydrothermal systems. *Appl. Environ. Microbiol.* **56:**1440–1447.
311. **Summit, M., and J. A. Baross.** 1998. Thermophilic subseafloor microorganisms from the 1996 North Gorda Ridge eruption. *Deep-Sea Research II* **45:**2751–2766.

312. **Summit, M., B. Scott, K. Nielson, E. Mathur, and J. Baross.** 1998. Pressure enhances thermal stability of DNA polymerase from three thermophilic organisms. *Extremophiles* **2:**339–345.
313. **Summit, M., A. Peacock, D. Ringelberg, D. C. White, and J. A. Baross.** Estimation of microbial biomass and community composition in hot, hydrothermally influenced sediments from Middle Valley, Juan de Fuca Ridge. *Proc. ODP Sci. Res. Leg 169*, in press.
314. **Takai, K., A. Inoue, and K. Horikoshi.** 1999. *Thermaerobacter marianensis* gen. nov., sp. nov., an anaerobic thermophilic marine bacterium from the 11,000-m deep Mariana Trench. *Intl. J. Syst. Bacteriol.* **49:**619–628.
315. **Takami, H., A. Inoue, F. Fuji, and K. Horikoshi.** 1997. Microbial flora in the deepest sea mud of the Mariana Trench. *FEMS Microb. Lett.* **152:**279–285.
316. **Taylor, C. D., and C. O. Wirsen.** 1997. Microbiology and ecology of filamentous sulfur formation. *Science* **277:**1483–1485.
317. **Trent, J. D., and A. A. Yayanos.** 1985. Pressure effects on the temperature range for growth and survival of the marine bacterium *Vibrio harveyi*. *Mar. Biol.* **89:**165–172.
318. **Trent, J. D., J. Osipiuk, and T. Pinkau.** 1990. Acquired thermotolerance and heat shock in the extremely thermophilic archaebacterium *Sulfolobus* sp. strain B12. *J. Bacteriol.* **172:**1478–1484.
319. **Trent, J. D., E. Nimmesgern, J. S. Wall, F.-U. Hartl, and A. L. Horwich.** 1991. A molecular chaperone from a thermophilic archaebacterium is related to the eukaryotic protein t-complex polypeptide-1. *Nature* **254:**490–493.
320. **Tunnicliffe, V.** 1992. Hydrothermal-vent communities of the deep sea. *Am. Sci.* **80:**336–349.
321. **Tunnicliffe, V., R. W. Embley, J. F. Holden, D. A. Butterfield, G. J. Massoth, and S. K. Juniper.** 1997. Biological colonization of new hydrothermal vents following an eruption on Juan de Fuca Ridge. *Deep-Sea Res. I* **44:**1627–1654.
322. **Tuovial, B. J., F. C. Dobbs, P. A. LaRock, and B. Z. Siegel.** 1987. Preservation of ATP in hypersaline environments. *Appl. Environ. Microbiol.* **53:**2749–2753.
323. **Turk, V., A. S. Rehnstam, E. Lundberg, and A. Hagstrom.** 1992. Release of bacterial DNA by marine nanoflagellates, an intermediate step in phosphorus regeneration. *Appl. Environ. Microbiol.* **58:**3744–3750.
324. **Turley, C. M.** 1993. The effect of pressure on leucine and thymidine incorporation by free-living bacteria and by bacteria attached to sinking oceanic particles. *Deep-Sea Res.* **40:**2193–2206.
325. **Turley, C. M., K. Lochte, and D. J. Patterson.** 1988. A barophilic flagellate isolated from 4500 m in the mid-North Atlantic. *Deep-Sea Res.* **35:**1079–1092.
326. **Tuttle, J. H., C. O. Wirsen, and H. W. Jannasch.** 1983. Microbial activities in the emitted hydrothermal waters of the Galapagos Rift vents. *Mar. Biol.* **73:**293–299.
327. **Van Dover, C. L., J. R. Cann, C. Cavanaugh, S. Chamberlain, J. R. Delaney, D. Janecky, J. Imhoff, J. A. Tyson, and the LITE Workshop Participants.** 1994. Light at deep sea hydrothermal vents. *Trans. Am. Geophys. Union* **75(4):**44–45.
328. **Van Dyk, T. K., W. R. Majarian, K. B. Konstantinov, R. M. Young, P. S. Dhurjati, and R. A. LaRossa.** 1994. Rapid and sensitive pollutant detection by induction of heat shock gene-bioluminescence gene fusions. *Appl. Environ. Microbiol.* **60:**1414–1420.
329. **Verati, C., P. Donato, D. Prieur, and J. Lancelot.** 1999. Evidence of bacterial activity from micrometer-scale layer analyses of black-smoker sulfide structures (Pito Seamount Site, Easter microplate). *Chem. Geol.* **158:**257–269.
330. **Vetriani, C., H. W. Jannasch, B. J. McGregor, D. A. Stahl, and A.-L. Reysenbach.** 1999. Population structure and phylogenetic characterization of marine benthic Archaea in deep-sea sediments. *Appl. Environ. Microbiol.* **65:**4375–4384.
331. **Vetter, Y.-A., and J. W. Deming.** 1994. Extracellular enzyme activity in the Arctic Northeast Water polynya. *Mar. Ecol. Prog. Ser.* **114:**23–34.
332. **Vetter, Y.-A., and J. W. Deming.** 1999. Growth rates of marine bacterial isolates on particulate organic substrates solubilized by freely released extracellular enzymes. *Microb. Ecol.* **37:**86–94.
333. **Vetter, Y.-A., J. W. Deming, P. A. Jumars, and B. B. Krieger-Brockett.** 1998. A predictive model of bacterial foraging by means of freely-released extracellular enzymes. *Microb. Ecol.* **36:** 75–92.

334. **Wang, C. L., P. C. Michels, S. C. Dawson, S. Kitisakkul, J. A. Baross, J. D. Keasling, and D. S. Clark.** 1997. Cadmium removal by a new strain of *Pseudomonas aeruginosa* in aerobic culture. *Appl. Environ. Microbiol.* **63:**4075–4078.
335. **Wang, Y., P. C. K. Lau, and D. K. Button.** 1996. A marine oligobacterium harboring genes known to be part of aromatic hydrocarbon degradation pathways of soil pseudomonads. *Appl. Environ. Microbiol.* **62:**2169–2173.
336. **Ward, D. M., R. Weller, and M. M. Bateson.** 1990. 16S rRNA sequences reveal uncultured inhabitants of a well-studied thermal community. *FEMS Microbiol. Rev.* **75:**105–116.
337. **Ward, D. M., M. M. Bateson, R. Weller, and A. L. Ruff-Roberts.** 1992. Ribosomal RNA analysis of microorganisms as they occur in nature. *Adv. Microb. Ecol.* **12:**219–286.
338. **Weichart, D., and S. Kjelleberg.** 1996. Stress resistance and recovery potential of culturable and viable nonculturable cells of *Vibrio vulnificus*. *Microbiology* **142:**845–853.
339. **Welch, T. J., A. Farewell, F. C. Neidhardt, and D. H. Bartlett.** 1993. Stress response of *Escherichia coli* to elevated hydrostatic pressure. *J. Bacteriol.* **175:**7170–7177.
340. **Wells, M. L., and E. D. Goldberg.** 1991. Occurrence of small colloids in seawater. *Nature* **353:** 342–344.
341. **Wells, M. L., and E. D. Goldberg.** 1994. The distribution of colloids in the North Atlantic and Southern Oceans. *Limnol. Oceanogr.* **39:**286–302.
342. **Wellsbury, P., K. Goodman, T. Barth, B. A. Cragg, S. P. Barnes, and R. J. Parkes.** 1997. Deep marine biosphere fueled by increasing organic matter availability during burial and heating. *Nature* **388:**573–576.
343. **Weyland, H., and E. Helmke.** 1989. Barophilic and psychrophilic bacteria in the Antarctic Ocean, p. 43–47. *In* T. Hattori, Y. Ishida, Y. Maruyama, R. Y. Morita, and A. Uchida (ed.), *Recent Advances in Microbial Ecology.* Proceedings of the 5th International Symposium on Microbial Ecology. Scientific Societies Press, Tokyo, Japan.
344. **Wiebe, W. J., W. M. Sheldon, Jr., and L. R. Pomeroy.** 1992. Bacterial growth in the cold: Evidence for an enhanced substrate requirement. *Appl. Environ. Microbiol.* **58:**359–364.
345. **Wiebe, W. J., W. M. Sheldon, Jr., and L. R. Pomeroy.** 1993. Evidence for enhanced substrate requirement by marine mesophilic bacterial isolates at minimal growth temperature. *Microb. Ecol.* **25:**151–159.
346. **Winn, C. D., J. P. Cowen, and D. M. Karl.** 1995. Microbes in deep-sea hydrothermal plumes, p. 255–273. *In* D. M. Karl (ed.), *Microbiology of Deep-Sea Hydrothermal Vents.* CRC Press, New York, N.Y.
347. **Wirsen, C. O., J. H. Tuttle, and H. W. Jannasch.** 1986. Activities of sulfur-oxidizing bacteria at the 21°N East Pacific Rise vent site. *Mar. Biol.* **92:**449–456.
348. **Wirsen, C. O., H. W. Jannasch, S. G. Wakeham, and E. A. Canuel.** 1987. Membrane lipids of a psychrophilic and barophilic deep-sea bacterium. *Current Microbiol.* **14:**319–322.
349. **Yager, P. L.** 1996. The microbial fate of carbon in high-latitude seas: impact of the microbial loop on oceanic uptake of CO_2. Ph.D. Thesis. University of Washington, Seattle.
350. **Yager, P. L., and J. W. Deming.** 2000. Pelagic microbial activity in the Northeast Water Polynya: testing for temperature and substrate interactions using a kinetic approach. *Limnol. Oceanogr.,* **44:** 1882–1893.
351. **Yager, P. L., A. R. M. Nowell, and P. A. Jumars.** 1993. Enhanced deposition to pits: a local food source for benthos. *J. Mar. Res.* **51:**209–236.
352. **Yano, Y., A. Nakayama, and K. Yoshida.** 1997. Distribution of polyunsaturated fatty acids in bacteria present in intestines of deep-sea fish and shallow-sea poikilothermic animals. *Appl. Environ. Microbiol.* **63:**2572–2577.
353. **Yayanos, A. A.** 1986. Evolutionary and ecological implications of the properties of deep-sea barophilic bacteria. *Proc. Natl. Acad. Sci. USA* **83:**9542–9546.
354. **Yayanos, A. A.** 1995. Microbiology to 10,500 meters in the deep sea. *Annu. Rev. Microbiol.* **49:** 777–805.
355. **Yayanos, A. A., and E. F. DeLong.** 1987. Deep-sea bacterial fitness to environmental temperatures and pressures, p. 17–32. *In* H. W. Jannasch, R. E. Marquis, and A. M. Zimmerman (ed.), *Current Perspectives in High Pressure Biology.* Academic Press, New York, N.Y.

356. **Yayanos, A. A., and A. S. Dietz.** 1982. Thermal inactivation of a deep-sea barophilic bacterium, isolate CNPT-3. *Appl. Environ. Microbiol.* **43:**1481–1489.
357. **Yayanos, A. A., and A. S. Dietz.** 1983. Death of a hadal deep-sea bacterium after decompression. *Science* **220:**497–498.
358. **Yayanos, A. A., A. S. Dietz, and R. Van Boxtel.** 1979. Isolation of a deep-sea barophilic bacterium and some of its growth characteristics. *Science* **205:**808–810.
359. **Yayanos, A. A., A. S. Dietz, and R. Van Boxtel.** 1981. Obligately barophilic bacterium from the Mariana Trench. *Proc. Natl. Acad. Sci. USA* **78:**5212–5215.
360. **Yayanos, A. A., A. S. Dietz, and R. Van Boxtel.** 1982. Dependence of reproduction rate on pressure as a hallmark of deep-sea bacteria. *Appl. Environ. Micobiol.* **44:**1356–1361.
361. **Zhao, H., H. G. Wood, F. Widdel, and M. P. Bryant.** 1988. An extremely thermophilic *Methanococcus* from a deep-sea hydrothermal vent and its plasmid. *Arch. Microbiol.* **150:**178–183.
362. **Zillig, W., K. O. Stetter, D. Pringishvilli, W. Shafer, S. Wunderl, D. Janekovic, I. Holz, and P. Palm.** 1982. Desulfurococcaceae, the second family of the extremely thermophilic anaerobic sulfur-respiring Thermoproteales. *Zentralbl. Bakteriol. Hyg. Abt. 1 Orig. Ser.* **C33:**304–317.
363. **ZoBell, C. E., and R. Y. Morita.** 1957. Barophilic bacteria in some deep-sea sediments. *J. Bacteriol.* **73:**563–568.

Nonculturable Microorganisms in the Environment
Edited by R. R. Colwell and D. J. Grimes

Chapter 11

Bacterial Viruses and Hosts: Influence of Culturable State

Frank T. Robb and Russell T. Hill

Many fundamental advances in molecular biology have resulted from the study of interactions between bacteriophages and their bacterial hosts. These advances include confirmation that DNA is the carrier of genetic information (16), the discovery of messenger RNA as the intermediate between DNA and protein (15), and the discovery of restriction endonucleases (2), a prerequisite for the growth of genetic engineering and biotechnology.

Several bacteriophages have been very intensively investigated at the molecular level, and interactions between these bacteriophages and their hosts are well understood under laboratory conditions. Indeed, the entire genomes of some bacteriophages have been sequenced, including bacteriophages λ (8), T7 (10), and M13 (53). However, bacteriophage ecology has received far less attention, and interactions between bacteriophages and hosts under natural conditions are less well understood. In fact, it has only recently become clear that bacteriophages may play an important ecological role, for example, in limiting populations of bacteria in aquatic ecosystems under some conditions (12a, 13, 43, 48a).

Approximately 2,100 bacteriophages were listed in 1981, and this number grows annually by about 100 (7) for a present total of significantly more than 3,000. One striking similarity in this otherwise extremely diverse assemblage is that almost all bacteriophages that have been described infect host cells that are growing logarithmically rather than hosts that are starved or in stationary phase. This is an important consideration, in view of the fact that bacteria inhabiting oligotrophic environments or high-nutrient habitats that are populated by the maximal density of microorganisms that can be supported are either in stationary-growth phase or else are proliferating at very slow rates.

Is infection of rapidly growing cells a general characteristic of bacteriophages, or have bacteriophages that infect stationary-phase or starved cells been overlooked as a consequence of methods used in phage isolation? Many methods have been

Frank T. Robb and Russell T. Hill • Center of Marine Biotechnology, University of Maryland Biotechnology Institute, Columbus Center, 701 East Pratt Street, Baltimore, MD 21202.

described for the isolation of phage from environmental samples (for example, see references 14, 25, 37, and 39). However, a common characteristic of standard methods that are available for isolation of phage is the use of plaque assays or dilution tubes. Methods for plaque assays generally specify the use of logarithmically growing host cells in liquid medium, to which environmental samples, phage enrichments, or phage concentrates are added. When the double-agar layer method of Adams (1) is used, host cells are inoculated onto solid media and undergo a period of rapid growth during which the presence of phage becomes evident by plaque formation. The dilution tube method is used primarily in cases where host bacteria do not grow on a solid substrate; in this procedure as well, exponentially growing host cells are inoculated into tubes. It is therefore evident that standard methods for phage isolation select for phage that infect logarithmically growing cells. We are unaware of attempts (besides our own) to specifically try to isolate phage that infect stationary-phase or starved cells.

It is important to distinguish between selection for phage that infect stationary-phase cells and those that infect starved and viable but nonculturable cells. Cells enter stationary phase in rich medium at high cell densities, either as a result of nutrient deprivation or buildup of toxic waste products. It has been postulated that this may be a programmed response to conditions of high cell density (38).

Obviously, the focus of laboratory studies on rapidly growing bacterial cells in nutrient-rich medium is not limited to the study of bacteriophage-host interactions. The physiology and molecular biology of bacteria are generally studied using these approaches, whereas bacteria in the natural environment are far more likely to be exposed to conditions of nutrient limitation. However, some significant advances have been made by researchers studying bacteria under more ecologically relevant conditions of nutrient limitation. The viable but nonculturable state (36) is now generally accepted as a response shown by many gram-negative bacteria under adverse environmental conditions. Cells entering the viable but nonculturable state are reduced in size, become ovoid, and retain very low levels of metabolic activity. Such cells, in contrast to starved cells, are not culturable on conventional media but retain viability as demonstrated by direct viable counting methods (20) or microautoradiography (18). A *Vibrio cholerae* mutant has been identified that showed an altered pattern of entry into the viable but nonculturable state, supporting the hypothesis that this state is a genetically controlled response to adverse conditions (32).

Response to starvation has been studied in some enteric and marine bacteria. Synthesis of new proteins occurs during starvation and alters cells to make them more efficient scavengers of scarce nutrients and to confer on them a more stress-resistant phenotype (24).

There is considerable evidence that there are profound differences between logarithmically growing cells and stationary-phase, starved, or viable but nonculturable cells. Here we outline work on phage infection of stationary-phase cells, propose experiments for isolation and study of phage that infect starved and viable but nonculturable cells, and speculate on the possible ecological significance of such phage.

PREVIOUS WORK ON THE INFLUENCE OF CULTURABLE STATE

Several previous observations suggest mechanisms by which the infection and propagation of bacteriophage may vary in response to the culturable state. The most obvious is lysogeny. In phage lambda, the decision to embark on the lytic cycle or enter the heritable symbiotic relationship with the host, namely lysogeny, is determined by a single protein, designated CII. The activity of CII is determined by environmental factors. The protein is unstable, and its level in the cell is determined by the relative rate that it is destroyed by bacterial proteases. Growth in rich medium activates proteases, and starvation has the opposite effect; therefore bacteriophage lambda more frequently lysogenizes starved cells (31).

In lysis inhibition, a phage-infected bacterium can be preserved for lengthy periods, up to 1 hour or three generations beyond the normal onset of lysis, by the superinfection of additional phage particles of the same type, for example by T-even phages (9). The burst size of the individual phage-infected cells is abnormally high, and the growth curves of the bacteria and bacteriophage are also affected by superinfection. In contrast, replication of three *Pseudomonas aeruginosa* bacteriophages was significantly altered in starved cells with an extended latent period and reduced burst size (21).

Phage Interactions with Stationary-Phase *Achromobacter*

The only well-studied case of phage infection of stationary-phase cells is that described initially by Woods (51). Bacterial strains with variable, growth-rate-dependent phage propagation responses were isolated (34). Wild-type *Achromobacter* was able to propagate bacteriophage growth in both rapidly growing and stationary, broth-grown cultures. In contrast, in *Achromobacter* strain 14, which was isolated after transduction experiments that may have resulted in the insertion of phage genomic material into the host bacterium, growth of phage was confined to stationary-phase cells. Mature colonies of *Achromobacter* strain 14 were sprayed with high-titer phage preparations, and after 54 h, lysis and miniplaque formation were observed within the colonies (33). Plaques of phage α3a growing on *Achromobacter* strain 14 appeared only after the lawns were at least 24 h old and presented with irregular outlines and continuous progression in size.

Kinetic Studies of Phage Development in *Achromobacter* sp. strain 2 Mutants

In experiments to determine the adsorption rates of phage to logarithmically growing and stationary-phase *Achromobacter* sp. strain 2 wt and strain 14 cells, all the cultures showed similar kinetics, with an initial rapid adsorption rate followed by slower adsorption after 30 min (33). In liquid culture, phage which were adsorbed irreversibly to strain 14 resulted in the emergence of phage bursts of average burst size of 709 per cell after 24 h, compared with 153 per cell in logarithmically growing cells. Individual bursts ranged from 8 plaques per cell to over 2,000 per cell for stationary-phase cells, whereas logarithmically growing cells ranged from 78 to 850 per cell (33). In this respect, the stationary-phase-specific growth in this system resembles lysis inhibition.

EFFECT OF AN *rpo* MUTATION, CONFERRING RIFAMPIN RESISTANCE, ON GROWTH PHASE-SPECIFIC PHAGE DEVELOPMENT

Stationary-phase-specific phage growth could be achieved by selecting a rifampin-resistant mutant of *Achromobacter* sp. strain 2. This mutation apparently mimics the effect of the genetic element present in *Achromobacter* strain 14. The RNA polymerase was isolated from both wild-type and Rif^r mutants and assayed for [^{3}H]UTP incorporation in vitro. An alteration in RNA polymerase, conferring rifampin resistance in vitro, and presumably in the β subunit, was found in the Rif^r strain (34). The correlation of phage growth phase specificity with an altered RNA polymerase is consistent with a model involving changed template specificity during the entry into stationary growth phase. A similar phenomenon, involving the generation of pleiotrophic, asporogenous Rif^+ mutants in *Bacillus subtilis* (23) has since proven to be the result of altered interaction of the core RNA polymerase with sporulation-specific factors.

Based on this evidence, it is tempting to speculate that the *Achromobacter* sp. strain 2 stationary-phase-specific pathway for phage development is coregulated with a growth-phase-specific factor. One possible candidate is the product of the *katF* (*rpoS*) gene which has been identified as a major regulator of the starvation response in *Escherichia coli* (38) and was recently shown biochemically to function as a sigma factor (42). The morphological description of phage α3a indicates that it is similar in size and appearance to the T-even coliphages, whose early developmental pathways are characterized by template switching, regulated by phage-encoded factors.

ECOLOGICAL RELEVANCE

Soil Ecosystems

Enumeration of phage in soil samples has been attempted by direct counts, using selected bacterial host systems. However, in many cases no phage can be detected unless prior enrichment is performed on samples. This could be a reflection of the low number of phage in soil. However, poor efficiency of extraction of phage from soil particles may result in the low counts being obtained (48). In fact, using optimized extraction procedures, Lanning and Williams (22) detected approximately 10^5 phage/g soil (dry weight) infecting certain *Streptomyces* hosts, suggesting that the total number of phage in soil was much greater than previously believed (22).

Williams et al. (48) reviewed a number of factors that may influence the replication of phage in soil, including adsorption of phage to soil colloids, temperature, pH, and the availability of active hosts. Many soil bacteria, such as *Bacillus* and *Streptomyces*, are gram positive and form resistant spores under adverse conditions. They would therefore become unavailable to phage predators. However, gram-negative bacteria such as pseudomonads would theoretically remain available for phage infection even under adverse conditions when these hosts are starved or have entered the viable but nonculturable state. Phage infection would not be as limited spatially and temporally if phage were present that could infect starved host cells.

Therefore, the potential ecological impact of phage, as a factor affecting bacterial populations, would be enhanced.

Aquatic Ecosystems

Initial estimates of the number of bacteriophages in freshwater and marine environments were obtained by direct counts, using host bacteria. Concentrations of coliphage and phage of other enteric bacteria were generally found to be between 10^3 and 10^5 phage per liter in freshwater environments (11). Numbers of phage in marine ecosystems were believed to be low, and almost all isolations of marine bacteriophages relied on enrichment cultures (26) and were therefore not useful for establishing phage numbers. One early indication that phage numbers may have been underestimated was obtained by using electron microscopy, which gave counts of virus particles of 10^7 viruses per liter in seawater samples (46). Recently, direct electron microscopic observations have shown abundances of 10^6 to 10^{11} viruses per liter in natural waters (3, 30). High viral abundances are widespread and have been found in estuaries (50), coral reef environments (28), and coastal and open ocean waters (for example, see references 6, 29, 40, and 47). It is likely that the vast majority of viruses observed in these studies were bacteriophages (29, 50), since viral abundance was generally found to be directly related to bacterial abundance, the virus particles observed morphologically resembled bacteriophages, and bacteria were the most abundant potential host.

These high viral abundances have resulted in renewed interest in the possible ecological role of bacteriophages in aquatic ecosystems. Current understanding of viruses as a cause of microbial mortality has been reviewed by Fuhrman and Suttle (13). Several factors warrant further investigation to assess more accurately the impact of phage on marine bacterial communities. Burst size is an important parameter when estimating mortality rates, and it would be useful to obtain burst sizes for more marine bacteriophages to refine estimates of average burst size. The proportion of hosts that contain lysogenic phage should be ascertained since these bacteria will be protected against lytic phage attack. Also of interest is the effect of the metabolic state of the host on phage infection. It is possible that starved or viable but nonculturable hosts may be in a "metabolic niche" where they are immune to phage attack. Alternatively, certain bacteriophages may have developed the capability to attack starved or viable but nonculturable cells in addition to rapidly growing hosts, or they may only be capable of attacking these less metabolically active cells. Moebus (26) posed the question, "How metabolically active must a cell be to become infected by adsorbed phage and produce new virions?" This question remains unanswered and is highly significant for understanding the ecology of bacteria in the marine environment, since these bacteria exist in an environment that contains very low nutrient concentrations, compared to soil or many freshwater aquatic systems.

The difficulties to be overcome in studying the influence of the metabolic state of host marine bacteria on their susceptibility to phage infection are considerable. Less than 1% of the marine bacteria observed by direct microscopy can be recovered on laboratory media (12, 19, 20). Many or most of these cells may be in the viable but nonculturable state, or other factors may prevent their growth in the laboratory. Furthermore, it is possible that many unusual groups of phage found in marine ecosystems have never been cultured in the laboratory. A striking example is the observation of exceptionally large marine virus-like particles with heads the size of small bacteria (5). Nevertheless, some phage host systems presently in laboratory culture can provide useful models for studying phage infection of both starved and viable but nonculturable cells.

SOME EXPERIMENTAL APPROACHES FOR ISOLATION OF BACTERIOPHAGES INFECTING STATIONARY-PHASE, STARVED, AND VIABLE BUT NONCULTURABLE CELLS

The approaches suggested here are applicable both to isolation of new phage host systems and to testing phage host systems in culture and those isolated on logarithmically growing cells for ability to infect cells in other culturable states.

Isolation of Phage Infecting Stationary-Phase Cells

For ease of conduct of the experiments, phage capable of infection of stationary-phase cells should be detectable using standard solid medium plaque isolation methods. A characteristic difference between these and phage that infect exponentially growing cells is that plaques will continue to grow after cells in bacterial lawns have entered stationary phase. In standard phage assay overlays, phage that infect only stationary-phase cells appear only after incubation for several days, once the host cells have entered stationary phase. Plaques that appear rapidly but continue to increase in size are usually indicative of phage that infect both rapidly growing and stationary-phase cells. Studies are in progress on one phage host system, isolated from oligotrophic Bahamian waters, that displays the characteristic of plaques that appear overnight but continue to increase in diameter for up to 4 days (Wommack, Hill, and Colwell, unpublished data). Infection of stationary-phase cultures in liquid medium can be monitored either by plaque assay or by transmission electron microscopy, based on the method of Børsheim et al. (4). Cells and cell debris are removed by centrifugation or filtration.

A method which has been used to modify an exponential phage-host system has been described and may be generally applicable for blocking exponential phage growth. An extremely halotolerant bacterial strain, initially identified as *Achromobacter* sp. strain 2, was isolated from salted ox hides and implicated in leather decay, since it is collagenolytic (45). Several lysogenic phages were isolated and characterized. The use of one of these phages in transduction experiments led to the isolation of a bacterial host strain with partial phage resistance, in which the pathway for development of phage in exponential cells was blocked (44). Interestingly, the phage developed plaques on mature bacterial lawns, and these infectious centers continued to grow with time over several days (51). This system provided a model for stationary-phase-specific growth in bacterial cells growing on rich medium, described in detail above.

Isolation of Phage Infecting Starved Cells

An approach based on the dilution tube method is most likely to be successful to isolate phage infecting starved cells. A sound strategy is the use of hosts and bacteriophage concentrates obtained from aquatic ecosystems since low-nutrient conditions frequently occur in these environments. Several procedures have been described for concentration of viruses from aquatic samples (29, 41, 49) and can be useful in obtaining phage concentrates to inoculate into dilution tubes. Successful infection of starved cells in these dilution tubes can be monitored by decrease in host cells (counted using the acridine orange epifluorescent microscopy procedure [17]) and/or by increase in phage counts (enumerated by using transmission electron microscopy [4]).

Isolation of Phage Infecting Viable but Nonculturable Cells

A wide range of gram-negative bacteria have been shown to enter the viable but nonculturable state, and suitable conditions have been established to facilitate entry (see, for example, references 27, 35, 36, 52). The approach for isolation of novel bacteriophages that infect viable but nonculturable bacteria should be similar to those described for starved cells. In addition, hosts presently known to be susceptible to infection by a particular phage can be tested to establish whether the hosts enter the viable but nonculturable state. Phage infection of these viable but nonculturable cells can be monitored as described above.

CONCLUSIONS

It is concluded that phage growth occurs on stationary-phase cells, with separate pathways for phage development operating in logarithmically growing and stationary-phase bacterial cells. Preliminary evidence indicates that transcriptional template selection may be critical in deciding which pathway is operative.

It is tempting to speculate that many oligotrophic environments, including oceanic marine waters, may yield evidence of similar pathways for phage development. Indeed, it would be surprising if alternative pathways (such as the bacteriophage α3a system) did not exist as an alternative to lysogeny, ensuring the propagation of phage under conditions of nutrient starvation and/or declining bacterial populations.

Acknowledgments. We express our gratitude and appreciation to David Woods. He was the first to observe the phenomenon of stationary-phase-specific phage growth, and his friendship and scientific acumen have been a great help during a collaboration that spans many years. F.T.R. was supported by grants from the Lucille P. Markey Foundation, the U.S. Department of Energy, the U.S. Department of Commerce ATP program and the National Science Foundation. R.J.H. was supported by a grant from the Wallenberg Foundation. This chapter is contribution number 251 from the Center of Marine Biotechnology.

REFERENCES

1. **Adams, M. H.** 1959. *Bacteriophages*. Interscience, New York, N.Y.
2. **Arber, W.** 1965. Host-controlled modification of bacteriophage. *Annu. Rev. Microbiol.* **19:**365–378.

3. **Bergh, O., K. Y. Borsheim, G. Bratbak, and M. Heldal.** 1989. High abundance of viruses found in aquatic environments. *Nature* (London) **340:**467–468.
4. **Børsheim, K. Y., G. Bratbak, and M. Heldal.** 1990. Enumeration and biomass estimation of planktonic bacteria and viruses by transmission electron microscopy. *Appl. Environ. Microbiol.* **56:** 352–356.
5. **Bratbak, G., O. H. Haslund, M. Heldal, A. Noess, and T. Roeggen**. 1992. Giant marine viruses? *Mar. Ecol. Prog. Ser.* **85:**201–202.
6. **Cochlan, W. P., J. Wikner, G. F. Steward, D. C. Smith, and F. Azam.** 1993. Spatial distribution of viruses, bacteria and chlorophyll a in neritic, oceanic and estuarine environments. *Mar. Ecol. Prog. Ser.* **92:**77–87.
7. **Coetzee, J. N.** 1987. Bacteriophage taxonomy, p. 45–85. *In* S.M. Goyal, C. P. Gerba, and G. Bitton (ed.), *Phage Ecology*. John Wiley and Sons, New York, N.Y.
8. **Daniels, D. L., J. L. Schroeder, W. Szybalski, F. Sanger, A. R. Coulson, G. R. Hong, D. F. Hill, G. B. Petersen, and F. R. Blattner.** 1983. Appendix II: Complete annotated lambda sequence, p. 519–676. *In* R. W. Hendrix, J. W. Roberts, F. W. Stahl, and R. A. Weisberg (ed.), *Lambda II*. Cold Spring Harbor Laboratory, Cold Spring Harbor, N.Y.
9. **Doermann, A. H.** 1948. Lysis and lysis inhibition with *Escherichia coli* bacteriophages. *J. Bacteriol.* **55:**257–275.
10. **Dunn, J. J., and F. W. Studier.** 1983. Complete nucleotide sequence of bacteriophage T7 DNA and the locations of T7 genetic elements. *J. Mol. Biol.* **166:**477–535.
11. **Farrah, S. R.** 1987. Ecology of phage in freshwater environments, p. 125–136. *In* S. M. Goyal, C. P. Gerba, and G. Bitton (ed.), *Phage Ecology*. John Wiley and Sons, New York, N.Y.
12. **Ferguson, R. L., E. N. Buckley, and A.V. Palumbo.** 1984. Response of marine bacterioplankton to differential filtration and confinement. *Appl. Environ. Microbiol.* **47:**49–55.
12a. **Fuhrman, J. A.** 1999. Marine viruses and their biogeochemical and ecological effects. *Nature* **399:** 541–548.
13. **Fuhrman, J. A., and C. A. Suttle.** 1993. Viruses in marine planktonic systems. *Oceanography* **6:** 51–63.
14. **Goyal, S. M.** 1987. Methods in phage ecology, p. 267–287. *In* S. M. Goyal, C. P. Gerba, and G. Bitton (ed.), *Phage Ecology*, John Wiley and Sons, New York, N.Y.
15. **Hall, B. D., and S. Spiegelman.** 1961. Sequence complementarity of T2 DNA and T2-specific RNA. *Proc. Natl. Acad. Sci. USA* **47:**137–146.
16. **Hershey, A. D., and M. Chase.** 1952. Independent functions of viral protein and nucleic acid in growth of bacteriophage. *J. Gen. Physiol.* **36:**39–56.
17. **Hobbie, J. E., R. J. Daley, and S. Jasper.** 1977. Use of nucleopore filters for counting bacteria by fluorescence microscopy. *Appl. Environ. Microbiol.* **33:**1225–1228.
18. **Hoppe, H. G.** 1976. Determination and properties of actively metabolizing heterotrophic bacteria in the sea, investigated by means of microautoradiography. *Mar. Biol.* **36:**291–302.
19. **Jannasch, H. W., and G. E. Jones.** 1959. Bacterial populations in seawater as determined by different methods of enumeration. *Limnol. Oceanogr.* **4:**128–139.
20. **Kogure, K., U. Simidu, and N. Taga.** 1979. A tentative direct microscopic method for counting living marine bacteria. *Can. J. Microbiol.* **25:**415–420.
21. **Kokjohn, T. A., G. S. Sayler, and R. V. Miller.** 1991. Attachment and replication of *Pseudomonas aeruginosa* bacteriophages under conditions simulating aquatic environments. *J. Gen. Microbiol.* **137:**661–666.
22. **Lanning, S., and S. T. Williams.** 1982. Methods for the direct isolation and enumeration of actinophages in soil. *J. Gen. Microbiol.* **128:**2063–2071.
23. **Linn, T., R. Losick, and A. L. Sonenshein.** 1975. Rifampicin resistance mutation of *Bacillus subtilis* altering the electrophoretic mobility of the beta subunit of ribonucleic acid polymerase. *J. Bacteriol.* **122:**1387–1390.
24. **Matin, A., E. A. Auger, P. H. Blum, and J. E. Schultz.** 1989. Genetic basis of starvation survival in nondifferentiating bacteria. *Ann. Rev. Microbiol.* **43:**293–316.
25. **Moebus, K.** 1980. A method for the detection of bacteriophages from ocean waters. *Helgolander Meeresunters.* **34:**1–14.

26. **Moebus, K.** 1987. Ecology of marine bacteriophages, p. 137–156. *In* S. M. Goyal, C. P. Gerba, and G. Bitton (ed.), *Phage Ecology*. John Wiley and Sons, New York, N.Y.
27. **Oliver, J. D., L. Nilsson, and S. Kjelleberg.** 1991. Formation of nonculturable *Vibrio vulnificus* cells and its relationship to the starvation state. *Appl. Environ. Microbiol.* **57:**2640–2644.
28. **Paul, J. H., J. B. Rose, S. C. Jiang, C. A. Kellogg, and L. Dickson.** 1993. Distribution of viral abundance in the reef environment of Key Largo, Florida. *Appl. Environ. Microbiol.* **59:**718–724.
29. **Paul, J. H., S. C. Jiang, and J. B. Rose.** 1991. Concentration of viruses and dissolved DNA from aquatic environments by Vortex Flow Filtration. *Appl. Environ. Microbiol.* **57:**2197–2204.
30. **Proctor, L. M., and J. A. Fuhrman.** 1990. Viral mortality of marine bacteria and cyanobacteria. *Nature* (London) **343:**60–62.
31. **Ptashne, M. A.** 1986. *Genetic Switch. Gene Control and Phage λ.* Blackwell Scientific Publications, Palo Alto, Calif.
32. **Ravel, J., R. T. Hill, and R. R. Colwell.** 1994. Isolation of a *Vibrio cholerae* transposon-mutant with an altered viable but nonculturable response. *FEMS Microbiol. Lett.* **120:**57–62.
33. **Robb, S. M., D. R. Woods, and F. T. Robb.** 1978. Phage growth characteristics on stationary phase *Achromobacter* cells. *J. Gen. Virol.* **41:**265–272.
34. **Robb, S. M., D. R. Woods, F. T. Robb, and J. K. Struthers.** 1977. Rifampicin-resistant mutant supporting bacteriophage growth on stationary phase *Achromobacter* cells. *J. Gen. Virol.* **35:**117–123.
35. **Rollins, D. M., and R. R. Colwell.** 1986. Viable but nonculturable stage of *Campylobacter jejuni* and its role in survival in the natural aquatic environment. *Appl. Environ. Microbiol.* **52:**531–538.
36. **Roszak, D. B., and R. R. Colwell.** 1987. Survival strategies of bacteria in the natural environment. *Microbiol. Rev.* **51:**365–379.
37. **Seeley, N. D., and S. B. Primrose.** 1982. The isolation of bacteriophages from the environment. *J. Appl. Bacteriol.* **53:**1–17.
38. **Siegele, D. A., and R. Kolter.** 1992. Life after log. *J. Bacteriol.* **174:**345–348.
39. **Suttle, C. A.** 1993. Enumeration and isolation of viruses, p. 121–134. *In* P. F. Kemp, B. F Sherr, E. B. Sherr, and J. J. Cole (ed.), *Aquatic Microbial Ecology*. Lewis Publishers, Boca Raton, Fla.
40. **Suttle, C. A., A. M. Chan, and M. T. Cottrell.** 1990. Infection of phytoplankton by viruses and reduction of primary productivity. *Nature* (London) **347:**467–469.
41. **Suttle, C. A., A. M. Chan, and M. T. Cottrell.** 1991. Use of ultrafiltration to isolate viruses from seawater which are pathogens of marine phytoplankton. *Appl. Environ. Microbiol.* **57:**721–726.
42. **Tanaka, K., Y. Takayanagi, N. Fujita, A. Ishihama, and H. Takahashi.** 1993. Heterogeneity of the principal s factor in *Escherichia coli*: the *rpoS* gene product, s38, is a second principal s factor of RNA polymerase in stationary-phase *Escherichia coli. Proc. Natl. Acad. Sci. USA* **90:**3511–3515.
43. **Thingstad, T. F., M. Heldal, G. Bratbak, and I. Dundas.** 1993. Are viruses important partners in pelagic food webs? *Trends Ecol. Evol.* **8:**209–213.
44. **Thompson, J. A., and D. R. Woods.** 1974. Bacteriophages and cryptic lysogeny in *Achromobacter. J. Gen. Microbiol.* **22:**153–157.
45. **Thompson, J. A., D. R. Woods, and R. L. Welton.** 1972. Collagenolytic activity of aerobic halophiles from hides. *J. Gen. Microbiol.* **70:**315–319.
46. **Torrella, F., and R. Y. Morita.** 1979. Evidence by electron micrographs for a high incidence of bacteriophage particles in the waters of Yaquina Bay, Oregon: ecological and taxonomical implications. *Appl. Environ. Microbiol.* **37:**774–778.
47. **Weinbauer, M. G., D. Fuks, and P. Peduzzi.** 1993. Distribution of viruses and dissolved DNA along a coastal trophic gradient in the northern Adriatic sea. *Appl. Environ. Microbiol.* **59:**4074–4082.
48. **Williams, S. T., A. M. Mortimer, and L. Manchester.** 1987. Ecology of soil bacteriophages, p. 157–179. *In* S. M. Goyal, C. P. Gerba, and G. Bitton (ed.), *Phage Ecology*. John Wiley and Sons, New York, N.Y.

48a. **Wommack, K. E., and R. R. Colwell.** 2000. Virioplankton: viruses in aquatic ecosystems. *Microbiol. Mol. Biol. Rev.* **64:**69–114.

49. **Wommack, K. E., R. T. Hill, and R. R. Colwell.** 1995. A simple method for the concentration of viruses from natural water samples. *J. Microbiol. Meth.* **22:**57–67.

50. **Wommack, K. E., R. T. Hill, M. Kessel, E. Russek-Cohen, and R. R. Colwell.** 1992. Distribution of viruses in the Chesapeake Bay. *Appl. Environ. Microbiol.* **58:**2965–2970.
51. **Woods, D. R.** 1976. Bacteriophage growth on stationary phase *Achromobacter* cells. *J. Gen. Virol.* **32:**45–50.
52. **Xu, H. S., N. Roberts, F. L. Singleton, R. W. Attwell, D. J. Grimes, and R. R. Colwell.** 1982. Survival and viability of nonculturable *Escherichia coli* and *Vibrio cholerae* in the estuarine and marine environment. *Microb. Ecol.* **8:**313–323.
53. **Yanisch-Perron, C., J. Vieira, and J. Messing.** 1985. Improved M13 phage cloning vectors and host strains: nucleotide sequences of the M13mp18 and pUC19 vectors. *Gene* **33:**103–119.

Nonculturable Microorganisms in the Environment
Edited by R. R. Colwell and D. J. Grimes

Chapter 12

The Importance of Viable but Nonculturable Bacteria in Biogeochemistry

D. Jay Grimes, Aaron L. Mills, and Kenneth H. Nealson

Biogeochemical cycles are critical parts of life itself—if elements do not cycle, life will grind to a halt, with essential elements being buried. Recycling of the elements allows for their reappearance in the food web to be used again and again, thereby allowing life to replenish itself. A simple example of this is shown in Fig. 1 for the carbon cycle, in which energy flow on Earth is diagrammatically linked to the carbon cycle. In this diagram we see the two major sources of energy on the planet (photic energy and geothermal energy) linked to life directly (through photosynthesis) and indirectly (through conversion of geothermal energy to reduced substrates that are used for lithotrophic metabolism). Both processes lead to the "fixation" or reduction of CO_2 to organic carbon—e.g., the conversion of carbon dioxide into biomass. The organic carbon is then recycled to the atmosphere via respiration by carbon-oxidizing bacteria and eukaryotes.

Similar biogeochemical cycles probably exist for all biologically important elements, although those that undergo redox transformation are perhaps the best understood, because they participate directly in redox reactions that yield energy for life. Such redox transformations are perhaps most vividly demonstrated by layered microbial communities that are found in physically stabilized environments (Fig. 2). Such layered communities provide evidence for biogeochemical cycling and the ability of microbes to catalyze a wide variety of redox transformations, taking advantage of environmental energy, e.g., nearly every redox transformation that yields energy has been "discovered" by the anaerobic microbes on Earth. Profiles like those shown in Fig. 2 provide direct field (chemical and physical) evidence for the importance and ubiquity of biogeochemical cycles on Earth, and

D. Jay Grimes • Institute of Marine Sciences, University of Southern Mississippi, P.O. Box 7000, Ocean Springs, MS 39566-7000. ***Aaron L. Mills*** • Laboratory of Microbial Ecology, Department of Environmental Sciences, Clark Hall, University of Virginia, Charlottesville, VA 22903. ***Kenneth H. Nealson*** • Division of Geology and Planetary Sciences, California Institute of Technology, MS 170-25, Pasadena, CA 90125.

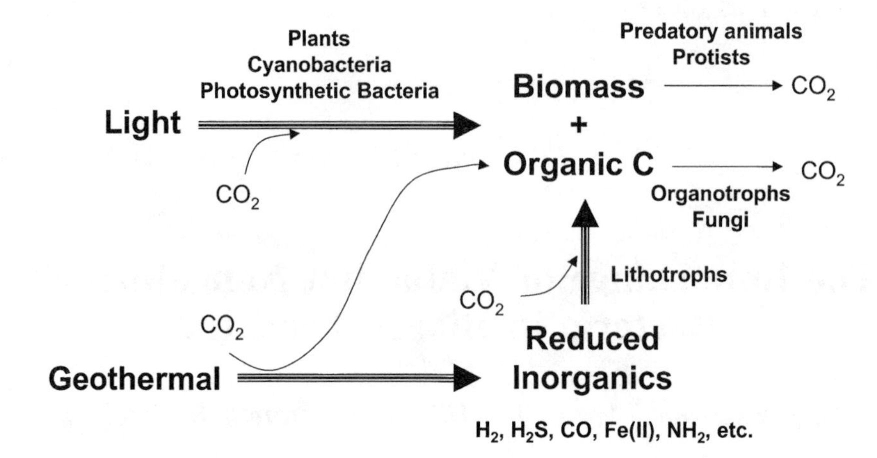

Figure 1. Energy flow and carbon cycling on Earth. This diagram represents the two major sources of energy on Earth, photic and geothermal, and how they relate to the cycling of organic carbon as mediated by the biota. It shows the connection between energy flow (photic energy being used directly, and geothermal energy being used indirectly via its conversion to reduced inorganics) and the cycle of carbon between inorganic (CO_2) and organic (fixed or reduced) forms.

the fact that they can be found at spatial scales ranging from hundreds of meters in the Black Sea (38) to micrometers in algal mats (29) attests to their importance.

As with many other areas of microbiology, however, it has become increasingly clear in the past few years that most of the organisms mediating the key chemical catalyses leading to the formation of layered communities are probably not yet cultivated and studied. A bulk of our knowledge is based on observations of actively growing axenic cultures of microorganisms in the laboratory. Various transformations, largely oxidation and reduction reactions, have been elucidated in the laboratory and this information has been used to infer role(s) in nature and to assemble model geochemical cycles. For the most part, these transformations have been carried out by pure cultures grown under ideal conditions that include nonlimiting nutrients. While such models provide starting points from which to explain how elements cycle in the biosphere, it is almost certain that the bulk of geochemical transformations are mediated by microorganisms (primarily bacteria) that are in one way or another under suboptimal conditions of growth and/or metabolism.

But do we really understand how this system works? Probably not, based on the discrepancies between the total number of microorganisms observed in a given environmental sample and the number counted indirectly by means of a plate count or most probable number (MPN). Such discrepancies have been the source of frustration and debate for decades. Typical of this discrepancy are the findings of Grimes et al. (20) showing that acridine orange direct counts (AODCs) of bacteria in seawater samples usually exceeded plate counts on heterotrophic plating media by 4 to 6 orders of magnitude and the lipid analyses of White (60) demonstrating that ocean sediments throughout the world contain 10^8 to 10^9 bacteria cm^{-1}. Data analysis presented by Meyer-Reil (35) also supports this observation and further

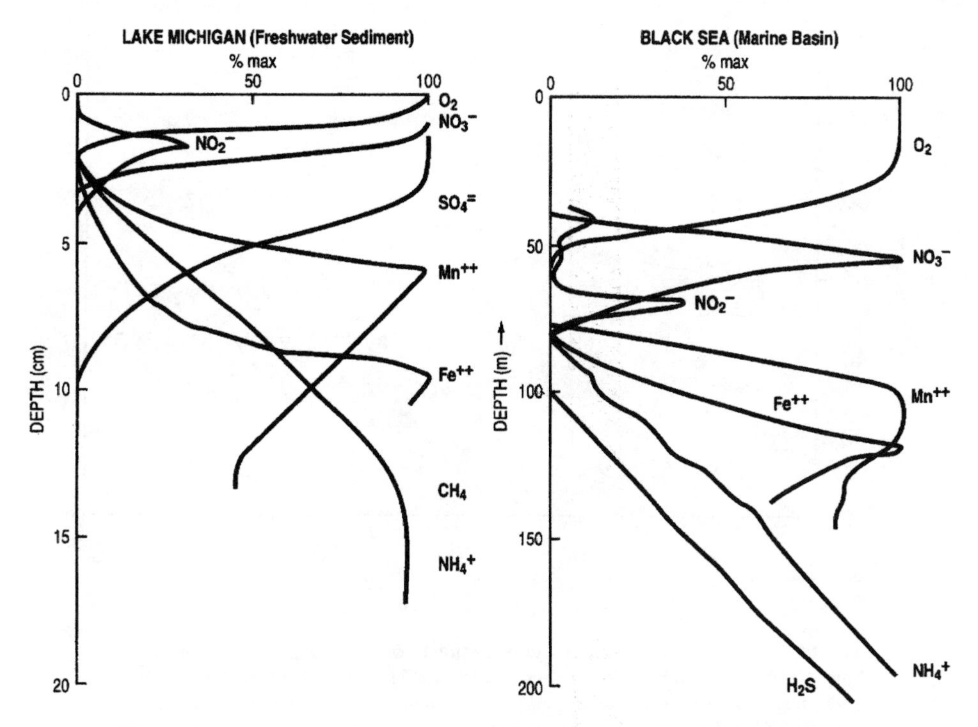

Figure 2. This figure shows what may be one of the most universal features of microbial communities on Earth, the layering of metabolites as a result of the redox chemistry of the microbes. In virtually every physically stabilized environment (sediments, soils, fjords, stratified lakes), we see such orderly progressions of redox chemistry that are the result of the consumption and/or production of nutrients at rates greater than diffusion. Such layered communities can occur on spatial scales of hundreds of meters as in the Black Sea, centimeters as in the Lake Michigan sediment, and much smaller as seen in bacterial mat communities or even biofilms. The relationship between biogeochemical cycling and microbial activities are perhaps nowhere better demonstrated than in such communities (38).

indicates the common failure of cultural-count data and direct-count data to show any correlation (see Fig. 3). Microbiologists began making this observation soon after the development of plate count methods, and reasons for the discrepancy were always stated to be one of the following: (i) the medium or conditions (e.g., temperature, pressure, salinity, pH, Eh, etc.) being used could not support the growth of all physiological types present in the sample and (ii) although still intact, many cells in the sample were moribund or dead and therefore incapable of further growth. Both of these hypotheses are still legitimate but for reasons not fully appreciated when they were first offered. Selection of growth medium clearly is important. There is no single medium that will support the growth of all heterotrophs, let alone autotrophs, be they phototrophic or lithotrophic. Even with a suite of media and conditions, many organisms will be overlooked. However, during the past decade, it has become apparent that viable bacteria that cannot be cultured

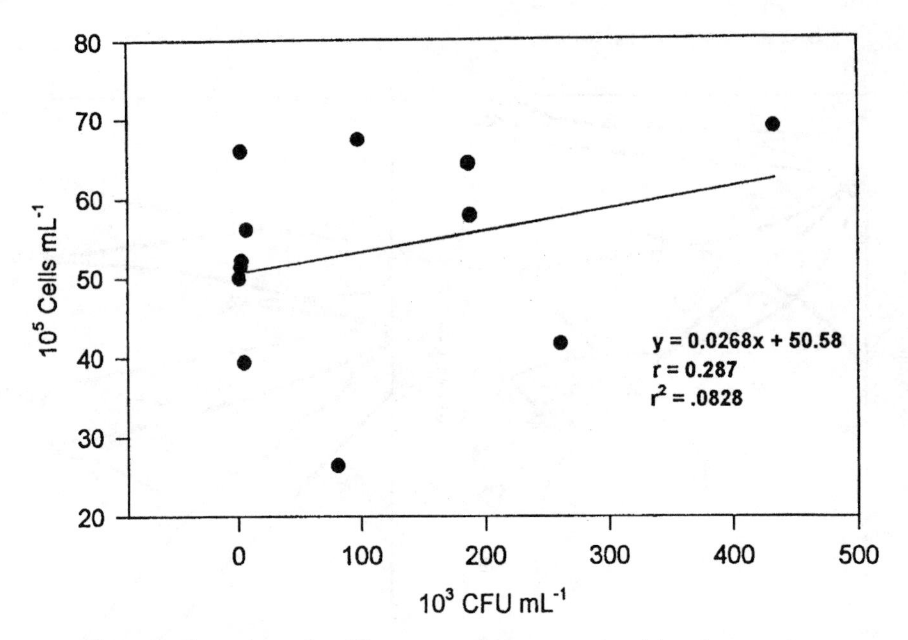

Figure 3. Comparison of acridine orange direct counts (y axis) and counts from spread plates (x axis) made from Kiel Fjord and Kiel Bight water samples. The line and equation represent the least squares regression of the data. The low values of r and r^2 show the weakness of the fit of the model.

also account for some fraction of the total count that will not grow under laboratory conditions. Both dormant organisms (i.e., somnicells) and organisms not yet cultured fail to grow on plating media and therefore contribute to the discrepancy. The second reason, intact moribund or dead cells, was long thought to be the principal reason for the discrepancy. However, while some cells in any given system are presumably moribund or dead, this factor can no longer be supported as the principal reason. For example, Tabor and Neihof (50) used a variety of direct methods (synthetically active bacteria, INT [2-(*p*-iodophenyl)-3-(*p*-nitrophenyl)-5-phenyltetrazolium chloride] reduction, and microautoradiographic determination of uptake-active organisms) to demonstrate that 25% to more than 85% of cells in Chesapeake Bay water samples (as determined by AODCs) were indeed viable (see Fig. 4). Clearly, viable but nonculturable bacteria can now be considered responsible for a major portion of the total count-plate count discrepancy in most systems.

Recent evidence suggests that there are countless "yet to be cultured" *Bacteria* and *Archaea* present in the biosphere, awaiting developments in microbial culture technology so that they too can be isolated and described. Bull et al. (5) estimated the total number of bacterial species to be 40,000. More recently, Tiedje (52) pointed out that different lines of evidence suggest that between 300,000 and 1 million species of bacteria inhabit the Earth, yet *Bergey's Manual of Systematic Bacteriology* (4 volumes containing 2,784 pages) lists only 3,100 species (26). Based on this hypothesis, *Bergey's Manual*, at least in contemporary format, could

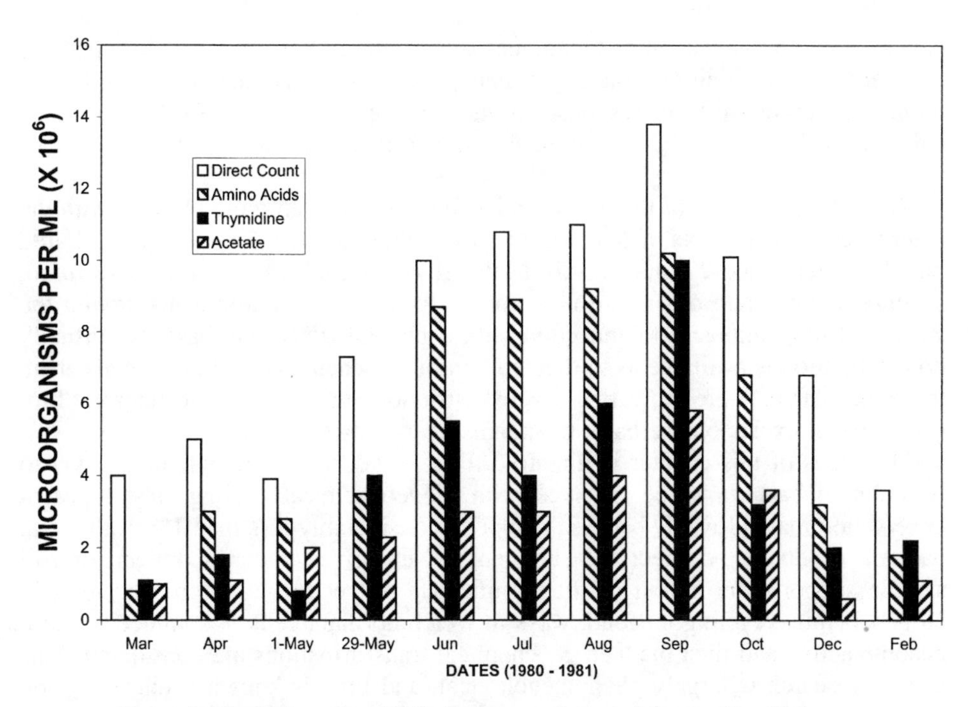

Figure 4. Distribution of total DC of microorganisms and the number of microorganisms active in the uptake of amino acids, thymidine, and acetate determined by microautoradiography for 11 Chesapeake Bay surface (1-m) water samples collected from 26 March 1980 through 9 February 1981. The boldfaced bar represents the total DC, and the bars to the right represent, in order, the numbers of organisms active in uptake of amino acids (AA), thymidine (Thd), and acetate (Aoc) determined for corresponding samples.

ultimately reach 400 to 1,300 volumes in length! Soil alone may contain 10,000 species per g, based on the heterogeneity of DNA extracted from soil samples (52). The oceans of the world contain countless picoplankton, many of which are yet to be cultured. For example, the cold water *Archaea*, first detected in seawater samples collected at depths ranging from 100 to 500 meters (11, 17), are now believed to comprise a substantial percentage of the total microbial biomass present in oceans throughout the world (12). These *Archaea* have not been isolated in pure culture, nor have they been grown and phenotypically described. They have been detected only by PCR amplification, characterization, and classification of their ribosomal RNA. Woese recently noted that approximately 300 archaeal species have been described in the literature. He believes that this number will expand significantly (C. R. Woese, personal communication, 1996). Boone has asserted that methanogens alone have accounted for 5 to 10 new species per year for the last 10 years (D. R. Boone, personal communication, 1996). It is doubtful that all of these yet to be cultured *Archaea* and *Bacteria* are dormant in their respective habitats; some must be active and involved in biogeochemical cycling in as yet unexplored ways. This hypothesis is supported by extensive circumstantial evidence, for example, data provided by vital staining techniques such as the AODC, direct viable counting (31), and INT staining (1). Based on their presumed relative abundance and community structure shifts in response to environmental disturbance (59), yet to be cultured *Archaea* and *Bacteria*, no doubt, transform vast quantities of geochemicals in the biosphere.

Recently, the extent of this situation has been made even more obvious with the emergence of major taxa of *Bacteria* for which there are few (or even no!) cultured members (reference 27 and Fig. 5). Major lineages such as the *Verrucomicrobia*, *Acidobacterium*, and others are now known to occur in environments around the world yet have almost no connection with a physiological data base, so virtually no information is available as to their role in biogeochemical cycling. Clearly then, the molecular data are indicating severe restrictions in our ability to interpret biogeochemical cycles on the basis of organisms already in culture.

The focus of this chapter is threefold. First, it addresses the potential of yet to be cultured bacteria to be involved with biogeochemical cycling. The repeated observation that cultivable bacteria comprise considerably less than 1% of the total bacteria observed by direct measurement of any given water, sediment, or soil sample supports the notion of the significance of yet to be cultured bacteria in biogeochemical cycling. Second, ways in which dormant cells or somnicells might become active and then mediate geochemical transformations are considered. This area of research is largely phenomenological, and little is known about activation mechanisms. Finally, it addresses the potential role of noncultivable bacteria in the emerging field of bioremediation.

NOT YET CULTURED MICROORGANISMS

One of the most exciting discoveries of modern microbial ecology has been made using the tools of biotechnology. PCR has been employed to amplify nucleic acids present in natural samples, followed by application of nucleic acid probes. Thus, it has been conclusively demonstrated that large numbers of bacteria exist which have not yet been cultured and consequently have not been characterized with regard to metabolic potential or ability. The elusive ocean *Archaea*, independently described by DeLong (11) and Fuhrman et al. (17), are a dramatic example of this situation. Since the majority of *Archaea* in culture are extremophiles (temperature, salinity, highly reducing environments, pH, etc.), it was a surprise to learn that archaeal types apparently comprise a substantial portion of the total microbial biomass in the oceans of the world (42)—in waters that average below 3°C, contain levels of oxygen that easily support aerobic growth, and exhibit mild alkalinity. In areas sampled by DeLong and coworkers, the ocean *Archaea* have been shown to comprise from 1.5 to 1.8 μg of carbon per liter of ocean water (12; Ed DeLong, personal communication). It is hard to imagine that such a dominant component of the oceanic microbial community does not play an active role in some aspect of marine biogeochemistry.

Another exciting example of marine microorganisms not yet cultivated was provided by Giovannoni and his colleagues (Giovannoni et al., 1990). Using the PCR to amplify 16S ribosomal RNA genes from Sargasso Sea picoplankton, Giovannoni

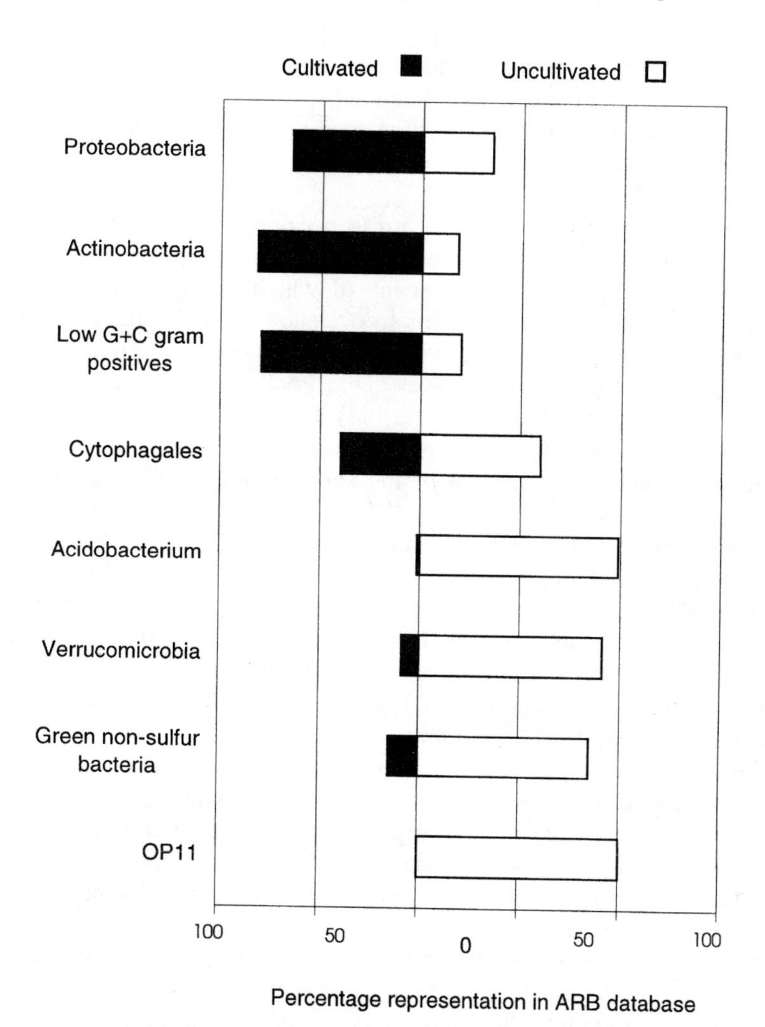

Figure 5. Percentage of bacteria of different groups that are presently uncultivated. This figure (from reference 27) shows that for some groups of bacteria, such as the *Proteobacteria*, *Actinobacteria*, and the low–G+C gram-positive bacteria, greater than 50% of the known sequences are represented by cultivated organisms. For others, only small percentages are cultivated, and for some (like the OP11 group) none have been cultivated. Apparently ubiquitous bacteria like the *Acidobacterium* and *Verrucomicrobia* groups have almost no cultivated representatives.

et al. (17a) discovered a novel bacterial group that accounted for 15% of the bacterial RNA in the Sargasso Sea. This group was designated SAR11 and was also found in coastal waters off Bermuda and Key Biscayne, Florida. More recently, Giovannoni's group found that SAR11 contains two subgroups that exhibit depth-specific distribution in samples collected from several Atlantic and Pacific Ocean sites (14a). Overall, the SAR11 cluster appears to account for 26% of the rDNAs recovered from seawater and about the same percentage of cells when counted by in situ hybridization (Giovannoni, personal communication, 1999). Unfortunately, even though Giovannoni and his students have invested considerable effort in trying to culture SAR11 cells, as of this writing they have been unsuccessful. The ubiquitous SAR11 is yet to be cultured, and there is no clue as to its function in the oceans of the world.

A further exciting finding, with regard to oceanic biota, is the recent discovery that marine subsurface zones contain a large community of bacteria (43), most of which have yet to be cultured, but many of which appear to be viable and are postulated to be important contributors to oceanic biogeochemical stability. Direct 16S rRNA gene sequence analysis of sediment samples from 1.6, 9.8, 78, and 503 m below the sea floor revealed diverse bacterial communities, including what seemed to be novel lineages of bacteria (43). Conservative estimates by Parkes et al. (43) suggest that oceanic subsurface bacteria could constitute as much as 10% of the total living biomass carbon of the surface biosphere. As with most other oceanic sediments, the numbers of total bacteria were on the order of 10^8 cm^{-3}, while viable anaerobic heterotrophs were only 10^5 to 10^6 cm^{-3}. The viable but as yet uncultivated population in oceanic sediments could conceivably constitute approximately 9% of the total biomass of the earth's crust, a staggering number. These data are supported by the lipid analyses of White (60), which also indicate the number of marine sediment bacteria to be 10^8 to 10^9 cm^{-3}.

Related to the ocean *Archaea*, at least in a phenomenological sense, are the oligotrophic bacteria. These ubiquitous microorganisms received considerable attention by many investigators in the 1970s (31a). Labeled oligobacteria by Button and his colleagues (6) these low-nutrient bacteria do not grow (at least in the laboratory) in the presence of high organic nutrient concentrations. Kuznetsov et al. (31a) defined oligotrophic bacteria as those bacteria capable of growth upon initial isolation on or in culture media containing 1 to 15 mg of C per liter, although some are inhibited even by this low amount of organic carbon. For example, Button et al. (6) observed that organic carbon (casein hydrolysate) additions of ≥5 mg of C per liter of seawater inhibited growth of oligotrophic bacteria. As discussed by Button et al. (6), it is clear that viability is an operational term. They suggest that one reason for low estimates of viability using traditional methods is that most marine bacteria reach stationary phase long before attaining visible turbidity. They go on to point out that even though epifluorescence microscopy and flow cytometry improve the sensitivity of detection by about 3 orders of magnitude, populations are truncated near these instrument sensitivity limits of 10^4 cells per ml. In other words, populations may grow (reproduce) in response to available nutrients, enter stationary phase, and presumably become dormant long before becoming detectable (6). It is highly probable, therefore, that at least some oligotrophic bacteria are members of the yet to be cultured class, and this was evidenced by the recent isolation of a new organism, *Oligobacterium* RB1 gen. nov. (D. K. Button, personal communication). Interestingly, this organism was isolated from oligotrophic seawater without hydrocarbon enrichment and was later found to be a toluene oxidizer. This is consistent with previous studies that have noted the presence of toluene in oligotrophic seawater and observed shifts in marine bacterial communities, including oligotrophic bacteria, in response to chronic hydrocarbon pollution (19).

Among the microorganisms not yet cultured is a category characterized by organisms that are thought to exist in large numbers but for which few representatives have been isolated. The assumption of large numbers is derived from the magnitude of an important process believed to be mediated by microorganisms. It is often easier to measure the rate and extent of a physiological reaction in environmental samples than it is to account for the microorganisms responsible, either qualitatively or quantitatively. Obviously, this situation is the exact opposite from the ocean *Archaea* of unknown biogeochemical function just described.

Take, for example, the use of acetate or other compounds as the carbon, energy, and electron sources for sulfate reduction. Prior to 1977, only a few compounds were believed to be important in the nutrition of the sulfate-reducing bacteria (SRB). Sulfate reduction was considered to be a process whereby lactate was split to form acetate and CO_2 (see for example Jørgensen [29]). In the late 1970s and early to mid-1980s, however, it became obvious that acetate was the preferred carbon source. The "type species" for sulfate reduction, *Desulfovibrio desulfuricans*, as well as a number of other common sulfate reducers, was isolated and cultured on media containing lactate as the carbon/energy source (44). Enumeration of sulfate reducers was often carried out using an MPN technique employing a lactate broth with high concentrations of SO_4^{2-}. A monograph on SRB (41) fails to list a method for cultural enumeration of the sulfate-reducing guild. Oddly enough, however, acetate-oxidizing SRB are not easily isolated (61, 62), much less enumerated as a guild. In a sense, these organisms fit perfectly the definition of viable but nonculturable bacteria, and this observation was recently confirmed by Ravenschlag et al. (45) when they found that 43% of the 353 rDNA clones isolated from Arctic Ocean sediments were potential SRB, based on dot blot hybridization with probes specific for SRB. None of these SRB were actually isolated or cultivated (45).

It is generally accepted that numbers of microorganisms should be positively correlated with levels of a given activity for which those microorganisms are assumed to be responsible. Thus, if the rate of ammonium oxidation, sulfate reduction, or iron reduction, for example, increases, we expect to find a concomitant increase in the numbers of ammonium-oxidizing, sulfate-reducing, or iron-reducing bacteria, respectively. The converse is also true, i.e., if we observe numbers of a specific guild of microbes to be elevated, compared with another location, we anticipate that the rate of the reaction of interest would be higher where there are more microorganisms. Of course, this further assumes that the material being examined limits microbial growth. In many cases, however, an activity might be carried out by a variety of microorganisms for which some other compound is the limiting factor. In the case of sulfate reduction, for example, the wide variety of substrates used as sources of energy and carbon by the SRB precludes the employment of a single medium for enumeration of all guild members simultaneously. Indeed, a literature review listed a large and diverse number of compounds that can be used as substrates for sulfate reduction (21). That list named two inorganics (H_2 and CO), the straight alkanes from C_{12} to C_{20}, 6 branched and 20 linear aliphatic fatty acids (from formate to eicosanoate), 7 dicarboxylic acids, 16 alcohols and glycols, 12 amino acids, 5 sugars, 34 aromatic compounds, and 10 or more miscellaneous compounds including several alicyclics. How could one possibly devise a medium with which to enumerate all SRB? For example, a medium employing acetate as the sole carbon source for the SRB should not be expected to be useful in the enumeration of lactate-oxidizing microbes or for those microorganisms using other volatile fatty acids, alcohols, or hydrocarbons as a source of energy. Use of an acetate-based medium, therefore, would render all the organisms using some other source of energy as viable, but nonculturable, at least for the purposes of that particular enumeration.

A great variety and number of yet to be cultured bacteria may also derive from the abundant symbioses that occur between marine *Eucarya* and *Bacteria*. In such cases, it is often possible to view readily by electron microscopy actively growing symbionts at cell concentrations of 10^9 ml^{-1} or more (51), when outside the host no growth has been demonstrated. An example of such an association is that of the many luminescent symbioses of bacteria with marine fishes and squids (22, 23, 30, 37). For example, it is known that the flashlight fish, *Photoblepharon palpabratus*, constantly inoculates the environment with bacteria from its light organ where they have grown, with doubling times on the order of 1 to several days (23). In a few days, a single fish can populate the water of a sizable aquarium with bacteria to a population of 10^4 ml^{-1}, which can be easily seen but not grown in culture (i.e., viable, but nonculturable). These bacteria are presumably the seed stock for the inoculation of young fish. To date, these bacteria have not been cultured, despite efforts by many different workers. Recent work by Haygood (22) has shown that the luminous bacteria from the light organs of anomalopid fishes have only one copy of the rRNA genes, which may be an adaptation to a symbiotic life style which results in slow growth of the bacteria in the light organ and perhaps extremely slow growth, or a state of pseudodormancy or dormancy outside the host. Such adaptations provide good molecular models to explain the existence of bacteria capable of surviving but unable to grow except under appropriate (e.g., symbiotic) conditions and, even then, slowly.

The metabolic symbioses of chemolithotrophic bacteria (sulfur oxidizers, ammonia oxidizers, and methane oxidizers) with vent and sediment *Eucarya* provide yet another example of easily visible but apparently uncultured bacteria (8, 9, 13, 14). These bacteria, which are the basis of the food chain in the deep hydrothermal vents and thus of central biogeochemical importance in this environment (28), clearly fall under the rubric of viable but nonculturable. Such observations may provide valuable clues to the study of environments where symbiosis is appreciated as a major mode of individuals and communities. Interestingly, it has also been reported that some of the vent symbionts show adaptations similar to those reported for the luminous symbionts—namely that the rRNA genes appear to be present in the cell in single copy (14). While this property does not define symbiotic bacteria, since some *Archaea* are also thought to contain single copies of rRNA genes (14), it is intriguing that two separate groups of nonculturable symbionts are now known to share this trait.

Of equal potential importance are the interactions of bacteria with other bacteria—microbial syntrophisms. The groups of organisms classified in the genera *Syntrophomonas* or *Syntrophobacter* provide a good example. These bacteria are key organisms in the anaerobic degradation of organic carbon, and they oxidize substrates such as butyrate with the concomitant release of molecular hydrogen. They are incapable of growth in the presence of molecular hydrogen and can only grow in coculture with hydrogen-consuming bacteria such as methanogens or sulfate reducers. The presence of such syntrophic bacteria can completely change the anaerobic flow of carbon and allows efficient functioning of carbon oxidation. Because of their sensitivity to hydrogen, a product of their own metabolism, these organisms are extremely difficult to obtain in pure culture; despite their abundance and biogeochemical importance, they remain for the most part uncultured or, at best, extremely difficult to culture. Given the close associations of bacteria in biogeochemical cycles, it is not hard to imagine that such interactions occur as part of many different biogeochemical cycles and that such examples may be the rule, rather than the exception, as knowledge of complex microbial communities expands (39, 48)

Most of the organisms described thus far are all aquatic *Bacteria* and *Archaea*. However, based on data of Tiedje (52) and others (2), many soil *Bacteria* and *Archaea* await discovery and characterization. Indeed, data being gathered by Schmidt and his students and colleagues suggest that several different phylogenetic clusters will be described in soil samples (4). In some cases, the isolation of new, novel organisms will be relatively easy, involving perhaps simple enrichment, as was the case with the *Sphingomonas* sp. (16), *Geobacter sulfurreducens* sp. nov. (7), and *Bacillus infernus* sp. nov. (2). In other cases, isolation will no doubt be arduous, following the example of the oceanic *Archaea* and SAR11. By whatever means they are finally cultured and described, new isolates will continue to redefine the importance of viable but nonculturable bacteria in biogeochemistry.

SOMNICELLS

Dormant bacteria are commonly thought of as those with specific mechanisms of dormancy, such as endospore-forming *Bacillus* and *Clostridium* species or the cyst-forming methanotrophs. However, as pointed out in chapter 1, another more subtle approach to the problem of long-term survival in nutrient-depleted or otherwise adverse habitats has evolved in several eubacteria—the ability to form somnicells (46). Somnicells have been the focus for most of the discussion in this monograph, and they most certainly play a central role in biogeochemistry. The newly discovered molecular mechanisms responsible for starvation survival in *Escherichia coli* (34), which are echoed in the marine vibrios (40), provide mechanisms for organisms to slow down their metabolism significantly and enter into a survival mode under conditions of starvation. Yet another (but probably related) mechanism has been offered by Chesbro et al. (10) in their description of the domains of slow growth for several bacteria. Given that many aquatic bacteria may enjoy a feast or famine type of life in and out of the stomachs of aquatic mammals, fishes, and invertebrates, it would not be unexpected to find starved and apparently dormant bacteria with mechanisms triggered by specific nutrient signals (perhaps supplied by digestive tract components) from higher organisms.

Factors that lead to induction of somnicells include temperature (58), nutrient stress (34, 40), and other factors (18, 20, 46), and are undoubtedly different for different bacteria. For example, do the thermophiles that were enriched by Vidal (54, 55) fall into the category of somnicells? Bacteria that tolerate or require high temperature (thermophiles) are present in seawater, probably throughout the oceans, but fail to grow at temperatures usually employed for culturing marine microorganisms. Vidal tested this hypothesis with a high-temperature heat exchanger located on the pier at Scripps Institute of Oceanography. After a few days, the exchanger became clogged with bacteria, most of which would not grow at less than 60°C! These thermophiles, with temperature requirements for 60°C and above for growth, were present and dormant and would have been missed by virtually every study of these waters, yet the enrichments indicated a population with remarkable ability to respond to high temperatures and grow rapidly once conditions were adjusted to their requirements.

In the case of the iron- and sulfur-oxidizing chemolithotroph *Thiobacillus ferrooxidans*, the definition of viable but nonculturable (47) fits appropriately. *T. ferrooxidans* is normally an acidophilic bacterium that does not oxidize iron at pH levels above about 4.5. Although the organism has been reported to oxidize sulfur at circumneutral pH, the importance of this process in the natural setting for the organism has not been determined. Cultural enumeration of the organism is usually carried out by employing the 9K medium of Silverman and Lundgren (49). This preparation consists of a high concentration of Fe(II) (about 9,000 mg liter^{-1}, hence the name 9K) and a low pH (usually around 3). When samples of circumneutral water are placed into tubes of 9K medium (because of the low pH and high concentration of dissolved solids the medium cannot be solidified with agar and is therefore only suitable for MPN assays), growth and oxidation of the iron do not occur. Does this mean that the organism is not present or, if present, is dead?

In their study of Lake Anna, Virginia, Baker and Mills (1) examined the distribution of *T. ferrooxidans* in an arm of an impoundment that receives most of its water from Contrary Creek, a large acid mine drainage stream. Results of several studies indicated that total (AODC) and heterotrophic bacteria (defined by spread plate methods on one-half strength nutrient agar) increased with distance from the mouth of the acid mine stream (36, 57), and further that the number of respiring organisms (INT-reducing bacteria) increased even more rapidly than the increase in total bacteria (Table 1). Furthermore, the numbers of *T. ferrooxidans* (enumerated

Table 1. Total and respiring bacterial cells in the Contrary Creek Arm of Lake Anna, Virginia[a]

Station	Cells ml^{-1}		% INT positive
	Total (AODC)	Respiring (INT)[b]	
C1	3.7×10^5	8.9×10^4	24
C3	4.0×10^5	2.4×10^5	61
A2	9.2×10^5	5.9×10^5	64

[a]The stations are listed in increasing distance from the source of the acid mine drainage which enters the lake at station C1. Station A2 is about 2 km from station C1, with station C3 about halfway between.
[b]INT-reduction assay.

by the 9K-MPN technique) dropped off drastically with distance from the acid water source, and no *T. ferrooxidans* could be measured at only a short distance from the mouth of the stream (36).

Baker and Mills (1) developed a fluorescent antibody (FA) assay against a strain of *T. ferrooxidans* with which they had been working, and counts made with this method showed that the cultural method underestimated the number of *T. ferrooxidans* cells in the circumneutral water of Lake Anna (Table 2). Still, identity of surface antigens on bacterial cells does not indicate viability, and the water samples showing positive FA reactions could not, therefore, be considered to contain viable cells. To overcome this difficulty, Baker and Mills (1) applied the INT reduction assay to determine what proportion of the cells had a functioning cytochrome system. Many of the FA-positive cells did not show viability by this approach, but a significant number did. When circumneutral water samples were placed directly into 9K medium, no *T. ferrooxidans* colonies were observed, but when such samples were amended with small amounts of acid over a period of about 10 hours (thereby gradually lowering the pH of the water to that of the 9K medium), after which they were transferred to the 9K medium, the cells grew with a concomitant oxidation of iron. In every case tested, presence of FAINT-positive cells indicated the ability to culture *T. ferrooxidans* through the gradual acidification technique. It was suggested that, as *T. ferrooxidans* moves from the acidic water to the more dilute and circumneutral waters of the lake, some cells survive exposure to the higher pH, although the cells may not be actively taking up substrate, as there was no evidence for either iron or sulfur oxidation.

There is significance to the findings of Baker and Mills (1) in that many have speculated on the initiation of acid mine drainage formation and the source of inoculum of *T. ferrooxidans* for the oxidation of Fe(II) in pyrite in coal and mineral deposits. If some fraction of *T. ferrooxidans* cells survive in circumneutral water for periods of time, the water itself could serve as the source of inoculum for newly exposed pyrites and other metal sulfides. The pH would be reduced via sulfur oxidation by *T. ferrooxidans* or other organisms with similar metabolic potential.

Table 2. *T. ferrooxidans* counts in water from acid mine drainage-contaminated and uncontaminated sites

Station	AMD	No. (mean) of *T. ferrooxidans* cells per ml determined by indicated technique[a]		
		FA	FAINT	MPN
C1	+[b]	7.7×10^4 (0.08)	5.6×10^4 (0.15)	33
C4	+/−	4.1×10^4 (0.19)	1.8×10^4 (0.24)	0
A2	+/−	2.7×10^3 (0.04)	4.1×10^2 (0.29)	0
F1	−	5.4×10^2 (0.50)	1.4×10^1 (1.0)	0
L1	−	4.6×10^3 (0.15)	5.4×10^1 (1.0)	0
PH	−	0	0	0

[a] Numbers of cells of *T. ferrooxidans* (FA), culturable *T. ferrooxidans* (MPN), and respiring *T. ferrooxidans* (FAINT) per milliliter in samples of water from acid mine drainage (AMD)-contaminated and uncontaminated sites. (Reproduced from reference 1.) Values for the direct counts are the means and coefficient of error obtained from three replicate water samples.

[b] +, presence of AMD; −, absence of AMD.

Luminescent marine bacteria can also form somnicells. An excellent example is the recent finding of Lee and Ruby (32) that the Hawaiian squid, *Euprymna scolopes*, inoculates the water around it constantly with luminous bacteria (*Vibrio fischeri*). Within a few days, these bacteria (which are estimated to be almost 10^3 cells per ml of seawater!) become nonculturable but retain the ability to establish a viable symbiosis with the light organ of newly hatched juvenile squid, an example of the viable but nonculturable state serving the bacterium in its transfer from the light organ of one squid to another and (more importantly) playing a role in the life cycle of its host species.

BIOREMEDIATION

Viable but nonculturable bacteria clearly function in bioremediation which, in natural habitats, can be defined as a planned and controlled biogeochemical elimination, reduction, or neutralization of toxic or harmful wastes. There are toxic waste sites in which metabolically competent bacteria exist, albeit in a dormant state, presumably because of a requirement for a nutrient or growth factor, without which their growth is limited. For example, trichloroethylene (TCE), one of the leading groundwater contaminants in the United States, does not quickly disappear from aquifers. However, in elucidating what has become a classic example of cometabolism, many investigators have demonstrated that a variety of methanotrophic and aromatic-oxidizing bacteria will cometabolize TCE when supplied with necessary substrates, such as methane and toluene (15, 24, 56). Presumably, the requisite bacteria exist in situ as somnicells and do not dehalogenate the TCE until they are supplied with a suitable substrate, e.g., methane, a scenario suggested by Bowman et al. (3). This type of in situ cometabolism was successfully demonstrated in large-scale field tests at the Savannah River Laboratory (Bowman et al. [3]) and resulted in a patent for the process (Hazen et al., U.S. Patent 5,384,048, assigned to the U.S. Department of Energy, January, 1995).

Recently, attention has been given to exploiting the small size of dormant bacteria to facilitate their transport to subsurface sites contaminated with toxic chemicals. Most bacteria become markedly smaller in size as they progress from vegetative cells to somnicells (20, 46). Costerton and his colleagues have proposed injecting these miniaturized cells into contaminated sites, and they have succeeded in reducing both cell size and extracellular polysaccharide (glycocalyx) volume for several biodegradative bacteria (25). Truex et al. (53) have also investigated the effects of starvation on bioremediation and demonstrated a 57% reduction in cell volume for *Pseudomonas cepacia* cells starved for 62 days. Interestingly, cells starved for more than 60 days demonstrated a more efficient degradation of quinoline than cells starved for only 2 days (53).

The above two examples represent an exciting opportunity to manipulate features of the viable but nonculturable phenomenon for the purpose of bioremediation. Toxic chemicals presumably are in contact with somnicells rather than active vegetative cells, when the chemicals first enter and contaminate the environment. Thereafter, both direct metabolism and cometabolism by activated or resuscitated somnicells are required to degrade toxic wastes. Subsurface transport represents a major roadblock to successful bioremediation. Therefore, mechanisms to transport both competent cells and nutrients to subsurface sites need to be devised. Clearly, additional research would be valuable in gaining an improved understanding of the physiology and ecology of somnicells, so that they can be used for in situ bioremediation of contaminated sites (Fig. 6).

A more global approach to bioremediation has been proposed for removal of anthropogenic CO_2 from the atmosphere to reduce greenhouse gases. One hypothesis holds that nutrients which limit photosynthesis (e.g., iron) can be supplied to the world oceans and thereby stimulate CO_2 uptake on a global scale. Martin et al.

Figure 6. Cartoon depicting bacteria in a subsurface waste site. Some bacteria are metabolically competent and fully active, capable of transforming organic and inorganic wastes that are present in subsurface waste sites. Other bacteria are metabolically incompetent and therefore cannot oxidize, reduce, or otherwise transform organic and/or inorganic wastes. And other bacteria (depicted by the "sleeping cell") are thought to be viable but nonculturable or dormant and therefore do not transform wastes whether they are metabolically competent or incompetent. Many metabolically competent bacteria probably transition between active and dormant states, as metabolizable wastes flux through the site and thereby provide utilizable nutrients. At question is, how do we interrogate conditions at the site?

(33) were able to demonstrate a doubling of plant biomass, a threefold increase in chlorophyll, and a fourfold increase in plant production, following addition of iron (injected as Fe^{+2}) to the equatorial Pacific Ocean. The predominant phytoplankton in the non-iron-enriched area were picoplankton (cells <5 μm in size comprised $>90\%$ of the chlorophyll biomass). Inside the iron-enriched area, biomass of all size classes increased, but the largest contributors to plankton biomass increases were cyanobacteria, especially *Synechococcus*. While Martin et al. (33) did not observe the amount of CO_2 drawdown that was expected, the experiment did demonstrate rapid in situ response by bacteria, probably dormant bacteria, to an increase in concentration of a limiting nutrient.

Much remains to be done before bioremediation becomes a useful, predictable, and utilized technology. Few contributing basic disciplines have amassed sufficient information to ensure successful applications. Genetics, physiology, ecology, and geology all make contributions to bioremediation research and development, as well as bioprocess engineering. In the area of genetics, little information is available on gene stability, expression, and exchange in natural environments, since most genetic research has been conducted in the laboratory under conditions unrepresentative of the natural environment. Research in microbial physiology has been neglected, and more information is needed on enzyme induction and repression, metabolic pathways, cometabolism, and other factors involved with metabolism of toxic chemicals in natural environments. Especially critical is the need for an understanding of how microorganisms function when presented with mixed wastes that may include chlorinated solvents, polyaromatic hydrocarbons, heavy metals, radionuclides, chelating agents, and fuel hydrocarbons (Fig. 6). Little is known about how microbial communities respond to wastes in situ. Metabolically competent bacteria may not respond to toxic wastes for many reasons, e.g., they may be weathered onto a substrate, exist in a dormant state, or both. They could be capable of degrading one substrate but remain inactive because of inhibition by other wastes. Presumably, in some situations, metabolically competent bacteria may be totally absent. Clearly, there is a need for better understanding of ecology (e.g., species interactions, bioavailability, attachment, taxis), geology (e.g., macro- to microheterogeneity, transport, weathering), and nondestructive, on-line methods for process monitoring and verification. With this additional information engineers can optimize processes and bring them to the marketplace to be used to clean up toxic waste sites, eliminate toxic chemicals from waste streams, and contribute to new waste-free production methods for industry and agribusiness, including aquaculture. An understanding of viable but nonculturable somnicells and how they can be managed for bioremediation would provide a major contribution to environmental health and management.

CONCLUSIONS

The ability to enumerate viable but nonculturable cells—either dormant or not yet cultured—can have a profound impact on the conclusions drawn from examination of microbiota in environmental systems. Traditional techniques for isolation and enumeration of microorganisms have served the scientific community well but are limited in what they can do. They will continue to provide invaluable information, but those cells which cannot be "seen," i.e., cultured, are enigmatic. Whether somnicells or yet to be cultured cells, these viable but nonculturable bacteria hold secrets valuable to understanding of and application in many facets of biogeochemistry. Ultimately, it should be possible to correlate cell numbers with reaction rates for a more careful definition and management of biogeochemical reactions.

REFERENCES

1. **Baker, K. H., and A. L. Mills.** 1982. Determination of the number of respiring *Thiobacillus ferrooxidans* cells in water samples by using combined fluorescent antibody-2-(*p*-iodophenyl)-3-(*p*-nitrophenyl)-5-phenyltetrazolium chloride staining. *Appl. Environ. Microbiol.* **43:**338–344.
2. **Boone, D. R., Y. Liu, Z. Zhao, D. L. Balkwill, G. R. Drake, T. O. Stevens, and H. C. Aldrich.** 1995. *Bacillus infernus* sp. nov., an Fe(III)- and Mn(IV)-reducing anaerobe from a deep terrestrial subsurface. *Int. J. Syst. Bacteriol.* **45:**441–448.
3. **Bowman, J. P., L. Jiménez, I. Rosario, T. C. Hazen, and G. S. Sayler.** 1993. Characterization of the methanotrophic bacterial community present in a trichloroethylene-contaminated subsurface groundwater site. *Appl. Environ. Microbiol.* **59:**2380–2387.
4. **Buckley, D. H., J. R. Graber, and T. M. Schmidt.** 1998. Phylogenetic analysis of nonthermophilic members of the kingdom *Crenarchaeota* and their diversity and abundance in soils. *Appl. Environ. Microbiol.* **64:**4333–4339.
5. **Bull, A. T., M. Goodfellow, and J. H. Slater.** 1992. Biodiversity as a source of innovation in biotechnology. *Annu. Rev. Microbiol.* **46:**219–252.
6. **Button, D. K., F. Schut, P. Quang, R. Martin, and B. R. Robertson.** 1993. Viability and isolation of marine bacteria by dilution culture: theory, procedures, and initial results. *Appl. Environ. Microbiol.* **59:**881–891.
7. **Caccavo, F., Jr., D. J. Lonergan, D. R. Lovley, M. Davis, J. F. Stolz, and M. J. McInerney.** 1994. *Geobacter sulfurreducens* sp. nov., a hydrogen- and acetate-oxidizing dissimilatory metal-reducing microorganism. *Appl. Environ. Microbiol.* **60:**3752–3759.
8. **Cary, C., C. R. Fisher, and H. Felbeck.** 1988. Mussel growth supported by methane as sole carbon and energy source. *Science* **240:**78–80.
9. **Cavanaugh, C. M., S. L. Gardiner, M. L. Jones, H. W. Jannasch, and J. B. Waterbury.** 1981. Prokaryotic cells in the hydrothermal vent tube worm *Riftia pachyptila* Jones: possible chemoautotrophic symbionts. *Science* **213:**340–342.
10. **Chesbro, W. R., M. Arbige, and R. Eifert.** 1990. When nutrient limitation places bacteria in the domains of slow growth: metabolic, morphologic and cell cycle behavior. *FEMS Microbiol. Ecol.* **74:**103–120.
11. **DeLong, E. F.** 1992. Archaea in coastal marine environments. *Proc. Natl. Acad. Sci. USA* **89:**5685–5689.
12. **DeLong, E. F., K. Y. Wu, B. B. Prézelin, and R. V. M. Jovine.** 1994. High abundance of Archaea in Antarctic marine picoplankton. *Nature* (London) **371:**695–697.
13. **Distel, D. L., and C. M. Cavanaugh.** 1994. Independent phylogenetic origins of methanotrophic and chemoautotrophic bacterial endosymbioses in marine bivalves. *J. Bacteriol.* **176:**1932–1938.
14. **Eisen, J. A., S. W. Smith, and C. M. Cavanaugh.** 1992. Phylogenetic relationships of chemoautotrophic bacterial symbionts of *Solemya velum* Say (Mollusca:Bivalvia) determined by 16S rRNA gene sequence analysis. *J. Bacteriol.* **174:**3416–3421.

14a. **Field, K. G., D. Gordon, T. Wright, M. Rappé, E. Urback, K. Vergin, and S. J. Giovannoni.** 1997. Diversity and depth-specific distribution of SAR11 cluster rRNA genes from marine planktonic bacteria. *Appl. Environ. Microbiol.* **63:**63–70.

15. **Folsom, B. R., P. J. Chapman, and P. H. Pritchard.** 1990. Phenol and trichloroethylene degradation by *Pseudomonas cepacia* G4: kinetics and interactions between substrates. *Appl. Environ. Microbiol.* **56:**1279–1285.

16. **Fredrickson, J. K., D. L. Balkwill, G. R. Drake, M. F. Romine, D. B. Ringelberg, and D. C. White.** 1995. Aromatic-degrading *Sphingomonas* isolates from the deep subsurface. *Appl. Environ. Microbiol.* **61:**1917–1922.
17. **Fuhrman, J. A., K. McCallum, and A. A. Davis.** 1992. Novel major archaebacterial group from marine plankton. *Nature* (London) **356:**148–149.
17a. **Giovannoni, S. J., T. B. Britschgi, C. L. Moyer, and K. G. Field.** 1990. Genetic diversity in Sargasso Sea bacterioplankton. *Nature* **345:**60–63.
18. **Gottschal, J. C.** 1990. Phenotypic response to environmental changes. *FEMS Microbiol. Ecol.* **74:** 93–102.
19. **Grimes, D. J., F. L. Singleton, and R. R. Colwell.** 1984. Allogenic succession of marine bacterial communities in response to pharmaceutical waste. *J. Appl. Bacteriol.* **57:**247–261.
20. **Grimes, D. J., R. W. Attwell, P. R. Brayton, L. M. Palmer, D. M. Rollins, D. B. Roszak, F. L. Singleton, M. L. Tamplin, and R. R. Colwell.** 1986. Fate of enteric pathogenic bacteria in estuarine and marine environments. *Microbiol. Sci.* **3:**324–329.
21. **Hansen, T. A.** 1993. Carbon metabolism of sulfate-reducing bacteria, p. 21–40. *In* J. M. Odom and R. Singleton, Jr. (ed.), *The Sulfate-Reducing Bacteria: Contemporary Perspectives.* Springer-Verlag, New York, N.Y.
22. **Haygood, M. G.** 1993. Light organ symbioses in fishes. *Crit. Rev. Microbiol.* **19:**191–216.
23. **Haygood, M. G., B. M. Tebo, and K. H. Nealson.** 1984. Luminous bacteria of a monocentrid fish (*Monocentris japonicus*) and two analopid fishes (*Photoblepharon palpabratus* and *Kryptophanaron alfredi*): population sizes and growth within the light organs and rates of release into the seawater. *Mar. Biol.* **78:**249–258.
24. **Henry, S. M., and D. Grbić-Galić.** 1991. Influence of endogenous and exogenous electron donors and trichloroethylene oxidation toxicity on trichloroethylene oxidation by methanotrophic cultures from a groundwater aquifer. *Appl. Environ. Microbiol.* **57:**236–244.
25. **Herman, D. C., and J. W. Costerton.** 1993. Starvation-survival of a *p*-nitrophenol-degrading bacterium. *Appl. Environ. Microbiol.* **59:**340–343.
26. **Holt, J. G. (ed.).** 1989. *Bergey's Manual of Systematic Bacteriology*, vol. 1–4. Williams & Wilkins, Baltimore, Md.
27. **Hugenholtz, P., B. M. Goebel, and N. R. Pace.** 1998. Impact of culture-independent studies on the emerging phylogenetic view of bacterial diversity. *J. Bacteriol.* **180:**4765–4774.
28. **Jannasch, H. W.** 1985. The chemosynthetic support of life and the microbial diversity at deep-sea hydrothermal vents. *Proc. R. Soc. London Ser. B.* **225:**277–297.
29. **Jørgensen, B. B.** 1980. Mineralization and the bacterial cycling of carbon, nitrogen, and sulphur in marine sediments, p. 239–251. *In* D. C. Ellwood, J. N. Hedger, M. J. Latham, J. M. Lynch, and J. H. Slater (ed.), *Contemporary Microbial Ecology.* Academic Press, Inc., (London) Ltd., London, United Kingdom.
30. **Kessel, M.** 1977. The ultrastructure of the relationship between the luminous organ of the teleost fish *Photoblepharon palpebratus* and its symbiotic bacteria. *Cytobiology* **15:**145–155.
31. **Kogure, K., U. Simidu, and N. Taga.** 1979. A tentative direct microscopic method for counting living marine bacteria. *Can. J. Microbiol.* **25:**415–420.
31a. **Kuznetsov, S. I., G. A. Dubinina, and N. A. Lapteva.** 1979. Biology of oligotrophic bacteria. *Annu. Rev. Microbiol.* **33:**377–387.
32. **Lee, K-H, and E. G. Ruby.** 1995. Symbiotic role of the viable but nonculturable state of *Vibrio fischeri* in Hawaiian coastal seawater. *Appl. Environ. Microbiol.* **61:**278–283.
33. **Martin, J. H., et al.** 1994. Testing the iron hypothesis in ecosystems of the equatorial Pacific Ocean. *Nature* (London) **371:**123–129.
34. **Matin, A.** 1991. The molecular basis of carbon-starvation-induced general resistance in *Escherichia coli. Mol. Microbiol.* **5:**3–10.
35. **Meyer-Reil, L. A.** 1978. Autoradiography and epifluorescent microscopy combined for the determination of number and spectrum of actively metabolizing bacteria in natural waters. *Appl. Environ. Microbiol.* **36:**506–512.
36. **Mills, A. L.** 1985. Acid mine waste drainage: Microbial impact on the recovery of soil and water ecosystems, p. 35–81. *In* R. L. Tate and D. Klein (ed.), *Soil Reclamation Processes.* Marcel Dekker, Inc., New York, N.Y.

37. **Nealson, K. H., and J. W. Hastings.** 1992. The luminous bacteria, p. 625–649. *In* A. Balows, H. G. Trueper, M. Dowrkin, W. Harder, and K. Schleifer (ed.), *The Prokaryotes.* Springer-Verlag, New York, N.Y.
38. **Nealson, K. H., and D. A. Stahl.** 1997. Microorganisms and biogeochemical cycles: what can we learn from layered microbial communities? *Rev. Mineral.* **35:**5–34.
39. **Nusslein, K., and J. M. Tiedje.** 1999. Soil bacterial community shift correlated with change from pasture vegetation in a tropical soil. *Appl. Environ. Microbiol.* **65:**3622–3626.
40. **Nystrom, T., N. H. Albertson, K. Flardh, and S. Kjelleberg.** 1990. Physiological and molecular adaptation to starvation and recovery from starvation by the marine *Vibrio* sp. S14. *FEMS Microbiol. Ecol.* **74:**129–140.
41. **Odom, J. M., and R. Singleton, Jr. (ed.).** 1993. *The Sulfate-Reducing Bacteria: Contemporary Perspectives.* Springer-Verlag, New York, N.Y.
42. **Olsen, G. J.** 1994. Archaea, archaea, everywhere. *Nature* (London) **371:**657–658.
43. **Parkes, R. J., B. A. Cragg, S. J. Bale, J. M. Getliff, K. Goodman, P. A. Rochelle, J. C. Fry, A. J. Weightman, and S. M. Harvey.** 1994. Deep bacterial biosphere in Pacific Ocean sediments. *Nature* (London) **371:**410–413.
44. **Postgate, J. R.** 1979. *The Sulphate-Reducing Bacteria.* Cambridge University Press, Cambridge, U.K.
45. **Ravenschlag, K., K. Scham, J. Pernthaler, and R. Amann.** 1999. High bacterial diversity in permanently cold marine sediments. *Appl. Environ. Microbiol.* **65:**3982–3989.
46. **Roszak, D. B., and R. R. Colwell.** 1987. Survival strategies of bacteria in the natural environment. *Microbiol. Rev.* **51:**365–379.
47. **Roszak, D. B., D. J. Grimes, and R. R. Colwell.** 1984. Viable but nonrecoverable stage of *Salmonella enteritidis* in aquatic systems. *Can. J. Microbiol.* **30:**334–338.
48. **Sekiguchi, Y., Y. Kamagata, K. Nakamura, A. Ohashi, and H. Harada.** 1999. Fluorescence in situ hybridization using 16S rRNA-targeted oligonucleotides reveals localization of methanogens and selected uncultured bacteria in mesophilic and thermophilic sludge granules. *Appl. Environ. Microbiol.* **65:**1280–1288.
49. **Silverman, M. P., and D. G. Lundgren.** 1959. Studies on the chemoautotrophic iron bacterium *Ferrobacillus ferrooxidans.* I. An improved medium and a harvesting procedure for securing high cell yields. *J. Bacteriol.* **77:**642–647.
50. **Tabor, P. S., and R. A. Neihof.** 1984. Direct determination of activities for microorganisms of Chesapeake Bay populations. *Appl. Environ. Microbiol.* **48:**1012–1019.
51. **Tebo, B. M., D. S. Linthicum, and K. H. Nealson.** 1979. Luminous bacteria and light-emitting fish: ultrastructure of the symbiosis. *Biosystems* **11:**269–281.
52. **Tiedje, J. M.** 1994. Microbial diversity: of value to whom? *ASM News* **60:**524–525.
53. **Truex, M. J., F. J. Brockman, D. L. Johnstone, and J. K. Fredrickson.** 1992. Effect of starvation on induction of quinoline degradation for a subsurface bacterium in a continuous-flow column. *Appl. Environ. Microbiol.* **58:**2386–2392.
54. **Vidal, F. V.** 1981. Microorganisms associated with areas of hydrothermal activity off Northern Baja California. Ph.D. Thesis. University of California, San Diego.
55. **Vidal, V. M. V., F. V. Vidal, and J. Isaacs.** 1978. Coastal submarine hydrothermal activity off Northern Baja California. *J. Geophys. Res.* **84:**1757–1774.
56. **Wackett, L. P., and D. T. Gibson.** 1988. Degradation of trichloroethylene by toluene dioxygenase in whole-cell studies with *Pseudomonas putida* F1. *Appl. Environ. Microbiol.* **54:**1703–1708.
57. **Wassel, R. A., and A. L. Mills.** 1983. Changes in water and sediment bacterial community structure in a lake receiving acid mine drainage. *Microb. Ecol.* **9:**155–169.
58. **Weichart, D., J. D. Oliver, and S. Kjelleberg.** 1992. Low temperature induced non-culturability and killing of *Vibrio vulnificus*. *FEMS Microbiol. Lett.* **100:**205–210.
59. **White, D. C.** 1993. *In situ* measurement of microbial biomass, community structure and nutritional status. *Phil. Trans. R. Soc. London Ser. A* **344:**59–67.
60. **White, D. C.** 1995. Chemical ecology: possible linkage between macro- and microbial ecology. *Oikos* **74:**174–181.
61. **Widdel, F.** 1987. New types of acetate oxidizing, sulfate-reducing *Desulfobacter* species, *D. hydrogenophilus* sp. nov., *D. latus* sp. nov., and *D. curvatus* sp. nov. *Arch. Microbiol.* **148:**286–291.
62. **Widdel, F., and N. Pfennig.** 1977. A new anaerobic, sporing, acetate-oxidizing, sulfate reducing bacterium, *Desulfotomaculum (emend.) acetoxidans*. *Arch. Microbiol.* **112:**119–122.

Nonculturable Microorganisms in the Environment
Edited by R. R. Colwell and D. J. Grimes

Chapter 13

Viable but Nonculturable Cells in Plant-Associated Bacterial Populations

Mark Wilson and Steven E. Lindow

Plants in terrestrial ecosystems are colonized by complex microbial communities composed of bacteria, yeasts, and filamentous fungi. These microbial communities have received much attention because of their effects on plant productivity. While these communities contain some deleterious organisms, such as phytopathogenic bacteria and fungi, they also contain beneficial organisms, such as nitrogen-fixing bacteria, bacteria capable of suppressing plant disease, and bacteria capable of promoting plant growth. Quantification of bacterial populations in these plant-associated microbial communities in epidemiological, pathological, and ecological studies has to date relied almost exclusively upon either plate counts using selective media (4, 48, 87) or upon quantitative immunofluorescence microscopy (13, 28). Only recently have techniques for the quantification of viable or metabolically active bacterial cells become available. Relatively few attempts have been made with plant-associated bacterial species to relate the population size of culturable cells, determined by plating on selective media, with the population size of viable or metabolically active cells (9, 10, 74, 107). The occurrence in these plant-associated microbial communities of a substantial proportion of viable or metabolically active cells which are not culturable would have serious implications for the disciplines of plant pathology, microbial ecology, and phytoremediation.

The viable but nonculturable state of a bacterial cell may be defined as the state in which a metabolically active cell is incapable of sustained cellular growth on media normally supporting the growth of that cell (66). Entry into the viable but nonculturable state may result from exposure to natural environmental stresses, such as temperature shock (66). Loss of culturability, however, may also be associated with the starvation-survival state observed in certain bacteria under oligotrophic conditions (61, 62, 67, 85) and with sublethal injury due to the effects of xenobiotic agents such as chlorine, antibiotics, or heavy metals (91, 92, 93). While

Mark Wilson • Biology Department, The Colorado College, Colorado Springs, CO 80903. *Steven E. Lindow* • Department of Plant and Microbial Biology, University of California, Berkeley, Berkeley, CA 94720.

cells which either are in the starvation-survival state or are sublethally injured do not conform to the sensu stricto definition of viable but nonculturable cells (66), they may nevertheless be both metabolically active and nonculturable on laboratory media. As this chapter is primarily concerned with the enumeration of bacteria in plant-associated microbial communities during epidemiological and ecological studies, we are for practical purposes adopting a sensu lato definition of viable but nonculturable cells which includes cells rendered nonculturable by exposure to natural environmental stresses, oligotrophic conditions, and xenobiotic agents.

QUANTIFICATION OF CULTURABLE, TOTAL, AND VIABLE BACTERIAL POPULATION SIZES

The quantification of culturable bacterial population size by the plate count method relies upon the use of selective media based on either nutritional requirements (87) or antibiotic-resistance markers (4, 48). Some antibiotics used in selective media, however, are not conducive to the growth of environmentally stressed cells (34, 63). Culturable population sizes of marked strains of *Rhizobium phaseoli*, *Agrobacterium tumefaciens*, and *Pseudomonas* sp. recovered from sterile soil were significantly greater on unamended media than on antibiotic-amended media (1). The antibiotic rifampin is frequently used for marking bacteria in ecological studies, but spontaneous rifampin-resistant mutants show pleiotropic mutations (78), which can affect cell survival (24). Further, rifampin-marked strains recovered from plants are not always culturable on rifampin-amended media (60), suggesting that rifampin may be antagonistic to the growth of environmentally stressed or adapted cells. In this discussion of the viable but nonculturable phenomenon, it must be borne in mind that culturability may be culture-medium specific and also that culturability of environmentally stressed or adapted bacteria may be affected by components of the medium required for their selective isolation from environmental samples.

The use of quantitative immunofluorescence microscopy permits the determination of the total population size of a given bacterial species in a plant-associated microbial community (13, 17, 20, 27, 28). Unfortunately, immunofluorescence microscopy detects intact nonviable cells as well as viable cells. Nonviable heat-killed rhizobial cells may persist in the soil environment for up to 2 weeks (12). The difference between the culturable population size and the total population size determined by immunofluorescence alone cannot therefore be assumed to represent viable but nonculturable cells.

Several techniques are now available for the quantification of viable cells, based on substrate responsiveness (50) or metabolically active cells, based on the reduction of artificial electron acceptors (66). These electron acceptors include 2-(*p*-iodophenyl)-3-(*p*-nitrophenyl)-5-phenyltetrazolium chloride (INT) (65); 5-cyano-2,3-ditolyl tetrazolium chloride (CTC) (82); and sodium 3′-{1-[(phenylamino) - carbonyl] - 3, 4 - tetrazolium} - bis(4 - methoxy - 6 - nitro) benzene sulfonic acid hydrate (XTT) (84). Comparisons of "viable" cell population sizes determined by these different methods may show significant differences (66, 79; C. Heijnen, Proc. 4th Int. Symp. on Bacterial Genetics and Ecol., 1993); hence, it must be borne in mind that determinations of viable population size are also technique specific (90). These viable cell techniques have not been applied extensively in the field of plant-associated microorganisms. One reason for this is the difficulty in dealing with environmental samples containing mixed microbial populations and interfering organic matter. The majority of studies of the viable but nonculturable phenomenon have been conducted in gnotobiotic microcosms (2, 44, 67, 86). Only a relatively few studies have used specific immunofluorescence protocols to determine the number of viable cells in nonsterile environmental samples (16, 42, 112). Methods for the quantification of substrate-responsive or metabolically active cells, combined with specific immunofluorescence techniques in rapid and reliable protocols, will make the problem of nonculturability of bacterial populations in plant-associated microbial communities more tractable.

Bearing these facts in mind, this chapter presents the evidence for the occurrence of viable but nonculturable cells in plant-associated bacterial populations and addresses the implications of this phenomenon for plant pathology, microbial ecology, and for the release of genetically engineered microorganisms (GEMs).

VIABLE BUT NONCULTURABLE BACTERIA IN THE PHYLLOSPHERE

Epiphytic bacteria colonizing aerial plant surfaces are subject to various environmental stresses, including osmotic and matric stress due to fluctuations in relative humidity, ultraviolet and visible radiations, and naturally occurring and applied bactericides. Epiphytic bacteria may also experience nutritional stress in the nutrient-limited phyllosphere (110). The characteristics of epiphytic bacteria that enable them to tolerate these stresses have been extensively reviewed (6, 30, 55), however, few attempts have been made to assess the culturability of the epiphytic populations subject to these stresses.

Epiphytic phytopathogenic bacteria in the phyllosphere microbial communities of host and nonhost plant species are of significance, because they provide inoculum for disease development during conducive environmental conditions (40, 51). Many epidemiological studies have been performed documenting the survival of phytopathogenic bacteria in the phyllosphere (38, 52, 88). These studies have documented declines in the population size of inoculated bacterial strains (11, 36, 94, 95, 96) and fluctuations in the population sizes of naturally occurring epiphytic bacteria, over time intervals of duration varying from a day (41) to a season (59, 83, 111). None of these epidemiological studies, however, determined whether the declines in culturable population size resulted from cell death or merely from loss of culturability. While differences between epiphytic bacterial population sizes determined by selective plating and immunofluorescence microscopy have been observed (73), these are not conclusive evidence for the existence of viable but nonculturable cells if nonviable cells remain intact in the phyllosphere environment for any period of time.

The effects of the environment on the culturability of epiphytic bacterial populations have only recently been addressed. Wilson and Lindow (107) compared the number of cells culturable on a selective medium with the number of viable cells in an epiphytic population of *Pseudomonas syringae* subject to desiccation in a controlled environment. Viability was defined as substrate responsiveness and was determined using a modified version of the yeast extract/nalidixic acid method (50) which allowed for specific detection of the applied *P. syringae* in an epiphytic community. When the epiphytic *P. syringae* population was subject to desiccation under conditions of low relative humidity, the culturable population size dropped, as would be observed following inoculation of beans in the field (108); however, the viable population size was not significantly different from the culturable population size. This suggests that the decline in population size of *P. syringae* inoculated onto bean plants in the field (108) is probably due to cell death and not due to a loss of culturability. Pedersen and Leser (73, 74), however, reported contrasting results. The population size of *Enterobacter cloacae* inoculated onto bean plants in the field was determined using selective plating and the yeast extract/nalidixic acid method (50), combined with quantitative immunofluorescence microscopy. At 14 days after inoculation the culturable population size was significantly less than the viable population size. In fact, less than 0.1% of the viable *E. cloacae* population was culturable on the selective medium used. Examined together, the findings of Wilson and Lindow (107) and Pedersen and Leser (73, 74) suggest either that the loss of culturability in epiphytic populations subject to desiccation stress is species specific or that the effects of desiccation under controlled conditions and under field conditions are not the same. Pedersen and Leser (73, 74) also observed a greater difference between the culturable population size and the total population size on bean than on barley, suggesting that the occurrence of viable but nonculturable cells in epiphytic bacterial populations may be host plant/bacterial species/environment specific.

Culturability of epiphytic bacterial populations may also be affected by the metabolic status of the cells. Following extended incubations of epiphytic *P. syringae* populations under constant environmental conditions, Wilson and Lindow (107) observed significant differences between the viable and culturable population sizes. In some samples, less than 25% of the viable *P. syringae* population was culturable on the selective medium. The loss of culturability of the epiphytic *P. syringae* population may have resulted from reduced metabolic activity due to low nutrient availability in the phyllosphere. In addition to epiphytic survival in the phyllosphere, certain phytopathogenic bacteria are also able to survive for extended periods in seeds and in dried plant debris (18, 26, 52, 88). In this "survival phase" the bacterial populations supposedly enter a state of reduced metabolic activity, or "hypobiosis" (52); however, there have been no published reports on the culturability levels of such bacterial populations. Models proposed to describe the growth and survival of epiphytic bacterial populations hypothesize a reversible transition between the survival phase and the multiplication phase (6, 7, 38, 49). The signal triggering the transition from the multiplication phase to the survival phase has been suggested to be low nutrient availability (6, 7). The loss of culturability observed in the epiphytic *P. syringae* populations may represent the onset of such a survival phase (107). It is possible, therefore, that epiphytic bacteria occasionally exhibit a starvation-survival phase similar to that which occurs in bacteria in oligotrophic aquatic environments (61, 62, 67, 85). Bacterial cells in the starvation-survival state are hypothesized to be cross-protected against other environmental stresses (66, 85). This could account for the stress tolerance of phytopathogenic bacteria in the survival phase.

There remain extensive opportunities for the study of the effects of environmental stress and metabolic status upon the culturability of epiphytic bacteria. Also, to our knowledge, there are no published reports on the culturability of sublethally injured epiphytic bacteria resulting from exposure to xenobiotic agents, such as antibiotics, and this area also, deserves to be examined.

VIABLE BUT NONCULTURABLE BACTERIA IN THE RHIZOSPHERE

Historically, there has been extensive interest in the survival of nitrogen-fixing rhizobia inoculated into the rhizosphere of crop plants in agroecosystems. In particular, with respect to survival under conditions of environmental stress, including desiccation (19, 35, 70, 102) and high soil temperature (70) but also in different soil types (25, 76). More recently, there has additionally been interest in the survival of saprophytic rhizobacteria, particularly pseudomonads, which may act as biological control agents of plant pathogens or promote plant growth (31, 43, 56, 89, 100). It has been recognized for some time that the population size of culturable soil bacteria determined using plate counts on laboratory media is significantly smaller than the total bacterial population size determined by microscopy (15). Such differences have been frequently attributed to the inability of oligotrophic species to grow on nutrient-rich isolation media (68) or clumping of cells on soil particles (81). Indirect evidence for the likely occurrence of viable but nonculturable cells in rhizobacterial populations has been provided by the application of quantitative immunofluorescence microscopy, where total bacterial population sizes exceed those determined by selective plating (73, 74, 77). The issue has also been raised in a review dealing with the enumeration of rhizobacterial populations (46). However, the problem so far has received only scant attention.

Many of the environmental stresses to which rhizosphere microbial communities are exposed, including desiccation and extremes of temperature and pH, could potentially affect culturability. Rhizobacterial populations experience both matric stress and osmotic stress during soil desiccation resulting from water uptake by the transpiring plant. Soil microcosms have been used to investigate the effects of matric and osmotic stress on culturability of plant-associated bacterial species. Pedersen and Jacobsen (72) recently compared the culturable and viable population sizes of *E. cloacae* and *Alcaligenes eutrophus* in soil microcosms subject to desiccation. Plate counts on selective media and direct viable counts, determined using quantitative immunofluorescence microscopy and the yeast extract/nalidixic acid method (50), indicated that *E. cloacae* was more desiccation tolerant than *A. eutrophus*. While in moist soil ($Y = -0.05$ MPa) the culturable population size and the viable population size were similar, in air-dried soil ($Y = -300$ MPa) the viable population size was significantly higher than the culturable population size. These findings are in agreement with the field observations that *E. cloacae* population sizes determined by selective plating and immunofluorescence microscopy were similar in inoculated top soil, except during periods of desiccation ($Y < -100$

MPa) (74). These data indicate the occurrence of viable but nonculturable cells in populations of both *E. cloacae* and *A. eutrophus* in response to desiccation.

As in the phyllosphere, the metabolic status of rhizobacterial populations may affect their culturability. When introduced into the rhizosphere or into bulk soil, bacterial populations usually undergo a slow decline in numbers even in the absence of environmental stress (56; Wei et al., Proc. Int. Cong. Plant Pathol., 1993), and these declines may reflect a loss of culturability. While bulk soil is generally thought to be oligotrophic, substantial carbon losses occur in the vicinity of actively growing roots (106). The growth of bacterial populations in the rhizosphere and surrounding bulk soil is dependent on carbon supply in excess of maintenance requirements of the population (3, 57). Only in the rhizosphere of an actively growing root is there sufficient exudation of carbon to exceed maintenance energy requirements and result in active growth. In the rhizosphere of aging roots or in bulk soil, energy availability is less than that required for maintenance, and the bacterial populations may enter a starvation-survival state, similar to that observed for microbes in oligotrophic aquatic environments (61, 62, 67, 85), which may result in reduced levels of culturability (101).

Norton and Firestone (65) used the artificial electron acceptor INT to quantify metabolically active cells in rhizosphere microbial communities of pine seedlings. The soil adjacent to roots contained greater numbers and a higher proportion of INT-active bacteria than the surrounding soil. Their data also suggested that many of the cells exhibiting metabolic activity might not be culturable on isolation media currently employed. A microcolony epifluorescence technique was used by Binnerup et al. (9) to detect viable but nonculturable *Pseudomonas fluorescens* cells introduced into soil microcosms. The microcolony assay permitted quantification of viable cells without the use of either an artificial electron acceptor or nalidixic acid. The culturable population size of the rhizosphere pseudomonad, determined by selective plating, declined over a period of 40 days, after which time only 0.02 to 0.35% of the initial inoculum was culturable. In contrast, the number of viable but nonculturable cells remaining at 40 days after inoculation represented 20% of the initial inoculum.

Heijnen (C. Heijnen, Proc. 4th Int. Symp. Bact. Genetics and Ecol., 1993) and Heijnen and van Elsas (37) compared the numbers of substrate-responsive cells and the number of metabolically active cells of *Flavobacterium* in rhizosphere and bulk soil. The number of substrate-responsive cells, determined using the yeast extract method and employing different nutritional substrates, and metabolically active cells, stained with the redox dye CTC, declined with time. After 9 to 13 days, more cells were substrate responsive in the rhizosphere than in the bulk soil. These results indicate that metabolic activity and hence culturability of rhizobacterial populations is likely to be different in the rhizosphere of young and old roots and in bulk soil. Such differences should be taken into account during epidemiological and ecological studies.

In a study using standard soil columns and field-scale lysimeters, the biocontrol strain *P. fluorescens* CHAO was shown to persist in the soil for extended time periods as viable but nonculturable cells (29, 45, 97, 98). The proportion of the soil population in the viable but nonculturable state was dependent upon the time period since inoculation, the soil temperature, the soil water activity and the abundance of roots in the soil (97, 98). Although this study did not indicate the presence of viable but nonculturable cells in association with the roots (99), a more recent study examining viability of *P. fluorescens* DR5-BN14 found that an average of 25% of cells in association with barley roots were in the viable but nonculturable state (64).

IMPLICATIONS FOR THE OCCURRENCE OF VIABLE BUT NONCULTURABLE CELLS IN PLANT-ASSOCIATED BACTERIAL POPULATIONS

The studies discussed in this chapter strongly suggest that viable but nonculturable cells do occur in both phyllosphere bacterial spp. (108) and rhizosphere bacterial spp. (37, 45, 58, 73, 74, 97; C. Heijnen, Proc. 4th Int. Symp. Bact. Genetics and Ecol., 1993). The magnitude of nonculturability, however, is probably dependent on the bacterial species (74), host plant species (73, 74, 97), environmental conditions (67), nature and rapidity of onset of the stress, and (in the case of inoculated cells) physiological status of the inoculum (47, 100, 108). The occurrence of viable but nonculturable cells in plant-associated bacterial populations has important implications for plant pathology, microbial ecology, and the environmental release of any microorganism, but particularly GEMs.

In plant pathology, the occurrence of a substantial proportion of viable but nonculturable cells in populations of epiphytic phytopathogenic bacteria would have serious implications for epidemiological studies, which usually are based on enumeration of culturable cells by selective plating. If viable but nonculturable epiphytic bacterial cells can return to a state of active growth, they could contribute significantly to epidemic disease development. Also in the field of plant pathology, it is frequently necessary to establish the absence of phytopathogenic bacteria contaminating seeds or other propagating material. The occurrence of viable but nonculturable cells in populations of phytopathogenic bacteria would dictate that alternative detection methods, such as PCR (39) and immunofluorescence (28) should be used. These techniques, however, may produce false positives, due to the detection of nonviable cells, which may result in unnecessary economic losses.

The dispersal, activity, and culturability of any bacterial population introduced into the environment are of significance (103, 104); however, these issues are of particular concern for GEMs released into the open environment (8, 22, 23, 109). Plant/soil microcosms are frequently used to model the survival of engineered organisms in advance of field releases (5, 53, 105). While survival of microorganisms in these microcosms is sometimes compared with survival in the field (14, 71), the problem of differential culturability under controlled and environmental conditions is usually ignored. Environmental stresses encountered in a microcosm and the field are unlikely to be identical, and the proportion of the population which loses culturability may not be the same in both environments (74, 107). Further, field experiments employing the parental strain may not be relevant because the effects of stress may not be the same in parental and engineered cells. Although some studies have found no significant differences between the survival of wild-type and engineered bacteria (32, 33, 75), other studies have revealed such differences (69), hence there may also be differences in the proportion of culturable cells. The behavior of a genetically engineered strain in the field might not therefore be accurately modeled by either microcosm studies with the engineered strain or field experiments with the nonengineered parental strain. Oliver et al. (67) even demonstrated that the proportion of culturable cells in a population of a recombinant organism differs, depending on whether the recombinant sequences are plasmid borne or integrated into the chromosome. The implications of these differences in culturability between wild types and recombinants for field releases are discussed in Chapter 14 (see Levin and Angle).

In conclusion, while there is sufficient evidence to indicate that viable but nonculturable cells occur in populations of plant-associated bacteria, in response to various environmental stresses, as yet there are insufficient data to assess the magnitude and hence the significance of the phenomenon. Thus a compelling need exists for further investigation of the viable but nonculturable phenomenon in plant-associated bacterial populations.

Acknowledgments. We acknowledge C. Heijnen for provision of unpublished research data and G. A. Beattie, E. A. Clarke, and J. C. Pedersen for critical review of the manuscript.

REFERENCES

1. **Acea, M. J., C. R. Moore, and M. Alexander.** 1988. Survival and growth of bacteria introduced into soil. *Soil Biol. Biochem.* **20:**509–515.
2. **Allen-Austin, D., B. Austin, and R. R. Colwell.** 1984. Survival of *Aeromonas salmonicida* in river waters. *FEMS Microbiol. Lett.* **21:**143–146.
3. **Anderson, T. H., and K. H. Domsch.** 1985. Determination of ecophysiological maintenance carbon requirements of soil microorganisms in a dormant state. *Biol. Fert. Soils* **1:**81–89.
4. **Andrews, J. H.** 1986. How to track a microbe, p. 14–34. *In* N. J. Fokkema and J. van den Heuvel (ed.), *Microbiology of the Phyllosphere.* Cambridge University Press, Cambridge, United Kingdom.
5. **Armstrong, J. L., G. R. Knudsen, and R. J. Seidler.** 1987. Microcosm method to assess survival of recombinant bacteria associated with plants and herbivorous insects. *Curr. Microbiol.* **15:**229–232.
6. **Beattie, G. A., and S. E. Lindow.** 1994. Epiphytic fitness of phytopathogenic bacteria: physiological adaptations for growth and survival, p. 1–28. *In* J. Dangle (ed.), *Bacterial Pathogenesis of Plants and Animals: Molecular and Cellular Mechanisms.* Springer-Verlag, Berlin, Germany.
7. **Beattie, G. A., and S. E. Lindow.** 1995. The secret life of foliar bacterial pathogens on leaves. *Annu. Rev. Phytopathol.* **33:**145–172.
8. **Beringer, J. E., and M. J. Bale.** 1988. The survival and persistence of genetically-engineered micro-organisms, p. 29–46. *In* M. Sussman, C. H. Collins, F. A. Skinner, and D. E. Stewart-Tull (ed.), *The Release of Genetically-Engineered Micro-Organisms.* Academic Press, London, United Kingdom.
9. **Binnerup, S. J., D. F. Jensen, H. Thordal-Christensen, and J. Sorensen.** 1993. Detection of viable, but nonculturable *Pseudomonas fluorescens* DF57 in soil using a microcolony epifluorescence technique. *FEMS Microbiol. Ecol.* **12:**97–105.
10. **Binnerup, S. J., and J. Sorensen.** 1993. Long term oxidant deficiency in *Pseudomonas aeruginosa* PA0303 results in cells which are nonculturable under aerobic conditions. *FEMS Microbiol. Ecol.* **13:**79–84.
11. **Blakeman, J. P.** 1993. Pathogens in a foliar environment. *Plant Pathol.* **42:**479–493.
12. **Bohlool, B. B., and E. L. Schmidt.** 1973. Persistence and competition aspects of *Rhizobium japonicum* observed in soil by immunofluorescence microscopy. *Soil Sci. Am. Proc.* **37:**561–64.
13. **Bohlool, B. B., and E. L. Schmidt.** 1980. The immunofluorescence approach in microbial ecology. *Adv. Microb. Ecol.* **4:**203–241.

14. **Bolton, H., Jr., J. K. Fredrickson, S. A. Bentjen, D. J. Workman, S. W. Li, and J. M. Thomas.** 1991. Field calibration of soil-core microcosms: fate of a genetically altered rhizobacterium. *Microb. Ecol.* **21:**163–173.
15. **Bowen, G. D., and A. D. Rovira.** 1976. Microbial colonization of roots. *Annu. Rev. Phytopathol.* **14:**121–144.
16. **Brayton, P. R., and R. R. Colwell.** 1987. Fluorescent antibody staining method for enumeration of viable environmental *Vibrio cholerae* 01. *J. Microbiol. Methods* **6:**309–314.
17. **Brlansky, R. H., and R. F. Lee.** 1990. Detection of *Xanthomonas campestris* pv. *citrumelo* from citrus using membrane entrapment immunofluorescence. *Plant Dis.* **74:**863–868.
18. **Bruehl, G. W.** 1987. Survival of bacteria in soil, p. 185–195. *In Soilborne Plant Pathogens.* Macmillan Publishers, Ltd., London, United Kingdom.
19. **Bushby, H. V. A., and K. C. Marshall.** 1977. Some factors affecting the survival of root-nodule bacteria on desiccation. *Soil Biol. Biochem.* **9:**143–147.
20. **Byrd, J. J., and R. R. Colwell.** 1992. Microscopy applications for analysis of environmental samples, p. 93–112. *In* M. A. Levin, R. J. Seidler, and M. Rogul (ed.), *Microbial Ecology, Principles, Methods and Applications.* McGraw-Hill Inc., New York, N.Y.
21. **Byrd, J. J., H.-S. Xu, and R. R. Colwell.** 1991. Viable but non-culturable bacteria in drinking water. *Appl. Environ. Microbiol.* **57:**875–878.
22. **Colwell, R. R., P. R. Brayton, D. J. Grimes, D. B. Roszak, S. A. Huq, and L. M. Palmer.** 1985. Viable but non-culturable *Vibrio cholerae* and related pathogens in the environment: implications for release of genetically engineered micro-organisms. *Bio/Technology* **3:**817–820.
23. **Colwell, R. R., C. Somerville, I. Knight, and W. Straube.** 1988. Detection and monitoring of genetically-engineered micro-organisms, p. 47–60. *In* M. Sussman, C. H. Collins, F. A. Skinner, and D. E. Stewart-Tull (ed.), *The Release of Genetically-Engineered Micro-Organisms.* Academic Press, London, United Kingdom.
24. **Compeau, G., B. J. Al-Achi, E. Platsouka, and S. B. Levy.** 1988. Survival of rifampin-resistant mutants of *Pseudomonas fluorescens* and *Pseudomonas putida* in soil systems. *Appl. Environ. Microbiol.* **54:**2432–2438.
25. **Crozat, Y., J. C. Cleyet-Marel, J. J. Giraud, and M. Obaton.** 1982. Survival rates of *Rhizobium japonicum* populations introduced into different soils. *Soil Biol. Biochem.* **14:**401–405.
26. **De Boer, S. H.** 1982. Survival of phytopathogenic bacteria in soil, p. 285–306. *In* M. S. Mount and G. H. Lacy (ed.), *Phytopathogenic Prokaryotes.* Academic Press, London, United Kingdom.
27. **De Boer, S. H.** 1984. Enumeration of two competing *Erwinia carotovora* populations in potato tubers by a membrane filter-immunofluorescence procedure. *J. Appl. Bacteriol.* **57:**517–522.
28. **De Boer, S. H.** 1990. Immunofluorescence for bacteria, p. 295–298 *In* R. Hampton, E. Ball, and S. De Boer (ed.), *Serological Methods for Detection and Identification of Viral and Bacterial Plant Pathogens.* APS Press, Minneapolis, Minn.
29. **Defago, G., C. Keel, and Y. Moenne-Loccoz.** 1996. Fate of introduced biocontrol agent *Pseudomonas fluorescens* CHAO in soil: biosafety considerations, p. 241–245. *In* T. Wenhua, R. J. Cook, and A. Rovira (ed.), *Advances in Biological Control of Plant Diseases.* China Agricultural University Press, Beijing, China.
30. **Dickinson, C. H.** 1986. Adaptations of micro-organisms to climatic conditions affecting aerial plant surfaces, p. 77–100. *In* N. J. Fokkema and J. van den Heuvel (ed.), *Microbiology of the Phyllosphere.* Cambridge University Press, Cambridge, United Kingdom.
31. **Dupler, M., and Baker.** 1984. Survival of *Pseudomonas putida*, a biological control agent, in soil. *Phytopathology* **74:**195–200.
32. **England, L. S., H. Lee, and J. T. Trevors.** 1993. Recombinant and wild-type *Pseudomonas aureofaciens* strains in soil: survival, respiratory activity and effects on nodulation of whitebean *Phaseolus vulgaris* L. by *Rhizobium* species. *Mol. Ecol.* **2:**303–313.
33. **England, L. S., H. Lee, and J. T. Trevors.** 1995. Recombinant and wild-type *Pseudomonas aureofaciens* strains introduced into soil microcosms: effect on decomposition of cellulose and straw. *Mol. Ecol.* **4:**221–210.
34. **Genthner, F. J., J. Upadhyay, R. P. Campbell, and B. R. S. Genthner.** 1990. Anomalies in the enumeration of starved bacteria on culture media containing nalidixic acid and tetracycline. *Microb. Ecol.* **20:**283–288.

35. **Griffin, G. W., and R. J. Roughley.** 1992. The effect of soil moisture potential on growth and survival of root nodule bacteria in peat culture and on seed. *J. Appl. Bacteriol.* **73:**7–13.
36. **Haas, J. H., and J. Rotem.** 1976. *Pseudomonas lachrymans* adsorption, survival, and infectivity following precision inoculation of leaves. *Phytopathology* **66:**992–997.
37. **Heijnen, C. E., and J. D. van Elsas.** 1994. Metabolic activity of bacteria introduced into soil, p. 187–189. *In* M. H. Ryder, P. M. Stephens, and G. D. Bowen (ed.), *Improving Plant Productivity with Rhizobacteria.* CSIRO Division of Soils, Adelaide, Australia.
38. **Henis, Y., and Y. Bashan.** 1986. Epiphytic survival of bacterial leaf pathogens, p. 252–268. *In* N. J. Fokkema and J. van den Heuvel (ed.), *Microbiology of the Phyllosphere.* Cambridge University Press, Cambridge, United Kingdom.
39. **Henson, J. M., and R. French.** 1993. The polymerase chain reaction and plant disease diagnosis. *Annu. Rev. Phytopathol.* **31:**81–109.
40. **Hirano, S. S., and C. D. Upper.** 1983. Ecology and epidemiology of foliar bacterial plant pathogens. *Annu. Rev. Phytopathol.* **21:**243–269.
41. **Hirano, S. S., and C. D. Upper.** 1989. Diel variation in population size and ice nucleation activity of *Pseudomonas syringae* snap bean leaflets. *Appl. Environ. Microbiol.* **55:**623–630.
42. **Hoff, K. A.** 1988. Rapid and simple method for double staining of bacteria with 4′,6-diamidino-2-phenylindole and fluorescein isothiocyanate-labeled antibodies. *Appl. Environ. Microbiol.* **54:**2949–2952.
43. **Howie, W. J., R. J. Cook, and D. M. Weller.** 1987. Effect of soil matric potential and cell motility on wheat root colonization by fluorescent pseudomonads suppressive to take-all. *Phytopathology* **77:** 286–292.
44. **Hussong, D., R. R. Colwell, M. O'Brien, E. Weiss, A. D. Pearson, R. M. Weiner, and W. D. Burge.** 1987. Viable *Legionella pneumophila* not detectable by culture on agar media. *Bio/Technology* **5:**947–950.
45. **Keel, C., M. Zala, J. Troxler, A. Natsch, H. A. Pfirter and G. Defago.** 1994. Application of biocontrol strain *Pseudomonas fluorescens* CHAO to standard soil-columns and field-scale lysimeters. I. Survival and vertical translocation, p. 255–257. *In* M. H. Ryder, P. M. Stephens, and G. D. Bowen (ed.), *Improving Plant Productivity with Rhizobacteria.* CSIRO Division of Soils, Adelaide, Australia.
46. **Kloepper, J. W., and C. J. Beauchamp.** 1992. A review of issues related to measuring colonization of plant roots by bacteria. *Can. J. Microbiol.* **38:**1219–1232.
47. **Klotz, M. G.** 1993. The importance of bacterial growth phase for *in planta* virulence and pathogenicity testing: co-ordinated stress response regulation in fluorescent pseudomonads. *Can. J. Microbiol.* **39:**948–957.
48. **Kluepfel, D. A.** 1993. The behavior and tracking of bacteria in the rhizosphere. *Annu. Rev. Phytopathol.* **31:**441–472.
49. **Knudsen, G. R.** 1991. Models for the survival of bacteria applied to the foliage of crop plants, p. 191–216. *In* C. J. Hurst (ed.), *Modeling the Environmental Fate of Microorganisms.* American Society for Microbiology, Washington, D.C.
50. **Kogure, K., U. Simidu, and N. Taga.** 1979. A tentative direct microscopic method for counting living marine bacteria. *Can. J. Microbiol.* **25:**415–420.
51. **Leben, C.** 1965. Epiphytic microorganisms in relation to plant disease. *Annu. Rev. Phytopathol.* **3:** 209–230.
52. **Leben, C.** 1981. How plant-pathogenic bacteria survive. *Plant Dis.* **65:**633–637.
53. **Liang, L. N., J. L. Sinclair, L. M. Mallory, and M. Alexander.** 1982. Fate in model ecosystems of microbial species of potential use in genetic engineering. *Appl. Environ. Microbiol.* **44:**708–714.
54. **Linder, K., and J. D. Oliver.** 1989. Membrane fatty acid and virulence changes in the viable but non-culturable state of *Vibrio vulnificus. Appl. Environ. Microbiol.* **55:**2837–2842.
55. **Lindow, S. E.** 1991. Determinants of epiphytic fitness in bacteria, p. 295–314. *In* J. H. Andrews and S. S. Hirano (ed.), *Microbial Ecology of Leaves.* Brock/Springer, New York, N.Y.
56. **Loper, J. E., C. Haack, and M. N. Schroth.** 1985. Population dynamics of soil pseudomonads in the rhizosphere of potato (*Solanum tuberosum* L.). *Appl. Environ. Microbiol.* **49:**416–422.
57. **Lynch, J. M.** 1990. Longevity of bacteria: considerations in environmental release. *Curr. Microbiol.* **20:**387–389.

58. **Manahan, S. H., and T. R. Steck.** 1997. The viable but nonculturable state in *Agrobacterium tumefaciens* and *Rhizobium meliloti*. *FEMS Microbiol. Ecol.* **22:**29–37.
59. **Mansvelt, E. L., and M. J. Hattingh.** 1988. Resident populations of *Pseudomonas syringae* pv. *syringae* on leaves, blossoms, and fruits of apple and pear trees. *J. Phytopathology* **121:**135–142.
60. **McInroy, J. A., G. Wei, G. Musson, and J. W. Kloepper.** 1992. Evidence for possible masking of rifampicin-resistance phenotype of marked bacteria *in planta*. *Phytopathology* **82:**1177 (abstract).
61. **Morita, R. Y.** 1985. Starvation and miniaturization of heterotrophs, with special emphasis on maintenance of the starved viable state, p. 111–130. *In* M. Fletcher and G. D. Floodgate (ed.), *Bacteria in Their Natural Environments*. Academic Press, London, United Kingdom.
62. **Morita, R. Y.** 1988. Bioavailability of energy and its relationship to growth and starvation survival in nature. *Can. J. Microbiol.* **34:**436–441.
63. **Mossel, D. A. A., and P. Van Netten.** 1984. Harmful effects of selective media on stressed microorganisms: nature and remedies, p. 329–369. *In* M. H. E. Andrew and A. D. Russell (ed.), *The Revival of Injured Microbes*. Academic Press, New York, N.Y.
64. **Normander, B., N. B. Hendrickson, and O. Nybroe.** 1999. Green fluorescent protein-marked *Pseudomonas fluorescens*: localization, viability and activity in the natural barley rhizosphere. *Appl. Environ. Microbiol.* **65:**4646–4651.
65. **Norton, J. M., and M. K. Firestone.** 1991. Metabolic status of bacteria and fungi in the rhizosphere of ponderosa pine seedlings. *Appl. Environ. Microbiol.* **57:**1161–1167.
66. **Oliver, J. D.** 1993. Formation of viable but non-culturable cells, p. 239–272. *In* S. Kjelleberg (ed.), *Starvation in Bacteria*. Plenum Press, New York, N.Y.
67. **Oliver, J. D., L. Nilsson, and S. Kjellberg.** 1991. Formation of nonculturable *Vibrio vulnificus* cells and its relation to the starvation state. *Appl. Environ. Microbiol.* **57:**2640–2644.
68. **Olsen, R. A., and L. R. Bakken.** 1987. Viability of soil bacteria: optimization of plate counting technique and comparison between total counts and plate counts within different size groups. *Microb. Ecol.* **13:**59–74.
69. **Orvos, D. R., G. H. Lacy, and J. Cairns, Jr.** 1990. Genetically engineered *Erwinia carotovora*: survival, intraspecific competition, and effects upon selected bacterial genera. *Appl. Environ. Microbiol.* **56:**1689–1694.
70. **Osa-Afiana, L. O., and M. Alexander.** 1982. Differences among cowpea rhizobia in tolerance to high temperature and desiccation in soil. *Appl. Environ. Microbiol.* **43:**435–439.
71. **Pedersen, J. C.** 1992. Survival of *Enterobacter cloacae*: field validation of a soil/plant microcosm. *Microb. Releases* **1:**87–93.
72. **Pedersen, J. C., and C. S. Jacobsen.** 1993. Fate of *Enterobacter cloacae* JP120 and *Alcaligenes eutrophus* AEO106(pRO101) in soil during water stress: effects on culturability and viability. *Appl. Environ. Microbiol.* **59:**1560–1564.
73. **Pedersen, J. C., and T. D. Leser.** 1992. Survival of *Enterobacter cloacae* on leaves and in soil detected by immunofluorescence microscopy in comparison with selective plating, p. 245–247. *In* D. E. S. Stewart-Tull and M. Sussman (ed.), *The Release of Genetically Modified Organisms*. Plenum Press, New York, N.Y.
74. **Pedersen, J. C., and T. D. Leser.** 1992. Survival of *Enterobacter cloacae* on leaves and in soil detected by immunofluorescence microscopy in comparison with selective plating. *Microb. Releases* **1:**95–102.
75. **Pillai, S. D., and I. L. Pepper.** 1991. Transposon Tn*5* as an identifiable marker in rhizobia: survival and genetic stability of Tn*5* mutant bean rhizobia under temperature stressed conditions in desert soils. *Microb. Ecol.* **21:**21–33.
76. **Postma, J., C. H. Hok-A-Hin, and J. H. Oude Voshaar.** 1990. Influence of the inoculum density on the growth and survival of *Rhizobium leguminosarum* biovar *trifolii* introduced into sterile and non-sterile loamy sand and silt loam. *FEMS Microbiol. Ecol.* **73:**49–58.
77. **Postma, J., J. D. van Elsas, J. M. Govaert, and J. A. van Veen.** 1988. The dynamics of *Rhizobium leguminosarum* biovar *trifolii* introduced into soil as determined by immunofluorescence and selective techniques. *FEMS Microbiol. Ecol.* **53:**251–260.
78. **Press, C. M., J. W. Kloepper, J. A. McInroy.** 1992. Pleiotropic mutations associated with spontaneous antibiotic-resistant mutants of rhizobacteria. *Phytopathology* **82:**1178 (abstract).

79. **Quinn, J. P.** 1984. The modification and evaluation of some cytochemical techniques for the enumeration of metabolically active heterotrophic bacteria in the marine environment. *J. Appl. Bacteriol.* **57:**51–57.
80. **Rattray, E. A., J. I. Prosser, L. A. Glover, and K. Killham.** 1992. Matric potential in relation to survival and activity of a genetically modified microbial inoculum in soil. *Soil Biol. Biochem.* **24:** 421–425.
81. **Richaume, A., C. Steinberg, and L. Jocteur-Monrozier.** 1993. Differences between direct and indirect enumeration of soil bacteria: the influence of soil structure and cell location. *Soil Biol. Biochem.* **25:**641–643.
82. **Rodriguez, G. G., D. Phipps, K. Ishiguro, and H. F. Ridgway.** 1992. Use of a fluorescent redox probe for direct visualization of actively respiring bacteria. *Appl. Environ. Microbiol.* **58:**1801–1808.
83. **Roos, I. M. M., and M. J. Hattingh.** 1986. Resident populations of *Pseudomonas syringae* on stone fruit tree leaves in South Africa. *Phytophylactica* **18:**55–58.
84. **Roslev, P., and G. M. King.** 1993. Application of a tetrazolium salt with a water-soluble formazan as an indicator of viability in respiring bacteria. *Appl. Environ. Microbiol.* **59:**2891–2896.
85. **Roszak, D. B., and R. R. Colwell.** 1987. Survival strategies of bacteria in the natural environment. *Microbiol Rev.* **51:**365–379.
86. **Roszak, D. B., D. J. Grimes, and R. R. Colwell.** 1984. Viable but nonrecoverable stage of *Salmonella enteriditis* in aquatic systems. *Can. J. Microbiol.* **30:**334–338.
87. **Rudolph, K., M. A. Roy, M. Sasser, D. E. Stead, M. Davis, and F. Gossele.** 1990. Isolation of bacteria, p. 43–94. *In* Z. Klement, K. Rudolph, and D. C. Sands (ed.), *Methods in Phytobacteriology.* Akedimiai Kado, Budapest, Hungary.
88. **Schuster, M. L., and D. P. Coyne.** 1974. Survival mechanisms of phytopathogenic bacteria. *Annu. Rev. Phytopathol.* **12:**199–221.
89. **Seong, K.-V., M. Hofte, J. Boelens, and W. Verstaete.** 1991. Growth, survival, and root colonization of plant growth beneficial *Pseudomonas fluorescens* ANP15 and *Pseudomonas aeruginosa* 7NSK2 at different temperatures. *Soil Biol. Biochem.* **23:**423–428.
90. **Servais, P., J. Vives-Rigo, and G. Billen.** 1992. Survival and mortality of bacteria in natural environments, p. 100–119. *In* J. C. Fry and M. J. Day (ed.), *Release of Genetically Engineered and Other Micro-Organisms.* Cambridge University Press, Cambridge, United Kingdom.
91. **Singh, A., and G. A. McFeters.** 1990. Injury of enteropathogenic bacteria in drinking water, p. 368–379. *In* G. A. McFeters (ed.), *Drinking Water Microbiology: Progress and Recent Developments.* Springer-Verlag, New York, N.Y.
92. **Singh, A., and G. A. McFeters.** 1990. Enumeration, occurrence, and significance of injured indicator bacteria in drinking water, p. 368–379. *In* G. A. McFeters (ed.), *Drinking Water Microbiology: Progress and Recent Developments.* Springer-Verlag, New York, N.Y.
93. **Singh, A., F.-P. Yu, and G. A. McFeters.** 1990. Rapid detection of chlorine-induced bacterial injury by the direct viable count method using image analysis. *Appl. Environ. Microbiol.* **56:**389–394.
94. **Sleesman, J. P., and C. Leben.** 1976. Bacterial desiccation: effects of temperature, relative humidity, and culture age on survival. *Phytopathology* **66:**1334–1338.
95. **Surico, G., B. W. Kennedy, and G. L. Ercolani.** 1981. Multiplication of *Pseudomonas syringae pv. glycinea* on soybean primary leaves exposed to aerosolized inoculum. *Phytopathology* **71:**532–536.
96. **Timmer, L. W., J. J. Marois, and D. Achor.** 1987. Growth and survival of xanthomonads under conditions nonconducive to disease development. *Phytopathology* **77:**1341–1345.
97. **Troxler, J., M. Zala, C. Keel, A. Natsch, and G. Defago.** 1994. Application of biocontrol strain *Pseudomonas fluorescens* CHAO to standard soil-columns and field-scale lysimeters. II. Detection and enumeration, p. 258–260. *In* M. H. Ryder, P. M. Stephens, and G. D. Bowen (ed.), *Improving Plant Productivity with Rhizobacteria.* CSIRO Division of Soils, Adelaide, Australia.
98. **Troxler, J., M. Zala, Y. Moenne-Loccoz, and G. Defago.** 1997. Predominance of nonculturable cells of the biocontrol strain *Pseudomonas fluorescens* CHAO in the surface horizon of large outdoor lysimeters. *Appl. Environ. Microbiol.* **63:**3776–3782.
99. **Troxler, J., M. Zala, A. Natsch, Y. Moenne-Loccoz, and G. Defago.** 1997. Autecology of the biocontrol strain *Pseudomonas fluorescens* CHAO in the rhizosphere and inside roots at later stages of development. *FEMS Microbiol. Ecol.* **24:**119–130.

100. **Vandenhove, H., R. Merckx, H. Wilmots, and K. Vlassak.** 1991. Survival of *Pseudomonas fluorescens* inocula of different physiological stages in soil. *Soil Biol. Biochem.* **23:**1133–1142.
101. **van Overbeek, L. S., L. Eberl, M. Givskov, S. Molin, and J. D. van Elsas.** 1995. Survival of, and induced stress resistance in, carbon-starved *Pseudomonas fluorescens* cells residing in soil. *Appl. Environ. Microbiol.* **61:**4202–4208.
102. **van Rensburg, H. J., and B. W. Strijdom.** 1980. Survival of fast- and slow-growing *Rhizobium* spp. under conditions of relatively mild desiccation. *Soil Biol. Biochem.* **12:**353–356.
103. **van Veen, J. A., and C. E. Heijnen.** 1994. The fate and activity of microorganisms introduced into soil, p. 63–71. *In* C. E. Pankhurst, B. M. Doube, V. V. S. R. Gupta, and P. R. Grace (ed.), *Soil Biota: Management in Sustainable Farming Systems.* CSIRO, Glen Osmond, Australia.
104. **van Veen, J. A., L. S. van Overbeek, and J. D. van Elsas.** 1997. Fate and activity of microorganisms introduced into soil. *Microbiol. Mol. Biol. Rev.* **61:**121–135.
105. **Walter, M. V., L. A. Porteous, V. J. Prince, L. Ganio, and R. J. Seidler.** 1991. A microcosm for measuring survival and conjugation of genetically engineered bacteria in rhizosphere environments. *Curr. Microbiol.* **22:**117–121.
106. **Williams, S. T.** 1985. Oligotrophy in soil: fact or fiction? p. 81–110. *In* M. Fletcher and G. D. Floodgate (ed.), *Bacteria in their Natural Environments.* Academic Press, London, United Kingdom.
107. **Wilson, M., and S. E. Lindow.** 1992. Relationship of total viable and culturable cells in epiphytic populations of *Pseudomonas syringae. Appl. Environ. Microbiol.* **58:**3908–3913.
108. **Wilson, M., and S. E. Lindow.** 1993. Effect of phenotypic plasticity on epiphytic survival and colonization by *Pseudomonas syringae. Appl. Environ. Microbiol.* **59:**410–416.
109. **Wilson, M., and S. E. Lindow.** 1993. Release of recombinant microorganisms. *Annu Rev. Microbiol.* **47:**913–944.
110. **Wilson, M., and S. E. Lindow.** 1994. Coexistence among epiphytic bacterial populations resulting from nutritional resource partitioning. *Appl. Environ. Microbiol.* **60:**4468–4477.
111. **Wimalajeewa, D. L. S., and J. D. Flett.** 1985. A study of populations of *Pseudomonas syringae* pv. *syringae* on stonefruits in Victoria. *Plant Pathology* **34:**248–254.
112. **Xu, H.-S., N. Roberts, F. L. Singelton, R. W. Atwell, D. J. Grimes, and R. R. Colwell.** 1982. Survival and viability of nonculturable *Escherichia coli* and *Vibrio cholerae* in the estuarine and marine environment. *Microb. Ecol.* **8:**313–323.

Nonculturable Microorganisms in the Environment
Edited by R. R. Colwell and D. J. Grimes

Chapter 14

Implications of the Viable but Nonculturable State in Risk Assessment Based on Field Testing of Genetically Engineered Microorganisms

Morris A. Levin and J. Scott Angle

The importance of the viable but nonculturable (VBNC) state, with respect to safety of environmental applications of genetically engineered microorganisms (GEMs), can best be appreciated in risk assessment. This can be achieved through knowledge of the survival of microorganisms released into specific environments, their characteristics expressed in response to the specific environment, and how these factors relate to the VBNC. From such data, it should be possible to determine whether a GEM will become VBNC and to estimate efficacy and probability of any adverse ecological and/or health effects.

Effects of application of microorganisms can be highly beneficial (e.g., degradation of a targeted pollutant), adverse (e.g., pathogenic for humans, plants, or animals), or neutral, (no obvious effect). Significance of an effect, (i.e., the risks and benefits) can be established by application of the principles of risk assessment as is done with both indigenous microorganisms and GEMs. If microorganisms become VBNC, assays of population densities for risk assessment may yield erroneous findings, and the potential for adverse effect, thereby, would be underestimated (13, 61, 75). To evaluate this potential, the role of the basic elements of risk assessment, namely hazard and exposure, must be measured in order to integrate the effect of VBNC microorganisms. Thus, two issues must be resolved: whether microorganisms entering the VBNC state increase or reduce potential hazard, and whether VBNC microorganisms increase or decrease exposure.

RISK ASSESSMENT

Many investigators (20, 36, 70), as well as national and international organizations such as the U.S. National Academy of Sciences (48, 49) and the World Health

Morris A. Levin • University of Maryland Biotechnology Institute, 701 E. Pratt St., Suite 231, Baltimore, MD 21202. ***J. Scott Angle*** • College of Agriculture and Natural Resources, University of Maryland, 1201 Symons Hall, College Park, MD 20742.

Organization (76) have described the process of risk assessment, including that associated with GEMs. The National Academy of Sciences reports (49, 50) stressed the importance of familiarity—the amount of background information available about the parent organisms and their host—with the conclusion that the more information available about the parent and GEM microorganisms, the greater confidence in predicting the behavior and overall effects of a GEM. Tiedje et al. (70) dissected this general statement into a series of attributes that should be considered, and they discussed the need for more scrutiny when less information is available about a given attribute. Thus, if the phenotypic characteristics of a host microorganism is well known and the inserted DNA is well characterized and/or from a closely related species, less concern is generated, and a lesser degree of scrutiny is required. Specific information requirements, such as how detailed the description of the organism or method of DNA insertion should be, are described in general terms.

The earlier National Academy of Sciences report (48) points out that determining the risk of employing a particular microorganism, naturally occurring or genetically engineered in any environmental setting requires a synthesis of potential adverse effects associated with the specific agent, i.e., hazard and the level of exposure of populations sensitive to the identified hazard. The probability and magnitude of the effect and options available to a regulatory body determine the risk management procedures that should be employed. However, the potential risk associated with use of a microorganism for an environmental application is related to the phenotype and genotype of the microorganism. Alteration of the genotype by one or a few genes may or may not affect the potential environmental outcome, which may be positive, negative, or neutral.

The VBNC issue can make a risk assessment more difficult if interactions occur, rendering the estimation of the hazard component or exposure component more difficult. For example, increased difficulty in determining colony-forming units (CFU) per gram of soil or milliliter of water, if the cells are in a VBNC state, would make estimating exposure more difficult. Also, in the VBNC state, microorganisms may or may not offer potential hazard, depending on which metabolic functions are operating. Upon returning to the culturable state, the microorganisms at least transiently may be more (or less) virulent, and estimates of hazard may be less certain. The objective of risk assessment is to determine the magnitude of an effect, based on integration of hazard and exposure. For example, Leser et al. (34) showed that the generation time and culturability of *Pseudomonas* sp. strain b13(fri) was reduced in a marine microcosm. After 3 days, the cells were unculturable, but PCR showed that the ribosomal content corresponded to a generation time of two hours. Similarly, Leung et al. (35) demonstrated a 24-fold increase in *lacZ* expression in *Pseudomonas aureofaciens* introduced into river water.

To determine the effect of VBNC microorganisms on the risk of using GEMs in environmental settings, it is necessary to determine if the culture undergoes the VBNC phenomenon and then which (if any) aspects affect the estimation of major components of risk assessment, namely hazard and exposure.

THE VBNC STATE

VBNC bacteria have been described by many authors (13, 25, 46, 61, 63, 64). The VBNC state is seen as a response to adverse environmental conditions, i.e., stress, often involving lack of nutrient (53, 56). It is accepted that the majority of bacteria can survive for long periods of time in the absence of a complete supply of nutrients (15, 56). It has also been demonstrated that aqueous suspensions of bacteria respire but do not multiply during starvation conditions (15, 56). Wilson and Lindow (75), working with *Pseudomonas syringae*, described the appearance of VBNC cells in relation to shifts in humidity. Pedersen and Jacobsen (55) concluded that different genera [e.g., *Enterobacter cloacae* JP120 and *Alcaligenes eutrophus* AEO106(pRO101)] exhibit different responses to desiccation, in terms of becoming VBNC. However, the loss (or, better, shutdown) of ability to replicate is not automatically reversed with the return of environmental conditions conducive to growth and multiplication (55, 65).

Luscombe and Gray (42) described four sequelae of exposure to limited nutrients: induction of better uptake mechanisms for the limited nutrient; reduced uptake of other nutrients; reducing bottlenecks due to specific growth limitations; and general modulation of synthesis of macromolecules "to allow balanced growth" as closely as possible. The response can be reversed by exposure of bacteria to carefully selected environments, such as low nutrient concentrations (26, 27), exposure to specific metabolites (24), and presence of specific metabolites in the medium and/or highly specific growth conditions (15, 30, 63, 64). Berry et al. (7) attribute the shift of *Escherichia coli* to the culturable phase from VBNC to exposure to heat shock. Postgate (56) pointed out that growth substrates can also accelerate death of populations exposed to a nutrient-poor environment.

Occurrence of the VBNC State

The ability to survive adversity takes many forms, depending on the bacterium involved and the environmental conditions. Responses range from the formation of spores by *Clostridium* or *Bacillus* species (1) to cysts formed by *Giardia* species (33). Aaronson (1) described the ability of bacteria to respond to environmental stress (presence of inhibitory chemicals or starvation) as the production of perennating cells. These are cells which are "appreciably more resistant to chemical and physical stress than vegetative cells"; they survive until conditions become favorable, usually on a cyclical schedule related to seasons.

Response to environmental conditions is influenced by both the condition to which the cells are exposed (53, 55, 75) and the history of the bacterial culture. Knowledge of the environmental conditions to which a microorganism will be exposed at the release site and the history of the culture is essential to predict response and effect (47).

The VBNC state has been reported in a variety of microorganisms. Nilsson et al. (51) reported the shift of *Vibrio vulnificus* to VBNC upon exposure to nutrient-limited seawater. Other workers (25, 29, 46, 71) have reported the ability to form VBNC cells in *Aeromonas*, *Alcaligenes*, *Shigella*, and *Salmonella* species. VBNC cell formation in marine barophilic bacteria has also been reported (44, 48). Bakhrouf et al. (4) report the "evolution" of *P. aeruginosa* to VBNC in sterile sea water. Byrd and Colwell (11) and Berry et al. (7) reported VBNC cell formation in *E. coli*. Brayton et al. (10) reported VBNC cell formation in *Vibrio cholerae*.

Recently, Manahan and Steck (1986), demonstrated the VBNC state in *Agrobacterium tumefaciens* and *Rhizobium meliloti* (43).

SIGNIFICANCE OF VBNC ORGANISMS IN RISK ASSESSMENT

Opinions and findings vary as to the significance of VBNC microorganisms with respect to public health. Studies with *Campylobacter jejuni* provide a good example (8, 45, 61, 65). Medema et al. (45) reported that all strains studied (seven) which became VBNC retained pathogenicity after returning to the culturable state. Saha et al. (65) found that 14% of the strains studied became VBNC and did not regain pathogenicity after returning to the culturable state. The latter report speculated that VBNC was not a risk factor in assessing public health risk of *Campylobacter* by bacterial density estimates. Medema et al. (45) demonstrated that all of the seven VBNC *Campylobacter* strains regained pathogenicity, suggesting that VBNC cells are important in estimating potential public health problems. Islam et al. (25), working with *Shigella*, also concluded that VBNC bacteria may be important in epidemiological studies. Rahman et al. (57) have demonstrated the potential virulence of VBNC *Shigella dysenteriae* type 1.

However, Rose et al. (62), working with *Aeromonas salmonicida*, declared that the VBNC state may be an artifact, since low numbers of small, rounded cells are present in a culture at all times. Working with cell suspensions in microcosms, they monitored CFUs per milliliter until no colonies were observed (27 to 30 days), allowed the microcosms to incubate for an additional 7 days at 22°C, and found that the microorganisms could be recovered 48 h after addition of 1% tryptone soy broth to the microcosm. No growth occurred in 1-ml samples taken from the microcosm to which nutrient had been added. They concluded that the bacteria were present in low numbers, possibly attached to the walls of the microcosm, but they did not do acridine orange direct counts or direct viable counts to verify their findings.

To gain a broader perspective on the significance of the effect of VBNC bacteria on risk assessment, at least two issues must be considered. The impact of the phenomenon on populations exposed to the particular agent should be known, and the methods used to calculate potential exposure and the impact of these factors on hazard determinations must also be critically estimated if accuracy and precision affected by cells in the VBNC state are to be measured.

Exposure Assessment and the VBNC Phenomenon

Exposure assessment involves determining bacterial population densities at specific locations. Of interest are locations where sensitive populations may be present, thus requiring knowledge of the transport and fate of the microorganism under consideration (GEM or naturally occurring). There are many established techniques for estimating viable microorganisms in laboratory or environmental settings (2,

39, 67). Much research on improved techniques to provide better exposure and hazard estimates for risk assessment has been done (38, 72; USDA International Symposium, 1990). Many methods for estimating CFUs in situ rely on plate counts and involve selective media, whereas other methods rely on molecular approaches, i.e., measuring RNA or DNA and relating the concentration detected to total number of cells and/or diversity of the population. Given the VBNC phenomenon, it is important to distinguish between "living" and "dead" cells. Postgate (56) viewed the distinction as a continuum and defined "living" as able to multiply when provided optimum conditions. However, optimum conditions for the same microorganism can vary, depending on the physiological state of the organism (26, 27, 42).

METHODS FOR MONITORING GEMs

Total Number of Cells

Techniques have been developed to monitor microorganisms that measure the total number of cells. PCR has been used by many workers to detect DNA in laboratory and environmental samples and thereby relate findings to cell numbers (5, 6, 18, 66). Microscopic techniques using fluorescent and electron microscopy have been developed to determine the total number of cells and in some cases to differentiate between living and dead cells (12).

Chaudhry et al. (14) used in vitro amplification of target DNA by PCR, followed by hybridizing the DNA with a specific DNA probe, to detect genetically modified *E. coli* in sewage and lake water. The method permitted detection of an amplified and unique marker (0.3-kb) DNA of the test GEM. DNA was inserted into a derivative of a 2,4-dichlorophenoxyacetic acid-degradative plasmid, pRC10, and transferred into *E. coli*. The genetically-altered microorganism was seeded into filter-sterilized lake and sewage water samples. Viable counts decreased over the test period of 14 days to an undetectable level. With PCR, it was possible to detect the amplified unique marker of the GEM for up to 10 to 14 days of incubation. The method requires only picogram amounts of DNA and "has an advantage over the plate count technique which can detect only culturable microorganisms (14)." The method can be used to monitor GEMs in complex environments, where discrimination between GEMs and indigenous microorganisms is either difficult or requires time-consuming tests. Working with *V. vulnificus*, Brauns et al. (9) reported using PCR to detect as little as 72 pg of DNA from culturable cells versus 31 pg of DNA from VBNC cells. Ribosomal content has been monitored by Leser (34) using *Pseudomonas* in marine microcosms.

Roszak and Colwell (63) used autoradiography to distinguish viable from nonviable cells. Deng and Oliver (17) reported that visualization of viable and nonviable *Giardia* cysts by low-voltage scanning electron microscopy yielded excellent correlation with estimated viability of the cysts determined by fluorescein diacetate-propidium iodide staining and determined by electron microscopy and morphology.

Viable Cell Counts

The use of the plate count to estimate bacterial populations in the laboratory has been well accepted for nearly a century (73) and is used to estimate populations of GEMs in the environment (3). Both standard and selective media are employed. Farrand (19) cites the need for selective media and for care in determining the titer of populations under study. Many variations have been implemented. Seidler and Hearn (67) reported the use of selective media, combined with follow-up analysis with a gene probe to identify a specific GEM. Recovery (or estimates of CFU) of microorganisms, using the viable plate count, is limited. The portion of the population finding the nutritional and physiological characteristics of the recovery medium and growth conditions satisfactory for growth and reproduction is often less than 1% of the total population. Despite this limitation, useful environmental information may still be obtained.

Techniques have been developed to demonstrate in environmental samples the presence of viable cells which do not produce colonies on routinely employed bacteriological media (12, 23, 32, 63). In general, these methods rely on visual differentiation between cells, for example, after exposure to nalidixic acid or a vital stain. The presence of the vital stain allows microscopic observation, and the inhibitor, such as nalidixic acid, permit differentiation, since VBNC cells continue to metabolize and will take up the stain. Nalidixic acid inhibits multiplication, resulting in filamentous forms, if the cells are viable.

Identifying and Estimating Persistence and Effect

Estimating efficacy, or a positive or desired effect, is straightforward if the effect is known and can be quantified. Assessment of an adverse effect, however, is perhaps the most problematical aspect of risk assessment in environmental applications of GEMs or any microorganism, for that matter. The assessor must specify the nature of the effect, and a method is required to estimate the effect quantitatively. Many studies have focused on defining adverse effects and developing methods to quantitate them and to determine their significance (21, 31, 37, 52).

There have been several reports of effects of environmental application of GEMs. For example, Bej et al. (5) found that introducing genetically engineered *Pseudomonas cepacia* AC1100 into soil microcosms resulted in elevated taxonomic diversity. This was determined at the phenotypic level by analyses of culturable isolates and at the genetic level by analysis of the heterogeneity of total microbial community DNA, using reannealing kinetics. The greatest impact occurred when the microorganism was introduced along with the herbicide 2,4,5-T, which *P. cepacia* AC1100 can degrade. Both a shift in population structure and genetic exchange occurred. Short et al. (69) reported a 400-fold decline in the number of fungal propagules, and a marked reduction in CO_2 evolution after soil, amended with 2,4-dichlorophenoxyacetate, was treated with a genetically engineered strain of *Pseudomonas putida*.

Persistence

Selenska and Klingmuller (68) studied the persistence of cells and specific DNA sequences in soil. Seventy days after inoculation of soil with *Enterobacter agglomerans* containing plasmid pEA9, the bacteria were no longer culturable on an agar medium. At the same time, specific *nif* sequences of the plasmid were detected in total DNA recovered from the soil. A modified method of DNA extraction from soil was used, which did not require amplification of the DNA sequences by PCR. Effects of pollutants on survival and the VBNC state of *E. coli* in fresh and marine water have been reported (28, 54), showing that pollutants may not affect production of exotoxins while fragments of the gene (detected by PCR probe) were still present.

Turpin et al. (71) developed an enzyme-linked immunosorbent assay (ELISA) and a microwell fluorescent antibody (FA) direct count method for monitoring *Salmonella* spp. in soil. When monitored by plate count, the survival of *Salmonella* spp. was greater in sterile than in nonsterile soil. Evidence was obtained for production of VBNC *Salmonella* cells in nonsterile soil; that is, the plate counts dropped rapidly with time, but FA direct counts and ELISA remained level. The *Salmonella* cells becoming progressively smaller and rounder with time. Dead *Salmonella* cells (autoclaved) introduced into soil rapidly disappeared, i.e., did not persist.

Many studies have been conducted to determine if genetic material inserted into a GEM genome can, or will, be transferred and expressed. Much of the background knowledge regarding gene exchange is derived from laboratory experimentation. The probability of gene transfer in aquatic (66), terrestrial (74), and wastewater (22) environments has been reviewed. Gene transfer in natural environments has been observed, and transfer frequencies measured. Gene transfer frequencies in nature are generally lower than those obtained in the laboratory. However, environmental parameters affecting gene transfer are not well understood and have been studied only to a limited extent (59). On the other hand, interspecies transfer of plasmids and rapid transfer of antibiotic resistance within microbial populations have been reported (22, 41).

Antibiotic resistance is a commonly used marker/selective mechanism in molecular biology and is known to spread within microbial populations. Expression of a transferred trait is the most obvious characteristic persisting after environmental application of GEMs, which may have a significant effect (60). The fact that gene transfer occurs readily (74) indicates that traits can be exchanged. Also, the fact that DNA can be incorporated in competent cells via transformation, after cell lysis (22), it suggests that traits can be acquired in the absence of energy production on the part of the donor. If exchange occurs between VBNC GEMs and other members of indigenous populations, characteristics of the recipient could be altered.

EFFECTS

As indicated above for *Campylobacter* spp., there may be an effect in terms of recovery of pathogenicity. Linder and Oliver (40) demonstrated that such conversion of *V. vulnificus* occurred after 24 days. They studied the occurrence of VBNC cells in *V. vulnificus* and, for comparison, *E. coli* in artificial seawater microcosms incubated at 5°C and found that, while total counts remained constant, a comparison of total counts with plate counts indicated nonculturability was reached by day 24. In contrast, direct viable counts indicated that the cells remained viable throughout the 32 days of incubation. As an indication of metabolic changes occurring as cells enter the state of nonrecoverability, membrane fatty acid analyses were performed. At the point of nonculturability of *V. vulnificus*, the major fatty acid species (C_{16}) decreased 57% from the start of the experiment and concomitant with the appearance of several short-chain acids. For *E. coli*, although cells were still culturable, a similar trend was observed. Mouse infectivity studies with *V. vulnificus* suggested that some loss of virulence occurred. Total cell counts were monitored by acridine orange epifluorescence microscopy, metabolic activity by direct viable counts, and culturability by plate counts on selective and nonselective media.

Rice and Oliver (58) noted that the marine barophile CNPT-3, after stress by starvation, exhibited a significant reduction in cell size and biovolume. The starved cells demonstrated a greater tendency to attach at in situ pressure (400 atm) and temperature (5°C) than at 1 atm, and the extent of attachment increased with increasing duration of starvation. However, Magarinos et al. (44) examined a variety of biochemical, physiological, and serological properties, including LD_{50} for *Pasteurella piscida*, and observed no such effect. The latter did not employ pressure as a factor.

SUMMARY AND CONCLUSIONS

The VBNC phenomenon is associated with changes in cell morphology and ability to multiply on routinely employed bacteriological culture media. VBNC cells appear to occur predominantly in gram-negative microorganisms. VBNC cells can reproduce if subjected to selected growth conditions and may return to culturability following biochemical and physiological measurement. Some of the initiators of the VBNC state have been identified and include starvation. Changes in cell morphology and fatty acid content after starvation have been identified for VBNC *V. vulnificus.*

Parental microorganisms, i.e., microorganisms to which DNA has been inserted (e.g., *E. coli*), have demonstrated the ability to enter the VBNC state. However, the underlying mechanism(s) and specific metabolic changes resulting from, or due to, VBNC are still not yet fully understood. Thus, there is no reason to assume that a GEM will not enter the VBNC state.

It is clear that if a GEM (or any microorganism) becomes VBNC it will become more difficult to detect and enumerate, especially if growth and multiplication on a standard bacteriological culture medium is the criterion, i.e., estimating CFUs. As a result, exposure assessment, conducted as a first step in risk assessment, becomes more difficult to perform and the results are less credible. Use of one of the methods for indicating the presence of VBNC cells in a bacterial population would aid in producing a more reliable estimate of the viable population, but one could not be entirely certain that the presence of VBNC cells was significant from a hazard perspective until further studies provide more data.

There is no argument as to the effect on a bacterial culture of passing through the VBNC state, i.e., on phenetic and/or genetic characteristics of the bacterium when and if it returns to culturability. VBNC cells can be considered to be a life cycle phase in which the microorganism awaits conditions suitable for growth and reproduction, at which time it will revert to active growth. It can equally well be considered as "clinging to life" and slowly approaching nonviability. In reality, there is most likely a continuum between these two extremes. For accurate estimation of environment effects, methods are needed to estimate better the status of microorganisms on this continuum.

VBNC organisms are known to retain cellular integrity longer than dead cells in the same environment, and therefore, during the VBNC state, it may be possible for the GEM to exchange DNA with other members of the microbial community. Although Gealt (22) points out the need for energy production to replicate the plasmid and for the exchange process itself, since exchange takes place when DNA is released to the environment as a function of cell lysis, genetic transformation may occur with minimal energy production.

In conclusion, VBNC cells can be thought of as a point on a continuum between readily culturable cells and dead cells, nonculturable even after using "heroic" techniques. Thus, the VBNC phenomenon creates difficulties in assessing risks associated with release of GEMs to the environment, including the following.

Use of viable count technique may not produce an accurate estimate of population density, adversely affecting exposure estimates. Inclusion of VBNC cells can result in overestimation of potential hazard since cells are not metabolically active. Including VBNC cells in a population estimate may overstate any hazard because the cells may not always revert to their original physiological status. Finally, inclusion of the VBNC portion of the population provides a worst-case estimate, suitable for risk assessment purposes. While many methods for enumerating VBNC cells have been developed, careful examination of the accuracy of such methods has not been conducted and needs to be done for risk analysis in the future.

REFERENCES

1. **Aaronson, S.** 1981. *Chemical Communication at the Microbial Level*, p. 30–33. CRC Press, Baton Rouge, Fla.
2. **American Public Health Association.** 1989. *Standard Methods for the Examination of Water and Wastewater.* American Public Health Association, Washington, D.C.
3. **Atlas, R.** 1992. Detection and enumeration of microorganisms based upon phenotype, p. 29–43. *In* M. A. Levin, R. J. Seidler, and M. Rogul (ed.), *Microbial Ecology: Principles, Methods and Applications.* McGraw-Hill, New York, N.Y.
4. **Bakhrouf, A., M. Jeddi, A. Bouddabous, and M. J. Gauthier.** 1989. Evolution of *Pseudomonas aeruginosa* cells towards a filterable stage in seawater. *FEMS Mic. Lett.* **59:**187–190.
5. **Bej, A. K., M. H. Mahbubani, and R. M. Atlas.** 1991. Detection of viable *Legionella pneumophila* in water by polymerase chain reaction and gene probe methods. *Appl. Environ. Microbiol.* **57:**597–600.
6. **Bej, A. K., M. Perlin, and R. M. Atlas.** 1991. Effect of introducing genetically-engineered microorganisms on soil microbial community diversity. *FEMS Microbiol. Lett.* **86:**(2)169–175.
7. **Berry, C., B. J. Lloyd, and J. S. Colbourne.** 1991. Effect of heat shock on recovery of *Escherichia coli* from drinking water, p. 85–88. *In* W. O. K. Grabow, R. Morris, and K. Botzenhart, (ed.), *Health Related Water Microbiology.* Int. Symp. On Health-Related Water Microbiology, Tuebingen (FRG), 1–6 April 1990. IAWPRC, Tuebingen, Germany.
8. **Beumer, R. R., J. De Vries, and F. M. Rombouts.** 1992. *Campylobacter jejuni* non-culturable coccoid cells. *Int. J. Food Microbiol.* **15:**153–163.
9. **Brauns, L. A., M. C. Hudson, and J. D. Oliver.** 1991. Use of the polymerase chain reaction in detection of culturable and nonculturable *Vibrio vulnificus* cells. *Appl. Environ. Microbiol.* **57:**2651–2655.

10. **Brayton, P. R., M. L. Tamplin, A. Huq, and R. R. Colwell.** 1987. Enumeration of *Vibrio cholerae* O1 in Bangladesh waters by fluorescent-antibody direct viable count. *Appl. Environ. Microbiol.* **53:** 2862–2865.
11. **Byrd, J. J., and R. R. Colwell.** 1990. Maintenance of plasmids pBR322 and pUC8 in nonculturable *Escherichia coli* in the marine environment. *Appl. Environ. Microbiol.* **56:**2104–2107.
12. **Byrd, J. J., and R. R. Colwell.** 1992. Microscopic applications for analysis of environmental samples, p. 607–623. *In* M. A. Levin, R. J. Seidler, and M. Rogul (ed.), *Microbial Ecology: Principles, Methods and Applications*. McGraw-Hill, New York, N.Y.
13. **Byrd, J. J., H. S. Xu, and R. R. Colwell.** 1991. Viable but nonculturable bacteria in drinking water. *Appl. Environ. Microbiol.* **57:**875–878.
14. **Chaudhry, G. R., G. A. Toranzos, and A. R. Bhatti.** 1989. Novel method for monitoring genetically engineered microorganisms in the environment. *Appl. Environ. Microbiol.* **55:**1301–1304.
15. **Dawes, E. A.** 1976. Endogenous metabolism and the survival of starved prokaryotes, p. 19–24. *In* T. R. G. Gray and J. R. Postgate (ed.), *The Survival of Vegetative Microbes.* Cambridge University Press, Cambridge, United Kingdom.
16. **Dawson, M. P., B. A. Humphrey, and K. C. Marshall.** 1981. Adhesion: a tactic in the survival strategy of a marine vibrio during starvation. *Curr. Microbiol.* **6:**195–199.
17. **Deng, M. Y., and D. O. Oliver.** 1992. Degradation of *Giardia lamblia* cysts in mixed human and swine wastes. *Appl. Environ. Microbiol.* **58:**2368–2374.
18. **Dockendorf, T. C., A. Breen, O. A. Oguntseitan, J. G. Packard, and G. S. Sayler.** 1992. Practical considerations of nucleic acid hybridization and reassociation techniques in environmental analysis, p. 286–311 and 393–419. *In* M. A. Levin, R. J. Seidler, and M. Rogul (ed.), *Microbial Ecology: Principles, Methods and Applications.* McGraw-Hill, New York, N.Y.
19. **Farrand, S. K.** 1992. Conjugal gene transfer on plants, p. 607–623. *In* M. A. Levin, R. J. Seidler, and M. Rogul (ed.), *Microbial Ecology: Principles, Methods and Applications.* McGraw-Hill, New York, N.Y.
20. **Fisk, J., and Covello, V. T.** 1986. *Biotechnology Risk Assessment*, p. 1–125. Pergamon Press, Elmsford, N.Y.
21. **Fredrickson, J. K., and C. Hagedorn.** 1992. Identifying ecological effects from the release of genetically engineered microorganisms and microbial pest control agents, p. 327–344. *In* M. A. Levin, R. J. Seidler, and M. Rogul (ed.), *Microbial Ecology: Principles, Methods and Applications.* McGraw-Hill, New York, N.Y.
22. **Gealt, M. A.** 1992. Gene transfer in wastewater, p. 327–344. *In* M. A. Levin, R. J. Seidler, and M. Rogul (ed.), *Microbial Ecology: Principles, Methods and Applications.* McGraw-Hill, New York, N.Y.
23. **Hobbie, J. E., R. J. Daley, and S. Jasper.** 1977. Use of nucleopore filters for counting bacteria by fluorescence microscopy. *Appl. Environ. Microbiol.* **33:**1225–1228.
24. **Heinmets, F., W. W. Taylor, and J. J. Lehman.** 1953. The use of metabolites on the restoration of the viability of heat and chemically inactivated *Escherichia coli. J. Bacteriol.* **67:**5–14.
25. **Islam, M. S., M. K. Hasan, M. A. Miah, G. C. Sur, A. Felsenstein, M. Venkatesan, R. B. Sack, and M. J. Albert.** 1993. Use of the polymerase chain reaction and fluorescent antibody methods for detecting viable but nonculturable *Shigella dysenteria* type 1 in laboratory microcosms. *Appl. Environ. Microbiol.* **59:**536–540.
26. **Jannasch, H. W.** 1979. Microbial ecology of aquatic low nutrient habitats, p. 243–260. *In* M. Shilo (ed.), *Strategies of Microbial Life in Extreme Environments.* Verlag Chemie, Weinheim, Germany.
27. **Jannasch, H. W., and G. E. Jones.** 1959. Bacterial populations in seawater by different methods of enumeration. *Limnol. Oceanogr.* **4:**128–139.
28. **Jolivet-Gougeon, J., A. S. Baux, F. Sauvager, M. Arturo-Schaan, and M. Cormier.** 1996. Influence of peracetic acid on *Escherichia coli* H10407 strain in laboratory microcosms. *Can. J. Microbiol.* **42:**60–65.
29. **Kandel, A., O. Nybroe, and O. F. Rasmussen.** 1992. Survival of 2,4-dichlorophenoxyacetic acid degrading *Alcaligenes eutrophus* AEO 106(pR0101) in lake water microcosms. *Microb. Ecol.* **24:** 291–303.
30. **Kjelleberg, S., B. A. Humphrey, and K. C. Marshall.** 1983. Initial phase of starvation and activity of bacteria at surfaces. *Appl. Environ. Microbiol.* **46:**978–984.

31. **Klein, D. A.** 1992. Measurement of microbial population dynamics: significance and methodology, p. 607–623. *In* M. A. Levin, R. J. Seidler, and M. Rogul (ed.), *Microbial Ecology: Principles, Methods and Applications.* McGraw-Hill, New York, N.Y.
32. **Kogure, K., U. Simidu, and N. Taga.** 1979. A tentative direct microscopic method for counting living marine bacteria. *Can. J. Microbiol.* **25:**415–420.
33. **LeChevallier, M. W., W. D. Norton, and R. G. Lee.** 1991. *Giardia* and *Cryptosporidium* spp. in filtered drinking water supplies. *Appl. Environ. Microbiol.* **57:**2617–2621.
34. **Leser, T. D., M. Boye, and N. B. Hendriksen.** 1995. Survival and activity of *Pseudomonas* sp. strain B13 (FRI) in a marine microcosm determined by quantitative PCR and an rRNA-targeting probe and its effect on the indigenous bacterioplankton. *Appl. Environ. Microbiol.* **61:**1201–1207.
35. **Leung, K., J. T. Trevors, and H. Lee.** 1995. Survival and lacZ expression in recombinant *Pseudomonas* strains introduced into river water microcosms. *Can. J. Microbiol.* **41:**461–469.
36. **Levin, M. A. and H. Strauss.** 1991. *Risk Assessment in Genetic Engineering*, p. 60–87. McGraw-Hill, New York, N.Y.
37. **Levin, S. A., and M. A. Harwell.** 1986. Potential ecological consequences of genetically engineered organisms. *Env. Management* **10:**495–513.
38. **Levin, M. A., R. J. Seidler, and M. Rogul. (ed.).** 1992. *Microbial Ecology: Principles, Methods and Applications.* McGraw-Hill, New York, N.Y.
39. **Levin, M. A., R. J. Seidler, A. R. Bourquin, J. Fowle III, and T. Barkay.** 1987. EPA developing methods to assess environmental release. *Bio/Technology.* **5:**38–45.
40. **Linder, K., and J. D. Oliver.** 1989. Membrane fatty acid and virulence changes in the viable but nonculturable state of *Vibrio vulnificus. Appl. Environ. Microbiol.* **55:**2837–2842.
41. **Lovins, K. W., J. S. Angle, J. L. Weber, and R. L. Hill.** 1993. Leaching of *Pseudomonas aeruginosa* and transconjugants through unsaturated, intact soil columns. *FEMS Microbiol. Ecol.* **13:** 105–112.
42. **Luscombe, B. M., and T. R. G. Gray.** 1971. Effect of varying growth rate on morphology of Arthobacter. *J. Gen. Microbiol.* **69:**433–448.
43. **Manahan, S. H., and T. R. Steck.** 1996. VBNC state in *Agrobacterium tumefaciens* and *Rhizobium meliloti. FEMS Microbiol. Lett.* **22(1):**39–48.
44. **Magarinos, B., J. L. Romalde, J. L. Bartha, and A. E. Torzano.** 1993. Evidence of a dormant but infective state of the fish pathogen *Pasteurella piscicida* in seawater and sediment. *Appl. Environ. Microbiol.* **60(1):**180–186.
45. **Medema, G. J., F. M. Schets, A. W. Giessen, and A. H. van de Havelaar.** 1992. Lack of colonization of 1 day old chicks by viable, nonculturable *Campylobacter jejuni. J. Appl. Bacteriol.* **72:** 512–516.
46. **Morgan, J. A. W., G. Rhodes, and R. W. Pickup.** 1993. Survival of nonculturable *Aeromonas salmonicida* in lake water. *Appl. Environ. Microbiol.* **59:**874–880.
47. **Morita, R. Y.** 1992. Survival and recovery of microorganisms from environmental samples, p. 607–633. *In* M. A. Levin, R. J. Seidler, and M. Rogul (ed.), *Microbial Ecology: Principles, Methods and Applications.* McGraw-Hill, New York, N.Y.
48. **National Academy of Sciences.** 1983. *Risk Assessment*, p. 83–111. National Research Council, Washington, D.C.
49. **National Academy of Sciences.** 1987. *Introduction of Recombinant DNA Engineered Organisms into the Environment*, p. 75–115. National Research Council, Washington, D.C.
50. **National Academy of Sciences.** 1989. *Field Testing of Genetically Engineered Organisms in the Environment*, p. 87–93. National Research Council, Washington, D.C..
51. **Nilsson, L., J. D. Oliver, and S. Kjelleberg.** 1991. Resuscitation of *Vibrio vulnificus* from the viable but nonculturable state. *J. Bacteriol.* **173:**5054–5059.
52. **Office of Technology Assessment.** 1986. *New Developments in Biotechnology. Field Testing Engineered Organisms*, p. 75–79. Office of Technology Assessment, Washington, D.C.
53. **Oliver, J. D., L. Nilsson, and S. Kjelleberg.** 1991. Formation of nonculturable *Vibrio vulnificus* cells and its relationship to the starvation state. *Appl. Environ. Microbiol.* **57:**2640–2644.
54. **Pathak, S. P., and J. W. Bhattacherjee.** 1994. Effect of pollutants on the survival of *Escherichia coli* in microcosms of river water. *Bull. Environ. Contam.* **53:**198–203.

55. **Pedersen, J. C., and C. S. Jacobsen.** 1993. Fate of *Enterobacter cloacae* JP120 and *Alcaligenes eutrophus* AE0106 (pR0101) in soil during water stress: effects on culturability and viability. *Appl. Environ. Microbiol.* **59:**1560–1564.
56. **Postgate, J. R.** 1976. Death in macrobes and microbes, p. 1–19. *In* T. R. G. Gray and J. R. Postgate (ed.), *The Survival of Vegetative Microbes.* Cambridge University Press, Cambridge, United Kingdom.
57. **Rahman, I., M. Shahamat, M. A. R. Chowdry, and R. R. Colwell.** 1996. Potential virulence of viable but nonculturable *Shigella dysenteriae* Type 1. *Appl. Environ. Microbiol.* **62:**115–120.
58. **Rice, S. A., and J. D. Oliver.** 1992. Starvation response of the marine barophile CNPT3. *Appl. Environ. Microbiol.* **58:**2432–2437.
59. **Richaume, A., J. S. Angle, and M. J. Sadowsky.** 1989. Influence of soil variables on in situ plasmid transfer from *Escherichia coli* to *Rhizobium fredii. Appl. Environ. Microbiol.* **55:**1730–1734.
60. **Rissler, J., and M. Mellon.** 1993. *Perils Amidst the Promise*, p. 4–23. Union of Concerned Scientists, Washington, D.C.
61. **Rollins, D. M., and R. R. Colwell.** 1986. Viable but nonculturable stage of *Campylobacter jejuni* and its role in survival in the natural aquatic environment. *Appl. Environ. Microbiol.* **52:**531–538.
62. **Rose, A. S., A. E. Ellis, and A. L. S. Munro.** 1990. Evidence against dormancy in the bacterial fish pathogen *Aeromonas salmonicida* subsp. *Salmonicida. FEMS Microbiol. Lett.* **68:**105–108.
63. **Roszak, D. B., and R. R. Colwell.** 1987. Metabolic activity of bacterial cells enumerated by direct viable count. *Appl. Environ. Microbiol.* **53:**2889–2893.
64. **Roszak, D. B., and R. R. Colwell.** 1987. Survival strategies of bacteria in the natural environment. *Microbiol. Rev.* **51:**365–379.
65. **Saha, S. K., S. Saha, and S. C. Sanyal.** 1991. Recovery of injured *Campylobacter jejuni* cells after animal passage. *Appl. Environ. Microbiol.* **57:**3388–3389.
66. **Saye, D. J., and S. B. O'Morchoe.** 1992. Evaluating the potential for genetic exchange in natural freshwater environments, p. 286–311. *In* M. A. Levin, R. J. Seidler, and M. Rogul (ed.), *Microbial Ecology: Principles, Methods and Applications.* McGraw-Hill, New York, N.Y.
67. **Seidler, R. J., and S. C. Hearn.** 1988. *EPA Special Report: the Release of Ice Minus Recombinant Bacteria at California Test Sites.* U.S. Environmental Protection Agency, Corvallis, Oreg.
68. **Selenska, S., and W. Klingmuller.** 1991. Direct detection of *nif*-gene sequences of *Enterobacter agglomerans* in soil. *FEMS Microbiol. Lett.* **80:**243–245.
69. **Short, K. A., J. D. Doyle, R. J. King, R. Seidler, G. Stotzky, and R. H. Olsen.** 1991. Effects of 2,4-dichlorophenol, a metabolite of genetically engineered bacteria, and 2,4-dichlorophenoxyacetate on some microorganism-mediated processes in soil. *Appl. Environ. Microbiol.* **57:**412–418.
70. **Tiedje, J. M., R. R. Colwell, Y. L. Grossman, R. E. Hodson, R. E. Lenski, R. E. R. N. Mack, and P. J. Regal.** 1989. The planned introduction of genetically engineered organisms. Ecological considerations and recommendations. *Ecology* **70:**103–120.
71. **Turpin, P. E., K. A. Maycroft, C. L. Rowlands, and E. M. H. Wellington.** 1993. Viable but non-culturable salmonellas in soil. *J. Appl. Bacteriol.* **74:**421–427.
72. **United States Department of Agriculture (USDA).** 1993. Biotechnology risk assessment research grants program. *Fed. Reg.* Jan 27, 1994. **58:**3978–3980.
73. **U.S. Environmental Protection Agency.** 1978. *Microbial Methods for Monitoring the Environment.* USEPA 600/8-78-017. U.S. Environmental Protection Agency, Washington, D.C.
74. **Walter, M. V., and R. J. Seidler.** 1992. Measurement of conjugal gene transfer in terrestrial ecosystems, p. 311–326. *In* M. A. Levin, R. J. Seidler, and M. Rogul (ed.), *Microbial Ecology: Principles, Methods and Applications.* McGraw-Hill, New York, N.Y.
75. **Wilson, M., and S. E. Lindow.** 1992. Relationship of total viable and culturable cells in ephiphytic populations of *Pseudomonas syringae. Appl. Environ. Microbiol.* **58:**3908–3913.
76. **World Health Organization.** 1983. *Biosafety Manual.* World Health Organization, Geneva, Switzerland.

Nonculturable Microorganisms in the Environment
Edited by R. R. Colwell and D. J. Grimes

Chapter 15

Chemical Disinfection and Injury of Bacteria in Water

Gordon A. McFeters and Mark W. LeChevallier

Exposure of enteric bacteria to sublethal levels of antibacterial agents and unfavorable conditions in many environments results in altered phenotypes within these organisms. Reports over the past five decades describe the failure of *Escherichia coli* to be recovered using commonly accepted media and incubation conditions following exposure to compounds such as phenolic antiseptics (33, 35) and chlorine (68). As reviewed by Harris (32) and Ray (75), these early observations of reduced bacterial recovery following exposure to a range of sublethal environmental stressors were interpreted as a form of cellular damage that reduced the culturability of allochthonous bacteria using methods that were not restrictive for freshly cultivated bacteria. Some of the earliest workers studying this phenomenon observed that such damaged bacteria became more fastidious in their nutritional requirements for subsequent growth (100). An additional facet of that phenomenon that was also recognized relatively early (32) was the capability of the stressed cells to regain their more robust phenotype through a "revival" process where the damage was repaired under favorable circumstances. This recovery typically requires between 1 and 3 h of incubation under nonrestrictive conditions. During this time the level of colony-forming units (CFU) on selective media increases to the constant concentration seen using nonselective media. As pointed out by Litsky (54), environmental conditions that stress enteric bacteria are a common feature of most aquatic systems, and it should not be surprising that damaged cells are incapable of colony formation on many harsh selective media that continue to be used in microbiological analyses. Litsky further argued that this scenario represents an error among the early environmental microbiologists since many of these analytical approaches were borrowed without modification from medical applications where the target bacteria are both more numerous and directly isolated from an environment within a patient or animal to which the bacteria are well adapted.

Gordon A. McFeters • Microbiology Department and Center for Biofilm Engineering, Montana State University, Bozeman, MT 59717. ***Mark W. LeChevallier*** • American Water Works Service Co. Inc., P.O. Box 1770, Voorhees, NJ 08043.

Bacterial injury has been widely recognized for many years within the context of indicator organisms used to determine the microbiological quality of foods. As in many other environments, enteric bacteria associated with most food products are exposed to potentially damaging chemical and physical conditions during processing and preservation that result in cellular debilitation. As a consequence, attempts to assess the microbiological quality of the food, which has been traditionally based on colony formation using selective media, can result in a significant underestimation of the actual number of viable indicator bacteria present. This circumstance is important in the food industry as it can lead to an overly optimistic estimation of the safety of the product because a high proportion of the injured indicator bacteria will not be detected while pathogens, such as spore-forming bacteria and viruses, might persist in high numbers. Food microbiologists have studied issues relating to bacterial injury in foods for some time, and the subject has been extensively discussed in earlier reviews (3, 13, 14, 38, 71, 74, 75, 76) and again more recently (77). The significance of bacterial injury is also commonly acknowledged by food microbiologists and agencies dealing with health concerns in the food industry (13). Furthermore, the detection of injured indicator bacteria in food products, using specially designed media and methods, is routine. However, although the occurrence, detection, and significance of injured bacteria in disinfected water has gained some recognition among water microbiologists over the past decade (45, 46), few laboratories or agencies dealing with the safety of potable water utilize comparable methods.

As in the case of many processed foods, the microbiological quality of treated water and wastewater is determined by the presence of enteric indicator bacteria that are, likewise, not well adapted to the ambient conditions in water where stressors (including disinfectants) are often present in low concentrations. In particular, the optimal concentration of disinfectant added as the treated water enters distribution systems rapidly decreases with flow time and distance due to reaction with biofilm components lining the pipes as well as with other materials. Therefore, it is not surprising that an average of 95% of the coliforms detected in three operating drinking water distribution systems in New England were injured and incapable of colony formation on the medium that was commonly used to determine water potability (60). These and other similar observations from drinking water circumstances have been reviewed (58, 59) and other reports have discussed the evolution of methods for the detection of injured coliform bacteria in water (45, 46). It has also been demonstrated that injured bacteria, not detected using accepted methods, penetrate drinking water treatment plants and enter municipal distribution systems (12). These observations collectively suggest that indicator organisms used to determine water contamination become unculturable, using accepted media and methods, following exposure to stressors such as disinfectants in drinking water. As in the case of foods, this occurrence brings into question the reliability of many of the methods that are commonly used in the assessment of water potability and safety.

The injury of waterborne bacteria may be defined by the same criteria used previously by food microbiologists (3, 14, 38, 71). Injury is generally regarded as the sublethal physiological and/or structural consequence(s) resulting from exposure to stressors, such as suboptimal concentrations of disinfectants, within aquatic environments. This is signified by the inability of injured cells to reproduce under conditions that are suitable for the active growth of uninjured or freshly cultured bacteria. The restrictive conditions under which the damaged cells are unculturable include selective or minimal media and elevated temperatures. Therefore, a medium and growth conditions that are nonrestrictive must be used as a reference in determining the extent of injury. This is illustrated in Fig. 1, where a suspension of *E. coli* was exposed in membrane diffusion chambers to stream water containing high metal concentrations (61). Increasing injury is seen in the EG5 location as a progressive difference in bacterial recovery between TSY (nonselective) (Trypticase soy broth supplemented with 0.3% yeast extract and 1.5% Bacto-Agar) and DLA (TSY plus 0.1% sodium desoxycholate) (selective) media (6). The working hypothesis used to interpret this type of data is that both injured and uninjured cells are culturable on the nonselective medium (TSY) while only the noninjured bacteria form colonies on the selective medium. The percentage of cells that are injured is usually calculated by dividing the difference of CFUs on the two media by the CFUs on the nonselective medium × 100. However, it should be emphasized that percent injury calculations must be viewed as relative values since various selective

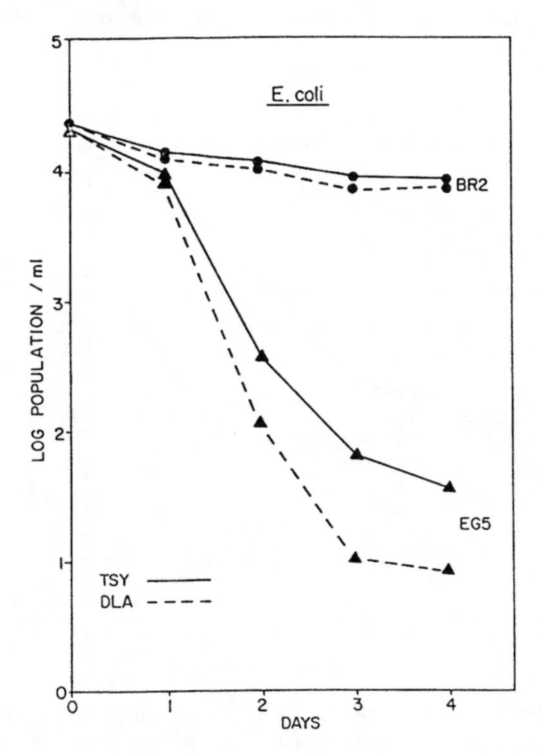

Figure 1. Comparative recovery of *E. coli* in membrane diffusion chambers located at sites BR2 (circles) and EG5 (triangles) in natural stream environments over a 4-day exposure period. Samples were surface-overlay plated using TSY (solid lines) and DLA (dashed lines) agar media.

media often detect bacteria with varying efficiencies (59), although this calculation can provide useful information describing the relative level of injury within a given population of waterborne bacteria. In that connection, Mossel and van Neeten (70) pointed out that such bacteria in natural systems respond to stress individually and are present in at least three states: those bacteria that are irreversibly inactivated or damaged, cells that are not injured, and those that are injured to varying degrees. Therefore, the restrictive medium or growth condition simply allows the discrimination of the stressed cells within the population that was initially undamaged, before exposure to the stressor(s). An additional criterion that is usually applied to injured bacteria is that they are capable of regaining their culturability under the selective conditions without growth. That concept is illustrated by the data presented in Fig. 2, where a suspension of injured *E. coli* was held in a nonrestrictive nutrient medium. It is noteworthy that there was no evidence of growth while the injured cells underwent repair during an extended lag phase.

The purpose of this chapter is to discuss the injury process of enteric bacteria following sublethal exposure to biocides. It is not intended that the review of the literature will be all inclusive, but much of the existing information concerning the causes and cellular consequences of this form of injury will be presented. The development of media to detect injured enteric bacteria in water will also be described with some information on the occurrence of stressed coliforms in disin-

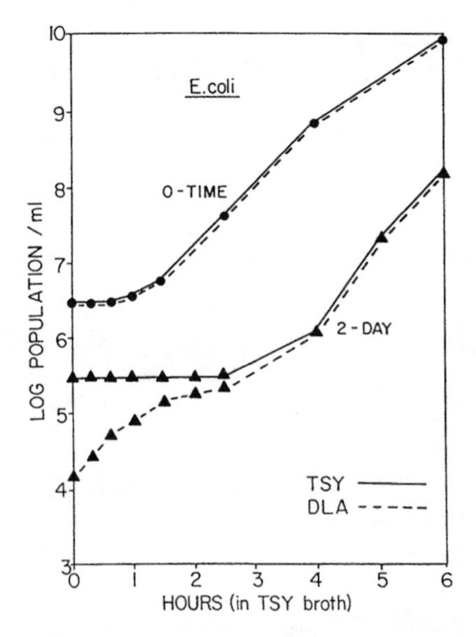

Figure 2. Repair of injury when *E. coli* was incubated in TSY broth following exposure in a natural stream environment (site EG6) using membrane diffusion chambers for 2 days (triangles). Control bacteria (0-time, circles) were freshly cultured. Both control and exposed bacterial suspensions were enumerated using TSY (solid lines) and DLA (dashed lines) on surface-overlay plates during the 6-hour incubation in TLY broth.

fected water. Finally, information will be presented and discussed concerning the significance of injured enteric bacteria in disinfected water systems.

DISINFECTANTS AS ENVIRONMENTAL STRESSORS

A range of chemical and physical stimuli have been implicated as causative factors in bacterial injury. These concepts have been reviewed in applications relating to both aquatic systems (58) and foods (38). However, sublethal concentrations of antimicrobial chemicals represent a predominant source of bacterial stress in a wide variety of systems where disinfectants occur at low concentrations. For example, the application of effective levels of disinfectant (i.e., ca. 1 mg/liter hypochlorous acid) at drinking water treatment facilities is followed by a steady decline to concentrations that are below the limit of detectability at the extremities of distribution networks (28, 73). As a consequence, potable water systems provide a range of environments with suboptimal concentrations of disinfectants that can cause injury.

Chlorine is the most commonly used biocide for the disinfection of potable water and wastewater streams. An early report by Mudge and Smith (72) described a reversible form of bacterial inactivation in chlorinated systems. Somewhat later, Schusner et al. (82) again reported the injury of *E. coli* by chlorine in foods, and the phenomenon was also demonstrated somewhat later in water (10) and wastewater (37). Those observations were further confirmed by Camper and McFeters (17) and McFeters and Camper (62) who demonstrated the very rapid occurrence of the injury process (i.e., less than 1 min) using hypochlorous acid under controlled laboratory conditions. The speed of this process distinguishes it from the phenomenon of viable but unculturable bacteria. That report also described a variety of physiological consequences including significant decreases in respiration and nutrient uptake within the affected bacteria. These reports highlight the need for growth media that would allow the recovery of injured coliforms from disinfected water since accepted media used in the analysis of water (i.e., m-Endo and m-FC) restricted the culturability and detection of indicator bacteria following sublethal stress (63). This observation was extended to chlorine-injured fecal streptococci by Lin (52) and Bissonnette (6). Others further demonstrated high levels of injured coliform bacteria in wastewater (29), and potable water (60) and in laboratory studies (93). The increasing use of chloramination as a disinfection strategy for the treatment of drinking water prompted a study demonstrating that exposure of enteric bacteria to monochloramine for 10 min also resulted in high levels (i.e., >90%) of injury (96). That finding was also observed when *Campylobacter jejuni* was exposed to monochloramine (9). In another study carried out within operating municipal drinking water treatment facilities, the majority of bacteria isolated from the treated water were injured both with and without chlorination (12). These findings suggest that factors in addition to disinfectants can lead to significant levels of injury in operating potable water systems. Early studies in our laboratory also suggested that influences other than disinfectants caused bacterial injury since suspensions of enteric bacteria exposed to natural stream water from different sources yielded reproducible levels of injury in the absence of added biocides (7). Alternative antimicrobial agents, such as ozone and chlorine dioxide, are being increasingly used to disinfect potable water. Of these, ozone has been shown to cause a reversible form of injury in *E. coli* that is manifest by the loss of culturability on m-FC medium, accentuated at 44.5°C (24). There is a need for additional studies on these and other alternative water disinfectants to determine if they likewise elicit bacterial injury.

Other potential sources of sublethal bacterial stress in the environment include metals, UV radiation, acidic pH, freezing and thawing, as well as biological factors. Early studies from our laboratory (6, 7) suggested that the stressors responsible for the injury observed in some natural streams were elevated concentrations of metals. That hypothesis is consistent with an earlier report demonstrating that metals including copper, zinc, and nickel effect damage in bacteria (57). Subsequent laboratory studies determined that low concentrations of copper caused stress, and an injury-concentration relationship was established (21). The work of Fujioka and Narikawa (25) as well as Kapuscinski and Mitchell (41) indicated that UV causes sublethal injury in bacteria suspended in seawater. The former of those reports indicated that fecal streptococci were much less susceptible to inactivation by sunlight than coliforms and that sewage bacteria diluted in seawater were more sensitive than when freshwater was used as the diluent. Others (69) found that following UV exposure in seawater, *E. coli* recovery was more efficient when a medium of lower nutrient concentration was used. Somewhat later, Bailey et al. (2) also described solar-radiation-induced injury in natural bacterial populations from the Chesapeake Bay. The antibacterial influence of acidic environments has been widely appreciated for many years, and a 1971 report described injury caused by low pH in *E. coli* (78). As is the case with other forms of injury, the affected bacteria were more sensitive to selective media. More recently, Hackney and Bissonnette (31), Double and Bissonnette (22), and Wortman and Bissonnette (99) demonstrated that coliforms exposed to acid mine water became injured and were more sensitive to media containing desoxycholate. These findings indicate that this environmental stressor also reduces the culturability of indicator bacteria in environments influenced by acid mine drainage. Temperature extremes have also been recognized as a causative factor in sublethal injury associated with bacteria in foods. Hurst (39) has reviewed the effects of sublethal heating while Mackey (56) discussed lethal and sublethal effects of refrigeration, freezing, and freeze-drying on microorganisms. Although these reports describe studies done with respect to food-related problems, it is likely that bacteria in aquatic environments such as heat exchangers respond in a similar manner because of exposure to alternating cycles of extreme heating, cooling, and temperatures below freezing. The possibility of biological interactions causing injury was also considered in aquatic systems since it is well established that high levels of heterotrophic plate count (HPC) bacteria suppress the detection of indicator bacteria in samples of drinking water (26). Subsequent laboratory studies revealed that when pseudomonads exceeded coliforms by a factor of 10^5, greater than 50% of the *E. coli* became injured and was reversibly unculturable (45). These reports indicate that a number of stressors that are ambient in many environments at sublethal levels can elicit injury and the reduced culturability of allochthonous bacteria.

OCCURRENCE OF INJURED BACTERIA IN AQUATIC SYSTEMS

Numerous reports of higher bacterial counts in waterborne communities when determined by most probable number (MPN) methods compared with approaches that relied on colony formation provided evidence of injured bacteria within aquatic systems (8, 27, 52, 53, 71, 84). Other early reports acknowledged injury as a reversible form of bacterial inactivation as a consequence of sublethal conditions of disinfection in water (34, 68). Later reports have shown that some of the difference was due to the positive statistical bias of the MPN calculation (55). The subsequent development and commercialization of a medium, designated m-T7, to specifically detect injured coliforms (43, 47), in addition to the inclusion of a discussion of this and other related methods in *Standard Methods for the Examination of Water and Wastewater*, Section 9212, (1), made available the techniques to test water for injured indicator bacteria. More recently, the performance of the commercially available enzyme detection media (Colisure, and ColiLert) in the enumeration of chlorine-injured coliforms and *E. coli* was evaluated in drinking water (61, 64–66). Naturally occurring coliforms in surface water were detected by these new media with efficiency comparable to that of the reference method prescribed by the U.S. Environmental Protection Agency (lauryl tryptose broth in an MPN configuration) when the bacteria were exposed to chlorine and diluted in drinking water. In addition, chlorine-stressed *E. coli* was detected with greater efficiency when using Colisure than with the reference method (64).

A renaissance of interest among aquatic microbiologists in the phenomenon of injury resulting from exposure to disinfectants was seen in the 1970s. Lin (52), Braswell and Hoadley (10) and Hoadley and Cheng (73) demonstrated that *E. coli* exposed to chlorine in sewage became unculturable on m-FC medium and also lost the ability to form gas from lactose, an important phenotype used to discriminate coliforms as indicator bacteria in water and wastewater (1). Those reports also described the progressive loss of culturability in a range of bacteria, with increasing exposure time in water. A survey of over 200 potable water samples in Montana and Massachusetts revealed that greater than 50% of the coliforms present were injured, as reviewed by LeChevallier and McFeters (46). A more detailed study was then conducted to survey disinfected water from several cities in northeastern U.S. using m-T7 and m-Endo media (60). Coliforms found in 71 samples of water from distribution systems and 46 samples of water leaving treatment plants were 97% (mean value) injured. Similarly high levels of injured coliforms were also found in backwash water from filters used in the treatment of drinking water and in samples following the repair of a broken pipe within a municipal distribution system. All 11 of the samples collected 1 week after the repair and rigorous disinfection of that broken pipe were found to be negative for coliforms on m-Endo while all were positive on m-T7 medium. That overall pattern was also seen in the results when a total of 1752 samples were compared from over 16 different locations across the U.S. and the Caribbean (59). As reported elsewhere (43), injury levels were high in most of the chlorinated systems while source and cistern water contained fewer injured coliforms. Again, the fraction of the chlorinated water samples in which coliforms were detected with m-T7 medium but not by m-Endo (the medium widely used for monitoring treated drinking water systems) is striking. A more recent study done in South Africa also compared the relative efficiency of m-Endo LES and m-T7 in the detection of coliform and HPC bacteria from chlorinated potable water (23). As in the previous reports, these workers demonstrated that greater than 50% of the viable coliforms failed to form colonies on m-Endo medium and that between 33 and 56% of the samples revealed 100% injury. They further demonstrated that the proportion of injured coliforms increased at times when disinfection appeared to be suboptimal. A parallel effect was noted in the detection efficiency of HPC bacteria on Plate Count Agar and R2A medium. These findings support the suggestion (15, 67) that populations of HPC bacteria in chlorinated water systems might likewise become injured and less culturable on some media. Therefore, established media used to determine water potability significantly underestimate the level of indicator organisms in disinfected water since injured bacteria become unculturable on selective media. These results underscore the significance of injured planktonic bacteria as a more sensitive indicator in the microbiological assessment of disinfected water and wastewater.

More recent studies have addressed the question of injury within biofilm communities exposed to biocides. Stewart et al. (88) demonstrated that biofilms of *Pseudomonas aeruginosa* and *Klebsiella pneumoniae* exposed to monochloramine retained respiratory and glucose transport activities to a much greater extent than their ability to form colonies on solid media. That same trend was also seen by Yu et al. (103) who showed that the culturability of bacteria within thin biofilms was at least ten times more susceptible to both hypochlorous acid and monochloramine than metabolic functions including respiration, RNA turnover, and transmembrane potential. That report also showed that the specific physiological processes and properties measured each responded somewhat differently to the biocides with time. These findings indicate that, although bacteria within interfacial communities are less susceptible to antimicrobial agents, they also undergo injury and become unculturable at a rate that exceeds their loss of various physiological activities. Watters et al. (96) found that bacteria injured sublethally following exposure to chloramine could repair the oxidative damage in biofilm communities. Presumably the fixed film provided a reduced environment capable of reversing partially oxidized cellular enzymes.

Caldwell and Morita (16) evaluated 600 samples from 10 groundwater-fed public water supplies in western Oregon using seven coliform methods. These methods included the presence-absence method and six membrane filter (MF) techniques: m-Endo LES, anaerobic incubation of m-Endo LES agar, lauryl sulfate (mLS), mTEC, modified HAB agar, and m-T7. These researchers found that m-T7 detected more coliforms than any of the methods evaluated and that coliform densities from the m-T7 test were 1.4 to 2.1 times greater than any of the other MF tests. Because 9 of the 10 systems were not chlorinated, the results illustrate that physiological changes can occur in systems that contain no added biocides.

Antimicrobial compounds are also known to cause injury in foods and food handling applications (14, 71). Typical of the early literature is a report by Schusner et al. (82) demonstrating that quaternary ammonium compounds and hypochlorite sanitizers cause injury in *E. coli*, *Staphylococcus aureus*, and *Streptococcus fae-*

calis. Mossel and Corry (71) reviewed evidence describing a wide range of antimicrobial agents used in the food industry as stressors for vegetative bacteria. More recently, Williams and Russell (97, 98) described sublethal injury in *Bacillus subtilis* spores following exposure to a range of biocides. Buazzi and Marth (11) also reported that exposure of *Listeria monocytogenes* to 8.5% sodium benzoate for 1 h resulted in over 99% injury. These reports are representative of a literature which supports the well-established concept within the food industry that exposure of bacteria to biocidal and antimicrobial compounds results in injury and the reversible loss of culturability when using selective media to enumerate bacteria in treated products.

Previous studies describing the injury of waterborne bacteria by antimicrobial compounds have largely focused on the indicator organisms. Although this information has been used to explain some persistent microbiological problems like those associated with unexplained occurrence of excessive levels of indicator bacteria in drinking water (discussed later), the vital consideration of injury in waterborne bacterial pathogens deserves to be addressed. The first question that needs to be resolved concerns the relative susceptibility of the indicator bacteria and pathogens to injury since the former are supposed to indicate health threats from the latter. Secondly, an important focus of interest in studies of pathogens is their ability to initiate infections and cause pathology in the infected host following environmental injury. In experiments done in our laboratory under controlled conditions, pathogens including *Yersinia enterocolitica*, *Salmonella enterica* serovar Typhimurium, and *Shigella* spp. were less susceptible to chlorine injury than the nonpathogenic coliforms and an enterotoxigenic *E. coli* (ETEC) strain (46). In addition, the virulence of these stressed populations was measured by the observation of 50% lethality levels when the bacteria were intraperitoneally injected into mice. The results of this study indicated that the virulence of the chlorine-stressed population was either virtually eliminated or required longer incubation for infection while viability was retained. In the case of *Y. enterocolitica*, the loss of virulence associated with chlorine-induced injury resulted from the acquired inability to invade while *S. enterica* serovar Typhimurium and ETEC became unable to attach to appropriate cell cultures. A parallel experiment with copper-induced stress yielded similar results. Walsh and Bissonnette (94) also demonstrated that sublethal concentrations of chlorine caused damage to the surface adhesins of ETEC and reduced their ability to attach to human leukocytes in an in vitro assay. However, the death and injury of some enteropathogenic bacteria like *C. jejuni* (9) are seen at lower concentrations of chlorine, and Terzieva and McFeters (89) demonstrated that sublethal chlorine exposure results in the reversible loss of virulence-related properties including attachment to and invasion of cultured HeLa cells. These are representative of findings suggesting that stressors such as chlorine affect different determinants involved in the virulence phenotype. Another set of experiments was done to follow the in vivo revival, growth, and pathogenicity of ETEC following chlorine-induced injury (85). Following injection into ligated ileal loops in mice, injured ETEC suspensions recovered within 4 h, with no increase in numbers. The same results were observed in vitro when chlorine-injured bacteria were incubated in saline containing homogenate of mouse intestinal mucosa but not in its absence.

The enterotoxigenicity of injured ETEC cells was also determined, by injection into ligated ileal loops in rabbits, to be undiminished. These and similar results from others (94) suggest that chlorine-injured ETEC can recover both in vitro and within the gut and that its toxigenic potential is retained. Singh and McFeters (86) demonstrated that when chlorine-injured ETEC were exposed to the additional stress of normal gastric acidity in orogastrically inoculated mice, virulence remained undiminished in the gut. Therefore, injury resulting from exposure to chlorine at levels approximating those found in drinking water reversibly reduces the enteropathogenic potential of the pathogens examined, and their virulence is recoverable within the mammalian gut following ingestion and passage through the stomach. The reversible loss of virulence in the pathogens might be considered analogous to the reversible loss of culturability in the indicator bacteria following sublethal exposure to chlorine. In addition, these results are similar to the viable but nonculturable state when a range of pathogens were exposed to simulated aquatic and marine environmental systems (80, 101). These findings provide additional evidence that allochthonous enteropathogenic bacteria retain both their viability and pathogenicity within various stressful aquatic environments.

The relationship between injured bacteria and those in the viable but nonculturable state has recently been addressed in a study by Smith et al. (87). Bacteria including ETEC, *Salmonella* serovar Typhimurium and *Y. enterocolitica* were exposed to pristine Antarctic seawater at the ambient temperature (−1.8°C) and tested for physiological activity as well as viability by colony formation and the direct viable count method. The optimal temperature for colony formation of the test bacteria became significantly decreased with exposure time. In addition, injured as well as viable but nonculturable subpopulations developed with environmental exposure. Therefore the bacterial phenomena of injury and viable but nonculturable can be concurrent processes in some environments and may be physiologically interrelated.

CELLULAR CONSEQUENCES OF INJURY

As already noted, the hallmark characteristic associated with injury is the reversible loss of culturability with exposure to sublethal levels of antimicrobials agents. However, reports in the literature also describe the cellular and physiological consequences of injury. Among various physiological responses to different stressors, functions associated with the membrane are frequently described as the focus of cellular damage (71), although the global cellular response to sublethal disinfectant exposure is likely to be the cumulative effect of a number of interrelated events (81).

The mechanism of chlorine damage has been investigated by a number of groups because of the importance of this agent in the disinfection of water and wastewater. Exposure to high concentrations of chlorine can affect a number of cellular alterations including disruption of protein synthesis (4), reactions with nucleic acids (83) and chromosomal aberrations (40). However, most workers studying the sublethal consequences of exposure to the low levels of chlorine common in drinking water systems agree that membrane-related functions are affected (3). Subsequent physiological studies were done to examine the hypothesis that membrane-related functions, such as respiration and nutrient transport, were compromised by chlorine (17). An immediate drop was observed in cellular ATP content following chlorine-induced injury while aldolase activity, a representative intracellular enzyme, remained unaffected. In addition, exposure to low concentrations of chlorine immediately inhibited the uptake of both radiolabeled glucose and amino acids as well as decreased oxygen use by 30 to 50% by the bacteria in contrast to the viable but unculturable state where uptake does not cease (79). These findings, which confirmed the original hypothesis regarding the action of chlorine on cytoplasmic membrane-related functions, are supported by the work of others (5, 30, 92). Therefore, physiological activities that are dependent upon membrane functions including respiration and nutrient transport are impaired in the chlorine-mediated injury process.

Studies to examine the recovery of disinfectant-damaged bacteria have also provided insights concerning the cellular consequences of the resulting injury. The development of a medium (m-T7) specifically for the improved recovery of coliform bacteria following exposure to free chlorine in clean water systems such as drinking water (43), was based on the earlier observation that many selective ingredients used in media for the detection of gram-negative bacteria are uniquely inhibitory for injured cells (63). Such findings support the hypothesis, already discussed from a physiological perspective, that sublethal exposure to chlorine causes detectable abnormalities in the bacterial membrane. Results obtained by Zaske et al. (104) provide direct confirmation of that premise (Fig. 3). More recently, Calabrese and Bissonnette (15) demonstrated that intracellular catalase activity is compromised with chlorine exposure and the culturability of chlorine-stressed coliform and HPC bacteria from wastewater was improved by the addition of enzymatically active catalase and pyruvate to a range of media. Those results also support the involvement of the cytoplasmic membrane in chlorine-induced stress since such injured cells become progressively sensitive to oxidants produced by aerobic respiration such as peroxide (15). That study further identified the insensitivity of m-T7 medium to recover coliform bacteria exposed to combined forms of chlorine, such as monochloramine, confirming the earlier observations of Watters et al. (96). Therefore, the approaches taken in the formulation of recovery media that have proved effective in the detection of chlorine-stressed coliforms support the involvement of the cell membrane in the injury process and indicate that different stressors, such as free and combined chlorine, elicit somewhat different physiological consequences.

For chloraminated bacteria, Watters et al. (96) found that addition of 0.1% sodium sulfite improved the recovery of cells with all media tested. This result was consistent with the theory that oxidation of sulfhydryl groups to disulfides by monochloramine is reversible. In contrast, the researchers found that sodium sulfite did not have a substantial effect when cells were stressed with free chlorine. The difference in culturability between free chlorine- and monochloramine-injured bacteria underscores the need to select the appropriate recovery medium commensurate to the type and physiological effect of the stressor.

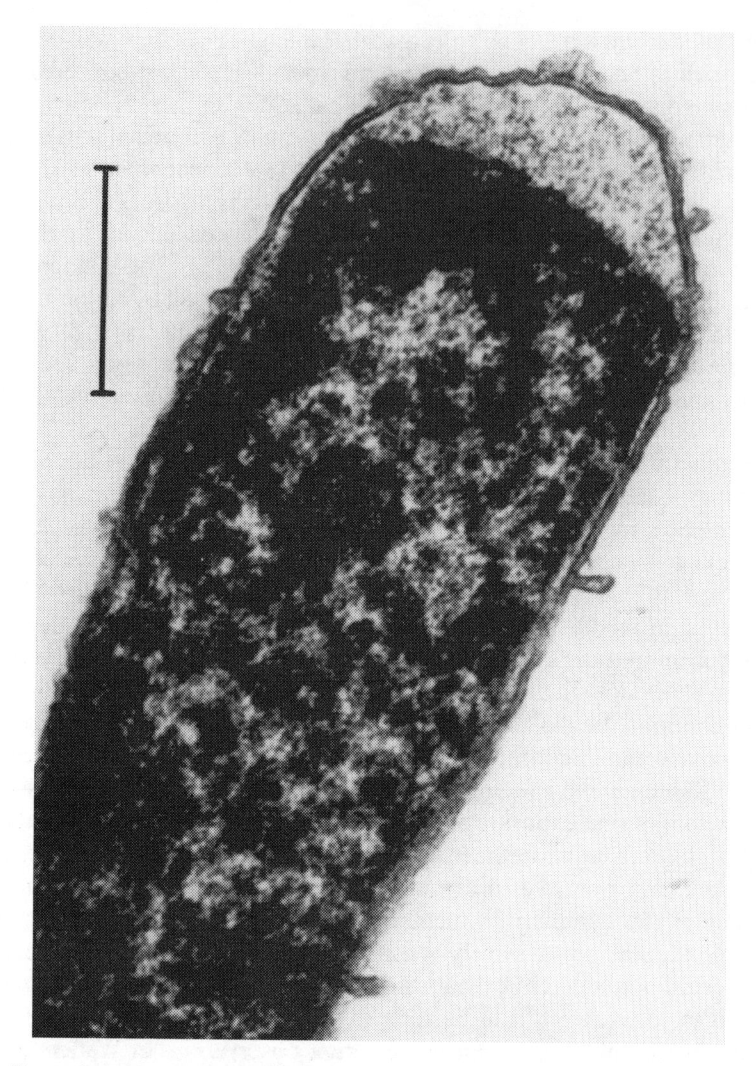

Figure 3. Electron micrograph of *E. coli* injured in reagent grade water. Blebs in the envelope were not seen in control cells. Bar, 0.4 μm. (Reproduced from reference 104 with permission.)

The emergence of knowledge concerning adaptive responses in bacteria following environmental stress and starvation (91) within the past decade raises the intriguing possibility that allochthonous bacteria can mount a similar reaction in response to sublethal disinfection. The activation of these sets of genes, termed regulons or stimulons, by a range of stressors results in the synthesis of a number of unique proteins not observed in actively growing, unstressed bacteria. Although little is known concerning the physiological or metabolic role of many of these proteins in such systems, it is clear that in some, like the oxidative stress response (20), a protective adaptation results. In another example, Leyer and Johnson (51) described the cross-protective effect of acid adaptation in serovar Typhimurium to other environmental stressors including heat, surface-active agents, and salt by the synthesis of unique proteins in the outer membrane. It is tempting to speculate that a similar protective adaptation occurs when gram-negative bacteria are exposed to chlorine since Doakes and Herson have shown that a unique set of proteins are synthesized when *Enterobacter cloacae* is injured by low levels of chlorine (D. A. Doakes and D. S. Herson, Abstr. 93rd Gen. Meet. Am Soc. Microbiol. 1993, p. 308).

A genetic basis for injury was suggested by LeChevallier et al. (48), who showed that genetically constructed strains (*recA* mutants) of *E. coli* had an increased sensitivity to deoxycholate. The results were consistent with the findings of Tessman and Peterson (90) who showed that many *recA* strains of *E. coli* were more sensitive to selective media and several drugs (crystal violet, kanamycin, mitomycin C, and chloramphenicol). Yoshida et al. (102) demonstrated that integration of an Hfr plasmid into the chromosome of *E. coli* made the strain more sensitive to sodium cholate. It is speculated that these chromosomal alterations can interfere with the integrity of the cell membrane.

Sublethal exposure to disinfectants like chlorine can influence virulence-related processes and phenotypes when pathogens become injured. As discussed previously in this chapter, such properties in a variety of enteric pathogens including ETEC, *Y. enterocolitica*, *Salmonella* serovar Typhimurium, *Shigella* spp., and *C. jejuni* are lost following chlorine injury. However, these processes and properties are regained upon recovery under suitable conditions including the succession of events accompanying ingestion into the mammalian digestive system in the waterborne route of infection. Therefore, the pathogenic genotype and potential of these bacteria persist following sublethal chlorination in water.

SIGNIFICANCE OF INJURED BACTERIA

The importance of detecting injured indicator bacteria in water can be justified from a number of perspectives. For example, data available from operating potable water systems indicate that between 50 and 90% of the viable coliforms may be injured and hence not detected using accepted media and methods. However, detection of the entire population of viable indicator bacteria, including injured cells, provides an increased margin of analytical sensitivity and safety in the early detection of emerging microbiological problems within a system. This kind of information allows the initiation of remedial action at an earlier stage and before legally mandated limits are exceeded that trigger notification of both regulatory agencies and the public that the water might constitute a hazard. Therefore, the use of more sensitive analytical methods capable of detecting injured indicator bacteria can help detect the source of problems within potable water systems and indicate when an appropriate solution has been employed.

For example, McFeters et al. (60) found that a high percentage (96.5%) of bacteria in treated drinking water were not detected using conventional Endo-type media because of injury. In this case, the occurrence of coliform bacteria in the distribution system could be explained by high levels (average 9.5 CFU/100 ml) of injured organisms passing through the treatment plant. Bucklin et al. (12) found coliform levels in filtered postbackwash water using m-T7 agar were as much as 151 times higher than those measured by m-Endo LES medium. Injured coliform bacteria (11 CFU/100 ml using m-T7 agar) were found after the replacement of a distribution system pipe that was disinfected for 24 h with 200 mg of free chlorine per liter and flushed before being placed back in service (60). The investigators also found coliforms using m-T7 agar in 11 samples, averaging 68 CFU/100 ml, 1 week after the rupture and repair of a distribution pipe. Significantly, all 11 samples were coliform negative by m-Endo LES. In all of the above examples, contamination of the distribution system—whether by passage through the treatment process, by main breaks, or by unsanitary installation of new pipes—could be detected using sensitive techniques designed to recover injured coliforms. This greater analytical sensitivity could also be useful in epidemiological investigations of waterborne disease outbreaks. Specifically, it is possible that outbreaks of waterborne morbidity, where coliforms are either undetected or detected in low numbers, might be explained by bacterial injury and/or viable but unculturable cells. Recognizing the source of the contamination permits the implementation of remedial measures.

Because conventional monitoring techniques are insensitive and do not detect the majority of viable cells present in a water supply, the water utility operator often does not know whether the remedial actions were effective. In the example of Hopewell, Virginia (Fig. 4), the occurrence of coliform bacteria enumerated by m-Endo LES was sporadic. Increased free chlorine residuals (between 2 to 3 mg/liter) in response to an episode of coliform bacteria in March, 1990, apparently resolved the problem. However, coliforms reappeared in June and again in November of 1991. Although no coliform bacteria were detected in December of 1991 or January through February of 1992, monitoring of the system using m-T7 agar found coliform bacteria in 10 to 30% of the samples (Fig. 4). In nearly 50% of the months during 1992 and the first half of 1993, coliforms were not detected using m-Endo LES agar (approximately 60 distribution system samples were collected each month), while injured coliform bacteria were routinely found using m-T7 agar. Because research suggested that chloramines may be more effective for controlling biofilm bacteria (49, 50), Hopewell converted the distribution system disinfectant to monochloramine (2 to 2.5 mg/l residual) during the first week of July, 1993. No coliform bacteria were detected in the distribution system using either m-T7 or m-Endo LES for the 7 months following the conversion to chloramines. The dramatic decrease in coliform levels measured by m-T7 agar (supplemented with 0.1% sodium sulfite) provides positive proof that the remedial efforts were completely successful.

Bacterial injury has additional significance in potable water because of the adoption of the presence-absence (PA) concept into the Federal drinking water regulations. This new regulatory approach is based on the frequency of coliform detection (presence) within systems, using any of several approved analytical approaches. This is different from the former practice of using estimates of coliform concentrations to determine if a water system has exceeded a maximal allowable limit of indicator bacteria. Caldwell and Morita (16) found that the distribution of coliform

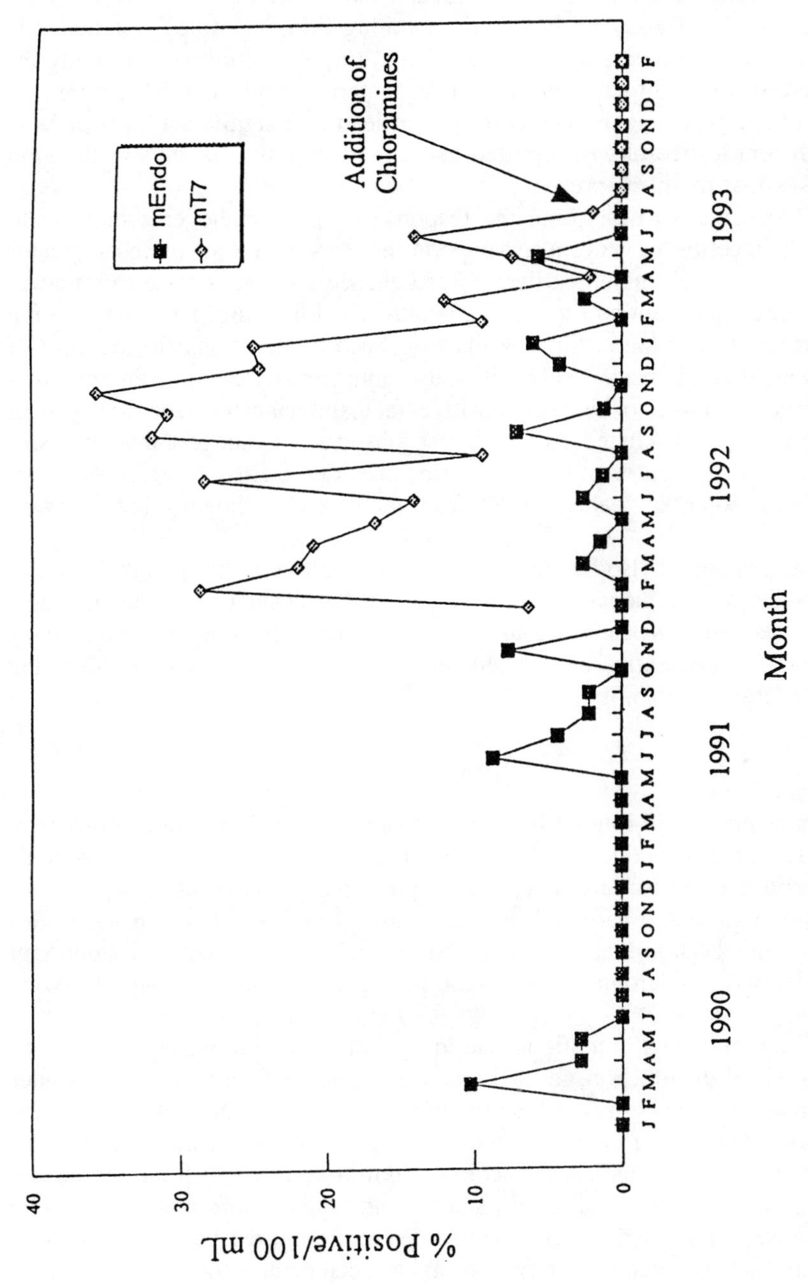

Figure 4. Incidence of coliform-positive samples within the Hopewell, Virginia, municipal drinking water system, 1990–1993. Samples were analyzed for coliform bacteria using both m-Endo LES and m-T7 media by membrane filtration.

results obtained from five of the six MF media tested fit the negative binomial model. However, the mean and the coefficient of aggregation parameters derived from the m-T7 data were almost twofold greater than those derived from the standard m-Endo results. Based on data from operating drinking water systems, it is clear that the accepted media and methods failed to detect coliforms in many instances where disinfectant-stressed but viable bacteria are present. Therefore, the application of the presence-absence concept can lead to a significant level of false negative coliform test results as reported earlier (59, 60) and the underestimation of health risks that might be present.

Published information describing the response of enteropathogenic bacteria to suboptimal disinfection in water also supports the importance of injured indicator bacteria in water. As discussed earlier, injured enteric pathogens regain virulence-related properties and continue to pose a health risk following ingestion into the mammalian gut. That relationship has likewise been demonstrated in the case of viable but nonculturable bacteria (18). It is also noteworthy that some enteric pathogens have been shown to be less sensitive to disinfectant-induced injury than coliform bacteria, as discussed earlier. These observations suggest that in disinfected water the detection of injured coliforms provides a greater degree of sensitivity to indicate potential health hazards from the enteric pathogens that are more resistant.

Injured bacteria might also be viewed as a manifestation of the sublethal cellular effects of biocides. This concept can be useful in the evaluation of disinfectants since it provides a more sensitive biological index when evaluating biocide efficacy within bacterial communities than when relying on the loss of culturability, the traditional criterion of viability.

SUMMARY

Some characteristics of injured bacteria are shared with those populations considered to be viable but nonculturable. Therefore, injured and viable but nonculturable bacteria may be partially overlapping phenomena and functionally related. The information presented here and in other chapters in this volume strongly supports the view that both injured and viable but nonculturable bacteria are important when considering the microbial ecology and public health microbiology of a wide range of aquatic systems.

Ambient environmental conditions within operating potable water systems, including suboptimal disinfection, result in the occurrence of injured bacteria. Under some circumstances, these stressed subpopulations, which are not culturable when using most of the accepted methods utilized in water quality monitoring, can represent the majority of the indicator bacteria within a system. Such an occurrence suggests that more sensitive and representative water quality information would be obtained if media or strategies were used to detect the injured yet viable subpopulation of indicator bacteria. Such information has been demonstrated to be of value in the detection of operational and mechanical problems within water treatment and distribution systems. Hence, the detection of injured bacteria affords an added measure of accuracy to assist in the early recognition of treatment deficiencies or contamination within operating water systems. Continuing efforts are needed to improve the detection of injured and nonculturable bacteria. More importantly, the adoption of methods to enhance the detection of injured and viable but nonculturable indicator bacteria within the drinking water industry as well as the appropriate regulatory agencies would help ensure the safety of potable water supplies in the U.S. and worldwide. The persistent incidence of waterborne morbidity in the U.S. (19, 36, 42) underscores the need for more sensitive and accurate indicators of potential microbiological health hazards in drinking water.

Acknowledgments. We are grateful to Cheryl Norton for collecting the data from Hopewell, Virginia, and to Barry H. Pyle for prepublication review of this manuscript as well as the many colleagues and former students who collaborated in the work reported here. We also acknowledge current support from the National Aeronautics and Space Administration, the National Science Foundation/Office of Polar Programs and the Center for Biofilm Engineering at Montana State University, an NSF-sponsored engineering center (cooperative agreement EEC-8907039).

REFERENCES

1. **APHA, AWWA, and WEF.** 1992. Stressed organisms. *In* pp. 924–926. A. E. Greenberg (ed.), *Standard Methods for the Examination of Water and Wastewater.* American Public Health Association, Washington, D.C.
2. **Bailey, C. A., R. A. Neihoff, and P. S. Tabor.** 1983. Inhibitory effect of solar radiation on amino acid uptake in Chesapeake Bay bacteria. *Appl. Environ. Microbiol.* **46:**44–49.
3. **Beuchat, L. R.** 1978. Injury and the repair of gram-negative bacteria with special consideration of the involvement of the cytoplasmic membrane. *Adv. Appl. Microbiol.* **23:**219–244.
4. **Benarde, M., W. B. Snow, V. P. Oliveri, and B. Davidson.** 1967. Kinetics and mechanism of bacterial disinfection by chlorine dioxide. *Appl. Microbiol.* **15:**257–265.
5. **Berg, J. D., P. V. Roberts, and A. Matin.** 1986. Effect of chlorine dioxide on selected membrane functions of *E. coli. J. Appl. Bacteriol.* **60:**213–220.
6. **Bissonnette, G. K.** 1974. Recovery characteristics of bacteria injured in the natural aquatic environment. PhD dissertation. Montana State University, Bozeman.
7. **Bissonnette, G. K., J. J. Jezeski, G. A. McFeters, and D. G. Stuart.** 1975. Influence of environmental stress on enumeration of indicator bacteria from natural waters. *Appl. Microbiol.* **29:**186–194.
8. **Bissonnette, G. K., J. J. Jezeski, G. A. McFeters, and D. G. Stuart.** 1977. Evaluation of recovery methods to detect coliforms from water. *Appl. Environ. Microbiol.* **33:**590–595.
9. **Blaser, M. J., P. F. Smith, W.-L. L. Wang, and J. C. Hoff.** 1986. Inactivation of *C. jejuni* by chlorine and monochloramine. *Appl. Environ. Microbiol.* **51:**307–311.
10. **Braswell, J. R., and A. W. Hoadley.** 1974. Recovery of *E. coli* from chlorinated sewage. *Appl. Microbiol.* **28:**328–329.
11. **Buazzi, M. M., and E. H. Marth.** 1992. Characteristics of sodium benzoate injury of *L. monocytogenes. Microbios* **70:**199–207.
12. **Bucklin, K. E., G. A. McFeters, and A. Amirtharaja.** 1991. Penetration of coliforms through municipal drinking water filters. *Water Res.* **25:**1013–1017.
13. **Busta, F. F.** 1976. Practical implications of injured microorganisms in food. *J. Milk Food Technol.* **39:**138–145.
14. **Busta, F. F.** 1978. Introduction to injury, p. 195–201. *In* D. Perlman (ed.), *Adv. Appl. Microbiol.* **23:**195–201. Academic Press, New York, N.Y.
15. **Calabrese, J. P., and G. K. Bissonnette.** 1990. Improved membrane filtration method incorporating catalase and sodium pyruvate for detection of chlorine-stressed coliform bacteria. *Appl. Environ. Microbiol.* **56:**3558–3564.

16. **Caldwell, B. A., and R. Y. Morita.** 1987. *Sampling Regimes and Bacteriological Tests for Coliform Detection in Groundwater.* EPA/60/2-87/083. U.S. Environmental Protection Agency, Cincinnati, Oh.
17. **Camper, A. K., and G. A. McFeters.** 1979. Chlorine injury and the enumeration of waterborne coliform bacteria. *Appl. Environ. Microbiol.* **37:**633–641.
18. **Colwell, R. R., P. R. Brayton, D. J. Grimes, and D. B. Rozak.** 1985. Viable but non-recoverable *V. cholerae* and related pathogens in the environment: implications for release of genetically engineered microorganisms. *Bio/Technology* **3:**817–820.
19. **Craun, G. F.** 1986. *Waterborne Diseases in the United States.* CRC Press, Boca Raton, Fla.
20. **Demple, B.** 1991. Regulation of bacterial oxidative stress genes. *Annu. Rev. Genet.* **25:**315–337.
21. **Domek, M. J., M. W. LeChevallier, S. C. Cameron, and G. A. McFeters.** 1984. Evidence for the role of copper in the injury process of coliforms in drinking water. *Appl. Environ. Microbiol.* **48:** 289–293.
22. **Double, M. A., and G. K. Bissonnette.** 1980. Enumeration of coliforms from streams containing acid mine water. *J. Water Pollut. Control Fed.* **52:**1947–1952.
23. **duPreez, M., and R. Kfir.** The extent of bacterial injury after chlorination. *Aqua*, in press.
24. **Finch, G. R., M. E. Stiles, and D. W. Smith.** 1987. Recovery of a marker strain of *E. coli* from ozonated water by membrane filtration. *Appl. Environ. Microbiol.* **53:**2894–2896.
25. **Fujioka, R. S., and O. T. Narikawa.** 1981. Effect of sunlight on enumeration of indicator bacteria under field conditions. *Appl. Environ. Microbiol.* **41:**690–696.
26. **Geldreich, E. E., M. D. Nash, D. J. Reasoner, and R. H. Taylor.** 1978 The necessity of controlling bacterial populations in potable waters: community water supply. *J. Am. Water Works Assoc.* **64:** 596–602.
27. **Geldreich, E. E., M. J. Allen, and R. H. Taylor.** 1978. Interferences to coliform detection in potable water supplies, p. 13–20. *In* C. W. Hendricks, (ed.), *Evaluation of the Microbiology Standards for Drinking Water.* (EPA-570/9-78-00C), USEPA, Washington, D.C.
28. **Goshko, M. A., W. O. Pipes, and R. R. Christian.** 1983. Coliform occurrence and chlorine residual in small distribution systems. *J. Am. Water Works Assoc.* **74:**372–378.
29. **Grahm, P. J., and G. R. Brenniman.** 1983. Enumeration of chlorine-damaged fecal coliforms in wastewater effluents. *J. Water Pollut. Control Fed.* **55:**164–169.
30. **Haas, C. W., and R. S. Englebrecht.** 1980. Physiological alterations of vegetative microorganisms resulting from chlorination. *J. Water Pollut. Control Fed.* **52:**1976–1989.
31. **Hackney, C. R., and G. K. Bissonnette.** 1978. Recovery of indicator bacteria in acid mine streams. *J. Water Pollut. Control Fed.* **50:**775–780.
32. **Harris, N. D.** 1963. The influence of recovery medium and the incubation temperature on the survival of damaged bacteria. *J. Appl. Bacteriol.* **26:**387–397.
33. **Harris, N. D., J. P. Richards, and M. Whitefield.** 1961. On the relationship between the viability of bacteria damaged by antiseptics and their responsiveness to modifications in the counting medium. *J. Appl. Bacteriol.* **24:**182–187.
34. **Heinmets, F., W. W. Taylor, and J. J. Lehman.** 1954. The use of metabolites in restoration of viability of heat and chemically inactivated *E. coli. J. Bacteriol.* **67:**5–12.
35. **Hershey, A. D.** 1939. Factors limiting bacterial growth: properties of *E. coli* surviving sublethal temperatures. *J. Bacteriol.* **38:**563–579.
36. **Herwaldt, B. L., G. F. Craun, S. L. Stokes, and D. D. Juranek.** 1992. Outbreaks of waterborne disease in the United States: 1989–1990. *J. Am. Water Works Assoc.* **84(April):**129–135.
37. **Hoadley, A. W., and C. M. Cheng.** 1974. Recovery of indicator bacteria on selective media. *J. Appl. Bacteriol.* **37:**45–57.
38. **Hurst, A.** 1977. Bacterial injury; a review. *Can. J. Microbiol.* **23:**936–944.
39. **Hurst, A.** 1984. Revival of vegetative bacteria after sublethal heating, p. 77–104. *In* M. H. E. Andrew and A. D. Russell (ed.), *The Revival of Injured Microbes.* Academic Press, London, United Kingdom.
40. **Ingols, R. S.** 1958. The effect of monochloramine and chromate on bacterial chromosomes. *Public Health Works* **89:**105–106.
41. **Kapuscinski, R. B., and R. Mitchell.** 1981. Solar radiation induces sublethal injury in *E. coli* in seawater. *Appl. Environ. Microbiol.* **41:**670–674.

42. **Kramer, M. H., B. L. Herwaldt, G. F. Craun, R. L. Calderon, and D. D. Juranek.** 1996. Waterborne disease: 1993 and 1994. *J. Am. Water Works Assoc.* **88:**66–80.
43. **LeChevallier, M. W., S. C. Cameron, and G. A. McFeters.** 1983. New medium for the recovery of coliform bacteria from drinking water. *Appl. Environ. Microbiol.* **45:**484–492.
44. **LeChevallier, M. W., A. Singh, D. A. Schiemann, and G. A. McFeters.** 1985. Changes in virulence of waterborne enteropathogens with chlorine injury. *Appl. Environ. Microbiol.* **50:**412–419.
45. **LeChevallier, M. W., and G. A. McFeters.** 1984. Recent advances in coliform methodology. *J. Environ. Health* **47:**5–9.
46. **LeChevallier, M. W., and G. A. McFeters.** 1985. Enumerating injured coliforms in drinking water. *J. Am. Water Works Assoc.* **77:**81–87.
47. **LeChevallier, M. W., P. E. Jakanoski, A. K. Camper, and G. A. McFeters.** 1984. Evaluation of m-T7 agar as a fecal coliform medium. *Appl. Environ. Microbiol.* **48:**371–375.
48. **LeChevallier, M. W., A. K. Camper, S. C. Broadaway, J. M. Henson, and G. A. McFeters.** 1987. Sensitivity of genetically engineered organisms to selective media. *Appl. Environ. Microbiol.* **53:**606–609.
49. **LeChevallier, M. W., C. D. Lowry, and R. G. Lee.** 1990. Disinfecting biofilms in a model distribution system. *J. Am. Water Works Assoc.* **82(7):**87–99.
50. **LeChevallier, M. W., C. D. Lowry, R. G. Lee, and D. L. Gibbon.** 1993. Examining the relationship between iron corrosion and the disinfection of biofilm bacteria. *J. Am. Water Works Assoc.* **85(7):** 111–123.
51. **Leyer, G. J., and E. A. Johnson.** 1993. Acid adaptation induces cross-protection against environmental stresses in *Salmonella typhimurium. Appl. Environ. Microbiol.* **59:**1842–1847.
52. **Lin, S.** 1973. Evaluation of coliform tests for chlorinated secondary effluents. *J. Water Pollut. Control Fed.* **45:**498–506.
53. **Lin, S.** 1976. Membrane filter method for recovery of coliforms in chlorinated sewage effluents. *Appl. Environ. Microbiol.* **32:**547–552.
54. **Litsky, W.** 1979. Gut critters are stressed in the environment, more stressed by isolation procedures, p. 345–347. *In* R. R. Colwell and J. Foster (ed.), *Aquatic Microbial Ecology.* A Maryland Sea Grant Publication, College Park, Md.
55. **Loyer, M. W., and M. A. Hamilton.** 1984. Interval estimation of the density of organisms using a serial-dilution experiment. *Biometrics* **40:**907–916.
56. **Mackey, B. M.** 1984. Lethal and sublethal effects of refrigeration and freeze-drying on microorganisms, p. 46–76. *In* M. H. E. Andrew and A. D. Russell (ed.), *Revival of Injured Microbes.* Academic Press, London, United Kingdom.
57. **MacLeod, R. A., S. C. Kuo, and R. Gelinas.** 1967. Metabolic injury to bacteria. II. Metabolic injury induced in distilled water or Cu in plating diluent. *J. Bacteriol.* **93:**961–969.
58. **McFeters, G. A.** 1989. Detection and significance of injured indicator and pathogenic bacteria in water, p. 179–210. *In* B. Ray (ed.), *Injured Index and Pathogenic Bacteria: Occurrence and Detection in Foods, Water and Feeds.* CRC Press, Boca Raton, Fla.
59. **McFeters, G. A.** 1990. Enumeration, occurrence and significance of injured indicator bacteria in drinking water, p. 478–492. *In* G. A. McFeters (ed.), *Drinking Water Microbiology: Progress and Recent Developments.* Springer-Verlag, New York, N.Y.
60. **McFeters, G. A., J. S. Kippin, and M. W. LeChevallier.** 1986. Injured coliforms in drinking water. *Appl. Environ. Microbiol.* **51:**1–5.
61. **McFeters, G. A., and D. G. Stuart.** 1972. Survival of coliform bacteria in natural waters; field and laboratory studies with membrane filter chambers. *Appl. Microbiol.* **24:**805–811.
62. **McFeters, G. A., and A. K. Camper.** 1983. Enumeration of indicator bacteria exposed to chlorine. *Adv. Appl. Microbiol.* **29:**177–193.
63. **McFeters, G. A., Cameron, S. C., and M. W. LeChevallier.** 1982. Influence of diluents, media and membrane filters on the detection of injured coliform bacteria. *Appl. Environ. Microbiol.* **43:** 97–103.
64. **McFeters, G. A., S. C. Brodaway, B. H. Pyle, M. Pickett, and Y. Egozy.** 1997. Comparative performance of Colisure. *J. Am. Water Works Assoc.* **89**(9)**:**112–120.
65. **McFeters, G. A., B. H. Pyle, S. J. Gillis, D. Ferrazza, and C. J. Acomb.** 1991. Effect of chlorine on the comparative performace of Colisure™, ColiLert™ and ColiQuick™ for the enumeration of

coliform bacteria and *E. coli* in drinking water, p. 493–495. *Technology Conference Proceedings*, American Water Works Association, Denver, Colo.
66. **McFeters, G. A., B. H. Pyle, S. J. Gillis, C. J. Acomb, and D. Ferrazza.** 1993. Chlorine injury and the comparative performace of Colisure™, ColiLert™ and ColiQuick™ for the enumeration of coliform bacteria and *E. coli* in drinking water. *Water Sci. Tech.* **27:**261–265.
67. **Means, E. A., L. Hanami, H. F. Ridgway, and B. H. Olson.** 1981. Evaluating mediums and plating techniques for enumerating bacteria in water distribution systems. *J. Am. Water Works Assoc.* **73(11):** 585–590.
68. **Milbauer, R., and N. Grossowicz.** 1959. Reactivation of chlorine-inactivated *E. coli*. *Appl. Microbiol.* **7:**67–70.
69. **Moss, S. K., and K. C. Smith.** 1981. Membrane damage can be a significant factor in the inactivation of *Escherichia coli* by near-ultraviolet radiation. *Photochem. Photobiol.* **33:**203–210.
70. **Mossel, D. A., and P. van Netten.** 1984. Harmful effects of selective media on stressed microorganisms: nature and remedies, p. 329–371. *In* M. H. E. Andrew and A. D. Russell (ed.), *The Revival of Injured Microbes.* Academic Press, London, United Kingdom.
71. **Mossel, D. A., and J. E. L. Corry.** 1977. Detection and enumeration of sublethally injured pathogenic and index bacteria in foods and water processed for safety. *Alimentia* **16**(Special Issue)**:**19–34.
72. **Mudge, C. S., and F. R. Smith.** 1935. Relation of action of chlorine to bacterial death. *Am. J. Public Health* **25:**442–447.
73. **O'Connor, J. T., L. Hash, and A. B. Edwards.** 1975. Deterioration of water quality in distribution systems. *J. Am. Water Works Assoc.* **67:**113–116.
74. **Ray, B.** 1979. Current methods to detect stressed microorganisms. *J. Food Protect.* **42:**346–355.
75. **Ray, B.** 1989. Enumeration of injured indicator bacteria from foods, p. 9–54. *In* B. Ray (ed.), *Injured Index and Pathogenic Bacteria: Occurrence and Detection in Foods, Water and Feeds.* CRC Press, Boca Raton, Fla.
76. **Ray, B.** 1989. Introduction, p. 1–8. *In* B. Ray (ed.), *Injured Index and Pathogenic Bacteria: Occurrence and Detection in Foods, Water and Feeds.* CRC Press, Boca Raton, Fla.
77. **Ray, B.** 1993. Sublethal injury, bacteriocins and food microbiology. *ASM News* **59:**285–291.
78. **Roth, L. A., and D. Leenan.** 1971. Acid injury of *E. coli*. *Can. J. Microbiol.* **17:**1005–1008.
79. **Roszak, D. B., and R. R. Colwell.** 1987. Survival strategies of bacteria in the natural environment. *Microbiol. Rev.* **51:**365–379.
80. **Roszak, D. B., D. J. Grimes, and R. R. Colwell.** 1984. Viable but nonrecoverable stage of *S. enteritidis* in aquatic systems. *Can. J. Microbiol.* **30:**334–338.
81. **Russell, A. D.** 1984. Potential sites of damage in microorganisms exposed to chemical or physical agents, p. 1–18. *In* M. H. E. Andrew and A. D. Russell (ed.), *The Revival of Injured Microbes.* Academic Press, London, United Kingdom.
82. **Schusner, D. L., F. F. Busta, and M. L. Speck.** 1971. Inhibition of injured *E. coli* by several selective agents. *Appl. Microbiol.* **21:**41–45.
83. **Shih, K. L., and J. Lederberg.** 1976. Effects of chloramine on *B. subtilis* deoxyribonucleic acid. *J. Bacteriol.* **125:**934–945.
84. **Shipe, E. L., and G. M. Cameron.** 1952. A comparison of the membrane filter with most probable number method for coliform determination from several waters. *Appl. Microbiol.* **2:**85–87.
85. **Singh, A., R. Yeager, and G. A. McFeters.** 1986. Assessment of in vivo revival, growth and pathogenicity of *E. coli* strains after copper and chlorine injury. *Appl. Environ. Microbiol.* **52:**832–837.
86. **Singh, A., and G. A. McFeters.** 1987. Survival and virulence of copper- and chlorine-stressed *Y. enterocolitica* in experimentally infected mice. *Appl. Environ. Microbiol.* **53:**1768–1774.
87. **Smith, J. J., J. P. Howington, and G. A. McFeters.** 1994. Survival, physiological response and recovery of enteric bacteria exposed to a polar marine environment. *Appl. Environ. Microbiol.* **60:** 2977–2984.
88. **Stewart, R. S., T. Griebe, R. Srinivasan, C.-I. Chen, F. P. Yu, D. deBeer, and G. A. McFeters.** 1994. Comparison of respiratory activity and culturability during monochloramine disinfection of binary population biofilms. *Appl. Environ. Microbiol.* **60:**1690–1692.

89. **Terzieva, S. I., and G. A. McFeters.** 1992. Effect of chlorine on some virulence-related properties of *C. jejuni. Int. J. Environ. Health Res.* **2:**24–32.
90. **Tessman, E. S., and P. Peterson.** 1985. Plaque color method for rapid isolation of novel *recA* mutants of *E. coli* K-12: new classes of protease-constitutive *recA* mutants. *J. Bacteriol.* **163:**677–687.
91. **VanBogelen, R. A., and F. C. Neidhardt.** 1990. Global systems approach to bacterial physiology: protein responders to stress and starvation. *FEMS Microb. Ecol.* **74:**121–128.
92. **Venkobacchar, C., L. Lyengar, and A. V. S. P. Rao.** 1977. Mechanism of disinfection: effect of chlorine on cell membrane functions. *Water Res.* **11:**727–729.
93. **Verville, K. M., and D. S. Herson.** 1989. The effect of free chlorine on *E. coli* populations. *Curr. Microbiol. 18:*235–241.
94. **Walsh, S. M., and G. K. Bissonnette.** 1983. Chlorine-induced damage to surface adhesins during sublethal injury to enterotoxigenic *E. coli. Appl. Environ. Microbiol.* **45:**1060–1065.
95. **Walsh, S. M., and G. K. Bissonnette.** 1987. Effect of chlorine injury on heat-labile enterotoxin production in enterotoxigenic *E. coli. Can. J. Microbiol.* **33:**1091–1096.
96. **Watters, S. K., B. H. Pyle, M. W. LeChevallier, and G. A. McFeters.** 1989. Enumeration of *E. cloacae* after chloramine exposure. *Appl. Environ. Microbiol.* **55:**3226–3228.
97. **Williams, N. D., and A. D. Russell.** 1992. Increased susceptibility of injured spores of *B. subtilis* to cationic and other stressing agents. *Lett. Appl. Microbiol.* **15:**253–255.
98. **Williams, N. D., and A. D. Russell.** 1992. The nature and site of biocide-induced sublethal injury in *B. subtilis* spores. *FEMS Microbiol. Lett.* **99:**277–280.
99. **Wortman, A. T., and G. K. Bissonnette.** 1985. Injury and repair of *E. coli* damaged by exposure to acid mine water. *Water Res.* **19:**1291–1297.
100. **Wright, J. H.** 1917. The importance of uniform culture media in the bacteriological examination of disinfectants. *J. Bacteriol.* **2:**315–346.
101. **Xu, H., N. Roberts, S. L. Singleton, R. W. Attwell, D. J. Grimes, and R. R. Colwell.** 1982. Survival and viability of nonculturable *E. coli* and *V. cholerae* in estuarine and marine environments. *Microb. Ecology* **8:**313–323.
102. **Yoshida, Y., N. Takamatsu, and M. Yoshikawa.** 1978. Preferential inhibitory action of sodium cholate on an *Escherichia coli* strain carrying a plasmid in an integrated state. *J. Bacteriol.* **133:** 406–408.
103. **Yu, F. P., and G. A. McFeters.** 1994. Physiological response of bacteria in biofilms to disinfection. *Appl. Environ. Microbiol.* **60:**2462–2466.
104. **Zaske, S. K., W. S. Dockins, and G. A. McFeters.** 1980. Cell envelope damage in *E. coli* caused by short-term stress in water. *Appl. Environ. Microbiol.* **40:**386–390.

Nonculturable Microorganisms in the Environment
Edited by R. R. Colwell and D. J. Grimes

Chapter 16

The Public Health Significance of Viable but Nonculturable Bacteria

James D. Oliver

A bacterium in the viable but nonculturable (VBNC) state is defined here as a cell which fails to grow on the routine bacteriological media on which it would normally grow and develop into a colony, but which is in fact alive and capable of metabolic activity. The term "nonculturable" seems to be a misnomer as, under the proper conditions, it appears that these cells are able to "resuscitate" to the metabolically active and culturable state (this point, along with a brief discussion on why cells enter this state of dormancy, is presented later in this chapter). In this review, the discussion of cells entering the VBNC state is limited to those cells which respond to a natural environmental stress (e.g., a temperature downshift) in such a manner. Thus, this review does not include a discussion of the detrimental effects of such agents as antibiotics, chlorine, heavy metals, or other chemicals to which cells may be exposed and which may result in cell injury or death (this area is reviewed in chapter 15). Similarly, this review does not describe cells that are most correctly termed "nonculturable," e.g., those animal symbionts which have never been cultured in the laboratory. Such cells are described in chapter 5. Instead, this chapter is limited to a discussion of human bacterial pathogens which are known to enter the VBNC state. Finally, I have for the most part selected to review only those studies which have employed such methods as the "direct viable count" originally described by Kogure et al. (38), *p*-iodonitrotetrazolium violet (INT) reduction (86), or 5-cyano-2,3-ditolyl tetrazolium chloride (CTC) hydrolysis (64) to demonstrate viability in cells no longer culturable. These methods have been described in an earlier review on the VBNC state (49), as well as in other chapters of this monograph. Despite these restrictions, bacterial cells from at least 16 different genera, mostly but not exclusively gram negative, comprising over 30 different species, have now been reported to enter the VBNC state (50).

James D. Oliver • Interdisciplinary Biotechnology Program, Department of Biology, University of North Carolina at Charlotte, 9201 University City Blvd., Charlotte, NC 28223.

HUMAN PATHOGENS ENTERING INTO THE VBNC STATE

Vibrio cholerae

Vibrio cholerae is an age-old human pathogen which is the cause of epidemic or Asiatic cholera. According to the Centers for Disease Control and Prevention, (CDC), the most recent pandemic (which began in Peru) has resulted in over 300,000 cholera cases and over 2,300 cholera-related deaths in the Western hemisphere in 1992. Transmission of cholera is typically through water, a result of poor sanitation practices or natural disasters (e.g., floods). The bacterium is able to penetrate the intestinal mucosa, resulting in colonization of the epithelial surface of the gut. Subsequent production of the cholera toxin leads to massive loss (up to 10 to 15 liters/day) of fluid and electrolytes.

While dormancy of a variety of bacteria has long been recognized (see Roszak and Colwell [66] for an early review), the first publication to present clear experimental evidence of a viable but nonculturable state was that of Xu et al. (83). These investigators were able to show that both *V. cholerae* and *Escherichia coli*, following incubation in artificial seawater (ASW), remained viable although they lost all ability to produce colonies on media routinely employed for their culture. The clinical isolate of *V. cholerae* O1 examined (strain ATCC 14035) was found to enter the VBNC state rapidly (6 to 9 days) in Chesapeake Bay water (11‰ salinity) at 4 to 6°C.

Following that study, it was realized that the VBNC state exhibited by this pathogen might well explain the seasonality and distribution of this organism in regions of the world where cholera is endemic. Subsequently, Colwell and coworkers (12, 31a, 84) showed that the organism was present in many waters from which it could not be cultured.

Later studies from Colwell's laboratory provided evidence that cells *of V. cholerae*, present in the VBNC state, may be capable of causing human cholera. Colwell et al. (20) inoculated cells of *V. cholerae* strain CA401, present in the VBNC state following incubation in Patuxent River water at 25°C, into rabbit ligated ileal loops. Positive virulence responses (distension, hemorrhage, and fluid accumulation) were observed in all injected loops, with fluid from the loops found to contain *V. cholerae* O1. In a later study (21), cells of an oral attenuated A−B+ strain (CVD 101) of classical *V. cholerae* were incubated at 4°C in phosphate-buffered saline microcosms. After becoming nonculturable, portions of the microcosm (containing ca. 10^8 cells/ml by direct count, and ca. 5×10^6 cells/ml by direct viable count [DVC]) were fed to two human volunteers. Approximately 48 h after challenge, one of the volunteers passed culturable *V. cholerae* O1 cells in his stool at a concentration of 2.9×10^3/g. Approximately 5 days postchallenge, the second volunteer also had cells of *V. cholerae* O1 in his stool. At no time did incubation of the cells in a nutrient medium reveal the presence of culturable cells present among the nonculturable population. A repeat of this study, employing cells which had been in the VBNC state for 1 month, did not yield viable *V. cholerae* O1 cells in the volunteers' stools. Nevertheless, this study supports the authors' hypothesis that cells of this bacterium, present in the VBNC state, are capable of resuscitating to the culturable, infectious state following human passage. Interestingly, Hasan et al. (J. A. K.

Hasan, abstr. D-138, Abstr. Annu. Meet. Am. Soc. Microbiol., 1992) reported PCR amplification of the cholera enterotoxin (CT) operon from VBNC cells and detected the cholera toxin gene in diarrheal stool samples from a patient with distinct clinical cholera symptoms but who was culture negative for both *V. cholerae* O1 and non-O1, as well as enterotoxigenic *E. coli*.

Vibrio vulnificus

Seafood is consumed in the United States at a rate of over 15 lb per person per year, a total of 3.7 billion lb per year. Whereas seafood is generally of excellent quality, a large variety of bacteria, viruses, and toxins may be present in seafood. Of the several bacterial infections which may result from consumption of raw shellfish, those caused by *V. vulnificus* account for 95% of all deaths in this country. The bacterium is a normal inhabitant of marine waters and occurs naturally in high numbers in molluscan shellfish. Fatality rates of over 60% are typical in those persons developing infection after consumption of raw or undercooked shellfish, especially oysters. The high mortality rate associated with these cases is further accentuated by the speed at which the infection develops. Patients dying within 2 to 24 h of hospital admission have been reported in several studies (40a, 48, 52).

More is known about the VBNC state of *V. vulnificus* than any other microorganism, and several reviews of the VBNC response in this pathogen have been published (49, 50). Linder and Oliver (40) showed that temperature is the factor that induces this state in *V. vulnificus*, which enters the VBNC state when incubated in artificial seawater at 5°C but not at room temperature. Cells incubated at 5°C enter the nonculturable state whether in nutrient-depleted artificial seawater (40, 47) or in heart infusion broth (53). Thus, nutrient level appears not to be a major factor in the nonculturable response of this bacterium. Wolf and Oliver (82) subsequently characterized in more detail the role of temperature as the inducer of the nonculturable state for this bacterium. Stationary-phase cells of *V. vulnificus* entered the VBNC state within 40 days at 5°C, whereas cells at 10, 15, 20, and 30°C remained culturable (at least 10^4 CFU/ml) throughout the 40-day study period. As significant variations had been observed regarding the time required for this bacterium to enter the VBNC state, Oliver et al. (54) examined eight factors, including incubation temperature, which might influence the nonculturable response of *V. vulnificus*. These included incubation temperature and physiological age of the inoculum, presence or absence of a nitrogen and phosphate source in the diluent, initial cell density, initial and subsequent temperature of the plating medium, and salt content of the plating medium. Of these parameters, only the physiological age of the inoculum was found to have a significant effect on the time required for the cells to enter the nonculturable state. Cells taken from the stationary phase generally required about twice as many days to become nonculturable at 5°C than did logarithmic-phase cells. This observation offered an explanation for the large differences in the time required to enter the VBNC state reported for some bacteria. Subsequent studies (Oliver, unpublished data) indicated that washing also reduces the time required for *V. vulnificus* to enter the VBNC state, apparently due to the removal of nutrients which are present when cells are not washed.

Morton et al. examined macromolecular synthesis in cells *of V. vulnificus* grown at 22°C in glucose-minimal medium and then transferred to 5°C (Morton et al., abstr. Q-260, Abstr. Annu. Meet. Am. Soc. Microbiol., 1992). At intervals pre- and posttransfer, the cells were monitored for DNA, RNA, and protein synthesis. Morton et al. observed a decrease in synthesis of all macromolecules within 15 min of the temperature downshift. These data suggest that rapid signal transduction occurs in *V. vulnificus* in response to a temperature downshift, since the cells had not reached 5°C until almost 3 h after the microcosm was transferred to that temperature. Such a response may represent a protective mechanism designed to prevent the cells from continued metabolism when nutrients are no longer available or their transport is no longer possible.

Subsequently, McGovern and Oliver (42) examined the induction of cold-responsive proteins in *V. vulnificus* following a temperature downshift from 23 to 13°C. They reported that while an immediate and sharp decline in protein synthesis occurred, some 40 proteins were induced by the temperature decrease. Such "cold shock" proteins are undoubtedly required for cell survival at such low temperatures and may be involved in the VBNC response (discussed further later in this chapter).

There is preliminary evidence that modifications in DNA may also be occurring as cells of *V. vulnificus* enter the VBNC state. Studies from this lab (77a) employing PCR technology have revealed marked changes in randomly amplified polymorphic DNA (RAPD) PCR banding patterns, as well as loss of banding, as cells are nutrient starved or exposed to low temperature. These studies are discussed in "The VBNC State in the Natural Environment," below.

The dramatic change in cell morphology exhibited by bacteria entering the nonculturable state suggests that significant cell envelope modifications must occur. Oliver (49) reported that the addition of ampicillin to *V. vulnificus* cells incubated at 5°C resulted in an extremely rapid loss of culturability. These data suggest that the nonculturable response requires active peptidoglycan synthesis, at least in its initial stages.

In one of only two studies published on changes in the membrane lipid composition occurring in cells entering the VBNC state, Linder and Oliver (40) found the major fatty acid species (C16, C16:1, C18) of nonculturable cells of *V. vulnificus* to be decreased almost 60% compared to culturable cells. The percentage of fatty acids with chain lengths less than C16 increased as the cells became nonculturable, and long-chain acids appeared which were not evident in culturable cells. While no fluidity measurements were made in this study, such changes presumably allow the cells to maintain proper membrane fluidity as they enter the nonculturable state.

Initial studies (40) examining the effect of nonculturability on virulence of *V. vulnificus* in mice suggested that VBNC cells lost virulence. However, a relatively low inoculum of *V. vulnificus* (5×10^4 cells based on DVC data) was employed. In a more recent study (51), we found that injections into mice of 0.5 ml of a population of nonculturable (<0.1 CFU/ml) *V. vulnificus* cells resulted in lethality. In our experimental protocol, less than 0.04 CFU of culturable *V. vulnificus* cells were injected, suggesting that in vivo resuscitation of the cells had occurred and that virulence was maintained.

Escherichia coli

Numerous strains of *E. coli* are known to be human pathogens, classified for example as enterotoxigenic, enteroinvasive, enteropathogenic, or enterohemorrhagic. Pathogenic strains are known to cause infections of the urinary bladder, renal pelvis, and kidney, a variety of intestinal (typically diarrheal) infections, and hemorrhagic colitis. That *E. coli* may be present but undetectable thus presents serious public health concerns. Another concern is that *E. coli* cells, along with other coliforms, have long been considered the "gold standard" for indicating the presence of fecal pollution (and therefore possibly other more serious human pathogens) in recreational and potable waters, as well as in milk and virtually all other foodstuffs. The presence of coliforms in the VBNC state, which would not be detected by routine microbiological analysis, would lead to the erroneous conclusion that fecal pollution had not occurred.

Early studies (29, 35) indicated that coliforms (including *E. coli*) rapidly lose culturability when incubated in seawater, and this observation was frequently used as justification to allow dumping of raw sewage into the ocean. It was not until a 1978 study by Dawe and Penrose (23) that this "die-off" was questioned. These authors demonstrated that while colony-forming units of sewage-contaminated seawater rapidly declined, no decline was observed in either ATP biomass or particle size distribution of radioactively labeled coliform particles.

The first laboratory study to provide direct evidence for the viable but nonculturable state in *E. coli* was that of Xu et al. (83) who found that cells "undergo a 'nonrecoverable' stage of existence, but remain viable" following incubation in saltwater microcosms. They speculated that "The usefulness of the coliform and fecal coliform indices for evaluating water quality for public health purposes may be seriously compromised . . ."

These preliminary studies have been verified and extended in numerous studies by a variety of investigators. Grimes and Colwell (31) observed total loss of culturability within 13 hours after placement of *E. coli* H10407 cells into seawater at 25°C, while direct viable counts showed only a minor decrease over a 112-h study. Despite the lack of culturability, these cells retained their virulence plasmid.

That cells are able to retain plasmids as they enter the VBNC state is of special concern, especially as such plasmids are often essential to the virulence of several *E. coli* strains. In addition to the study of the enteropathogenic *E. coli* strain H10407 cited above, Byrd and Colwell (14) found plasmids pBR322 and pUC8, present in *E. coli* JM83 and JM101, respectively, to be maintained after the cells had entered the VBNC state following incubation in sterile ASW at 15°C for 21 days. These same authors subsequently reported (15) that several enteropathogenic strains maintained plasmids after over 3 years of incubation at 15 and 25°C, during which time the cells had demonstrated a long-term starvation-survival response. *E. coli* has also been reported (73) to retain R+ factors following incubation in natural seawater. Not only does the maintenance of plasmids in VBNC or starved cells provide additional cause for concern, but such plasmids may be a major factor in determining how a bacterial strain responds to various environmental temperatures (see below).

A complication sometimes encountered by researchers studying the VBNC state is that different strains of the same species may demonstrate dramatically different responses to the same environmental perturbation. This problem is perhaps best illustrated by those studies involving *E. coli*. For example, Gauthier et al. (28) found the seven strains of *E. coli* they studied to demonstrate significant individual variation in their survival in seawater. Most dramatic may be the response to temperature by this bacterium. While the strains of *E. coli* examined in the studies cited above by Byrd and Colwell (15) remained culturable for over 3 years at 15°C in ASW, the same researchers (14) have reported *E. coli* strains JM83 and JM101 to rapidly (6 and 21 days, respectively) become nonculturable when incubated under the same conditions. Other authors have reported the opposite response of *E. coli* to temperature. Martinez et al. (41) found strain ATCC 10536 to rapidly lose culturability at 20°C in seawater, and Fujioka et al. (26) reported fecal coliforms to decrease over 4 log units within 3 days at 24°C. Similarly, Wait and Sobsey (D. A. Wait and M. D. Sobsey, abstr. Q-94, Abstr. Annu. Meet. Am. Soc. Microbiol., 1980) found the *E. coli* strain they studied to become "inactivated" more rapidly at 28°C (0.5 days) than at 6°C (32 days). In contrast and like those studies reported above by Colwell and coworkers as well as Garcia-Lara et al. (27) and others, we have reported *E. coli* to enter the VBNC state only at low temperature (40, 63). Interestingly, Korhonen and Martikainen (39) monitored the survival of *E. coli* in eutrophic lake water and found little difference in culturability at either 4 or 25°C, regardless of whether the water was raw, filtered, or filtered and autoclaved. Indeed, these authors reported *E. coli* to remain fully culturable for over 40 days at 4°C in filtered (eutrophic) lake water. Such variations in survival may be explained, at least in part, by variations in the "seawater" diluent. Cornax et al. (22) incubated cells of *E. coli* at 18°C in either saline, artificial seawater, untreated natural seawater, or filtered seawater. They found the natural seawater microcosms, especially those which were unfiltered, to result in a much faster loss of culturability than either saline or ASW.

Another factor affecting variations in cellular response may be the presence of plasmid DNA. Oliver et al. (55) observed that entry into the VBNC state may be dramatically affected by the possession of plasmids. Figure 1A indicates that *Pseudomonas fluorescens* 10586, which lacks any plasmids, enters the VBNC state within 4 days at 37°C, while remaining fully culturable at 5°C. In contrast, the same strain, when harboring the plasmid pFAC510, responds in exactly the opposite manner (Fig. 1B). When a spontaneous insertion of the entire plasmid into the chromosome occurred, the resultant strain responded to these temperatures identical to that of the parent strain lacking any plasmid DNA. Thus, plasmid carriage (and copy number?) may have a dramatic effect on culturability and the VBNC state of bacteria.

Other factors, e.g. illumination of the culture, may also be important. Cornax et al. (22) found cells of *E. coli* to become nonplateable (a >6 log unit decrease) within 7 days at 18°C in ASW microcosms which were illuminated with white light. In contrast, cells in microcosms maintained in the dark remained fully culturable (10^6 to 10^7) over the same time period. Barcina and coworkers (1, 4, 5) have studied this phenomenon in detail in *E. coli* and found that the production of

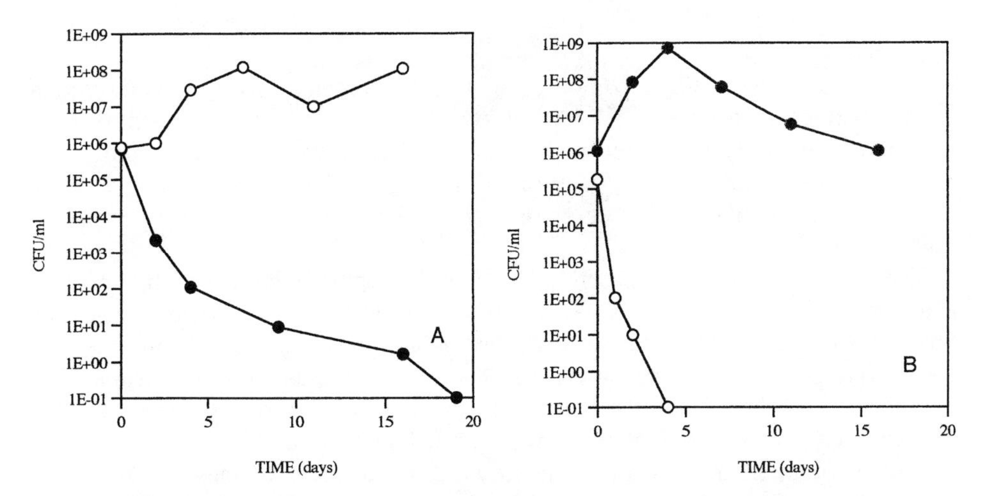

Figure 1. Role of plasmid DNA in entrance of *P. fluorescens* into the VBNC state. Parent strain (*P. fluorescens* 10586) with no plasmid DNA (○) and *P. fluorescens* 10586 pFAC510 (●). Cells were incubated in filtered and autoclaved river water at 5°C (A) or 37°C (B).

photoproducts (e.g., hydrogen peroxide) by cells when illuminated dramatically affects entry into the VBNC state. Indeed, such reactive oxygen species may be critical to the development of the VBNC state, as discussed in the last section of this chapter.

Colwell et al. (20) have reported resuscitation and evidence of continued virulence of *E. coli* when these nonculturable cells were introduced into ligated rabbit ileal loops. Similarly, when cells of *E. coli*, following 4 days of nonculturability in natural seawater, were concentrated by centrifugation and introduced into ligated rabbit ileal loops, Grimes and Colwell (31) reported successful reisolation of the cells after 36 hours. Confirmation of the identity of the cells was accomplished through plasmid characterization. Although they were not directly tested, the authors considered the nonculturable cells to have retained virulence, as the plasmids harbored by this *E. coli* strain encoded colonizing factor antigen I and both the heat-stable and heat-labile toxin subunits of the *E. coli* enterotoxin.

Recently, Pommepuy et al. (60) reported on the retention of pathogenicity of *E. coli* H10407 after sunlight-induced entry into the VBNC state. They found such cells to produce enterotoxin when incubated in rabbit intestinal loops, as indicated by ganglioside-enzyme-linked immunosorbent assay, and thus to be of potential public health concern.

The significance of the findings that virulence may be retained by *E. coli* cells present in the VBNC state is accentuated as new pathogenic strains of this species emerge. A case in point are the O157:H7 (enterohemorrhagic; EHEC) strains, which we have recently shown to enter the VBNC state when subjected to temperature downshifts (63). Our findings may explain the reduced detection of the organism during colder months.

Salmonella

The genus *Salmonella* is composed of two species, and six subspecies which can be distinguished into over 1,500 serotypes. These are a primary agent of food infection (commonly associated with contaminated poultry and poultry products), of which over 40,000 cases were reported in the United States alone in 1992 (17). In addition, some 400 to 500 cases of typhoid fever are reported in this country each year.

The first report on the "die-off" of *Salmonella* spp. may have been that of Wait and Sobsey (D. A. Wait and M. D. Sobsey, abstr. Q-94, Abstr. Annu. Meet. Am. Soc. Microbiol., 1980), who reported *Salmonella enterica* serovar Typhi (then referred to as *S. typhi*) to demonstrate a 90% loss of culturability when incubated in fresh seawater. They found the rate of inactivation to be inversely proportional to the water temperature (with cells decreasing much faster at 28°C than at 6°C). These authors did not conduct any direct viability assays, however, to show entry into the VBNC state.

Roszak et al. (68) were the first to actually demonstrate entry into the VBNC state by a *Salmonella* species, in this case by *Salmonella enteritidis* 13-1BB. The factor inducing this loss of culturability was apparently nutrient limitation, as cells introduced into sterile Potomac River water underwent loss of plateability within 48 h at 25°C. These nonculturable cells could be resuscitated to a level of culturability similar to that observed before nutrient deprivation by the addition of nutrient to the cells maintained in the nonculturable state for as long as 21 days.

In a later study, Roszak and Colwell (67) reported the same strain to become nonculturable within 20 days when incubated in water from the Chesapeake Bay. Evidence that the cells were in the VBNC state was provided not only by direct viable counts but by microautoradiography using radiolabeled glutamic acid or thymidine. This study also reported that the cells, when inoculated into ligated rabbit ileal loops, could be recovered on standard culture media. Further, and as has been shown for *E. coli*, the cells maintained their original plasmid profiles.

Similar to the conflicting studies on the VBNC state of *E. coli*, studies by other researchers have reported significantly different findings with *S. enteritidis*. When Chao et al. incubated this species in either groundwater or river water, no entry into the VBNC state was observed (W. L. Chao et al., abstr. N-49, Abstr. Annu. Meet. Am. Soc. Microbiol., 1987). Instead, they recorded a rapid initial decline, with the surviving population (105) either then declining at a much slower rate, or remaining constant. Chmielewski and Frank (18) reported *S. enteritidis* to exhibit complete loss of culturability after 5 weeks when incubated in potassium phosphate buffer at 7°C but to retain prolonged culturability when starved at 21°C.

Cornax et al. (22) reported the effects of various seawater diluents on the culturability of *Salmonella paratyphi*. At 18°C, these cells exhibited a rapid decline in CFU in both natural and filtered seawater, whereas little decreases were seen in ASW or saline solutions.

Roszak et al. (68) were apparently the first to study resuscitation of a pathogen from the VBNC state. In their in vitro study, they reported that cells of *S. enteritidis*, made nonculturable (<1 CFU/ml) in sterile river water, resuscitated to nearly the original culturable cell density through the addition of nutrients (heart infusion broth) to subsamples taken from the microcosm. A period of 25 h was required following nutrient addition before plateable cells appeared. These results were obtained when cells had been nonculturable 4 days; attempts to resuscitate cells after 21 days of nonculturability were unsuccessful. The authors suggested that longer periods of "dormancy" would require conditions other than simple nutrient addition to restore culturability.

Shigella

Shigella species are strict pathogens that cause bacillary dysentery, and shigellosis is a major health problem throughout the world. The organisms are highly virulent, with fewer than 100 cells able to initiate infection in even healthy persons (70). In some areas it accounts for the highest mortality rate of any of the diarrheal illnesses (36).

Colwell et al. (20) found *S. sonnei* to enter the VBNC state within 2 to 3 weeks when incubated in Chesapeake Bay water (salinity of 15‰) at 25°C. *S. flexneri* was reported to become nonculturable at a rate three times faster than *S. sonnei*.

Islam et al. (33) examined a type 1 strain of *S. dysenteriae*, inoculated into 25°C microcosms composed of autoclaved pond, lake, river, or drain water taken in Bangladesh. Using fluorescent antibody techniques, these authors observed this strain to enter the VBNC state within 2 to 3 weeks. These nonculturable cells could still be detected after 6 weeks by PCR. Believing that nonculturability is a part of the normal life cycle of a bacterium and a strategy for survival, the authors concluded that the VBNC state may be important in the epidemiology of shigellosis. Rahman et al. (61) studied several virulence factors of *S. dysenteriae* type 1 after induction into the VBNC state. They found that this virulent strain retained the *stx* shiga toxin gene, maintained biologically active Shiga toxin ShT, and remained able to adhere to intestinal epithelial cells. However, the ability of these cells when in the VBNC state to invade intestinal cells could not be demonstrated. Nevertheless, the authors concluded that VBNC cells of type 1 *S. dysenteriae* remained potentially virulent and posed a public health problem.

Campylobacter

Campylobacter spp. are important pathogens for a variety of animals, including several of agricultural concern. These include ovine and bovine animals, as well as cats, dogs, and birds. In addition, campylobacteriosis caused by *C. jejuni* is now recognized as one of the most common enteric diseases of humans, with rates of isolation exceeding those of *Salmonella* spp. (24, 72). Major sources of human infection are poultry, with raw milk and water also implicated (7).

Rollins and Colwell (65) examined the entry of *C. jejuni* strain HC of biotype 2 (originally isolated from a human campylobacteriosis patient) into the VBNC state when incubated in filter-sterilized stream water at various temperatures. It was observed that cells held in water at low (4°C) temperature maintained viability for over 4 months, whereas cells incubated at 25 or 37°C demonstrated a relatively rapid decline in culturability, entering the VBNC state within 28 and 10 days, respectively. The same conclusion was reached by Korhonen and Martikainen (39) who studied a biotype 1 strain of *C. jejuni* incubated at these same temperatures in lake water which was untreated, filtered, autoclaved, or supplemented with 0.9% NaCl. Interestingly, Rollins and Colwell (65) found that shaking the microcosms resulted in an immediate decline in culturability in 3 days, about 3 to 4 times faster than static cultures. This observation brings into question the clinical practice of 37°C shaking-incubation of this species.

Cells of *C. jejuni* have long been known to produce coccal forms, generally in conditions unfavorable to growth (62). In a comparative study, Moran and Upton (45) concluded that the coccoid forms were ". . a degenerate cell form which is undergoing cellular degradation," and which were nonviable. In their study, Rollins and Colwell (65) observed cells of *C. jejuni* to undergo significant morphological changes, from the typical spiral morphology to a predominance of coccoid forms, as they entered the VBNC state. While these cells maintained an intact cytoplasmic membrane, a condensed cytosol was observed, an observation similar to that reported for a variety of other bacteria entering the VBNC state (3, 20, 49). Although no data were shown, the authors reported that animal passage indicated that these VBNC cells retained viability.

Following up on the observations of Rollins and Colwell, Beumer et al. (7) confirmed, through ATP measurements and direct viable counting, that cells of *C. jejuni* develop into coccoid cells as they enter into the VBNC state. However, these investigators did not observe resuscitation of coccoid cells developed from bovine, porcine, avian, or human strains following their introduction into simulated gastric, ileal, or colon environments. Further, oral administration to laboratory animals and volunteers caused no typical symptoms of campylobacteriosis, nor could the organism be demonstrated in stool samples.

Medema et al. (43) employed seven strains of *C. jejuni* taken from human, chicken, and water sources to study the VBNC state. They incubated cells in filter-sterilized and pasteurized surface water at either 15 or 25°C, with monitoring by plate counts, acridine orange counts, and direct viable counts. They observed a rapid (2 log units/day) decline for all strains when incubated at 25°C, with total loss of plateability by 3 days. In contrast, survival at 15°C was extended to over 30 days. At each temperature, however, direct viable counts closely paralleled plate counts, and six of the seven strains failed to indicate a prolonged VBNC state. Only one strain (CB258) from a chicken showed a "typical" VBNC response, with DVC values remaining high throughout a 13-day study. The ability of this strain to colonize day-old chick intestines or to cause death following the oral administration of cells at a level ca. 700 to 7,000 times that of the LD_{50} of freshly cultured cells was also tested. The authors were unable to isolate culturable cells from the ceca of the chicks after 7 days of incubation, nor did passage through the allantoic fluid of embryonated eggs yield culturable cells. In that the infectious oral dose for cultured cells in this model was stated to be as low as 26 CFU, these authors questioned the significance of VBNC cells in the transmission of campylobacters. They further suggested that the microcosms employed by Rollins and Colwell (65) in their infectivity studies may have included the presence of a few culturable cells in the population. Such criticism has generally followed all studies which have attempted to prove the virulence of cells in the VBNC state and remains one of the major concerns of this area of research.

As noted above, Medema et al. (43) found only one of the seven strains they examined to enter the VBNC state. Similarly, in one of the most comprehensive studies performed to date on the physiological traits of any bacterium entering the VBNC state, Tholozan et al. (76) found only 3 of the 36 strains of *C. jejuni* they examined entered the VBNC state. Those strains became nonculturable in approximately 15 days when incubated at 4°C in surface water. These authors recorded an increase in cell volume of VBNC cells, with significantly lower potassium content and membrane potentials compared to cells in the culturable state. All these traits were suggested to be strategies to minimize cell maintenance requirements.

Campylobacter coli is less frequently isolated from the environment. Jacob et al. (34) questioned whether this was due to a faster die-off of this species under environmental conditions or whether it more readily enters into the VBNC state. They studied *C. coli* CK205 isolated from swine feces, employing an incubation in sterile distilled water at either 4 or 37°C. They reported transition to the VBNC state within 48 h at 37°C, whereas (as has been reported for *C. jejuni*) *C. coli* required 14 days at 4°C. Like Rollins and Colwell (65), they found a conversion to coccoid forms, some with a slightly enlarged periplasmic space or membrane budding. In their study, which is one of the very few to report on biochemical changes occurring in any bacterium on entry into the VBNC state, Jacob et al. (34) found no changes in whole cell protein or lipooligosaccharide patterns as the cells became nonculturable. They concluded that coccoid forms may be considered dormant forms that would not be detected in water by conventional microbiological methods.

Legionella

Legionella pneumophila was first isolated following a widely publicized outbreak of pneumonia in Philadelphia in 1976. Members of the genus *Legionella* are now realized to be an important cause of pneumonia in humans, typically accounting for over 1,000 reported cases each year (17). As might be expected for bacteria which have an aquatic habitat (they can be isolated from freshwater lakes and ponds), and which are now routinely isolated from man-made aquatic habitats (e.g., air conditioning cooling towers and showerheads), these organisms appear to have also developed an aquatic survival strategy.

Paszko-Kolva et al. (57) monitored the culturability of an environmental isolate of *L. pneumophila* at 20°C in filtered and unfiltered samples of seawater, river water, and tap water, using plate counts on buffered charcoal yeast extract agar. At the same time, viability was determined through [^{3}H]thymidine labeling (employed in place of "direct viable counts" using nalidixic acid, as *Legionella* spp. are resistant to this antibiotic), acridine orange counts, and direct fluorescent microscopy. Their results indicated that this species is capable of surviving in aquatic habitats in a VBNC state. They observed little change in culturability, regardless of water source, during the first 5 days of incubation. After that period, plateability decreased rapidly only for cells incubated in either filtered or unfiltered seawater. However, [^{3}H]-

thymidine uptake studies, as well as acridine orange direct counts, indicated the cells to be in the VBNC state. This is one of the few investigations which has employed natural, unfiltered water samples containing their normal protozoan flora. As a result, the authors were able to determine that grazing likely accounts for some, but not all, of the difference between culturable and nonculturable counts observed in natural waters.

Atlas et al. (2) used PCR-gene probe, fluorescent antibody microscopic, and plate count detection methods to determine the presence of *Legionella* spp. in water samples taken from 28 dental facilities in six U.S. states. In some samples, plate counts failed to detect any *Legionella* spp. even when both PCR and the antibody methods detected >1,000/ml of *Legionella* spp. The PCR-gene probe method detected *Legionella* spp. in 68% of the dental unit water samples, with *L. pneumophila* detected in 8%. Hussong et al. (32) employed VBNC assay techniques to survey environmental samples from sources implicated in an epidemic of Legionnaires' disease. While none of these samples revealed culturable cells of *Legionella* spp., fluorescently labeled antibodies against L. pneumophila revealed numerous antibody-positive cells. Laboratory microcosm studies employing autoclaved tap water indicated that the loss of one log-unit of culturability by *L. pneumophila* strain Philadelphia 1 required 29 days at 4°C, whereas only 13 days were required at 37°C. When bacteria taken from these microcosms were injected into embryonated eggs, greater mortality was observed than could be accounted for by the number of nonculturable cells which were injected. These results indicated to the authors that the nonculturable cells present in the microcosm resuscitated to culturable cells, which were then capable of significant multiplication. In contrast, Yamamoto et al. (85) were unable to demonstrate resuscitation of VBNC cells following 100 days of starvation, and they found rRNA subunit degradation had occurred. However, whenever lack of resuscitation is observed when such studies are being performed, it must be considered whether or not the appropriate conditions for resuscitation are being employed. Recently, Steinert et al. (74) reported that VBNC cells of *L. pneumophila* did resuscitate when cells of *Acanthamoeba castellanii* were added to the dormant bacterial cells. *A. castellanii* is one of several protozoan species which support the intracellular growth of *L. pneumophila* and which are thought to provide the intracellular environment required for growth of this bacterium.

Yamamoto et al. (85) used PCR to detect cells of *L. pneumophila* and reported successful amplification even after 300 days of nonculturability. Bej et al. (6), however, found that this method would detect cells only when viable cells could be detected by culture methods or when a direct viable count method (such as INT reduction) indicated the presence of viable cells. Thus, whether PCR has promise in detecting such cells is not clear (see also "The VBNC State in the Natural Environment," below).

Helicobacter

Helicobacter pylori has been recently implicated as the causative agent of gastritis (13) and peptic ulcers (59), and evidence exists associating this bacterium with stomach cancer. Despite these findings, the mode of transmission has not been elucidated (75). While *H. pylori* has yet to be isolated from animal or environmental reservoirs, a correlation between human infection and drinking water sources has been documented (37).

Evidence that this organism may exist in a VBNC state may have been first presented by West et al. (79, 80). Shahamat et al. (71) confirmed these studies, using autoradiography ([^{3}H]thymidine uptake) to demonstrate the continued metabolic activity of *H. pylori* cells in the VBNC state. These authors employed both autoclaved river water and distilled water as the laboratory microcosms, with incubation at 4, 15, 22, or 37°C. Cells at 4°C were found to undergo a decline, from 10^8 CFU/ml to nonculturability, within 20 to 25 days. During this time, total direct counts remained constant, and the authors stated further that total numbers had not declined for over 2 years. In contrast, cells incubated at higher temperatures rapidly lost culturability, with the time to nonculturability decreasing with increased temperature. Cells at 37°C entered the VBNC state within 24 h, although again the total counts remained constant.

These same authors found that cells of *H. pylori* underwent a morphological change, from curved rod to horseshoe-shaped or coccoid forms, during incubation in water. Such coccoid cells, present in the VBNC state, continued to uptake thymidine, although at a much reduced rate. Recently, Gribbon and Barer (30) observed that 99% of nonculturable cells of *H. pylori* produced by incubation at 4°C retained oxidative activity for at least 250 days. Taken together, these studies suggest that *H. pylori* may persist in water in the VBNC state and thus may be of public health significance.

H. pylori may also be capable of remaining viable within the host while in a VBNC state. Perkins et al. (58) reported that 2 weeks after treatment with antibiotics, six of six cats were negative for *H. pylori* by culture. However, PCR determined that the majority of the cats had gastric fluid which was positive *for H. pylori*. The authors speculated that a failure of the antibiotics to successfully eradicate *H. pylori* might allow subsequent recolonization of the stomach after cessation of therapy. The authors also suggested that the antibiotics may have been the inducer of the VBNC coccoid state of this bacterium.

Other Bacterial Pathogens

Klebsiella pneumoniae is another cause of bacterial pneumonia, typically in persons with depressed immune systems. Clinically, pneumonia cased *by K. pneumoniae* cannot be distinguished from that caused by *Streptococcus pneumoniae*. Byrd et al. (16) found *K. pneumoniae* to rapidly (<1 day) enter the VBNC state when incubated in sterile-filtered drinking (tap) water at 25°C. Despite this precipitous decline in plateability, the cells retained a high level of viability (10^5 to 10^6/ml) for at least 10 days, as indicated by direct viable counts.

Thibault et al. have suggested that the asymptomatic latent form of melioidosis, caused by *Burkholderia pseudomallei* (formerly *Pseudomonas pseudomallei*) might be explained by entry of this bacterium into the VBNC state (F. M. Thibault et al., abstr. D-115, Abstr. Annu. Meet. Am. Soc. Microbiol., 1997, abstr. D-166, p. 228, 1997).

THE VBNC STATE IN THE NATURAL ENVIRONMENT

Despite the increasing number of studies which have reported entry of cells into the VBNC state, very few have actually demonstrated this phenomenon in the natural environment (as opposed to laboratory microcosms) or the resuscitation of cells from the VBNC state in the natural environment. (Note: McFeters and co-workers have extensively studied survival of bacterial cells following their suspension in membrane diffusion chambers into various natural environments. These studies, which are on survival and injury of the cells as opposed to their entry into or resuscitation from the VBNC state, are described in chapter 15.)

Grimes and Colwell (31) suspended cells of *E. coli* in seawater 100 miles off the Bahama coast. The cells were present in a membrane chamber which allowed containment of the cells and hence their response to this environment. They observed total loss of culturability within 13 h after introduction of the cells into the seawater, while the number of cells which reacted with a fluorescently labeled antibody remained constant. Direct viable counts also showed only a minor decrease over the 112-h study. Such data clearly indicate entrance into the VBNC state by these cells in a natural environment.

We (Oliver et al. [56]) have found that cells of *V. vulnificus*, when placed into membrane diffusion chambers in coastal waters of North Carolina, remain culturable when water temperatures are >15°C. In contrast, when the water temperature was below 15°C, cells entered into the VBNC state, as indicated by CFUs and direct viable counts (Fig. 2A). When cells, which were made to enter the VBNC state in the laboratory by incubation in ASW at 5°C, were placed into chambers in warm coastal water (average temperature of 25°C), the cells resuscitated to the original level of culturability (Fig. 2B). This study, which to our knowledge is the first to report resuscitation in a natural environment, indicates that the VBNC state is not merely a laboratory phenomenon and likely accounts for the seasonality observed in many bacterial pathogens (e.g., *V. cholerae* and *V. vulnificus*).

It should be pointed out here that the use of PCR technology, which would appear to be ideally suited for the detection of cells present in the environment in the VBNC state, may in fact not be an effective method for this purpose. We have observed in both *V. vulnificus* and in *E. coli* that, when cells are induced into the VBNC state, changes in and/or loss of amplification patterns occurs. Using PCR to amplify the hemolysin gene of *V. vulnificus*, we (Brauns et al. [11]) found that approximately 500 times more DNA from cells in the VBNC state was required for signal detection than for cells in the culturable state. This was true whether whole-cell lysates or extracted DNA was employed for gene amplification. Subsequently, Coleman and Oliver (19) compared the conditions for amplification of this gene and again found that at least 100 times more cells from the VBNC state as compared to the culturable state were required. In that study, we pointed out the need to optimize PCR conditions using not culturable but nonculturable cells when performing PCR on cells in the VBNC state. In more recent studies, we have

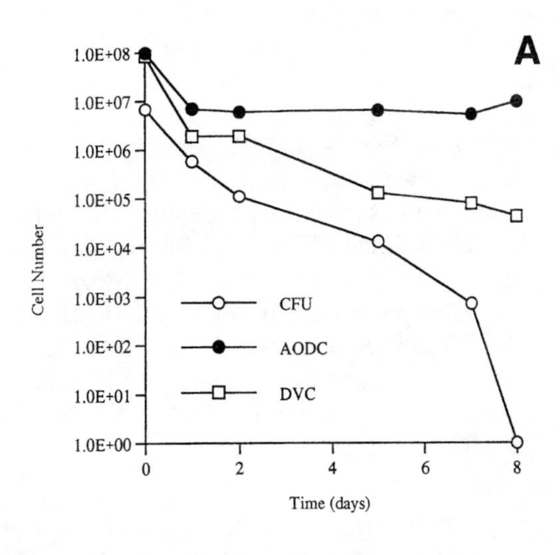

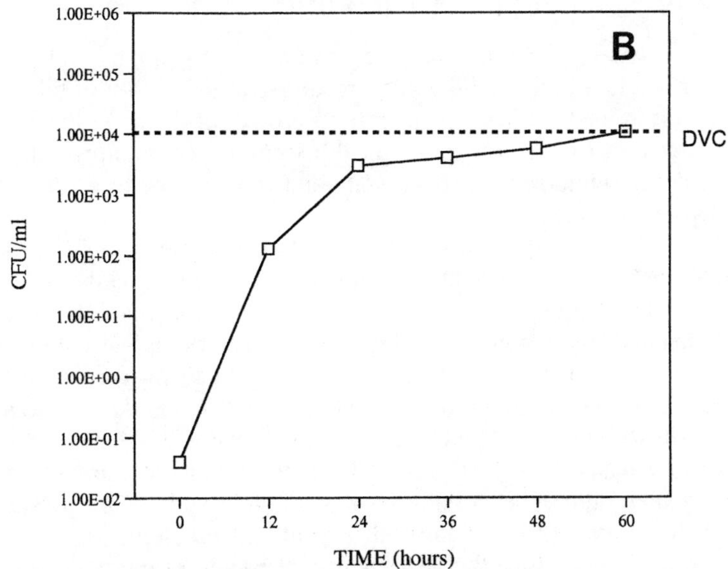

Figure 2. Entrance into and resuscitation from the VBNC state by *V. vulnificus* in a natural estuarine environment. Cells were incubated in membrane diffusion chambers. (A) Culturable cells (strain CVD713t in stationary phase) at ca. 10^8 CFU/ml in the estuarine water at a temperature of 10 to 13°C. Assays were plate counts (○), acridine orange direct counts (●), and direct viable counts (□) by the method of Kogure et al. (38). Cells entered the VBNC state (<1 CFU/ml) within 8 days. (B) Cells in the VBNC state (<1 CFU/ml) were placed in the estuarine water at an average temperature of 25°C. In these studies, full resuscitation to the direct viable count number (dotted line) occurred within 3 to 5 days.

employed the more sensitive randomly amplified polymorphic (RAPD) PCR method to monitor changes in the DNA of cells of *V. vulnificus* as these cells entered the VBNC state (J. M. Warner and J. D. Oliver, abstr. N-173, Abstr. Annu. Meet. Am. Soc. Microbiol., 1997). We observed banding changes (suggestive of genomic rearrangements?) and a gradual loss of amplification, with total loss typically occurring within 7 days. We speculated that these changes might be caused by the production of cold-induced stress proteins which also have a DNA-binding role. These changes could be reversed by subjecting the cells to a temperature upshift, resulting in their resuscitation. The same observation has been reported by Bej et al. (A. K. Bej et al., abstr. Q-177, Abstr. Gen. Meet. Am. Soc. Microbiol., 1997). We observed a similar phenomenon when enterohemorrhagic strains of *E. coli* were employed (S. Mickey et al., abstr. P-78, Abstr. Gen. Meet. Am. Soc. Microbiol., 1997), as have Graves and Bej in a study of biocide-treated *Salmonella enterica* serovar Typhimurium cells (S. Graves and A. K. Bej, abstr. Q-259, Abstr. Annu. Meet. Am. Soc. Microbiol., 1994).

RESUSCITATION, AND PROBLEMS IN STUDYING VIRULENCE OF CELLS IN THE VBNC STATE

Resuscitation has been defined as a reversal of the metabolic and physiologic processes that resulted in nonculturability, resulting in the ability of the cells to be culturable on those media normally supporting growth of the cell (49). For cells of pathogenic bacteria to remain of public health significance requires both that they (i) retain virulence while in the VBNC state and (ii) are able to resuscitate to the metabolically active state.

Oliver et al. (47) have pointed out the potential problem with in vitro resuscitation studies that rely on nutrient addition to nonculturable cells. Whether true resuscitation of the VBNC cells occurs, or whether a few culturable cells which may have remained undetected following normal bacteriological culture methods multiplied as a result of the added nutrient is difficult to prove. As an example, Roszak et al. (68), in examining the resuscitation of *Salmonella enteritidis*, determined their cells to be in the VBNC state when they were at a density of <1 CFU/ml. However, because nutrient was added to 50-ml subsamples for their resuscitation experiments, the possibility of culturable cells being present cannot be excluded. In an attempt to circumvent the nutrient problem, we (Nilsson et al. [47]) reported for the first time the resuscitation of bacteria from ASW microcosms without nutrient addition. In these studies, cells of *V. vulnificus* became nonculturable (<0.1 CFU/ml) after 27 days at 5°C. The determination of such a low level of culturability was accomplished through filtration of 10-ml samples of the microcosm, with the filters being placed on a nonselective (LB agar) medium. When the nonculturable cells were subjected to a temperature upshift (to room temperature), the original bacterial numbers were detectable by plate counts within 3 days. No increase in total cell count was observed during the temperature-induced resuscitation, suggesting that the plate count increases were not due to growth of a few culturable cells. In these studies, the subsamples removed for the resuscitation studies were of a 10-ml volume, which should have contained, on average, <1 culturable cell of *V. vulnificus*. That resuscitation was not a result of a few culturable cells remaining present in the microcosm was further suggested by microscopic examination which revealed no cocci present after 48 hours (suggesting that the cocci which were present prior to the temperature upshift had resuscitated to rods). As we subsequently reported however (78), while no exogenous nutrient had been added to achieve resuscitation, the increases in cell numbers observed by Nilsson et al. (47) could have been due to regrowth of a few culturable cells, employing nutrients released by moribund or dead cells present in the VBNC population.

The "resuscitation versus regrowth" dilemma appears to have been satisfied when Whitesides and Oliver (81) recently reported studies which argue strongly that "true resuscitation," at least in the case of *V. vulnificus*, does occur following removal of the VBNC-inducing stress. Whereas cells in the VBNC state would not form colonies on heart infusion (HI) agar, even when incubated at room temperature in a moist chamber for 30 days, a simple temperature upshift to 22°C for 24 h in the total absence of exogenous nutrient resulted in their resuscitation to culturable levels at or near those at the initiation of the study. Using extensive dilutions prior to a temperature upshift, we observed the appearance of culturable cells taken from ASW microcosms which were diluted so that while 10^3 VBNC cells were present, <0.0001 culturable cell could have remained (Fig. 3). Further, in another study resuscitation from $<3.3 \times 10^1$ to 5×10^4 CFU/ml was observed to occur within 1 h of incubation at 22°C (Fig. 4). This increase in culturability in such a short time frame would have required a generation time of ca. 6 min if culturability had been due to regrowth of undetected culturable cells. Thus, the appearance of cul-

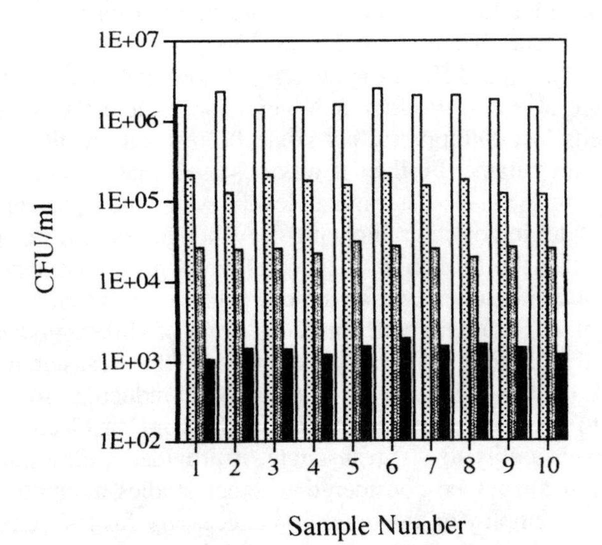

Figure 3. Resuscitation of *V. vulnificus* from the VBNC state (<0.1 CFU/ml) as observed in 10 individual 1-ml aliquots of an undiluted sample (open bars) and of samples diluted 10^{-1} (lightly stippled bars), 10^{-2} (heavily stippled bars), and 10^{-3} (solid bars). Resuscitation was at room temperature for 24 h, with plating onto HI agar. Taken from Whitesides and Oliver (81).

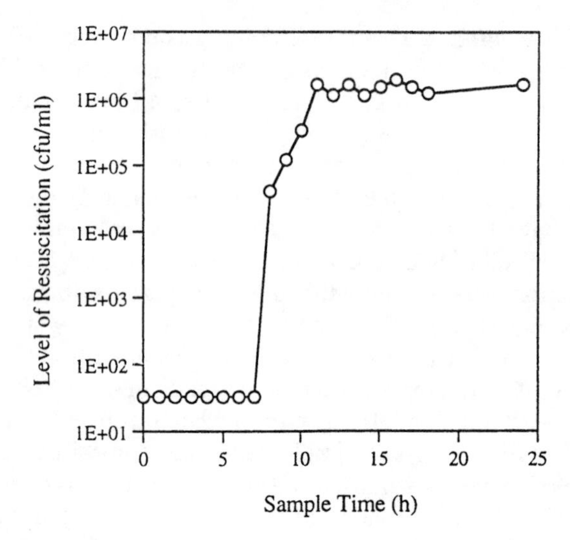

Figure 4. Time required for resuscitation of VBNC *V. vulnificus* cells. Cells from a VBNC microcosm (<3.3 × 10^1 CFU/ml) were shifted to room temperature, and aliquots were removed at hourly intervals and plated onto HI agar. Taken from Whitesides and Oliver (81).

turable cells must have been due to true resuscitation of the VBNC population and not to regrowth of undetected culturable cells. Studies by Mukamolova et al. (46) have also employed dilution studies to demonstrate resuscitation as opposed to regrowth in the case of *Micrococcus luteus* cells.

Roth et al. (69) used a different approach to show that nonculturable cells of *E. coli* were capable of true resuscitation. When exposed to an osmotic stress (0.8 M NaCl), these cells lost culturability but could be resuscitated through additions of the osmoprotectant betaine. Further, it was observed that betaine restored colony-forming ability (to 80% of the original level) even when protein synthesis was inhibited by the addition of chloramphenicol. Although nutrient was present in their system, the fact that the formation of new cells was prevented appears to rule out the possibility of cell proliferation as the cause of resuscitation.

Binnerup et al. (8), studying a denitrifying strain of *P. aeruginosa*, employed an immobilized cell technique to directly examine whether resuscitation or regrowth of cells occurred following removal of the VBNC-inducing stress. This method allowed them to conclude that "individual cells from an electron starved batch culture regain their ability to . . . resuscitate, if provided with a pulse of nitrate."

Finally, it must always be considered whether studies attempting resuscitation of VBNC cells are employing the conditions necessary for resuscitation. A simple reversal of the inducing stress, which is successful in the case of *V. vulnificus*, may not be sufficient for other bacteria. A case in point is *L. pneumophila*, which enters the VBNC state in the laboratory in response to nutrient depletion. Addition of nutrient has generally not been found to allow resuscitation of this pathogen, whereas Steinert et al. (74) recently found the addition of certain amoebae, which are natural hosts to this bacterium, resulted in resuscitation (see section in this chapter on *Legionella*). In the case of *V. cholerae*, Wai et al. (77) found that resuscitation of cells which had been in a low-temperature-induced VBNC state for 86 days could be achieved when the cells were plated following a heat shock of 45°C for 1 min. Why such a heat shock might induce resuscitation is not known, but the authors suggested that heat-shock-induced stress proteins might be involved.

It may also be the case that resuscitation may sometimes occur but not be evident. Ekweozor et al. (25) observed transient increases in culturability when cells of *C. jejuni* were entering the VBNC state. These increases, which occurred as cells were incubated at low temperatures in human fecal emulsions, could not be attributed to sampling errors, cell clumping, or the influence of fluctuations in experimental conditions. The authors concluded that "*C. jejuni* is capable of either growth at low temperatures or to transition temporarily between nonculturable and culturable states."

ARE VBNC CELLS "NONCULTURABLE"?

Might there be conditions under which pathogenic cells in the VBNC state can be directly cultured, without subjecting the cells to a resuscitation step? If so, then the phrase "viable but nonculturable" may not be fully appropriate. Instead of stating that cells are viable but nonculturable, it may be better to describe them as being in a viable but nonculturable state. This would suggest a reversible physiological state into which cells might enter in response to some environmental stress, but from which they can also exit (resuscitate). Although a great deal of information on the VBNC state has appeared in recent years, we do not yet understand why cells enter this state. In fact, nonculturability may be an indirect consequence of entrance into this state. Bloomfield et al. (9) recently proposed that nonculturability may result from the rapid transport of nutrient into a cell which prior to plating was in a state of dormancy. These authors suggested that such an "explosive" transport would result in an imbalance in metabolism and the production of potentially lethal reactive oxygen species (ROS). Thus, "nonculturability" may only be observed when the cells are placed in/on high nutrient media, as a result of the cells' attempts to transport and metabolize organics. This result would be less dramatic in broth media, as any ROS compounds produced would tend to diffuse away from the cell in contrast to cells which are plated, where the toxic ROS would remain close to the cells. Indeed, our results with *V. vulnificus* (81) indicated that the nutrient in heart infusion broth is bacteriostatic, and not bactericidal, to these cells. If this hypothesis is correct, then it should be possible to neutralize these ROS substances as they are produced, thus allowing direct culture of the cells. Indeed, Mizunoe et al. (44) have recently shown that restoration of culturability of *E. coli* cells in the VBNC state is possible through the use of ROS-degrading compounds. The authors were able to culture these cells using media containing catalase, sodium pyruvate, or α-ketoglutaric acid. Similarly, we have successfully resuscitated cells of *V. vulnificus*, present in the VBNC state for over 5 years, through plating on media containing catalase as well as other anti-ROS compounds (M. F. Hite, T. Bates, and J. D. Oliver, unpublished data). Further studies such as these, coupled with a greater understanding of the molecular events which accompany the entrance of cells into this state of dormancy, may ultimately permit our culture of cells which would otherwise remain "nonculturable" as a result of some environmental stress.

CONCLUSIONS

The VBNC state appears to be common to many bacteria, especially those which have aquatic habitats, and may represent a mechanism to survive an adverse environmental factor (e.g., low temperature) or a means of inducing "cross-protection" against other adverse factors (49). Among those bacteria entering this state are many significant human pathogens and "indicator" bacteria of these pathogens. The studies conducted to date on the virulence of such bacteria in the VBNC state strongly indicate that such cells may represent a public health hazard, although epidemiological studies need to be done to establish to what extent such cells are a factor in human health and/or disease.

Acknowledgments. This study was supported in part by grants from the National Institutes of Health (AI31216-01A1), the U.S. Department of Agriculture (91-37201-6877), the National Marine Fisheries Service (NA36FD0271), and the North Carolina Biotechnology Center (9313-ARG-0413).

REFERENCES

1. **Arana, I., A. Muela, J. Iriberri, L. Egea, and I. Barcina.** 1992. Role of hydrogen peroxide in loss of culturability mediated by visible light in *Escherchia coli* in a freshwater ecosystem. *Appl. Environ. Microbiol.* **58:**3903–3907.
2. **Atlas, R. M., J. F. Williams, and M. K. Huntington.** 1995. *Legionella* contamination of dental-unit waters. *Appl. Environ. Microbiol.* **61:**1208–1213.
3. **Baker, R. M., F. L. Singleton, and M. A. Hood.** 1983. Effects of nutrient deprivation on *Vibrio cholerae*. *Appl. Environ. Microbiol.* **46:**930–940.
4. **Barcina, I., J. M. González, J. Iriberri, and L. Egea.** 1989. Effect of visible light on progressive dormancy of *Escherichia coli* cells during the survival process in natural fresh water. *Appl. Environ. Microbiol.* **55:**246–251.
5. **Barcina, I., J. M González, J. Iriberri, and L. Egea.** 1990. Survival strategy of *Escherichia coli* and *Enterococcus faecalis* in illuminated fresh and marine systems. *J. Appl. Bacteriol.* **68:**189–198.
6. **Bej, A. K., M. H. Mahbubani, and R. M. Atlas.** 1991. Detection of viable *Legionella pneumophila* in water by polymerase chain reaction and gene probe methods. *Appl. Environ. Microbiol.* **57:**597–600.
7. **Beumer, R. R., J. de Vries, and F. M. Rombouts.** 1992. *Campylobacter jejuni* non-culturable coccoid cells. *Int. J. Food Microbiol.* **15:**153–163.
8. **Binnerup, S. J., O. Hojberg, and D. Gerlif.** 1995. Resuscitation demonstrated in a mixed batch of culturable and non-culturable *Pseudomonas aeruginosa* PAO303. *Int. Symp. Microb. Ecol.* **7:**1–5.3, 103.
9. **Bloomfield, S. F., G. S A. B. Stewart, C. E. R. Dodd, I. R. Booth, and E. G. M. Power.** 1998. The viable but nonculturable phenomenon explained? *Microbiology* **144:**1–3.
10. **Brauns, L. A., and J. D. Oliver.** 1994. Polymerase chain reaction of whole cell lysates of *Vibrio vulnificus*. *Food Biotechnol.* **8:**1–6.
11. **Brauns, L. A., M. C. Hudson, and J. D. Oliver.** 1991. Use of the polymerase chain reaction in detection of culturable and nonculturable *Vibrio vulnificus* cells. *Appl. Environ. Microbiol.* **57:**2651–2655.

12. **Brayton, P., M. Tamplin, A. Huq, and R. Colwell.** 1987. Enumeration of *Vibrio cholerae* O1 in Bangladesh waters by fluorescent-antibody direct viable count. *Appl. Environ. Microbiol.* **53:**2862–2865.
13. **Buck, G. E.** 1990. *Campylobacter pylori* and gastroduodenal disease. *Clin. Microbiol. Rev.* **3:**1–12.
14. **Byrd, J. J., and R. R. Colwell.** 1990. Maintenance of plasmids pBR322 and pUC8 in nonculturable *Escherichia coli* in the marine environment. *Appl. Environ. Microbiol.* **56:**2104–2107.
15. **Byrd, J. J., and R. R. Colwell.** 1993. Long-term survival and plasmid maintenance of *Escherichia coli* in marine microcosms. *FEMS Microbiol. Lett.* **12:**9–14.
16. **Byrd, J. J., H.-S. Xu, and R. R. Colwell.** 1991. Viable but nonculturable bacteria in drinking water. *Appl. Environ. Microbiol.* **57:**875–878.
17. **Centers for Disease Control.** 1993. *Summary of Notifiable Diseases, United States, 1992.* Massachusetts Medical Society, Waltham, Mass.
18. **Chmielewski, R. A. N., and J. F. Frank.** 1995. Formation of viable but nonculturable *Salmonella* during starvation in chemically defined solutions. *Lett. Appl. Microbiol.* **20:**380–384.
19. **Coleman, S. S., and J. D. Oliver.** 1996. Optimization of conditions for the polymerase chain reaction amplification of DNA from culturable and nonculturable cells of *Vibrio vulnificus*. *FEMS Microbiol. Ecol.* **19:**127–132.
20. **Colwell, R. R., P. R. Brayton, D. J. Grimes, D. B. Roszak, S. A. Huq, and L. M. Palmer.** 1985. Viable but non-culturable *Vibrio cholerae* and related pathogens in the environment: implications for the release of genetically engineered microorganisms. *Biotechnology* **3:**817–820.
21. **Colwell, R. R., P. Brayton, D. Herrington, B. Tall, A. Huq, and M. M. Levine.** 1996. Viable but nonculturable *Vibrio cholerae* O1 revert to a culturable state in the human intestine. *World J. Microbiol. Biotechnol.* **12:**28–31
22. **Cornax, R., M. A. Moriñigo, P. Romero, and J. J. Borrego.** 1990. Survival of pathogenic microorganisms in seawater. *Curr. Microbiol.* **20:**293–298.
23. **Dawe, L. L., and W. R. Penrose.** 1978. "Bactericidal" property of seawater: death or debilitation? *Appl. Environ. Microbiol.* **35:**829–833.
24. **Dirksen, J., and P. Flagg.** 1988. Pathogenic organisms in dairy products; cause, effects and control. *Food Sci. Technol. Today* **2:**41–43.
25. **Ekweozor, C. C., C. E. Nwoguh, and M. R. Barer.** 1998. Transient increases in colony counts observed in declining populations of *Campylobacter jejuni* held at low temperature. *FEMS Microbiol. Lett.* **158:**267–272.
26. **Fujioka, R. S., H. H. Hashimoto, E. B. Siwak, and R. H. F. Young.** 1981. Effect of sunlight on survival of indicator bacteria in seawater. *Appl. Environ. Microbiol.* **41:**690–696.
27. **Garcia-Lara, J., P. Menon, P. Servais, and G. Billen.** 1991. Mortality of fecal bacteria in seawater. *Appl. Environ. Microbiol.* **57:**885–888.
28. **Gauthier, M. J., P. M. Munroe, and S. Mohajer.** 1987. Influence of salts and sodium chloride on the recovery of *Escherichia coli* from seawater. *Curr. Microbiol.* **15:**5–10.
29. **Greenberg, A. E.** 1956. Survival of enteric organisms in sea water. *Public Health Rep.* **71:**77–86.
30. **Gribbon, L. T., and M. R. Barer.** 1995. Oxidative metabolism in nonculturable *Helicobacter pylori* and *Vibrio vulnificus* cells studied by substrate-enhanced tetrazolium reduction and digital image processing. *Appl. Environ. Microbiol.* **61:**3379–3384.
31. **Grimes, D. J., and R. R. Colwell.** 1986. Viability and virulence of *Escherichia coli* suspended by membrane chamber in semitropical ocean water. *FEMS Microbiol. Lett.* **34:**161–165.

31a. **Huq, A., R. R. Colwell, R. Rahman, A. Ali, M. A. R. Chowdhury, S. Parveen, D. A. Sack, and E. Russek-Cohen.** 1990. Occurrence of *Vibrio cholerae* in the aquatic environment measured by fluorescent antibody and culture method. *Appl. Environ. Microbiol.* **56:**2370–2373.

32. **Hussong, D., R. R. Colwell, M. O'Brien, A. D. Weiss, A. D. Pearson, R. M. Weiner, and W. D. Burge.** 1987. Viable *L. pneumophila* not detectable by culture on agar media. *Biotechnology* **5:**947–950.
33. **Islam, M. S., M. K. Hasan, M. A. Miah, G. C. Sur, A. Felsenstein, M. Venkatesan, R. B. Sack, and M. J. Albert.** 1993. Use of the polymerase chain reaction and fluorescent-antibody methods for detecting viable but nonculturable *Shigella dysenteriae* type 1 in laboratory microcosms. *Appl. Environ. Microbiol.* **59:**536–540.

34. **Jacob, J., W. Martin, and C. Holler.** 1993. Characterization of viable but nonculturable state of *Campylobacter coli*, characterized with respect to electron-microscopic findings, whole cell protein and lipooligosaccharide (LOS) patterns. *Zentbl. Mikrobiol.* **148:**3–10.
35. **Ketchum, B. H., J. C. Ayers, and R. F. Vaccaro.** 1952. Processes contributing to the decrease of coliform bacteria in a tidal estuary. *Ecology* **33:**247–258.
36. **Khan, M. U., G. T. Curlin, and M. I. Huq.** 1979. Epidemiolgy of *Shigella dysenteriae* type 1 infections in Dacca [sic] urban area. *Trop. Geogr. Med.* **31:**213–223.
37. **Klein, P. D., D. Y. Graham, A. Gaillor, A. R. Opekun, and E. O. Smith.** 1991. Water source as risk factor for *H. pylori* infection in Peruvian children. *Lancet* **337:**1503–1505.
38. **Kogure, K., U. Simidu, and N. Taga.** 1979. A tentative direct microscopic method for counting living marine bacteria. *Can. J. Microbiol.* **25:**415–420.
39. **Korhonen, L. K., and P. J. Martikainen.** 1991. Survival of *Escherichia coli* and *Campylobacter jejuni* in untreated and filtered lake water. *J. Appl. Bacteriol.* **71:**379–382.
40. **Linder, K., and J. D. Oliver.** 1989. Membrane fatty acid and virulence changes in the viable but nonculturable state of *Vibrio vulnificus*. *Appl. Environ. Microbiol.* **55:**2837–2842.
40a. **Linkous, D. A., and J. D. Oliver.** 1999. Pathogenesis of *Vibrio vulnificus*. *FEMS Microbiol. Lett.* **174:**207–214.
41. **Martinez, J., J. Garacia-Lara, and J. Vibes-Rego.** 1989. Estimation of *Escherichia coli* mortality in seawater by the decrease in ^{3}H-label and electron transport system activity. *Microb. Ecol.* **17:** 219–225.
42. **McGovern, V. P., and J. D. Oliver.** 1995. Induction of cold-responsive proteins in *Vibrio vulnificus*. *J. Bacteriol.* **177:**4131–4133.
43. **Medema, G. J., F. M. Schets, A. W. van de Giessen, and A. H. Havelaar.** 1992. Lack of colonization of 1 day old chicks by viable, non-culturable *Campylobacter jejuni*. *J. Appl. Bacteriol.* **72:** 512–516.
44. **Mizunoe, Y., S. N. Wai, A. Takade, and S.-I. Yoshida.** 1999. Restoration of culturability of starvation-stressed and low-temperature-stressed *Escherichia coli* 0157 cells using H_2O_2-degrading compounds. *Arch. Microbiol.* **172:**63–67.
45. **Moran, A. P., and M. E. Upton.** 1986. A comparative study of the rod and coccoid forms of *Campylobacter jejuni* ATCC 29428. *J. Appl. Bacteriol.* **60:**103–110.
46. **Mukamolova, G. V., A. S. Kaprelyants, and D. B. Kell.** 1995. Secretion of an antibacterial factor during resuscitation of dormant cells in *Micrococcus luteus* cultures held in an extended stationary state. *Antonie van Leeuwenhoek* **67:**289–295.
47. **Nilsson, L., J. D. Oliver, and S. Kjelleberg.** 1991. Resuscitation of *Vibrio vulnificus* from the viable but nonculturable state. *J. Bacteriol.* **173:**5054–5059.
48. **Oliver, J. D.** 1989. *Vibrio vulnificus*, p. 569–600. *In* M. P. Doyle (ed.), *Foodborne Bacterial Pathogens.* Marcel Dekker, Inc., New York, N.Y.
49. **Oliver, J. D.** 1993. Formation of viable but nonculturable cells, p. 239–272. *In* S. Kjelleberg (ed.), *Starvation in Bacteria.* Plenum Press, New York, N.Y.
50. **Oliver, J. D.** 1995. The viable but non-culturable state in the human pathogen, *Vibrio vulnificus*. *FEMS Microbiol. Lett.* **133:**203–208.
51. **Oliver, J. D., and R. Bockian.** 1995. In vivo resuscitation, and virulence towards mice, of viable but nonculturable cells of *Vibrio vulnificus*. *Appl. Environ. Microbiol.* **61:**2620–2623.
52. **Oliver, J. D., and J. B. Kaper.** 1997. *Vibrio* species. *In* M. P. Doyle, L. R. Beuchat, and T. J. Montville (ed.), *Food Microbiology: Fundamentals and Frontiers.* ASM Press, Washington, D.C.
53. **Oliver, J. D., and D. Wanucha.** 1989. Survival of *Vibrio vulnificus* at reduced temperatures and elevated nutrient. *J. Food Safety* **10:**79–86.
54. **Oliver, J. D., L. Nilsson, and S. Kjelleberg.** 1991. Formation of nonculturable *Vibrio vulnificus* cells and its relationship to the starvation state. *Appl. Environ. Microbiol.* **57:**2640–2644.
55. **Oliver, J. D., D. McDougald, T. Barrett, L. A. Glover, and J. I. Prosser.** 1995. Effect of temperature and plasmid carriage on nonculturability in organisms targeted for release. *FEMS Microbiol. Ecol.* **17:**229–238.
56. **Oliver, J. D., F. Hite, D. McDougald, N. L. Andon, and L. M. Simpson.** 1995. Entry into, and resuscitation from, the viable but nonculturable state by *Vibrio vulnificus* in an estuarine environment. *Appl. Environ. Microbiol.* **61:**2624–2630.

57. **Paszko-Kolva, C., M. Shahamat, H. Yamamoto, T. Sawyer, J. Vives-Rego, and R. R. Colwell.** 1991. Survival of *Legionella pneumophila* in the aquatic environment. *Microb. Ecol.* **22:**75–83.
58. **Perkins, S. E., L. L. Yan, Z. Shen, A. Hayward, J. C. Murphy, and J. G. Fox.** 1996. Use of PCR and culture to detect *Helicobacter pylori* in naturally infected cats following triple antimicrobial therapy. *Antimicrob. Agents Chemother.* **40:**1486–1490.
59. **Peterson, W. L.** 1991. *Helicobacter pylori* and peptic ulcer disease. *N. Engl. J. Med.* **324:**1043–1048.
60. **Pommepuy, M., M. Butin, A. Derrien, M. Gourmelon, R. R. Colwell, and M. Cormier.** 1996. Retention of enteropathogenicity by viable but nonculturable *Escherichia coli* exposed to seawater and sunlight. *Appl. Environ. Microbiol.* **62:**4621–4626.
61. **Rahman, I., M. Shahamat, M. A. R. Chowdhury, and R. R. Colwell.** 1996. Potential virulence of viable but nonculturable *Shigella dysenteriae* type 1. *Appl. Environ. Microbiol.* **62:**115–120.
62. **Rhoades, H. E.** 1954. The illustration of the morphology of *Vibrio fetus* by electron microscopy. *Am. J. Vet. Res.* **15:**630–633.
63. **Rigsbee, W., L. M. Simpson, and J. D. Oliver.** 1997. Detection of the viable but nonculturable state in *Escherichia coli* O157:H7. *J. Food Safety* **16:**255–262.
64. **Rodriguez, G. G., D. Phipps, K. Ishiguro, and H. F. Ridgway.** 1992. Use of a fluorescent redox probe for direct visualization of actively respiring bacteria. *Appl. Environ. Microbiol.* **58:**1801–1808.
65. **Rollins, D. M., and R. R. Colwell.** 1986. Viable but nonculturable stage of *Campylobacter jejuni* and its role in survival in the natural aquatic environment. *Appl. Environ. Microbiol.* **52:**531–538.
66. **Roszak, D. B., and R. R. Colwell.** 1987. Survival strategies of bacteria in the natural environment. *Microbiol. Rev.* **51:**365–379.
67. **Roszak, D. B., and R. R. Colwell.** 1987. Metabolic activity of bacterial cells enumerated by direct viable count. *Appl. Environ. Microbiol.* **53:**2889–2983.
68. **Roszak, D. B., D. J. Grimes, and R. R. Colwell.** 1984. Viable but nonrecoverable stage of *Salmonella enteritidis* in aquatic systems. *Can. J. Microbiol.* **30:**334–338.
69. **Roth, W. G., M. P. Leckie, and D. N. Dietzler.** 1988. Restoration of colony-forming activity in osmotically stressed *Escherichia coli* by betaine. *Appl. Environ. Microbiol.* **54:**3142–3146.
70. **Ryan, K. J.** 1990. Enterobacteriaceae, p. 357–383. *In* J. C. Sherris (ed.), *Medical Microbiology: an Introduction to Infectious Diseases,* 2nd ed. Appleton & Lange, Publishers, Norwalk, Conn.
71. **Shahamat, M., U. Mai, C. Paszko-Kolva, M. Kessel, and R. R. Colwell.** 1993. Use of autoradiography to assess viability of *Helicobacter pylori* in water. *Appl. Environ. Microbiol.* **59:**1231–1235.
72. **Skirrow, M. B.** 1989. *Campylobacter* perspectives. *PHLS Microbiol. Digest* **6:**113–117.
73. **Smith, P. R., E. Farrell, and K. Dunican.** 1974. Survival of R+ *Escherichia coli* in sea water. *Appl. Microbiol.* **27:**983–984.
74. **Steinert, M., L. Emody, R. Amann, and Jorg Hacker.** 1997. Resuscitation of viable but nonculturable *Legionella pneumophila* Philadelphia JR32 by *Acanthamoeba castellanii. Appl. Environ. Microbiol.* **63:**2047–2053.
75. **Taylor, D. N., and M. J. Blaser.** 1991. The epidemiology of *Helicobacter pylori* infection. *Epidemiol. Rev.* **13:**42–59.
76. **Tholozan, J. L., J. M. Cappelier, J. P. Tissier, G. Delattre, and M. Federighi.** 1999. Physiological characterization of viable-but-nonculturable *Campylobacter jejuni* cells. *Appl. Environ. Microbiol.* **65:**1110–1116.
77. **Wai, S. N., T. Moriya, K. Kondo, M. Hiroyasu, and K. Amako.** 1996. Resuscitation of *Vibrio cholerae* O1 strain TSI-4 from a viable but nonculturable state by heat shock. *FEMS Microbiol. Lett.* **136:**187–191.
77a. **Warner, J. M., and J. D. Oliver.** 1998. Randomly amplified polymorphic DNA analysis of starved and viable but nonculturable *Vibrio vulnificus* cells. *Appl. Environ. Microbiol.* **64:**3025–3028.
78. **Weichart, D., J. D. Oliver, and S. Kjelleberg.** 1992. Low temperature induced non-culturability and killing of *Vibrio vulnificus*. *FEMS Microbiol. Lett.* **100:**205–210.
79. **West, A. P., M. R. Millar, and D. S. Tompkins.** 1990. Survival of *Helicobacter pylori* in water and saline. *J. Clin. Pathol.* **43:**609.
80. **West, A. P., M. R. Millar, and D. S. Tompkins.** 1992. Effect of physical environment on survival of *H. pylori*. *J. Clin. Pathol.* **45:**228–231.

81. **Whitesides, M. D., and J. D. Oliver.** 1997. Resuscitation of *Vibrio vulnificus* from the viable but nonculturable state. *Appl. Environ. Microbiol.* **63:**1002–1005.
82. **Wolf, P. W., and J. D. Oliver.** 1992. Temperature effects on the viable but nonculturable state of *Vibrio vulnificus. FEMS Microbiol. Ecol.* **101:**33–39.
83. **Xu, H.-S., N. Roberts, F. L. Singleton, R. W. Attwell, D. J. Grimes, and R. R. Colwell.** 1982. Survival and viability of nonculturable *Escherichia coli* and *Vibrio cholerae* in the estuarine and marine environment. *Microb. Ecol.* **8:**313–323.
84. **Xu, H.-S., N. C. Roberts, L. B. Adams, P. A. West, R. J. Siebeling, A. Huq, M. I. Huq, R. Rahman, and R. R. Colwell.** 1984. An indirect fluorescent antibody staining procedure for detection of *Vibrio cholerae* serovar O1 cells in aquatic environmental samples. *J. Microbiol. Methods* **2:**221–231.
85. **Yamamoto, H., Y. Hashimoto, and T. Ezaki.** 1996. Study of nonculturable *Legionella pneumophila* cells during multiple-nutrient starvation. *FEMS Microbiol. Ecol.* **20:**149–154.
86. **Zimmerman, R., R. Iturriaga, and J. Becker-Birck.** 1978. Simultaneous determination of the total number of aquatic bacteria and the number thereof involved in respiration. *Appl. Environ. Microbiol.* **36:**926–935.

Nonculturable Microorganisms in the Environment
Edited by R. R. Colwell and D. J. Grimes

Chapter 17

Epidemiological Significance of Viable but Nonculturable Microorganisms

Anwarul Huq, Irma N. G. Rivera, and Rita R. Colwell

Although the term "viable but nonculturable" (VBNC) has been used in the literature during the past two decades to describe a survival strategy of microorganisms, notably human pathogens in the aquatic environment, there is an abundance of philosophical musings on this subject in the microbiological literature. For example, Radsimosky in 1930 (119) noted a significant difference in the number of autotrophic and organotrophic bacteria in water samples. Direct microscopic enumeration, used in water bacteriology at the time, yielded counts 200 to 5,000 times higher than culture counts on bacteriological plates (16, 17). The difference between results of bacterial enumeration by direct observation and subsequent measurement of oxygen demand led Butkevitch and Butkevitch (18) to conclude that a significant portion of a given bacterial population, which did not appear as colonies on plates, must be present in a resting stage. Knaysi (82) and Jennison (70) were able to demonstrate metabolic activity of organisms observed by direct microscopy to be present in a sample but failing to grow, hence not appearing as colonies on plates. ZoBell (156) reconfirmed earlier findings that plate culture counts of seawater, although the widely used method at the time (and to this day), yield only a small percentage of the bacteria actually present in a given sample.

These findings of many investigators, including Valentine and Bradfield (146), Postgate (118), Jannasch (69), Stevenson (142), Kurath and Morita (85), and others, provide an underpinning for the concept of a "resting stage" for nonsporulating bacteria. A variety of terms have been used to describe this phenomenon. Dawe and Penrose (43) provided evidence of viability by determining ATP levels, suggesting coliforms present in seawater did not die off, as had long been thought. Colwell and associates (153) studied this phenomenon, both in *Vibrio cholerae* and

Anwarul Huq and Rita R. Colwell • Center of Marine Biotechnology, Columbus Center, University of Maryland Biotechnology Institute, 701 E. Pratt St., Baltimore, MD 21202, and Department of Cell and Molecular Biology, University of Maryland, College Park, MD 20742. ***Irma N. G. Rivera*** • Center of Marine Biotechnology, Columbus Center, University of Maryland Biotechnology Institute, 701 E. Pratt St., Baltimore, MD 21202, and Department of Microbiology, Biomedical Science Institute, São Paulo University, São Paulo, S.P. Brazil, CEP 05508-900.

Escherichia coli, and presented convincing experimental evidence of a VBNC state of *E. coli* in seawater, with viability of the organism determined by the nalidixic acid method of Kogure et al. (83). That is, the cells were unable to grow on conventional culture media, but responded to nalidixic acid treatment. Xu et al. (153) concluded that the VBNC phenomenon represents a state of dormancy, allowing survival and persistence of bacterial cells in the natural or host environment. Bacterial cells in this state maintain viability and metabolic activity while ceasing multiplication, that is, they do not form visible colonies on bacteriological culture media. Thus, ecologists and epidemiologists studying pathogenic and infectious agents, especially those that are autochthonous members of the aquatic environment, such as *V. cholerae*, will need to reevaluate the methods routinely employed for counting bacteria in samples, whether environmental or clinical. In fact, intact and live bacterial cells (i.e., those maintaining metabolic activity and not recoverable by conventional culture methods employing routine bacteriological media) have been found in both terrestrial and aquatic environments, as well as in clinical specimens (28, 38, 64, 95, 144).

The definition of living, as opposed to dead, bacteria is complex and difficult to enunciate. Even very recent publications still conclude that bacterial cells not growing on culture plates are dead (137). However, most authors, including Gonzalez et al. (48), suggest that bacterial cells should be considered dead only when they lose both culturability and cellular integrity. Major changes have been observed in cellular morphology during the process of conversion from culturable to nonculturable cells in laboratory microcosms, where cells become coccoid and smaller in size, with the central region compressed and surrounded by denser cytoplasm (8, 11, 38, 84, 127, 153; S. Chaiyanan et al., abstr. no. Q-66, Abstr. 97th Gen. Meet. Am. Soc. Microbiol, 1997). Several factors, such as temperature, salinity, osmotic pressure, pH, and nutrient concentration, may be involved in the induction of this state (7, 34, 38, 57, 67, 109, 136, 150).

Several investigators have reported the clinical importance of the VBNC phenomenon, notably for vibrios and related enteric pathogens (40, 91). In fact, the VBNC state may allow pathogenic *Vibrio* to be more resistant to pressure and conventional food processing methods. Thus, over-reliance on conventional methods for microbiological analysis of seafood carries the risk of generating false negatives for process-resistant VBNC *Vibrio* (7). The number of published studies of clinically important bacteria demonstrating the VBNC phenomenon has increased significantly, mainly because of its significance for public health.

V. cholerae O1, the causative agent of cholera, and many other enteric pathogens adapt to conditions in the environment that may be inimical to cell growth and division. The response may be manifested in various ways. When traditional culture media are employed to isolate an organism from environmental samples, injury to the cells may occur during sample processing, in addition to adverse effects of incomplete or inhibitory nutrient concentration (32). A rapid change in air temperature can cause significant fluctuation in surface water temperature, which in turn can affect microorganisms significantly, especially under unfavorable nutritional conditions (90, 108). It has been suggested that cells of *V. cholerae* enter into the VBNC state to persist under such adverse conditions (130). Under low nutrient conditions, cells may alter metabolic pathways and enzyme syntheses for efficient use of available nutrients. Furthermore, bacteria in nutrient-rich environments, such as bodies of water receiving sewage and farm runoff, may not be able to maintain their normal rate of metabolism and therefore enter into the VBNC state (130, 138). The survival of *V. cholerae* O1 and non-O1 in seawater, sewage, and freshwater microcosms has been evaluated, and survival was longer in seawater for both groups (134).

The time required for different species to enter into the nonculturable state may vary from several hours to months (50, 56, 152). It is speculated that there may be one or more factor(s) in the environment responsible for converting or inducing a normal culturable cell to the VBNC state. Some investigators have reported a dramatic decrease in nutrient uptake and respiration when VBNC cells are exposed to visible light (5, 116), while others did not observe any effect of light on culturability of those cells (56). Different strains of the same bacterial species have been reported to require longer or shorter periods of time to enter into the VBNC state (153). Bacterial strains of *Vibrio vulnificus* and related organisms have been reported to require different times of exposure to conditions adverse to growth and division, e.g., reduced temperature, before entry into the VBNC state (57, 90, 124, 150, 152). A unity in the explanation of conditions inducing this phenomenon has not yet been achieved, perhaps because of microeffects, either chemical and/or physical, in the surrounding environment. Thus, the time required to enter into the VBNC state will vary, even for the same strain, e.g., *Vibrio salmonicida*, tested at different times when microcosm conditions are otherwise held constant (56).

VBNC PATHOGENS

Of those microorganisms comprising the flora of the natural environment, only 2 to 3% are considered to be harmful to humans. Organisms are considered pathogenic when they demonstrate ability to infect and cause disease (45). The term "infection" is used when successful introduction and multiplication of a pathogen on or within the host occur, and the term "disease" is used to describe an infection causing significant overt damage to the host. However, organisms isolated from infected bodies may not be pathogenic, but can be considered potential pathogens, unless and until a definite role of disease-causing ability is determined.

In some cases, symptoms of a particular disease are expressed, while implication of a specific organism cannot be established by traditional bacteriological culture methods. Often, they are considered nonbacterial infections, without further elaborate investigation. Such symptoms, if bacterial by cause, may be associated with the presence of a pathogenic bacterium in the VBNC state. Because of the lack of knowledge of the occurrence of VBNC microorganisms, causative agents of many diseases have yet to be identified or appropriate diagnosis and treatment provided.

Until recently, most studies of the VBNC phenomenon in bacteria were limited to aquatic pathogens. Early studies were done on *E. coli* and *V. cholerae*, two important waterborne pathogens demonstrated to survive in the VBNC state (153). During the past decade, several investigations were reported on the existence of the VBNC state in other microbial pathogens, regardless of habitat. A list of human pathogens for which the VBNC state has been observed and/or investigated is provided in Table 1.

Although growth-limiting chemical factors and physical factors are known to play an important role in the survival of bacterial cells, the most common is temperature. A brief description of conditions triggering entry to the VBNC state in selected pathogenic organisms is provided below.

Aeromonas spp.

Aeromonas hydrophila causes severe wound infections and occasionally gastritis in humans exposed to natural water in which the bacterium is present. *Aeromonas* spp. are considered to be aquatic in origin (68) and have been shown to infect divers, especially professional divers who, because of their duties, must dive and/or swim in contaminated waters, e.g., police and rescue divers, sewage pipe repairmen, etc. (37). Although aeromonads are mesophiles, based on results of laboratory microcosm studies, it has been found that they will enter into the VBNC state within 10 days at 35°C and 30 days at 10°C (104; J. A. K. Hasan et al., abstr. Q-139, Abstr. 91st Gen. Meet. Am. Soc. Microbiol., 1991).

Table 1. Pathogens and related references for which the VBNC phenomenon has been studied and/or observed

Pathogen	Reference(s)
Aeromonas spp.	2, 104; J. A. K. Hasan et al., abstr. Q-139, Abstr. 91st Gen. Meet. Am. Soc. Microbiol., 1991
Campylobacter coli	57
Campylobacter jejuni	8, 21, 75, 99, 102, 127, 133, 144
Escherichia coli	5, 19, 50, 115, 116, 153
Helicobacter pylori	11, 31, 93, 105, 106, 114, 135; M. Shahamat et al., abstr. I-101, Abstr. 90th Gen. Meet. Am. Soc. Microbiol., 1990
Klebsiella pneumoniae	20
Legionella pneumophila	6, 13, 64, 67, 111, 112, 141
Pseudomonas putida	85, 103
Salmonella enteritidis	129, 145
Salmonella enterica serovar Typhimurium	145
Shigella sonnei	34
Shigella flexneri	34
Shigella dsysenteriae	66, 120, 121; I. Rahman et al., abstr. Q-4, Abstr. 93rd Gen. Meet. Am. Soc. Microbiol., 1993
Vibrio cholerae O1	7, 14, 34, 36, 38, 39–41, 51–53, 60–63, 84, 94, 123, 124, 132, 134, 138, 153, 154; S. Chaiyanan et al., abstr. no. Q-66, Abstr. 97th Gen. Meet. Am. Soc. Microbiol., 1997
Vibrio cholerae O139	Xu et al., unpublished data
Vibrio mimicus	24, 152; Xu et al., unpublished data
Vibrio parahaemolyticus	25, 73, 76, 110, 152
Vibrio vulnificus	54, 90, 108, 109, 148, 149, 150, 151, 152
Pasteurella piscida	92

Campylobacter spp.

Campylobacter jejuni causes a food- or waterborne enteric disease. The natural reservoir of *Campylobacter* spp. is the intestine of warm-blooded animals. *C. jejuni* can grow at temperatures of 35°C to 42°C under microaerophilic conditions. In a laboratory microcosm study, using stream water at 4°C, *C. jejuni* was induced to enter the VBNC state (127). The effects of temperature, pH, and NaCl concentration on the VBNC state have been studied (21), with only AMP detected after 30 days of incubation in microcosm water (144).

Escherichia coli

E. coli can cause both gastrointestinal and extraintestinal infections. Survival of various strains has been extensively investigated, and the failure to recover *E. coli* by culture methods after exposure to environmental water systems was ascribed in very early studies to rapid "die-off" (98). Subsequently, in extensive studies of the survival and viability of *E. coli*, it was found that under low temperature and nutrient-limited conditions, the cells do not die, but lose their ability to grow on conventional culture media (153). Furthermore, the cells remain metabolically active and viable (28). *E. coli* was observed to enter the VBNC state when exposed to filter-sterilized Chesapeake Bay water at 4 to 6°C and 11‰ salinity (153). *E. coli* has long been used as an indicator for water quality and an index of the level of pollution, by enumerating culturable cells. However, these methods must be reconsidered in light of the possibility that VBNC cells may be present but not enumerated by culture methods.

Helicobacter pylori

Helicobacter pylori is a fastidious, microaerophilic human pathogen associated with type B gastritis (15) and/or peptic ulcer (113). This species was originally assigned to the genus *Campylobacter* (49), because its growth conditions and habitat are similar to those of *C. jejuni*. In developed countries, *H. pylori* seropositivity is found in 40 to 50% of the population aged 20 to 60 years (86), but the mode of spread remains unclear. However, it was proposed that the coccoid forms are ingested orally, after which the coccoid cells outgrow to helical bacteria and colonize the stomach (11). This pathogen can survive in natural water microcosms, entering the VBNC state in water at 4°C (81, 93; M. Shahamat et al., abstr. I-101, Abstr. 90th Gen. Meet. Am. Soc. Microbiol., 1990). Subsequent studies have shown that VBNC *H. pylori* cells are metabolically active (135). The coccoid forms of VBNC *H. pylori* retain urease activity, reduce tetrazolium salts, exhibit a protein pattern similar to that of the helical bacteria, and synthesize DNA (11, 106). Narikawa et al. (105) showed that the coccoid form of *H. pylori* had impaired genomic DNA, the cells being still alive and able to increase in volume.

Klebsiella pneumoniae

In addition to causing gastrointestinal infections, *Klebsiella* spp. account for a significant proportion of urinary tract, bloodstream, and other extraintestinal infections. This waterborne pathogen, when tested for ability to grow and maintain viability in drinking water, demonstrated a short survival time as culturable cells but entered into the VBNC state within 4 days after inoculation into filtered drinking water incubated at 25°C (20), retaining viability despite nonculturability.

Legionella pneumophila

Legionella is the causative agent of Legionnaires' pneumonia. This potential microbiological hazard can arise from dispersed aerosols from cooling towers or evaporative condensers. In natural and man-made environments, free-living amoebae serve as host organisms in which legionellae multiply intracellularly (13). Conditions for inducing the VBNC state have been studied, and *L. pneumophila* was reported to become nonculturable in tapwater microcosms at 37°C (64) and in low-nutrient environments (112). The pathogen may also enter the VBNC state at 4°C but requires a longer incubation time to do so at this temperature. Exposure to cooling tower water to which hypochlorite has been added may also induce the VBNC state (6). Steinert et al. (141) demonstrated that *L. pneumophila* entered the VBNC state following starvation, from which it could be resuscitated by the addition of *Acanthamoeba castellanii.*

Pseudomonas putida

Pseudomonas putida is a common inhabitant of soil and water and is associated with plants. It has been recovered from various organs of patients (viz., skin, ear, respiratory tract) and is believed to be responsible for extraintestinal illness (89). Among the pseudomonads, *P. putida* (synonym, *Pseudomonas ovalis*) is one of only a few species in which induction of the VBNC state has been studied under environmental conditions. Although no attempt was made to detect VBNC cells, the number of cells declined to undetectable levels when released into sterile lake water (103).

Salmonella spp.

Salmonellae are causative agents of various diseases, including self-limited gastroenteritis, with mild clinical symptoms of short duration, severe gastritis, with or without bacteremia, and typhoid fever which can cause debilitating and potentially life-threatening illness. In the genus *Salmonella*, *S. enteritidis* was the first species to be reported to enter a VBNC state (129). *Salmonella enterica* serovar Typhimurium was reported to exhibit similar behavior (145). *Salmonella* spp. grow optimally at 37°C under aerobic conditions. Experiments using autoclaved river water microcosms maintained at 25°C showed that most of the cells enter the VBNC state by day 4 of exposure; after 21 days of continued incubation, all cells became VBNC (129).

Shigella spp.

Shigella species cause classical bacillary dysentery, characterized by severe, cramping abdominal pain and bloody mucoid diarrhea. These organisms possess the potential to invade mucosal cells, causing death and sloughing of the cells into the lumen of the bowel (89), and grow at ca. 37°C. Laboratory microcosm experiments, using filter-sterilized Chesapeake Bay water at 15‰ salinity and 25°C, showed that *Shigella sonnei* and *Shigella flexneri* enter the VBNC state within 21 days of exposure to Chesapeake Bay water (34). Further studies showed that these species can survive in the VBNC state for more than 6 months (120). Earlier studies also reported the entry of *Shigella dysenteriae* into the VBNC state (I. Rahman et al., abstr. Q-4, Abstr. 93rd Gen. Meet. Am. Soc. Microbiol., 1993).

Vibrio spp.

Most of the pathogenic vibrios are primarily waterborne (26, 33–35, 38, 61) and can cause a variety of human infections, both gastrointestinal and extraintestinal (89). Among all the described vibrios, *V. cholerae* is considered to have the highest epidemic potential. This species was reported to enter into the VBNC state by Xu et al. in 1982 (153), and subsequent studies were carried out by Colwell et al. (34). *V. cholerae* is a mesophile and grows optimally at a temperature of ca. 37°C. Entry into the VBNC state when cultured in filter-sterilized Chesapeake Bay water (11‰ salinity) and instant ocean water (10‰ salinity) broth at 4°C (123, 153) has been demonstrated. The bacteria maintained metabolic activities, detected using radiolabeled ([^{14}C]glucose/acetate and [^{3}H]thymidine) substrate uptake studies (28). However, if vibrios are not challenged with a factor(s) responsible for inducing nonculturability, *V. cholerae* will remain culturable (152).

A novel strain of *V. cholerae* which does not agglutinate with *V. cholerae* O1 antiserum was reported and described as *V. cholerae* O139 Bengal (122). This new serotype is responsible for epidemics of cholera in several Asian countries (22, 23). Recent microcosm studies showed *V. cholerae* O139 enters the VBNC state within 7 days at 25 and 4°C (B. Xu, A. Huq, and R. R. Colwell, unpublished data).

Vibrio mimicus is a species of the genus *Vibrio* with the potential to cause enteric and extraintestinal illness in humans (27, 42). Like *V. cholerae* (33), *V. mimicus* is also an autochthonous member of the aquatic environment (24), and most of the outbreaks associated with the pathogen are known to be water- and/or foodborne (42). Laboratory studies of the VBNC state in *V. mimicus* have not been extensive. An attempt by Wolf and Oliver (152) was unsuccessful in inducing nonculturability in this pathogen. In a recently completed study, *V. mimicus* was found to enter into the VBNC state within 7 days when maintained in filter-sterilized natural water and instant ocean water at 25°C and 4°C (Xu et al., unpublished data). Environmental studies in different geographical regions of the world showed lack of success in culturing the organism from the water column during winter months (24), most probably due to conversion to the VBNC state. A similar response has also been reported for *Vibrio parahaemolyticus*, a food- and waterborne pathogen associated with gastritis, which also disappears from the water column during unfavorable seasons (25, 76). However, laboratory microcosm studies by Wolf and Oliver (152) failed to determine conditions that would induce the VBNC state of these organisms. Jiang and Chai (73) found that the morphological changes of *V. parahae-*

molyticus were dependent on the strains and the temperatures at which the cells were maintained.

V. vulnificus is a known causative agent of seafood-associated human illness and is responsible for septicemia, wound infections, and occasionally gastroenteritis (9, 74). Like other vibrios, disappearance of *V. vulnificus* from the water column during winter months has been observed by several investigators (80, 107), suggesting that *V. vulnificus* enters into a nonculturable state in the environment. In fact, laboratory microcosm studies have demonstrated the VBNC state in *V. vulnificus*; i.e., this organism enters into the VBNC state in artificial seawater at 5°C within 24 days of incubation (90). Furthermore, temperature appears to be critical for formation of nonculturable, coccoid cells of *V. vulnificus* (152). However, resuscitation of VBNC cells occurs when the temperature stress is removed (151).

Vibrio fischeri is symbiotic when in the VBNC state (88). During the past several years, microbial ecologists and medical microbiologists have attempted to address the questions of how, when, and why the microorganisms are pathogenic. For example, *V. cholerae* is known to be an autochthonous microorganism in the aquatic environment and to play an important role in the balance of the aquatic ecosystem, specifically as a detritus consumer (39). The ecological evidence has been supported by taxonomic and epidemiology data, results of survival studies, and isolation from samples collected worldwide.

Cholera occurs seasonally in Bangladesh, with two annual peaks. During the inter-epidemic period, cells convert to the nonculturable state and the number of culturable cells decreases (61, 63). When *V. cholerae* is present outside of its natural ecosystem, various factors or conditions, including pH, temperature, anaerobic environment, and osmolarity, can affect a wide variety of its metabolic functions (100, 101) and induce the VBNC state (63). The relationship of *V. cholerae* to its host appears to be affected similarly by environmental parameters.

DETECTION METHODS AND THEIR APPLICATIONS

To study VBNC bacteria, the methods used must differentiate injured cells, starved cells, and dormant cells. However, some of the criteria used to determine VBNC cells, with respect to public health and epidemiology significance, are discussed below.

Identification or detection of pathogenic, virulent organisms with epidemic potential in clinical or environmental specimens is essential for proper mobilization and initiation of action by public health officials to prevent spread of the disease, especially in developing countries. Nonculturable cells can pose significant risk as they will be overlooked if only conventional culture methods are employed. The absence of culturable bacteria in a specimen may not necessarily reflect the state of the original population. Methods developed during the past decade have been successful and are now being widely used since the public health implications of VBNC organisms have been recognized. Quantification of bacterial numbers and biomass helps to determine the role of the bacteria in the environment. Examples of methods used to detect or demonstrate VBNC bacterial cells are provided in Table 2.

Table 2. Examples of methods for detection of VBNC cells

Method	Reference(s)
Respiration	77, 130, 155
Fluorochrome (AODC)	143
Erythrosine staining	1
Urea	146
Membrane potential	4
INT	155
Intracellular volume	131
Nalidixic acid enrichment	83
Fluorochrome-based DAPI	117
Indirect fluorescent antibody	154
DFA-DVC	14, 29
INT-DVC	J. Hasan et al., abstr. P-212, U.S.-Japan Cholera Conf. 1991
PCR-cholera toxin	137
Redox dye staining	126
Coagglutination	36
Fluorescent-oligo-probe	54
Autoradiography	135
Colorimetrically based immunoassay	51
Luminescence	44
Direct fluorescent antibody (DFA)	53
PCR	30

More than half a century ago, Karsinkin and Kusnetsov (77) and later Alfimov (1) used erythrosine stain to distinguish live cells from dead cells. Strugger (143) used an acridine orange stain employing 3,6-bis(dimethylamino)acridinium chloride to differentiate live cells from dead cells in soil. This method of detection, now popularly known as the acridine orange (AO) direct counting procedure (AODC), is probably best recognized for use in enumeration of total bacteria in environmental samples (12, 46). Unfortunately, dead, metabolically inactive cells cannot be distinguished from viable living cells by AO staining (78). Another fluorochrome-based staining method, using 4′,6-diamidino-2-phenylindole (DAPI), was introduced by Porter and Fieg (117); this method performs more efficiently. A review of the use of AO and DAPI indicates that the latter method is currently more widely used than AO (78).

The most commonly employed and more persuasive method for detection of VBNC bacterial cells is the direct viable count (DVC), developed by Kogure et al. (83). In this method, a small amount of yeast extract is added as nutrient with addition of nalidixic acid; the latter inhibits DNA replication and protein synthesis for cell division, resulting in elongated large cells easily distinguished from nonresponsive cells. Zimmerman et al. (155), demonstrated the viability of bacterial cells using soluble 2-4-iodophenyl-3-(4-nitrophenyl)-5-phenyltetrazolium chloride (INT) which is reduced by live metabolizing cells, producing insoluble INT formazan, a dark precipitation in the cell membrane easily recognized by microscopy. Wolf and Oliver (152) demonstrated excellent correlation between DVC and INT counts.

A very useful and practical method was developed by Brayton et al. (14), combining the DVC method of Kogure et al. (83) and the indirect fluorescent antibody method of Xu et al. (154) for direct detection of VBNC cells of *V. cholerae* O1. Brayton et al. (14) used a fluorescent monoclonal antibody in an antibody and direct viable count procedure (FA-DVC) to detect cells of *V. cholerae* O1 present in natural water from the aquatic environment.

Intracellular ATP has been used as an index of viability. However, inconsistent results were obtained. For example, 10^5 culturable cells of *C. jejuni* that were nondetectable by culture, within 4 days after suspensions were prepared yielded constant ATP measurements for more than 3 weeks thereafter (8).

By combining microscopic determination of substrate-responsive cells by DVC and individual respiring microorganisms (INT) by detection of active electron transport system (ETS), the viable and respiring cells of nonculturable *A. hydrophila* could be detected (J. Hasan et al., Abstr. P-212, USA-Japan Cholera Conf., 1991). In the combined INT-DVC method, substrate-responsive, enlarged cells with intracellular opaque deposits (indicating an active ETS) made possible enumeration of both culturable and VBNC organisms. The INT-DVC method has also been applied successfully to *S. dysenteriae* (120). In another method for testing viability of bacterial cells, Rodriguez et al. (126) used the redox dye 5-cyano-2,3-ditolyl tetrazolium chloride (CTC). The dye is colorless and nonfluorescent, but upon reduction via electron transport activity it produces red-fluorescing CTC-formazan in the cell. This method allows fluorescing CTC-formazan to be easily visualized in cells, even when the size of the cells is reduced (109). A luminescence-based detection of VBNC bacterial cells, reported to be highly sensitive (44), involving incubation of VBNC cells and using Kogure's nalidixic acid method (83), causes cells to increase in volume, with luminescence directly proportional to the size of the bacteria and measurable by luminometer.

An improved fluorescent monoclonal antibody staining kit for direct detection of *V. cholerae* O1 was developed, employing a monoclonal antibody labeled with fluorescein isothiocyanate (FITC) (53). This method is a modification of the indirect fluorescent antibody (IFA) technique described by Brayton et al. (14). CholeraScreen, based on coagglutination, for detection of *V. cholerae* O1, has proven to be useful in the field (36) because it does not require culture or specialized equipment, i.e., it can be used to detect the presence of a bacterial species even when the cells are in the nonculturable state, with results obtained within minutes. A colorimetric immunoassay, SMART, a sensitive membrane antigen detection test, has also been designed to detect bacterial cells in the nonculturable state. This rapid test kit was originally developed to diagnose *Neisseria gonorrhoeae* in clinical specimens. Using SMART technology, Hasan et al. (51) optimized CholeraSMART for detection of *V. cholerae* O1. The three test kits, Cholera DFA, CholeraSMART, and CholeraScreen, do not require culture, which makes them highly useful in clinical diagnosis as well as environmental monitoring, especially when a large number of bacteria are in the nonculturable state (36, 51, 53, 94).

Autoradiography has been used to show metabolic activity of VBNC cells; in one report it was shown to confirm the viability of *H. pylori* in microcosms maintained for more than 2 years, i.e., at least 26 months (135). While studying the metabolic activity and detection of living cells by measuring the production of organic acids, researchers observed that the pattern changed significantly at lower temperatures for *Campylobacter coli* (57). In addition, the internal potassium content and the membrane potential were significantly lower in the VBNC state than in the culturable state for *C. jejuni* (144).

With the advent of molecular methods and their application in bacteriological detection, impetus has been given to understanding the ecology of VBNC pathogens. These methods also are used to confirm the presence of VBNC bacteria in both clinical and environmental specimens. For identification purposes, the analysis of 16S and 23S rRNA sequences is referred to as the "gold standard," a powerful means for scanning microbial niches (3). PCR, the most common and frequently used method, detects the specific segment of a gene. Oligonucleotide primers were designed for PCR detection of VBNC cells of *V. cholerae* O1, based on the *ctxAB* gene of cholera toxin (*ctx*) (52). Chun et al. (30) compared similarity between 16S-23S rRNA intergenic spacer regions of *V. cholerae* and *V. mimicus* and found a common region for both bacteria that could be used as primers or probes in their identification and quantification. Additionally, for cholera epidemiological surveillance it could be applied to detect the natural population as enterobacterial repetitive intergenic consensus (ERIC)-PCR, which allows differentiating pathogenic and nonpathogenic populations (125).

PCR was used to detect a virulence plasmid in microcosm samples of VBNC *S. dysenteriae* type 1 (66). PCR is especially useful when the number of bacterial cells is extremely low, as in environmental samples. Sommerville et al. (139) and Fuhrman et al. (47) developed methods for extracting DNA directly from water samples, whereby the DNA is concentrated and treated on membrane filters (6). An additional step in preparation of DNA extracted from environmental water samples before PCR, restriction enzyme analysis, and hybridization studies has been described (10).

Enumeration of *V. vulnificus* collected on membrane filters, after hybridization with a fluorescent oligonucleotide eubacterial probe complementary to 16S rRNA, has been successful (54). Good correlation was observed when fluorescent oligonucleotide direct counts were compared with AODC counts, suggesting the potential of this test for other organisms (54, 55). A fluorescent molecular probe, LIVE/DEAD *Bac*Light, is available commercially (Molecular Probes, Inc., Eugene, Oreg.) in a kit form for detection of live bacteria with intact plasma membranes, which fluoresce green, readily distinguished from dead bacteria with compromised membranes, which fluoresce red (140). The kit has subsequently been modified for flow cytometry (71, 72, 96). A kinetic luminometric method for measuring the membrane-perturbing activity of complement based on genetically engineered luminescent bacteria has been described (147). Lebaron et al. (87) compared seven blue nucleic acid dyes for flow cytometric enumeration and concluded that SYBR-II is the best candidate for bacterial cell counting in aquatic ecosystems.

It was interesting that RNA levels declined similarly to DNA levels in the coccoid forms, suggesting that the decreased nucleic acid content was associated with the loss of cytoplasm by autolysis, as described for *C. jejuni* (102), *H. pylori* (105, 114), and *V. vulnificus* (150). A randomly amplified polymorphic DNA (RAPD)

PCR protocol was developed for *V. vulnificus* detection in VBNC state, but after 4 h of starvation the cells were not detectable by this method (148).

VIRULENCE POTENTIAL OF VBNC PATHOGENS

A pathogen is an agent capable of causing disease, and pathogenicity refers to the ability of organisms to induce harmful physiological or anatomical changes in a host. When organisms are "pathogenic" or "nonpathogenic," a relative virulence of the organism is implied (i.e., capability to produce disease under appropriate conditions). A pathogen may be virulent for one host, while avirulent for another. A pathogenic microorganism may cause disease if it is sufficiently virulent to overcome defense mechanisms and enter the host cell. If a pathogenic bacterium is not able to penetrate defense mechanisms of its host, disease will not be the outcome, even if pathogenic factors are present. Ability to multiply is also a significant factor in disease. An important question is whether nonculturable pathogenic bacteria maintain sufficient virulence to cause disease. Fluid accumulation in ligated ileal loops of rabbits inoculated with VBNC *E. coli* provided the first evidence that virulence was retained (34). The fact that VBNC cells remain metabolically active (28) and virulence plasmids are maintained in VBNC *E. coli* led to the conclusion that pathogenicity is retained in VBNC bacterial pathogens (19, 50). Strains of *E. coli,* after exposure to sunlight and entering the VBNC state, as well as culturable *E. coli*, retained pathogenicity as demonstrated by GM1-enzyme-linked immunosorbent assay (ELISA) (115). Also, VBNC cells of *L. pneumophila* injected into chick embryos showed retention of pathogenicity (64).

In 1990, a human volunteer study was carried out using VBNC cells of an attenuated strain of *V. cholerae* O1. Two volunteers each drank a suspension of ca. 10^8 cells. One of the two volunteers developed clinical symptoms of cholera, shedding culturable cells of *V. cholerae* O1 at a concentration of ca. 3×10^3 cells/g of stool. The stools of the second volunteer yielded colonies of *V. cholerae* 5 days after ingestion of the suspension, when plated on TCBS agar (35). It is important to note that entry into the VBNC state and restoration of the culturable state are likely to yield different results with different hosts. Sack et al. (132) evaluated an inoculum of *V. cholerae* directly from vials of frozen samples for volunteer studies and concluded that challenge with frozen bacteria results in a reproducible illness similar to that induced by freshly harvested bacteria.

The time elapsing before transition to the VBNC state varies, and occasionally cultures will not enter the VBNC state if specific conditions are not met (152). Similarly, infection is influenced by the ability of the host to prevent invasion and subsequent injury. Thus, just like viable and culturable cells, VBNC cells of pathogens may be virulent for one host and avirulent for another. Thus, the accumulated evidence strongly points to retention of virulence and in vivo transition to culturability in both laboratory animals and human volunteers (35, 41).

An initial study conducted by Linder and Oliver (90), wherein mice were injected with VBNC *V. vulnificus*, failed to demonstrate virulence. However, in a subsequent study, it was found that a small inoculum of VBNC cells of *V. vulnificus* (5×10^4 cells) injected into mice was sufficient to kill the mice (109).

Demonstration of resuscitation of frozen-thawed, VBNC cells of *C. jejuni* and virulence following repeated passage through the rat gut was reported by Saha et al. (133). The results indicated that nonculturable *C. jejuni* reverted to a culturable and fully virulent state by passaging through a susceptible host (133) and was capable of infecting mice; however, this property may differ between strains (75). In contrast, Medema et al. (99) were unable to demonstrate colonization of young chick intestine by oral introduction of VBNC cells of *C. jejuni*.

VBNC cells of a virulent strain of *S. dysenteriae* type 1 maintained a low level of metabolic activity, determined by ^{35}S-labeled methionine uptake, and proved cytopathic for cultured HeLa cells (120). Subsequent studies of the virulence potential of VBNC cells of *S. dysenteriae* showed that the gene coding for Shiga toxin production, detected by PCR of genomic DNA, was retained and toxin production continued even when the cells were nonculturable (66, 121). Shiga toxin was produced 45 days after nonculturability, with toxin detection achieved by ELISA using affinity-purified polyclonal antibody against the B subunit of the toxin. In addition, nonculturable cells also maintained a capacity for adhesion to Henle 407 cells (121).

Pace et al. (110) suggested that a bile-acid-containing environment found in the human host favors growth of virulent strains of *V. parahaemolyticus* and that bile acids enhance the expression of virulence factors. This is not surprising, since various factors or conditions are known to affect a variety of bacterial functions. Additionally, factors specific to the host, such as bile acids, may serve as signals.

Reentry into the culturable state appears to be unique for each organism and is influenced by a multitude of physical and chemical parameters (120). Maintenance of virulence by nonculturable pathogens varies in different species and is influenced by the conditions applied. Nonculturable cells of *C. jejuni* and *V. vulnificus* retain infectivity for mice (109, 127). Whether virulence of *Aeromonas salmonicida* persists when it is in the VBNC state is controversial (104). It was noted that the coccoid form of *H. pylori*, in contrast to the spiral form, binds poorly to gastric epithelial cells and induces little, if any, interleukin-8 secretion (31).

It is concluded from evidence gathered to date that pathogens, even though VBNC, maintain virulence genes and remain virulent. It is also concluded that in vivo passage can trigger VBNC cells to revert to the culturable, virulent state. Therefore, pathogens, even those in the VBNC state, pose a public health threat.

PUBLIC HEALTH AND EPIDEMIOLOGICAL SIGNIFICANCE

Environmental and clinical samples no longer can be considered free of pathogens if culturing yields negative results. Direct detection is required in instances of extraordinary host susceptibility, such as immunocompromised individuals and patients undergoing chemotherapy. For the general public, the presence of VBNC pathogens in water and food may be related to low-grade infections or so-called "aseptic" infections.

The morphology of bacterial cells in the VBNC state commonly is altered, and their size is significantly reduced (84; S. Chaiyanan et al., abstr. no. Q-66, Abstr. 97th Gen. Meet. Am. Soc. Microbiol., 1997). The most common occurrence is reduction in size and morphological conversion to an ovoid shape (40). Nonculturable pathogens in river or brackish water used for drinking as well as washing fruits and vegetables and utensils pose serious risk of infection. For example, results of microcosm studies clearly indicate that *V. cholerae* remains viable, even though nonrecoverable by plating onto agar media or culturing in broth (153).

Results of studies carried out in Bangladesh showed that 44% of surface water sources tested culture positive for *V. cholerae* in communities with reported cholera (58). In another study 10 years later, >63% of plankton samples collected from 10 ponds and river sites during a 3-year period were positive for *V. cholerae* O1 by fluorescent antibody direct detection (61). However, none of the plankton samples was positive by culture. More importantly, these results provided an explanation for the sudden appearance of cholera and the apparent disappearance of *V. cholerae* O1 from the environment in Bangladesh after epidemics receded. Huq et al. (61) demonstrated that *V. cholerae* O1 was, in fact, present in the environment year round, mostly in the nonculturable state, in pond and river water. The fluorescent monoclonal antibody method provided powerful evidence for the autochthonous presence of *V. cholerae* O1 in the aquatic environment. These water sources, regularly used for domestic purposes including drinking, posed a risk of infection especially during periods of plankton blooms (58). Earlier, Huq et al. (60) had reported that *V. cholerae* survives for longer periods of time when incubated with live, healthy copepods. Thus, when conditions are not favorable for growth and multiplication, *V. cholerae* O1 transforms to the nonculturable state in association with crustacean copepods (59, 60), as a mechanism for survival in the environment (38). *V. cholerae* O1 has been reported also to persist in the nonculturable state in laboratory microcosms of phytoplankton for more than 15 months (65). Persistence of *V. cholerae* in water in the VBNC state is an important public health factor, since detection will not be successful if only conventional culture methods are employed.

In summary, it is important to recognize that nonculturable bacteria are capable of producing disease. The first evidence of pathogenicity of nonculturable cells was the demonstration of fluid accumulation in rabbit ileal loops by VBNC *V. cholerae* O1 (34), followed by human volunteer experiments reported by Colwell et al. (35, 41). Additional experiments employing VBNC *E. coli*, wherein culturable cells were reisolated after passage through rabbit ileal loops 4 days postinoculation (50) and chick embryos died when injected with nonculturable cells of *L. pneumophila* (64), led to the conclusion that VBNC pathogens remain potentially pathogenic. Stool samples from patients symptomatically confirmed to be cholera cases, but culture negative when tested by cholera toxin (CT)-PCR, proved positive for *V. cholerae* (52). Three such culture-negative but CT-PCR-positive stool samples, further tested by fluorescent monoclonal antibody (14) and CholeraScreen (36), yielded positive results. Shirai et al. (137) described four stool samples from acute cholera patients who were negative by culture but positive by PCR for *ctx* gene. The four stool specimens were also examined for the presence of other enteric organisms and were reported negative. These clinical findings suggest a significant implication of VBNC pathogens for public health and a role in epidemiology, particularly at the onset of epidemics when patients may harbor nonculturable cells and yet remain undiagnosed since culture is the most common method for diagnosis in laboratories throughout the world.

Shigellosis caused by *Shigella* spp. is responsible for the largest number of deaths from all diarrheal illness among hospitalized patients (79) and has even been reported in the United States to result from swimming in contaminated water (128). Survival of VBNC *S. dysenteriae* type 1 cells for more than 6 months in microcosms (120) suggests a public health risk because of the extremely low infectious dose of *Shigella* for humans, i.e., as few as 10 bacteria. Earlier, Colwell et al. (34) reported that *S. sonnei* became VBNC when exposed to Chesapeake Bay water, remaining viable even though nonculturable for more than 21 days. The VBNC *S. dysenteriae* maintained virulence, as determined by retention of cytopathogenicity for HeLa cells (120).

VBNC *A. hydrophila* (J. A. K. Hasan et al., abstr. Q-139, Abstr. 91st Gen. Meet. Am. Soc. Microbiol., 1991) will not be detected if the diagnosis is made based only on culturing. It has been shown that divers diving in water contaminated with pathogens including *V. cholerae*, *A. hydrophila*, and *Pseudomonas aeruginosa* are colonized (throat, ear, and nose) without clinical symptoms being manifested. An acute infection was confirmed by detection of elevated serum antibody titers in blood samples collected 30 days after the dive, compared with predive blood samples (91). In this study, elevated antibody titers against *V. cholerae* in divers after diving at sites where VBNC cells of *V. cholerae* were detected were also recorded (62). In June 1991, VBNC cells of *V. cholerae* O1 were detected in plankton samples during field studies in the Black Sea. None of the water or plankton samples collected from the Black Sea was positive by culture (62), but the plankton samples were positive by fluorescent antibody. Four weeks after the field work was completed, a local newspaper in Odessa, Ukraine, reported an outbreak of cholera in that Black Sea coast city, where the beaches are used for recreational swimming and bathing: yet another example of VBNC cells playing a role in disease outbreaks. Such outbreaks can be averted through public health awareness, with a system in place for monitoring by appropriate methods other than culture alone.

Unfortunately, environmental studies will produce misleading results if only culture methods are employed. For example, Xu et al. (153) showed that a "die-off" of *E. coli* would be reported if plate count methods were used. In seawater of 25‰ salinity at 25°C, the total number of *E. coli* determined by direct counting methods remained the same while plate counts declined. Thus, under certain environmental conditions, seawater samples containing large numbers of *E. coli*, if tested by culture methods alone, would indicate few or no *E. coli* when large numbers of VBNC cells may, in fact, be present. In the case of shellfish harvesting areas, if coliform or fecal coliform counts are used to assess contamination by sewage or presence of potential human pathogens, the public health problem could be very serious, yet undetected (153).

In our laboratory, water and plankton samples are currently being analyzed by PCR using specific primers for *V. cholerae* based on the 16S-23S rRNA intergenic spacer region (30). Both qualitative and quantitative results can be obtained in a short period of time using this highly specific probe. Samples collected to date from the Chesapeake Bay were ToxR positive, cholera toxin (*ctxA*) negative, and toxin-coregulated pilus (*tcpA*) negative.

In conclusion, the evidence gathered over the past 20 years is sufficient to say that VBNC human pathogens cannot be considered dead but rather need to be viewed as a potential public health hazard. As better understanding of the genetic regulation of this phenomenon is gained, the risk to public health can be more precisely determined.

REFERENCES

1. **Alfimov, N. M.** 1954. Comparative evaluation of methods for determination of the bacterial count in the sea water. *Microbiology* **23:**693.
2. **Allen-Austin, D., B. Austin, and R. R. Colwell.** 1984. Survival of *Aeromonas salmonicida* in river water. *FEMS Microbiol. Lett.* **21:**143–146.
3. **Amann, R., W. Ludarg, and K. Schleifer.** 1994. Identification of uncultured bacteria: a challenging test for molecular taxonomists. *ASM News* **60:**360–365.
4. **Bakker, E. P., H. Rottenberg, and S. R. Caplan.** 1976. An estimation of the light-induced electrochemical potential difference of protons across the membrane of *Halobacterium halobium*. *Biochim. Biophys. Acta* **440:**557–572.
5. **Barcina, I., J. M. Gonzalez, J. Iniberri, and L. Egea.** 1989. Effect of visible light on progressive dormancy of *E. coli* cells during the survival process in natural fresh water. *Appl. Environ. Microbiol.* **55:**246–251.
6. **Bej, A. K., M. H. Mahbubani, J. L. D. Cesare, and R. M. Atlas.** 1991. PCR-gene probe detection of microorganisms using filter concentrated samples. *Appl. Environ. Microbiol.* **57:**3529–3534.
7. **Berlin, D. L., D. S. Herson, D. T. Hicks, and D. G. Hoover.** 1999. Response of pathogenic *Vibrio* species to high hydrostatic pressure. *Appl. Environ. Microbiol.* **65:**2776–2780.
8. **Beumer, R. R., J. De Vries, and F. M. Rombouts.** 1992. *Campylobacter jejuni* non-culturable coccoid cells. *Int. J. Food Microbiol.* **15:**153–163.
9. **Blake, P. A., M. H. Merson, R. E. Weaver, D. G. Hollis and P. C. Hueblein.** 1979. Disease caused by a marine vibrio: clinical characteristics and epidemiology. *N. Engl. J. Med.* **300:**1–5.
10. **Bocuzzi, V. M., W. L. Straube, J. Ravel, R. R. Colwell, and R. T. Hill.** 1998. Preparation of DNA extracted from environmental water samples for PCR amplification. *J. Microbiol. Methods* **31:**193–199.
11. **Bode, G., F. Mauch, and P. Melfertheiner.** 1993. The coccoid forms of *Helicobacter pylori*. Criteria for their viability. *Epidemiol. Infect.* **111:**483–490.
12. **Bowden, W. B.** 1977. Comparison of two direct-count techniques for enumerating aquatic bacteria. *Appl. Environ. Microbiol.* **33:**1229–1232.
13. **Bozue, J. A., and W. Johnson.** 1996. Interaction of *Legionella pneumophila* with *Acanthamoeba castellani*: uptake by coiling phagocytosis and inhibition of phagosome-lysosome fusion. *Infect. Immun.* **64:**668–673.
14. **Brayton, P., M. L. Tamplin, A. Huq, and R. R. Colwell.** 1987. Enumeration of *Vibrio cholerae* O1 in Bangladesh waters by fluorescent-antibody direct viable count. *Appl. Environ. Microbiol.* **53:** 2862–2865.
15. **Buck, G. E.** 1990. *Campylobacter pylori* as gastroduodenal disease. *Clin. Microbiol. Rev.* **3:**1–12.
16. **Butkevitch, V. S.** 1932. Zur Methodik der bacteriologischen Meresuntersuchungem und einige Augaben über die Verteilung dur Bakteriea im Wasser und inden Boden des Barends Meeres. *Trans. Oceanogr. Inst. Moscow* **2:**5–39.
17. **Butkevitch, V. S.** 1938. On the bacterial population of Caspian and Azov Seas. *Microbiology* (*Moscow*) **7:**1005–1021.
18. **Butkevitch, N. V., and V. S. Butkevitch.** 1936. Multiplication of sea bacteria depending on the composition of the medium and temperature. *Microbiology* (Moscow) **5:**3223.
19. **Byrd, J. J., and R. R. Colwell.** 1990. Maintenance of plasmids pBR322 and pUC8 in noncultivable *Escherichia coli* in the marine environment. *Appl. Environ. Microbiol.* **56:**2104–2107.

20. **Byrd, J. J., H. S. Xu, and R. R. Colwell.** 1991. Viable but non-culturable bacteria in drinking water. *Appl. Environ. Microbiol.* **57:**875–878.
21. **Cappelier, J. M., and M. Federighi.** 1998. Demonstration of viable but nonculturable state for *Campylobacter jejuni. Rev. Med. Vet.* **149:**319–326
22. **Cholera Working Group, International Center for Diarrhoeal Diseases Research, Bangladesh.** 1993. Large epidemic of cholera-like disease in Bangladesh caused by *Vibrio cholerae* O139 synonym Bengal. *Lancet* **342:**387–390.
23. **Chongsa-nguan, M., W. Chaicumga, P. Moolarsart, P. Kandhasingha, T. Shinada, H. Kurazono, and Y. Takeda.** 1993. *Vibrio cholerae* O139 Bengal in Bangkok. *Lancet* **342:**430–431.
24. **Chowdhury, M. A. R., H. Yamanaka, S. Miyoshi, K. M. S. Aziz, and S. Shimoda.** 1989. Ecology of *Vibrio mimicus* in aquatic environments. *Appl. Environ. Microbiol.* **55:**2073–2078.
25. **Chowdhury, M. A. R., H. Yamanaka, S. Miyoshi, and S. Shimoda.** 1990. Ecology and seasonal distribution of *Vibrio parahaemolyticus* in aquatic environments of a temperate region. *FEMS Microbiol. Ecol.* **74:**1–9.
26. **Chowdhury, M. A. R., S. Miyoshi, H. Yamanaka, and S. Shimoda.** 1992. Ecology and distribution of toxigenic *V. cholerae* in aquatic environments of a temperate region. *Microbios* **72:**203–213.
27. **Chowdhury, M. A. R., R. T. Hill, and R. R. Colwell.** 1994. A gene for the enterotoxin zonula occludens toxin is present in *Vibrio mimicus* and *Vibrio cholerae* O139. *FEMS Microbiol. Lett.* **119:** 377–380.
28. **Chowdhury, M. A. R., R. T. Hill, A. Huq, and R. R. Colwell.** 1995. Physiology and molecular genetics of viable but nonculturable microorganisms, p. 105–122. *In* M. Levin, C. Grim, and J. S. Angle (ed.), *Biotechnology and Risk Assessment.* Univ. of Maryland Biotechnology Inst., Baltimore, Md.
29. **Chowdhury, M. A. R., B. Xu, R. Montilla, J. A. K. Hasan, A. Huq, and R. R. Colwell.** 1995. A simplified immunofluorescence technique for detection of viable cells of *Vibrio cholerae* O1 and O139. *J. Microbiol. Methods* **24:**165–170.
30. **Chun, J., A. Huq, and R. R. Colwell.** 1999. Analysis of 16S–23S rRNA intergenic spacer regions of *Vibrio cholerae* and *Vibrio mimicus. Appl. Environ. Microbiol.* **65:**2202–2208.
31. **Cole, S. P., D. Cirillo, M. F. Kagnoff, D. G. Guiney, and L. Eckmann.** 1997. Coccoid and spiral *Helicobacter pylori* differ in their abilities to adhere to gastric epithelial cells and induce interleukin-8 secretion. *Infect. Immun.* **65:**843–846.
32. **Collins, V. G., and C. Kipling.** 1957. The enumeration of waterborne bacteria by a new direct count method. *J. Appl. Bacteriol.* **20:**257–264.
33. **Colwell, R. R., J. Kaper, and S. W. Joseph.** 1977. *Vibrio cholerae, Vibrio parahaemolyticus* and other vibrios: occurrence and distribution in Chesapeake Bay. *Science* **198:**394–396.
34. **Colwell, R., P. Brayton, D. Grimes, D. Roszak, S. Huq, and L. Palmer.** 1985. Viable, but nonculturable *Vibrio cholerae* and related pathogens in the environment: implications for release of genetically engineered microorganisms. *Bio/Technology* **3:**817–820.
35. **Colwell, R. R., M. L. Tamplin, P. R. Brayton, A. L. Gauzens, B. D. Tall, D. Harrington, M. M. Levine, S. Hall, A. Huq, and D. A. Sack.** 1990. Environmental aspects of *V. cholerae* in transmission of cholera, p. 327–343. *In* R. B. Sack and Y. Zinnaka (ed.), *Advances in Research on Cholera and Related Diarrhoeas,* 7th ed. K. T. K. Scientific Publishers, Tokyo, Japan.
36. **Colwell, R. R., J. A. K. Hasan, A. Huq, L. Loomis, R. J. Siebling, M. Torres, S. Galvez, S. Islam, and D. Bernstein.** 1992. Development and evaluation of a rapid, simple sensitive monoclonal antibody-based co-agglutination test for direct detection of *V. cholerae* O1. *FEMS Microbiol. Lett.* **97:**215–220.
37. **Colwell, R. R., A. Huq, K. A. Cunningham, and G. Losonsky.** 1992. Prospective study of diving-associated illnesses, p. 63–70. *Proc. Int. Symp. on Hazards of Diving in Polluted Waters.* Maryland Sea Grant College Publication No. UM-SG-TS-92-02. University of Maryland, College Park, Md.
38. **Colwell, R. R., and A. Huq.** 1994. Vibrios in the environment: viable but nonculturable *Vibrio cholerae,* p. 117–133. *In* I. K. Wachsmuth, O. Olsvik, and P. A. Blake (ed.), *Vibrio cholerae and Cholera: Molecular to Global Perspectives.* ASM Press, Washington, D.C.
39. **Colwell, R. R., and A. Huq.** 1999. Global microbial ecology: biogeography and diversity of *Vibrios* as a model. *J. Appl. Microbiol. Symp. Suppl.* **85:**134S–137S.

40. **Colwell, R. R., and W. M. Spira.** 1992. The ecology of *Vibrio cholerae*, p. 107–127. *In* D. Barua and W. B. Greenough III (ed.), *Cholera* Plenum Medical Book Company, New York, N.Y.
41. **Colwell, R. R., P. Brayton, D. Herrington, B. Tall, A. Huq, and M. M. Levine.** 1996. Viable but non-culturable *Vibrio cholerae* O1 revert to a cultivable state in the human intestine. *World J. Microbiol. Biotechnol.* **12:**28–31.
42. **Davis, B. R., G. R. Fanning, J. M. Madden, A. G. Steigerwalt, H. B. Bradford, Jr., H. L. Smith, Jr., and D. J. Breuner.** 1981. Characterization of biochemically atypical *Vibrio cholerae* strains and designation of a new pathogenic species, *Vibrio mimicus. J. Clin. Microbiol.* **14:**631–639.
43. **Dawe, L. L., and W. R. Penrose.** 1978. Bactericidal property of seawater: death or deliberation? *Appl. Environ. Microbiol.* **35:**829–833.
44. **Duncan, S., L. A. Glover, K. Killham, and J. I. Prosser.** 1994. Luminescence-based detection of activity of starved and viable but non-culturable bacteria. *Appl. Environ. Microbiol.* **60:**1308–1316.
45. **Finlay, B., and S. Falkow.** 1997. Common themes in microbial pathogenicity revisited. *Microbiol. Mol. Biol. Rev.* **61:**136–169.
46. **Fry, J. C.** 1990. Direct methods and biomass estimation. *Methods Microbiol.* **22:**41–85.
47. **Fuhrman, J. A., D. E. Comeau, A. Hagstrom, and A. M. Cham.** 1988. Extraction from natural planktonic microorganisms of DNA suitable for molecular biological studies. *Appl. Environ. Microbiol.* **54:**1426–1429.
48. **Gonzalez, J. M., J. Iriberri, L. Egea, and I. Barcina.** 1992. Characterization of culturable protistan grazing and death of enteric bacteria in aquatic ecosystem. *Appl. Environ. Microbiol.* **58:**998–1004.
49. **Goodwin, C. S., and J. A. Armstrong.** 1990. Microbiological aspects of *Helicobacter pylori* (*Campylobacter pylori*). *Eur. J. Clin. Microbiol. Infect. Dis.* **9:**1–13.
50. **Grimes, D. J., and R. R. Colwell.** 1986. Viability and virulence of *Escherichia coli* suspended by membrane chamber in semi-tropical ocean water. *FEMS Microbiol. Lett.* **34:**161–165.
51. **Hasan, J. A. K., A. Huq, M. L. Tamplin, R. Siebeling, and R. R. Colwell.** 1994. A novel kit for rapid detection of *V. cholerae* O1. *J. Clin. Microbiol.* **32:**249–252.
52. **Hasan, J. A. K., M. A. R. Chowdhury, M. Shahabuddin, A. Huq, L. Loomis, and R. R. Colwell.** 1994. Polymerase chain reaction for the detection of cholera. Toxin genes in viable but non-culturable *V. cholerae* O1. *World J. Microbiol. Biotechnol.* **10:**568–571.
53. **Hasan, J. A. K., D. Bernstein, A. Huq, L. Loomis, M. L. Tamplin, and Rita R. Colwell.** 1994. Cholera DFA: an improved direct fluorescent monoclonal antibody staining kit for rapid detection and enumeration of *Vibrio cholerae* O1. FEMS Microbiol. Lett. **120:**143–148.
54. **Heidelberg, J. F., K. R. O'Neill, D. Jacobs, and R. R. Colwell.** 1993. Enumeration of *Vibrio vulnificus* on membrane filters with a fluorescently labeled oligonucleotide probe specific for kingdom-level 16S rRNA sequences. *Appl. Environ. Microbiol.* **59:**3474–3476.
55. **Heidelberg, J. F., M. Shahamat, M. A. Levin, I. Rahman, and R. R. Colwell.** 1997. Effect of aerosolization on culturability and viability of gram-negative bacteria. *Appl. Environ. Microbiol.* **63:** 3585–3588.
56. **Hoff, K. A.** 1989. Survival of *Vibrio anguillarum* and *Vibrio salmonicida* at different salinity. *Appl. Environ. Microbiol.* **55:**1775–1786.
57. **Höller, C., D. Witthuhn, and B. Janzen-Blunck.** 1998. Effect of low temperatures on growth, structure, and metabolism of *Campylobacter coli* SP10. *Appl. Environ. Microbiol.* **64:**581–587.
58. **Hughes, J. M., J. M. Boyce, R. J. Levine, M. U. Khan, K. M. A. Aziz, M. I. Huq, and G. T. Curlin.** 1982. Epidemiology of El Tor cholera in rural Bangladesh: importance of surface water in transmission. *Bull. W.H.O.* **60:**395–404.
59. **Huq, A., E. B. Small, P. A. West, M. I. Huq, R. Rahman, and R. R. Colwell.** 1983. Ecological relationships between *Vibrio cholerae* and planktonic crustacean copepods. *Appl. Environ. Microbiol.* **45:**275–283.
60. **Huq, A., E. Small, P. West, and R. R. Colwell. 1984.** The role of planktonic copepods in the survival and multiplication of *Vibrio cholerae* in the environment, p. 521–534. *In* R. R. Colwell (ed.), *Vibrios in the Environment*. John Wiley & Sons, New York, N.Y.
61. **Huq, A., R. R. Colwell, R. Rahman, A. Ali, M. A. R. Chowdhury, S. Parveen, D. A. Sack, and E. Russek-Cohen.** 1990. Detection of *V. cholerae* O1 in the aquatic environment by fluorescent monoclonal antibody and culture method. *Appl. Environ. Microbiol.* **56:**2370–2373.

62. **Huq, A., J. A. K. Hasan, G. Losonsky, V. Diomin, and R. R. Colwell.** 1994. Colonization of professional divers by toxigenic *Vibrio cholerae* O1 and *V. cholerae* non-O1 at dive sites in the United States, Ukraine and Russia. *FEMS Microbiol. Lett.* **120:**137–142.
63. **Huq, A., and R. R. Colwell.** 1995. A microbiological paradox: viable but nonculturable bacteria with special reference to *Vibrio cholerae*. *J. Food Protect.* **59:**96–101.
64. **Hussong, D., R. R. Colwell, M. O'Brien, E. Weiss, A. D. Pearson, R. M. Weiner, and W. D. Burge.** 1987. Viable *Legionella pneumophila* not detectable by culture on agar media. *Bio/Technology* **5:**947–952.
65. **Islam, M. S., B. S. Drasar, and D. J. Bradley.** 1990. Long-term persistence of toxigenic *V. cholerae* O1 in the mucilagenous sheath of a blue-green alga, *Anabaena variabilis*. *J. Trop. Med. Hyg.* **93:** 133–139.
66. **Islam, M. S., M. K. Hasan, M. A. Miah, G. C. Sur, A. Felsenstein, M. Venkatesan, R. B. Sack, and M. J. Albert.** 1993. Use of the polymerase chain reaction and fluorescent antibody methods for detecting viable but nonculturable *Shigella dysenteriae* type 1 in laboratory microcosms. *Appl. Environ. Microbiol.* **59:**536–540.
67. **James, B. W., W. S. Mauchline, P. J. Dennis, C. W. Keevil, and R. Wait.** 1999. Poly-3-hydroxibutyrate in *Legionella pneumophila*, an energy source for survival in low-nutrient environments. *Appl. Environ. Microbiol.* **65:**822–827.
68. **Janda, J. M., E. J. Botton, and M. Reltano.** 1983. *Aeromonas* species in clinical microbiology: significance, epidemiology, and specification. *Diagn. Microbiol. Infect. Dis.* **1:**221–228.
69. **Jannasch, H. W.** 1969. Estimations of bacterial growth rates in natural waters. *J. Bacteriol.* **99:** 156–160.
70. **Jennison, M. W.** 1937. Relations between plate counts and direct microscopic counts of *E. coli* during logarithmic growth period. *J. Bacteriol.* **33:**461–469.
71. **Jentsch, T. J., A. M. Garcia, and H. F. Lodish.** 1989. Primary structure of a noted 4-acetamido-4′isothiocyanostilbene-2-2′disulphonic acid (SITS)-binding membrane protein lights expressed in Torpedo California electroplax. *Biochem. J.* **261:**155.
72. **Jepras, R. I., J. Carter, S. C. Pearson, F. E. Paul, and M. J. Wilkinson.** 1995. Development of a robust flow cytometric assay for determining numbers of viable bacteria. *Appl. Environ. Microbiol.* **61:**2696–2701.
73. **Jiang, X., and T. Chai.** 1996. Survival of *Vibrio parahaemolyticus* at low temperatures under starvation conditions and subsequent resuscitation of viable, nonculturable cells. *Appl. Environ. Microbiol.* **62:**1300–1305.
74. **Johnston, J. M., S. F. Becker, and L. M. McFarland.** 1986. Gastroenteritis in patients with stool isolates of *Vibrio vulnificus*. *Am. J. Med.* **80:**336–338.
75. **Jones, D. M., E. M. Sutcliffe, and A. Curry.** 1991. Recovery of viable non-culturable *Campylobacter jejuni*. *J. Gen. Microbiol.* **137:**2477–2482.
76. **Kaneko, T., and R. R. Colwell.** 1973. Ecology of *Vibrio parahaemolyticus* in Chesapeake Bay. *J. Bacteriol.* **113:**24–32.
77. **Karsinkin, G. S., and S. J. Kusnetsov.** 1931. Neue Methoden in der Limnologie. *Arb. Limnol. Sta. Kossino.* **13:**47–48. (In Russian, with German summary.)
78. **Kepner, R. L., Jr., and J. R. Pratt.** 1994. Use of fluorochromes for direct enumeration of total bacteria in environmental samples: past and present. *Microbiol. Rev.* **58:**603–615.
79. **Khan, M. U., G. T. Curlin, and M. I. Huq.** 1979. Epidemiology of *Shigella dysenteriae* type 1 infections in Dacca urban area. *J. Trop. Geogr. Med.* **31:**213–223.
80. **Kim, Y. M., B. H. Lee, S. H. Lee, and T. S. Lee.** 1990. Distribution of *Vibrio vulnificus* in seawater of Kwangan Beach, Pusan, Korea. *Bull. Korean Fish. Soc.* **22:**385–390.
81. **Klein, P. D., D. Y. Graham, A. Gaillour, A. R. Opekum, and E. O. Smith.** 1991. Water source as risk factor for *H. pylori* infection in Peruvian children. *Lancet* **337:**1503–1505.
82. **Knaysi, G.** 1935. A microscopic method of distinguishing dead from living cells. *J. Bacteriol.* **30:** 193–206.
83. **Kogure, K., U. Simidu, and N. Taga.** 1979. A tentative direct microscopic method for counting living marine bacteria. *Can. J. Microbiol.* **25:**415–420.
84. **Kondo, K., A. Takade, and K. Amako.** 1994. Morphology of the viable but non-culturable *Vibrio cholerae* as determined by the freeze fixation technique. *FEMS Microbiol. Lett.* **123:**179–184.

85. **Kurath, G., and Y. Morita.** 1983. Starvation survival and physiological studies of a marine *Pseudomonas* spp. *Appl. Environ. Microbiol.* **45:**1206–1211.
86. **Lambert, J. R., S. K. Lin, and J. Aranda-Michel.** 1995. *Helicobacter pylori. Scand. J. Gastroenterol.* **30**(Suppl. 208)**:**33–46.
87. **Lebaron, P., N. Parthuisot, and P. Catala.** 1998. Comparison of blue nucleic acid dyes for flow cytometric enumeration of bacteria in aquatic systems. *Appl. Environ. Microbiol.* **64:**1725–1730.
88. **Lee, K., and E. G. Ruby.** 1995. Symbiotic role of the viable but non-culturable state of *V. fischeri* in Hawaiian coastal waters. *Appl. Environ. Microbiol.* **61:**278–283.
89. **Lennette, E. H., A. Balows, W. J. Hausler, Jr., and H. J. Shadomy (ed.).** 1985. *Manual of Clinical Microbiology*, 4th ed. American Society of Microbiology, Washington, D.C.
90. **Linder, K., and J. D. Oliver.** 1989. Membrane fatty acid and virulence changes in the viable but nonculturable state of *Vibrio vulnificus*. *Appl. Environ. Microbiol.* **55:**2837–2842.
91. **Losonsky, G. A., J. A. K. Hasan, A. Huq, S. Kaintuch, and R. R. Colwell.** 1994. Serum antibody responses of divers to waterborne pathogens. *J. Clin. Diagnos. Lab. Immun.* **1:**182–185.
92. **Magarinos, B., J. L. Romalde, J. L. Barja, and A. E. Toranzo.** 1994. Evidence of a dormant but infective state of the fish pathogen *Pasteurella piscicida* in seawater and sediment. *Appl. Environ. Microbiol.* **60:**180–186.
93. **Mai, U. E. H., M. Shahamat, and R. R. Colwell.** 1990. Survival of *Helicobacter pylori* in the aquatic environment, p. 90–96. *In* H. Menge, M. Gregor, G. N. J. Tytgat, B. J. Marshal, and C. A. M. McNulty (ed.), *Proceedings of the 2nd International Symposium on Helicobacter pylori*, August 25–26, Bad Nauheim, Berlin. Springer-Verlag, Berlin, Germany.
94. **Martins, M. T., P. S. Sanchez, M. I. Z. Sato, P. R. Brayton, and R. R. Colwell.** 1993. Detection of *Vibrio cholerae* O1 in the aquatic environment in Brazil employing direct immunofluorescence microscopy. *World J. Microbiol. Biotechnol.* **9:**390–392.
95. **Martins, M. T., I. G. Rivera, D. L. Clark, and B. H. Olson.** 1992. Detection of virulence factors in culturable *Escherichia coli* isolates from water samples by DNA probes and recovery of toxin-bearing strains in minimal *o*-nitrophenol-beta-D-galactopyranoside-4-methylumbelliferyl-beta-D-glucoronide media. *Appl. Environ. Microbiol.* **58:**3095–3100.
96. **Mason, J. D., L. A. R. Allman, J. M. Stark, and D. Lloyd.** 1995. The ability of membrane potential dyes and calcofluor white to distinguish between viable and nonviable bacteria. *J. Appl. Bacteriol.* **78:**309–315.
97. **Maurelli, A. T., A. E. Hromockyj, and M. L. Bernardini.** 1992. Environmental regulation of *Shigella* virulence. *Curr. Top. Microbiol. Immunol.* **180:**95–116.
98. **McFeters, G. A., and D. G. Stuart.** 1972. Survival of coliform bacteria in natural waters: field and laboratory studies with membrane filter chambers. *Appl. Environ. Microbiol.* **24:**805–811.
99. **Medema, G. J., F. M. Schets, A. W. van de Giessen, and A. H. Haveljar.** 1992. Lack of colonization of 1 day chicks by viable non-culturable *Campylobacter jejuni*. *J. Appl. Bacteriol.* **72:** 512–516.
100. **Mekalanos, J. J.** 1992. Environmental signals controlling expression of virulence determinants in bacteria. *J. Bacteriol.* **174:**1–7.
101. **Miller, J. F., J. J. Mekalanos, and S. Falkow.** 1989. Coordinate regulation and sensory transduction in the control of bacterial virulence. *Science* **243:**916–922.
102. **Moran, A. P., and M. E. Upton.** 1987. Factors affecting production of coccoid forms by *Campylobacter jejuni* on solid media during incubation. *J. Appl. Bacteriol.* **62:**527–537.
103. **Morgan, J. A. W., C. Winstanley, R. W. Pickup, J. A. Jones, and J. R. Saunders.** 1989. Direct phenotypic and genotypic detection of a recombinant pseudomonal population released into lake water. *Appl. Environ. Microbiol.* **55:**2537–2544.
104. **Morgan, J. A. W., G. Rhodes, and R. W. Pickup.** 1993. Survival of nonculturable *Aeromonas salmonicida* in lake water. *Appl. Environ. Microbiol.* **59:**874–880.
105. **Narikawa, S., S. Kawai, H. Aoshima, O. Kawamata, R. Kawaguchi, K. Hikiji, M. Kato, S. Iino, and Y. Mizushima.** 1997. Comparison of the nucleic acids of helical and coccoid forms of *Helicobacter pylori*. *Clin. Diagn. Lab. Immunol.* **4:**285–290.
106. **Nilius, M., A. Ströhle, G. Bode, and P. Malfertheiner.** 1993. Coccoid like forms (CLF) of *Helicobacter pylori*. Enzyme activity and antigenicity. *Int. J. Med. Microbiol. Virol. Parasitol. Infect. Dis.* **280:**259–272.

107. **O'Neill, K. R., S. H. Jones, and D. J. Grimes.** 1992. Seasonal incidence of *Vibrio vulnificus* in the Great Bay Estuary of New Hampshire and Maine. *Appl. Environ. Microbiol.* **58:**3257–3262.
108. **Oliver, J. D., and D. Wanucha.** 1989. Survival of *V. vulnificus* at reduced temperatures and elevated nutrients. *J. Food Safety* **10:**79–86.
109. **Oliver, J. D.** 1993. Formation of viable but non-culturable cells, p. 239–272. *In* S. Kjelleberg (ed.), *Starvation in Bacteria.* Plenum Press, New York, N.Y.
110. **Pace, J. L., T. Chai, H. A. Rossi, and X. Jiang.** 1997. Effect of bile on *Vibrio parahaemolyticus*. *Appl. Environ. Microbiol.* **63:**2372–2377.
111. **Paszko-Kolva, C., M. Shahamat, H. Yamamoto, T. Sawyer, J. Vives-Rego, and R. R. Colwell.** 1991. Survival of *Legionella pneumophila* in the aquatic environment. *Microbiol. Ecol.* **22:**75–83.
112. **Paszko-Kolva, C., M. Shahamat, and R. R. Colwell.** 1992. Long-term survival of *Legionella pneumophila* serogroup 1 under low nutrient conditions and associated morphological changes. *FEMS Microbiol. Ecol.* **102:**45–55.
113. **Peterson, W. L.** 1991. *Helicobacter pylori* and peptic ulcer disease. *N. Engl. J. Med.* **324:**1043–1048.
114. **Phadnis, S. H., M. H. Parlow, M. Levy, D. Ilver, C. M. Caulkins, J. B. Connors, and B. E. Dunn.** 1996. Surface localization of *Helicobacter pylori* urease and a heat shock protein homolog requires bacterial autolysis. *Infect. Immun.* **64:**905–912.
115. **Pommepuy, M., M. Butin, A. Derrien, M. Gourmelon, R. R. Colwell, and M. Cormier.** 1996. Retention of enteropathogenicity by viable but nonculturable *Escherichia coli* exposed to seawater and sunlight. *Appl. Environ. Microbiol.* **62:**4621–4626.
116. **Pommepuy M., L. Fiksdal, M. Gourmelon, H. Melikechi, M. L. Caprais, M. Cormier, and R. R. Colwell.** 1996. Effect of seawater on *Escherichia coli* β-galactosidase activity. *J. Appl. Bacteriol.* **81:**174–180.
117. **Porter, K. G., and Y. S. Fieg.** 1980. The use of DAPI for identifying and counting aquatic microflora. *Limnol. Oceanogr.* **25:**943–948.
118. **Postgate, J. R.** 1969. Viable counts and viability, p. 611–628. *In* J. R. Norris and D. W. Ribbons (ed.), *Methods in Microbiology.* Academic Press, Inc., London, United Kingdom.
119. **Radsimosky, R.** 1930. Vorlantige Augaben über die Dichtigheit der berkteriellen Besiedlung einiger Gewasser. *Trav. Sta. Biol. Dmiepre.* **5:**385–402.
120. **Rahman, I., M. Shahamat, P. A. Kirchman, E. Russek-Cohen, and R. R. Colwell.** 1994. Methionine uptake and cytopathogenicity of viable but nonculturable *Shigella dysenteriae* type 1. *Appl. Environ. Microbiol.* **60:**3573–3578.
121. **Rahman, I., M. Shahamat, M. A. R. Chowdhury, and R. R. Colwell.** 1996. Potential virulence of viable nonculturable *Shigella dysenteriae* type I. *Appl. Environ. Microbiol.* **62:**115–120.
122. **Ramamurthy, T., R. Garg, S. K. Sharma, G. B. Nair, T. Shimada, T. Takeda, T. Karasawa, H. Kuraziano, A. Pal, and Y. Takeda.** 1993. Emergence of novel strains of *V. cholerae* with epidemic potential in Southern and Eastern India. *Lancet* **341:**703–705.
123. **Ravel, J., R. T. Hill, and R. R. Colwell.** 1994. Isolation of a *Vibrio cholerae* transposon-mutant with an altered viable but nonculturable response. *FEMS Microbiol. Lett.* **120:**57–62.
124. **Ravel, J., I. T. Knight, C. E. Monahan, R. T. Hill, and R. R. Colwell.** 1995. Temperature-induced recovery of *Vibrio cholerae* from the viable but nonculturable state: growth or resuscitation?. *Microbiology* **141:**377–383.
125. **Rivera, I. G., M. A. R. Chowdhury, A. Huq, D. Jacobs, M. T. Martins, and R. R. Colwell.** 1995. Enterobacterial repetitive intergenic consensus sequences and the PCR to generate fingerprints of genomic DNA from *Vibrio cholerae* O1, O139, and non-O1. *Appl. Environ. Microbiol.* **61:**2898–2904.
126. **Rodriguez, G. G., D. Phipps, K. Ishiguro, and H. F. Ridgway.** 1992. Use of a fluorescent redox probe for direct visualization of actively respiring bacteria. *Appl. Environ. Microbiol.* **58:**1801–1808.
127. **Rollins, D. M., and R. R. Colwell.** 1986. Viable but non-culturable stage of *Campylobacter jejuni* and its role in survival in the natural aquatic environment. *Appl. Environ. Microbiol.* **52:**531–538.
128. **Rosenberg, M. L., K. K. Hazlet, J. Schaefer, J. C. Wells, and R. C. Pruneda.** 1976. Shigellosis from swimming. *JAMA* **236:**1849–1852.

129. **Roszak, D. B., D. J. Grimes, and R. R. Colwell.** 1984. Viable but non-recoverable stage of *Salmonella enteritidis* in aquatic systems. *Can. J. Microbiol.* **30:**334–338.
130. **Roszak, D. B., and R. R. Colwell.** 1987. Survival strategies of bacteria in the natural environment. *Microbiol. Rev.* **51:**365–379.
131. **Rottenberg, H.** 1979. The measurement of membrane potential and ΔpH in cells, organelles, and vesicles. *Methods Enzymol.* **55:**547–569.
132. **Sack, D. A., C. O. Tackt, M. B. Cohen, R. B. Sack, G. A. Losonsky, J. Shimko, J. P. Nataro, R. Edelman, M. M. Levine, R. A. Giannella, G. Schiff, and D. Lang.** 1998. Validation of a volunteer model of cholera with frozen bacteria as the challenge. *Infect. Immun.* **66:**1968–1972.
133. **Saha, S. K., S. Saha, and S. C. Sanyal.** 1991. Recovery of injured *Campylobacter jejuni* cells after animal passage. *Appl. Environ. Microbiol.* **57:**3388–3389.
134. **Sato, M. I. Z., P. S. Sanchez, I. G. Rivera, and M. T. Martins.** 1995. Survival of culturable *Vibrio cholerae* O1 and non-O1 in seawater, freshwater and wastewater and effect of the water environmental on enterotoxin production. *Rev. Microbiol.* (Brazil) **26:**83–89.
135. **Shahamat, M., U. Mai, C. Paszko-Kolva, M. Kessel, and R. Colwell.** 1993. Use of autoradiography to assess viability of *Helicobacter pylori* in water. *Appl. Environ. Microbiol.* **59:**1231–1235.
136. **Shiba, T., R. T. Hill, W. L. Straube, and R. R. Colwell.** 1995. Decrease in culturability of *Vibrio cholerae* caused by glucose. *Appl. Environ. Microbiol.* **61:**2583–2588.
137. **Shirai, H., M. Nishibuchi, T. Ramamurthy, S. K. Bhattacharya, S. C. Pal, and Y. Takeda.** 1991. Polymerase chain reaction for detection of cholera enterotoxin operon of *V. cholerae*. *J. Clin. Microbiol.* **29:**2517–2521.
138. **Sleightholme, V., and D. Roberts.** 1994. Viable but non-culturable *V. cholerae* O1: a short review. *Public Health Lab. Serv. Microbiol. Digest* **11:**77–80.
139. **Sommerville, C. C., I. T. Knight, W. L. Straube, and R. R. Colwell.** 1989. Simple rapid method for direct isolation of nucleic acid from aquatic environment. *Appl. Environ. Microbiol.* **55:**548–559.
140. **Sorscher, E. J., C. M. Fuller, and R. J. Bridges.** 1992. Identification of a membrane protein from T84 cells using antibodies made against a DIDS-binding peptide. *Am. J. Physiol.* **262:**C136.
141. **Steinert, M., L. Emödy, R. Amann, and J. Hacker.** 1997. Resuscitation of viable but nonculturable *Legionella pneumophila* Philadelphia JR32 by *Acanthamoeba castellanii*. *Appl. Environ. Microbiol.* **63:**2047–2053.
142. **Stevenson, L. H.** 1978. A case for bacterial dormancy in aquatic systems. *Microb. Ecol.* **4:**127–133.
143. **Strugger, S.** 1948. Fluorescence microscope examination of bacteria in soil. *Can. J. Res. Sect.* **26:** 188–193.
144. **Tholozan, J. L., J. M. Cappelier, J. P. Tissier, G. Gelattre, and M. Federighi.** 1999. Physiological characterization of viable-but-nonculturable *Campylobacter jejuni* cells. *Appl. Environ. Microbiol.* **65:**1110–1116.
145. **Turpin, P. E., K. A. Maycroft, C. L. Rowlands, and E. M. H. Wellington.** 1993. Viable but non-culturable salmonellas in soil. *J. Appl. Bacteriol.* **74:**421–427.
146. **Valentine, R. C., and J. R. G. Bradfield.** 1954. The urea method for bacterial viability counts with electron microscope and its relation to other viability counting methods. *J. Gen. Microbiol.* **11:**349–357.
147. **Virta, M., M. Karp, S. Rönnemaa, and E. M. Lilius.** 1997. Kinetic measurements of the membranolytic activity of serum complement using bioluminescent bacteria. *J. Immunol. Methods* **201:** 215–221.
148. **Warner, J. M., and J. D. Oliver.** 1998. Randomly amplified polymorphic DNA analysis of starved and viable but nonculturable *Vibrio vulnificus* cells. *Appl. Environ. Microbiol.* **64:**3025–3028.
149. **Weichart, D., J. D. Oliver, and S. Kjelleberg.** 1992. Low temperature induced non-culturability and killing of *Vibrio vulnificus*. *FEMS Microbiol. Lett.* **100:**205–210.
150. **Weichart, D., D. McDougald, D. Jacobs, and S. Kjelleberg.** 1997. In situ analysis of nucleic acids in cold-induced nonculturable *Vibrio vulnificus*. *Appl. Environ. Microbiol.* **63:**2754–2758.
151. **Whitesides, M. D., and J. D. Oliver.** 1997. Resuscitation of *Vibrio vulnificus* from the viable but nonculturable state. *Appl. Environ. Microbiol.* **63:**1002–1005.

152. **Wolf, P. W., and J. D. Oliver.** 1992. Temperature effects on the viable but nonculturable state of *V. vulnificus*. *FEMS Microbiol. Ecol.* **101:**33–39.
153. **Xu, H. S., N. Roberts, F. L. Singleton, R. W. Attwell, D. J. Grimes, and R. R. Colwell.** 1982. Survival and viability of nonculturable *Escherichia coli* and *Vibrio cholerae* in the estuarine and marine environment. *Microb. Ecol.* **8:**313–323.
154. **Xu, H. S., N. C. Roberts, L. B. Adams, P. A. West, R. J. Seibeling, A. Huq, M. I. Huq, R. Rahman, and R. R. Colwell.** 1984. An indirect fluorescent antibody staining procedure for detection of *Vibrio cholerae* serovar O1 cells in aquatic environmental samples. *J. Microb. Methods* **2:** 221–231.
155. **Zimmerman, R., R. Iturriaga, and J. Becker-Birck.** 1978. Simultaneous determination of the total number of aquatic bacteria and the number thereof involved in respiration. *Appl. Environ. Microbiol.* **36:**926–935.
156. **ZoBell, C. E.** 1946. Marine Microbiology: a Monograph on *Hydrobacteriology*, p. 41–58. Carnica Botanica Co., Waltham, Mass.

Nonculturable Microorganisms in the Environment
Edited by R. R. Colwell and D. J. Grimes

Chapter 18

Bacterial Death Revisited

Rita R. Colwell

The subject of anabiosis or "latent life" has long intrigued microbiologists and philosophers. It is well established that some microorganisms, notably the gram-positive spore formers, can under certain conditions be deprived of all visible signs of life, and yet these organisms are not dead. When their original conditions are restored, they can return to normal life and activity (61). This state of an organism has been referred to in the older literature variously as "viable lifelessness," suspended animation, viability, latent life, and even by the not very suitable term "anabiosis," meaning "latent life." This last term was used to describe the state of an organism when its metabolic activity is at lowest ebb, reaching a hardly measurable value and, in some cases, the physiological and biochemical processes being reversibly arrested for varying periods of time.

Not long ago, R. J. Cano (California Polytechnic State University at San Luis Obispo) reported reviving bacteria, Lazarus-like, from a 30-million-year-old dormancy (94). Cano and Borucki (23) reportedly recovered bacterial spores of *Bacillus sphaericus* from the stomachs of ancient bees embedded in amber. Subsequently, T. Hamamoto and K. Horikoshi, Institute of Physical and Chemical Research, Wako, Japan (48), claimed to have isolated *Bacillus subtilis* from ancient amber.

A debate over putative microorganisms in a Martian meteorite collected in Antarctica has also contributed to the controversy about survival and longevity of microorganisms in the environment. At a press conference on August 7, 1996, NASA scientists announced that they had evidence of past microbial life on Mars (81). The possibility of life on Mars was strongly supported by Percival Lowell (76), founder of Lowell Observatory, Flagstaff, Arizona. However, the purported lack of water on Mars is the strongest argument against life on that planet. The conclusion drawn by McKay and colleagues is based on results of studies of a small rock, "Martian meteorite ALH84001" (43). The meteorite landed in Antarctica ca. 13,000 years ago and contains carbonate "globules" approximately 3.6

Rita R. Colwell • Center of Marine Biotechnology, Columbus Center, University of Maryland Biotechnology Institute, 701 E. Pratt St., Baltimore, MD 21202, and Department of Cell and Molecular Biology, University of Maryland, College Park, MD 20742.

billion years old. The globules contain polycyclic aromatic hydrocarbons (PAHs), detectable at levels of parts per million. Scanning electron micrographs reveal small, regularly shaped ovoid and elongated forms 20 to 100 nm in length, resembling fossilized filamentous bacteria, but of an order of magnitude smaller in size than terrestrial bacteria microfossils. However, there is a great deal of skepticism about the NASA claims (4), with an inorganic explanation offered in place of these forms being microorganisms (43). These recent events apply to the discussion of bacterial life and death as will be seen in the discussion that follows.

The study of anabiosis traces to 1702, with Antonie van Leeuwenhoek's observations recorded in a letter, "On certain Animalcules found in the sediment in gutters of the roofs of houses." In his letter, he dealt with "animalcules," a category including rotifers and tardigrades, the latter being the most resistant creatures known. Tardigrades are animals less than a millimeter long and have the nicknames of "water bears" and "moss piglets." They are by far the toughest animals on earth, having the ability to shut down their metabolism completely while maintaining their cellular structure (129). The Italian scientist Lazaro Spallanzani (123, 124, 125), who repeated van Leeuwenhoek's experiments in 1776, named some of the animals he saw "tardigrades" because of their peculiar gait. The tardigrades survive freezing, heating to 125°C for several minutes, X-ray doses 250 times stronger than those that would kill mammals, and pressures of 6,000 atm; they will even survive after having their picture taken in a scanning electron microscope! Under these conditions, the tardigrade curls up to reduce surface area and slow the rate of water loss. Once shut down, it ceases respiration, i.e., stops metabolism totally. Crowe and Cooper (31) showed that tardigrades in the tun (dormant) state contain trehalose and protect cell structure by replacing water in their membranes with the trehalose. Tardigrades have been shown to survive in the tun state as long as 120 years. Surviving freezing is achieved by producing proteins that raise the freezing point of the animal to ensure that it freezes quickly, so that only very small crystals form. Tardigrades, based on rRNA analysis, are considered to be an offshoot of crustaceans, originating in the sea and moving to brackish water and then to freshwater in their evolutionary history (130).

The reaction of Lazaro Spallanzani (123), Professor of Natural History of the University of Pavia, to van Leeuwenhoek's discovery and the subsequent work of J. T. Needham (92) on microscopic eel worms was to deny the animal nature of the dried filaments of eel worms in blighted wheat grains. Spallanzani denied the phenomenon of reviviscence described by Needham (92) and Baker (6), prompting Needham to abandon his original view. Subsequently D. Maurice Roffredi (112, 113, 114) and Felice Fontana (41) each independently unraveled the main stages of the life history of eel worms and demonstrated without doubt the animal nature of the material discovered by Needham in blighted wheat grains (92). Moreover, Fontana (41) repeated Needham's observations and showed that the lifeless eel worms (although they were so desiccated that the mere touch of the point of a hair or a needle made them crumble to powder) could be revived within a few minutes when they were brought into contact with water, demonstrating the phenomenon of reviviscence. In their papers, both Roffredi (112, 113) and Fontana (41) attacked Needham, severely and with sarcasm, for abandoning his original correct view as to the true animal nature of his desiccated and revived eel worms. In a letter to Abbé Rozier (editor of the *Journal de Physique*), Needham (93) paid generous tribute to the remarkable achievement of Roffredi but expressed resentment at the criticisms directed against his view. This historical anecdote is detailed nicely by Keilin (61) and illustrates how little has changed in science over the last 200 years. We are still arguing the concepts of dormancy, nonculturability, and other survival strategies of organisms and with equally sarcastic criticism!

Spallanzani went on to publish "Observations and experiments upon some singular animals which may be killed and revived" (125). Spallanzani, unlike his predecessors, was not satisfied with the mere observations of these animalcules during desiccation and reviviscence. He submitted them to different temperatures, vacuum, electric shock, and treatment with chemical agents. Nevertheless, the view of Spallanzani about "real resurrection after death" was not generally accepted. The problem of resuscitation of microorganisms was continued by his contemporaries and generations of biologists up to now. Some workers confirmed his results, others rejected them. Ehrenberg (38) was one of the strongest opponents of the concept of resuscitation of lifeless desiccated microorganisms, espousing instead the hypothesis of greatly slowed-down life processes. His opposition and the discrepancy in results of various workers (the latter now recognized to have been due to species differences and inadequacy of methods to measure the degree of desiccation) led to a decline in the interest in the problem of the resuscitation of microorganisms. Interestingly, there followed a loss of confidence in microscopic observation, as well; the microscope was discredited as a tool of serious research for a period of time.

The subjects of resuscitation and spontaneous generation provided stormy and dramatic discussions between 1858 and 1864 during a revival of interest in the subjects that occurred in that period. The term anabiosis or "return to life" was introduced by Wilhelm Preyer (107, 108) for the phenomenon of resuscitation. That is, death was the lifeless and not viable state, whereas anabiosis was "viable lifelessness." Keilin (61) introduced the term "cryptobiosis," that is, "hidden life," to describe the state of an organism when it shows no visible signs of life and when its metabolic activity becomes hardly measurable or comes reversibly to a standstill. The conclusions of Keilin (61) serve well even today. Namely, "in a living active organism, the state of many of its constituents is the result of dynamic equilibrium between the reactions involved in their constant degradations and regenerations. The organism must constantly provide the energy for the upkeep of its complex structure, which has a tendency to collapse. The stability of such an organism is of a dynamic nature."

When it is in this condition, no energy can be supplied by the organism for the upkeep of its complex structure which, nevertheless, remains intact. The stability of such an organism is of a purely static nature, resembling the stability of biological material in vitro, obtained experimentally or from palaeobiochemical data. As long as the structure of these organisms remains intact, they retain the ability to return to normal active life. The concept of life applied to an organism in the state of anabiosis (cryptobiosis, dormancy, etc.) becomes synonymous with that of its structure, which supports all the components of its catalytic systems. Only when the structure is damaged or destroyed does the organism pass from the state of anabiosis (or latent life) to that of death.

In microbiological studies, the ability to culture microorganisms on artificial media in the laboratory is accepted as constituted proof of viability (53, 62, 106). Depending on the efficiency and/or selectivity of the media employed, interpretation of the viability of bacteria in environmental samples varies, with terms such as "live" cells, "dead" cells, "vegetative" cells, "viable" cells, "nonviable" cells, "stressed" cells, "injured" cells, "moribund" cells, and "static" cells being applied, sometimes quite ambiguously (11, 62, 79, 80, 87, 118). The terms used to describe the metabolic and reproductive states for the bacteria in the early literature included "anabiosis" and "cryptobiosis" (see above).

Survival, "maintenance of viability under adverse circumstances," includes response to stresses such as dormancy, as well as formation of spores and cysts, the latter characteristic of soil microorganisms. However, there are microorganisms that neither sporulate nor encyst, yet can be isolated from unfavorable environments. Such microorganisms persist as vegetative cells but use up their energy reserves slowly, as a result of lowered metabolic activity (89, 95).

Kurath and Morita (70), in studies of starvation survival of marine bacteria, recognized the "live" status of respiring cells, as previously described by Valentine and Bradfield (128). Many years earlier, however, Postgate and Hunter (105) had suggested cells were dead if they did not divide, acknowledging that nondividing bacteria may, in some sense, be alive because they retained their osmotic barrier. Postgate (104) further described the transient state between viability and death when bacterial cells exposed to starvation were incapable of multiplication but maintained metabolic function. Thus, according to Postgate (104), the major criteria for the viability of cells were multiplication and formation of colonies. Roszak and Colwell (117, 118) coined the term "viable but nonculturable" to describe *Salmonella enteriditis* cells detected by direct viable count (67) that were not culturable on bacteriological media. It was evident that those viable but nonculturable organisms maintained metabolic activity, indicated by uptake of various metabolic substrates (116–118). During the past decade, the term "viable but nonculturable" has been widely used in the literature and applied to a wide range of bacterial species of both clinical and environmental origin. The list of these is constantly growing, e.g., *Aeromonas salmonicida* (3); *Campylobacter jejuni* (60, 115); enterotoxigenic *Escherichia coli* (122); *Klebsiella pneumoniae* (133); *Salmonella enteritidis* (116); *Salmonella paratyphi, Staphylococcus* spp. (30), and *Salmonella enterica* serovar Typhimurium (10, 24); *Vibrio cholerae* O1 (27); *Vibrio vulnificus* (98); and *Yersinia enterocolitica* (121).

The viable but nonculturable phenomenon has engendered intense debate, ranging from total rejection on the basis that such cells are not really unculturable to simply evidence of the failure to provide appropriate conditions to support growth; in effect, the term was concluded to be an oxymoron (12). Barer and Harwood (13) reviewed the relationships between viability and culturability in bacteria, making a fine distinction between temporarily nonculturable bacteria and those that have never been cultured (62), and provided an extensive discussion of definitions and descriptions. The conclusion that Barer and Harwood (13) and Bogosian (17) come to is that culturability is the only secure operational definition of viability. Barer and Harwood (13) further split hairs by offering the terms "temporarily nonculturable" and "not immediately culturable" (NIC) for cells obviously still metabolizing. Bogosian (17), however, flatly concludes that all nonculturable cells are dead and that the ability to recover organisms on agar medium is the cornerstone of microbiology, without exception; i.e., if it does not grow on agar, the organism is dead.

The difficulty with such a rigid definition is that in microbial ecology the heterotrophic plate count has been shown to be inefficient for enumerating viable bacteria, especially those bacteria in marine and estuarine systems. Large differences have consistently been reported between results of plate counts and total microscopic counts, the latter accomplished easily with fluorochrome staining of physically intact bacteria. Comparison of results of plate counting, direct microscopic enumeration, and indirect activity measurements shows that the number of bacteria capable of forming colonies on a solid medium is always less than the number actually present and metabolically active, often by several orders of magnitude, in freshwater, marine environments, and soil (10, 45, 79, 85, 101, 117, 118).

Dodd et al. (35) proposed that some of the lethal effects associated with inimical processes (physical or chemical agents) result from "self-destruction" by the cells themselves (a suicide response), that is, self-destruction in exponentially growing cells caused by an oxidative burst which occurs when cells are growth arrested following such treatment. Protection against self-destruction can be provided by reducing the oxygen tension or by adaptive responses associated with the stationary phase, which protect the cell against DNA damage, free-radical damage, and protein denaturation. That is, a sublethal injury becomes lethal because, via growth arrest, cell division is decoupled from metabolism (16). The conclusion of Dixon (34) and Bloomfield et al. (16) is that there is every reason to accept the practical existence of a VBNC state, but better understanding of the biochemistry and physiology of the interactions between growth, respiration, and oxidative damage is needed.

Guelin et al. (46) observed the "rounding up" phenomenon, as well as concomitant decrease in cell volume at initial exposure to organic compound-free conditions. However, the "round body" phenomenon in *Vibrio* spp. was first described and reported by Felter et al. (39, 40, 64) and subsequently further described by Baker et al. (7) for starving *V. cholerae* cells. Indeed, bacteria in the viable but nonculturable state undergo a series of events, often indicated by the morphology of the cells. Xu et al. (132) reported changes in the shape and size of *E. coli* cells entering the viable but nonculturable state. Kondo et al. (69), using a freeze-fixation technique and electron microscopy, demonstrated morphological changes occurring in viable but nonculturable cells of *V. cholerae*. A sequence of gradual morphological changes in cell structure has been documented in our laboratory using scanning electron microscopy (see Fig. 1 through 3). In turn, morphological changes have been correlated with altered nutritional and physicochemical conditions of the environment. An autoradiographic study of tritium-labeled *Helicobacter pylori* (119) revealed aggregation of silver grains in the cells associated with uptake of

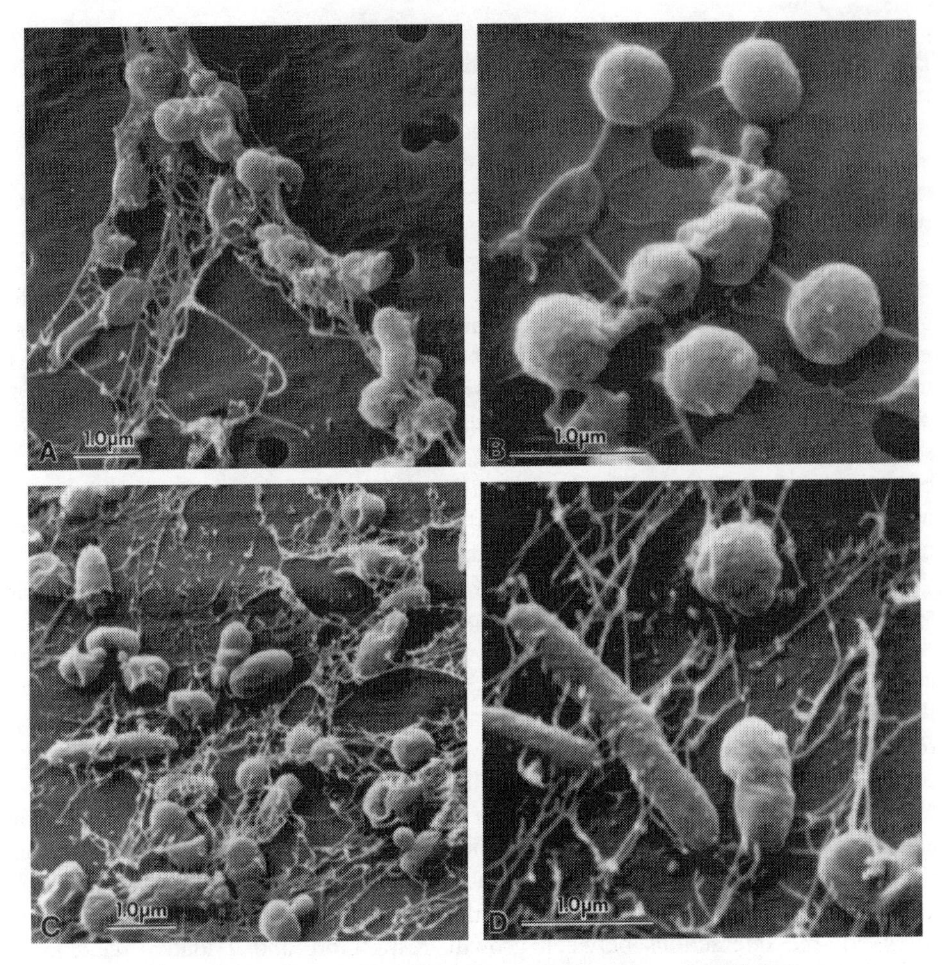

Figure 1. Viable but nonculturable *V. cholerae* O1 viewed by transmission electron microscopy (Chaiyanan and Colwell, unpublished data). (A) *V. cholerae* O1, 2 months old; (B) *V. cholerae* O139, 2 months old; (C) *V. cholerae* O1 (with 1% yeast extract), 2 months old; (D) *V. cholerae* O139 (with 1% yeast extract), 2 months old.

radiolabeled substrate, evidence of continuing metabolic activity as described earlier by Roszak and Colwell (118).

Kondo et al. (69) applied the freeze-fixation technique to study viable but nonculturable *V. cholerae*. They determined the size to be ca. two-thirds of the growing cell. Culturability of *V. cholerae* TS1-4 and *V. cholerae* O1 classical biotype was determined in phosphate-buffered saline. Thin sectioning, with rapid freezing and fixation and shadow-casting methods, revealed actively growing cells of *V. cholerae* O1 with a typical outer membrane, cell membrane, and uniform distribution of ribosomes in the cytoplasm. Viable but nonculturable cells reveal the same intracellular structures as growing cells (see Fig. 1), such as intact outer membrane,

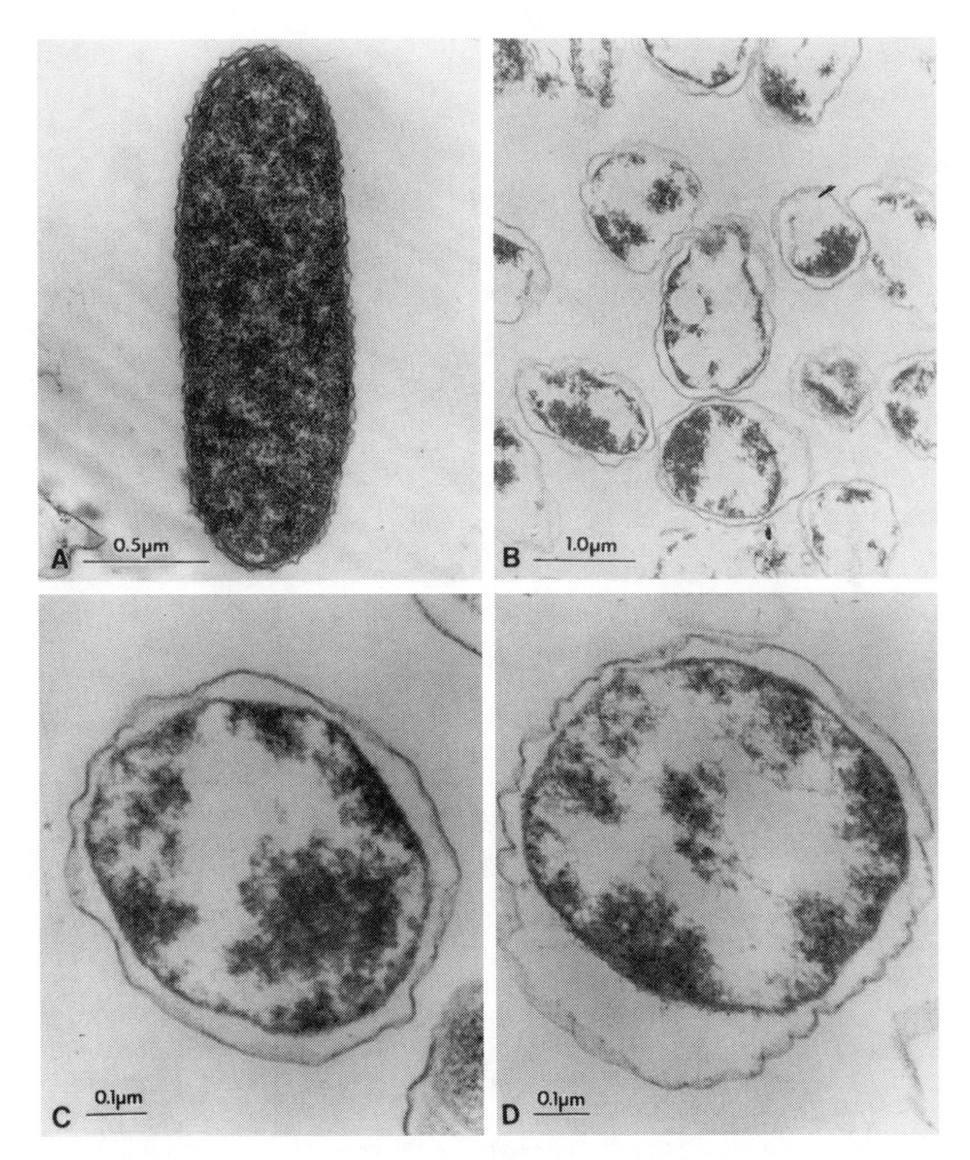

Figure 2. (A) *V. cholerae* O139 NT 330, culturable; (B) 2 months old, viable but nonculturable; (C) 2 months old, viable but nonculturable; (D) 6 months old, viable but nonculturable (Chaiyanan and Colwell, unpublished data).

cell membrane, and a cytoplasm with ribosomes. The overall shape of the cells is either a short rod or spherical, and they are smaller than cells in a growing culture, i.e., about two-thirds the size.

Under high magnification, the cell envelope shows some unique structures. The entire cell surface is covered with delicate fibrous layers not found in growing cells. Also, a thick electron-dense layer is found in the periplasmic space, possibly

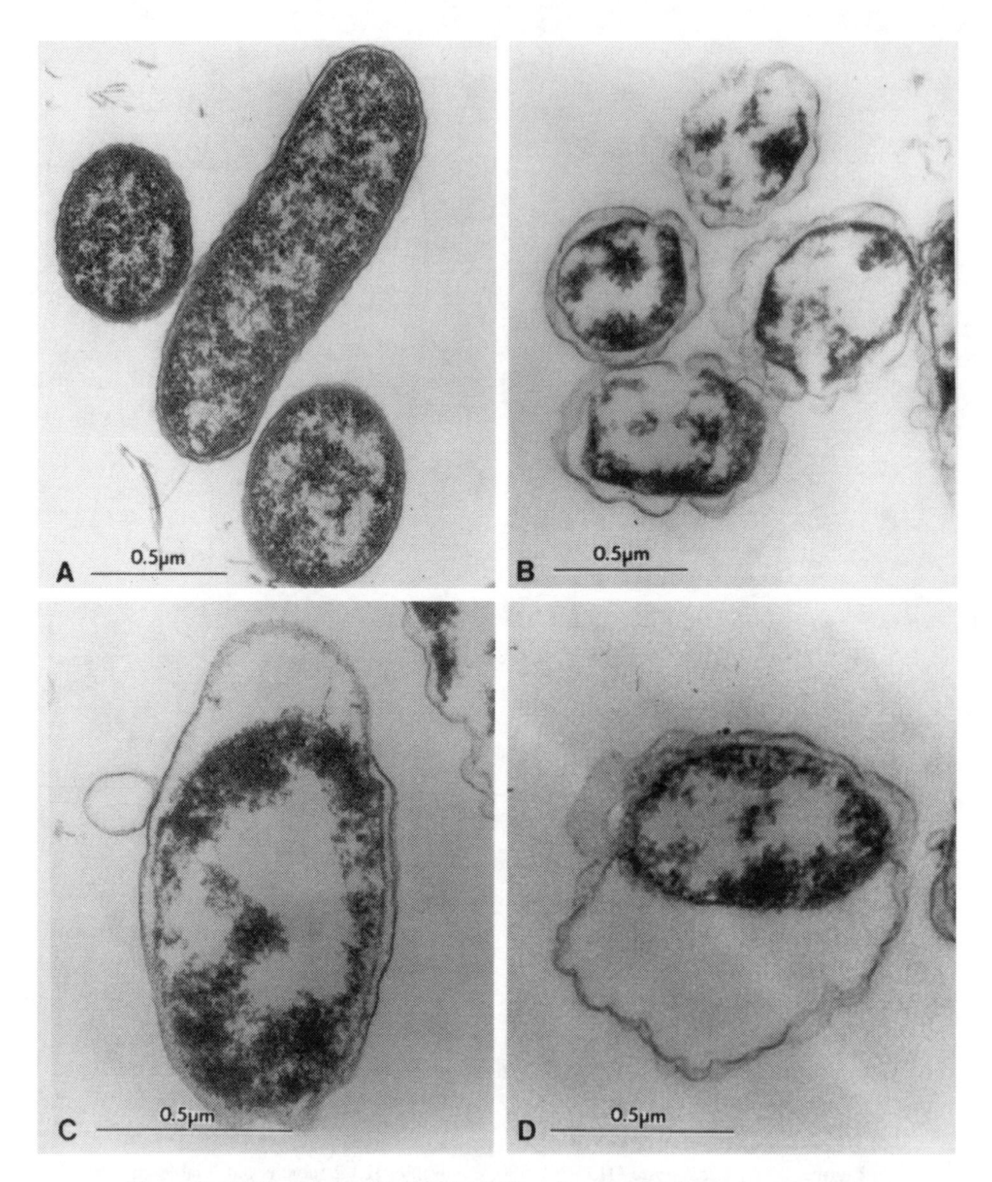

Figure 3. (A) *V. cholerae* O1 ATCC 14035, culturable; (B) 2 months old, viable but nonculturable; (C) 2 months old, viable but nonculturable; (D) 6 months old, viable but nonculturable (Chaiyanan and Colwell, unpublished data).

a peptidoglycan layer and more than three times thicker than that in actively growing cells. Finally, the outer membrane is frequently heavily undulated.

Along with changes in the cytoplasmic membrane, there are changes in proteins found in the periplasmic space. Holmquist and Kjelleberg (55, 56) found three dominant carbon starvation-induced periplasmic proteins produced by *Vibrio* sp.

strain S14 not produced during nitrogen or carbon starvation. No periplasmic proteins or starvation-specific antigens were found in *Vibrio* sp. strain DW1 (2). Therefore, production of periplasmic proteins may be species specific.

In the early 1950s, Bisset (15) claimed that nearly all bacteria possess resting or specialized distributive stages, which he termed "microcysts." These microcysts were considered to be a survival stage in the life cycle of certain *Vibrio* spp., serving to perpetuate the species through periods of extreme adversity (47). The morphological changes that occur within the cell when some cells "round up" (7, 33, 46, 64) and decrease in volume ca. 15-fold (46) to 300-fold (83) have been reported to be typical physiological responses of some bacteria upon exposure to organic nutrient-free conditions (78). Stevenson (126), in reviewing the occurrence of dormancy in bacterial cells, maintained that the small ovoid forms of bacteria normally observed in natural systems by direct microscopy represent exogenously dormant forms in delayed development, responding to unfavorable chemical or physical conditions in the environment. Roszak and Colwell (117) suggested that this is a spore-like or "somnicell" stage for non-spore-forming bacteria. Recently, dormancy and resuscitation in nonsporulating bacteria have been reported to be associated with secondary metabolite or pheromone production (63).

Roszak and Colwell (118) published a list of methods for differentiating living from dead cells and summarized definitions of "viable," other than multiplying and forming colonies (i.e., signs of viability, such as respiration, when cells are unable to divide under prevailing conditions). Different staining methods, such as acridine orange staining as introduced by Strugger in 1949 (127), were used to differentiate live cells from dead cells. However, the acridine orange method is used mainly to enumerate total bacteria in environmental samples (54, 102) since the color distinction between live and dead cells is not very clear (19, 42). In light of current knowledge of viable but nonculturable bacteria, there are several sensitive and specific methods of direct detection that have become available during the past few years. A fluorochrome-based staining method employing 4′,6-diamidino-2-phenylindole (DAPI) was introduced by Porter and Fieg (102). This staining procedure has become more popular during recent years, and it is the method preferred by environmental microbiologists since it estimates the proportion of total bacterial cells within a specific serotype or taxon when fluorescent DNA probes or fluorochrome-labeled antibodies are combined with a general fluorochrome, e.g., DAPI (65). Methods for detection have been reviewed by Oliver (97) and are presented in chapter 16 of this volume. Also useful are papers by Lloyd (74), McFeters et al. (80), Monfort et al. (88), Porter et al., (103), and Bernander et al. (14).

Of the several methods developed during the past 20 years for detection of viable but nonculturable *V. cholerae* and *E. coli*, the direct viable count (DVC) developed by Kogure et al. (67) is perhaps the most convincing and widely used to detect viable cells, irrespective of culturability. In this procedure, viable cells are easily recognized because of enlarged, elongated morphology. This test has been further optimized for detection of *V. cholerae* by combining it with fluorescent antibody (FA) staining. Such a procedure initially employed polyclonal antibody and, subsequently, monoclonal antibody (20, 21, 132). Using monoclonal antibody and the FA method, Huq et al. (57) successfully demonstrated the presence of *V. cholerae*

O1 cells in the natural aquatic environment year-round in Bangladesh, where cholera is endemic. The FA method, combined with the DVC method (21), was further developed to permit direct fluorescent antibody-direct viable counting (DFA-DVC), which is now used as a more convenient method for detection of viable but nonculturable cells of *V. cholerae* (25). The DFA-DVC method can be used to detect very small numbers of organisms in food and water samples by concentrating bacteria on filters (52). Also, the DVC has been used to assess survival of *E. coli* in marine environments (37). The use of soluble *p*-iodonitrotetrazolium (INT) to demonstrate viable bacterial cells was introduced by Zimmerman et al. (134), combined with the Kogure et al. (67) DVC method, and described as the INT-DVC method by Hasan (51). This method enables detection of viable but nonculturable bacteria, including *Aeromonas hydrophila* (51) and *Shigella dysenteriae* (110), in water samples.

Cytometry has proven particularly valuable in describing physiological functions (88), providing rapid analysis in liquid systems (8, 86).

There are good reasons for believing that the energization of the bacterial membrane is a critical factor in determining survival and that death ensues when endogenous sources of energy-yielding substrates are exhausted. The integrity of the membrane is not only important for preserving the organism's internal milieu against its surroundings, but also for promoting the uptake of nutrients via appropriate transport systems when these substrates become available again in the environment (32, 87). Changes in composition of cell wall and/or membrane have been reported by Linder and Oliver (73) and Oliver (97). Other interesting findings are the induction of β-galactosidose activity in individual nonculturable cells by quantitative cytological assay by Nwoguh et al. (96).

Using transposon mutagenesis, Ravel et al. (111) produced mutants showing altered viable but nonculturable response. One mutant consistently demonstrated more rapid entry into the viable but nonculturable state compared to wild type. Thus, genetic regulation of the phenomenon of viable but nonculturability is clearly implied. In the ultimate analysis, survival of a bacterial species in nature depends on preservation of the genome of a single organism. Although there may be many means adopted by different bacteria to combat their possible extinction, there are significant gaps remaining in our overall understanding of the nature of this phenomenon. Recently, the entire genome of *V. cholerae* has been sequenced (Heidelberg et al., unpublished data). These data strongly suggest an active role of the *rpoS* "pathway."

The gene *rpoS* (71) controls starvation-induced expression of many genes, affecting near-UV light resistance, acid phosphatase production, and HP II catalase production and overall stationary-phase gene expression. Thus, the gene plays a role in activating and repressing the synthesis of ca. 20 proteins at the onset of starvation (68).

Continuing from the early observations of anabiosis, microbial survival in the environment has long been of interest to physicians and veterinarians. Mitscherlich and Marth (84) compiled data on the survival of bacteria in the environment, tabulating the influence of factors such as species of bacteria, age, stage of growth, and nutritional conditions during growth in an 802-page volume. They included the substrate in or on which a given bacterium was found or grown, as well as temperature, irradiation, drying, pressure, and water activity. The data on vibrios, predominantly on *V. cholerae*, are interesting and go back to the turn of the century. For example, when *V. cholerae* was described by Koch (66), there was much more interest in its survival in the environment than in later decades.

The number of culturable *V. cholerae* in water depends on a variety of factors. Water temperature is critical, and seasonality also is important, since *V. cholerae* can be isolated in Chesapeake Bay in the late fall and early in spring, when the water temperature is 15 to 20°C. During winter, when the water temperature is 0 to 10°C, *V. cholerae* strains are not isolated (26, 79). Generally, culturability after suspension in river water lasts up to 12 days; in seawater, culturable cells can be recovered up to 60 days and 285 days or longer (91) after addition to a seawater microcosm.

Bacterial cells in the viable but nonculturable state can retain virulence. The first report of this potential in *E. coli* and *V. cholerae* was demonstrated by the accumulation of fluid in the ileal loop of rabbits after inoculation with viable but nonculturable cells (27). These findings were corroborated by the observation that virulence plasmids are maintained in *E. coli* when the cells are in the viable but nonculturable state (8, 22, 44, 101). Other studies yielded similar results: clinical symptoms of cholera were manifested in human volunteers after ingestion of viable but nonculturable cells of *V. cholerae* O1 (28, 29). Subsequent studies were done with *Shigella dysenteriae* type 1 (109, 110). Viable but nonculturable cells of *V. vulnificus* have been shown to cause death in mice (73, 97, 98). Excellent work on *S. enterica* serovar Typhimurium was done similarly by Baleux et al. (9) and Caro et al. (24) and reviewed by Monfort and Baleux (87). Virulence of viable but nonculturable cells of *C. jejuni* has been demonstrated in rats (121). Some reports suggest that virulence is lost upon entry of some cells into the viable but nonculturable state (82). Lazaro et al. (72), however, showed survival of *C. jejuni* at 4 and 20°C, with intact DNA content after 116 days, along with cellular integrity and respiring cells. Viable but nonculturable or dormant microorganisms have been implicated in interstitial cystitis (IC) (36), an inflammatory disease of the urinary bladder that has no known etiology. Urine cultures are usually negative. However, presence of bacterial 16S rRNA genes in bladder biopsies was demonstrated in 29% of patients with IC, but not from control patients with other urological diseases. A sensitive and nested PCR method was used to identify the presence of bacterial DNA. Cloning and sequencing of the 16S rRNA gene fragments showed the genes to be derived from genera representing gram-negative bacteria. Also, 0.22-μm-filterable forms were isolated in culture from biopsy tissue of 14 of 14 IC patients and 1 of 15 controls. The amplified fragments were most similar to *E. coli* and bacteria of the genus *Pseudomonas*.

Recently, coexistence of both *V. cholerae* O1 and O139 was reported in plankton samples in Bangladesh; the direct detection method was used because the culture method was not always successful (59). A field study was conducted in Bangladesh in which stool samples were collected from patients who were symptomatically confirmed as suffering from cholera. Even these samples did not always yield a positive culture. Furthermore, 8% of the stools which were negative by culture turned out to be positive by Cholera DFA, CholeraScreen, a coagglutination test, and a fluorescent antibody test, all of which are direct detection tests now commercially available (49). Some of these culture-negative, DFA-positive stools were further tested by PCR and found to be positive (50).

A cholera outbreak, resulting in one death, occurred in California in 1993 among airline passengers arriving from Argentina (1). Stool samples from these passengers were subjected to culture, with less than 100% success. However, vibriocidal antibody titer against *V. cholerae* was significantly high in the blood collected from those patients whose stool samples were culture negative but positive by CholeraScreen (1).

A study conducted by Colwell and colleagues demonstrated that infections resulting from diving in contaminated water do not always manifest clinical symptoms. In this study, infection was determined by detection of raised antibody titers against *Pseudomonas aeruginosa* in post-dive blood samples collected 30 days after a dive and compared with pre-dive blood samples taken when the organism was isolated from the diving site (75). Similarly, elevated antibody titer was detected against *V. cholerae* O1 in post-dive blood samples collected from American and Russian divers after diving in waters of their respective countries (58). Culture of *V. cholerae* was not always successful, but the organisms were readily detected by using the DFA method and employing monoclonal antibody. During a field trip to Russia, some of the water and plankton samples collected from the Black Sea were positive for *V. cholerae* O1 by culture, and most of the samples revealed the presence of nonculturable cells of *V. cholerae* O1, as determined by the fluorescent antibody method. Subsequently, an outbreak of cholera occurred in Odessa, a coastal area where swimming and bathing were permitted. The report appeared in a local Ukrainian newspaper (58). From investigations reported to date, it appears that most pathogens maintain virulence, suggesting that the careful analysis of food and water samples beyond bacteriological culture alone is needed to assess public health safety (see earlier chapters in this volume).

Some of the most interesting information about viable but nonculturable bacteria comes from studies of gamma irradiation-resistant microorganisms (99, 100). Ionizing-radiation-resistant microorganisms are exemplified by *Deinococcus radiodurans* (5), which shows no loss of viability at doses up to 5 kGy (90). Microorganisms in the reactor core after the Three Mile Island disaster grew at dose rates of 10 Gy/h (18).

Studies of indigenous microbiota in rock from the Yucca Mountain, Nevada Test Site revealed that irradiated bacteria enter the viable but nonculturable state (99). The ability to recover microorganisms after long-term (2 months) incubation at 4°C has implications for microbial-induced corrosion, structural degradation, and transport of radionuclides (77).

Kolter et al. (68) has focused on the stationary phase of the bacterial life cycle, pointing out that in the natural environment, bacterial growth is characterized by long periods of nutritional deprivation punctuated by short periods that allow fast growth, perhaps a feast-or-famine life style. The "starvation" of gram-negative bacteria involves, for some species, the ability to form dormant cells or multicellular aggregates as a response. Kolter et al. (68) describe a developmental program yielding less metabolically active and more resistant cells during starvation or stress. The differentiation is more gradual; the slower the growth rate of the culture, the more growing cells resemble starved cells (68). It is interesting that the starved, gram-negative cells possess many of the characteristics of gram-negative spores, i.e., they behave as if they were spores.

In summary, the long and sometimes painful journey leading to an understanding of the viable but nonculturable phenomenon appears to be coming to a conclusion. Clearly, both entry into and exit from stationary phase are communal activities of bacteria (131). The viable but nonculturable and not immediately culturable states are, therefore, parts of the complicated strategy of bacterial survival in an ever-challenging environment.

REFERENCES

1. **Abbott, S. L. and M. Janda.** 1993. Rapid detection of acute cholera in airline passengers by coagglutination assay. *J. Infect. Dis.* **168:**797–799.
2. **Albertson, N. H., G. W. Jones, and S. Kjelleberg.** 1987. The detection of starvation-specific antigens of two marine bacteria. *J. Gen. Microbiol.* **133**:2225–2232.
3. **Allen-Austin, D., B. Austin, and R. R. Colwell.** 1984. Survival of *Aeromonas salmonicida* in river water. *FEMS Microbiol. Lett.* **21:**143–146.
4. **Anders, E.** 1996. Evaluating the evidence for past life on Mars. *Science* **274:**2119–2121.
5. **Anderson, A. W., H. C. Nordan, R. F. Cain, G. Parrish, and D. Duggan.** 1956. Studies on a radioresistant *Micrococcus.* I. Isolation, morphology, cultural characteristics, and resistance to gamma radiation. *Food Technol.* **10:**575–577.
6. **Baker, H.** 1753. *Employment for the Microscope* (see Part II, chapter IV, Eels in blighted wheat, **250:**60). London.
7. **Baker, R. M., F. L. Singleton, and M. A. Hood.** 1983. Effects of nutrient deprivation on *Vibrio cholerae*. *Appl. Environ. Microbiol.* **46:**930–940.
8. **Baleux, B., and P. Got.** 1996. Apport de l'observation microscopique couplée à l'analyse d'images dans l'evaluation de la qualité bactériologique des eaux: approche cellulaire globale. *TSM* **6:**430–436.
9. **Baleux, B., A. Caro, J. Lesne, P. Got, S. Binard, and B. Delpeuch.** 1998. Survie et maintien de la virulence de *Salmonella* Typhimurium VNC exposée simultanément à trois facteurs stressants expérimentaux. *Oceanol. Acta* **21:**939–950.
10. **Barcina, I., P. Lebaron, J. Vives-Rego.** 1997. Survival of allochtonous bacteria in aquatic systems: a biological approach. *FEMS Microbiol. Ecol.* **23:**1–9.
11. **Barer, M. R., L. T. Gribbon, C. R. Harwood, and C. E. Nwoguh.** 1993. The viable but nonculturable hypothesis and medical bacteriology. *Rev. Med. Microbiol.* **4:**183–191.
12. **Barer, M. R.** 1997. Viable but nonculturable and dormant bacteria: time to resolve an oxymoron and a misnomer? *J. Med. Microbiol.* **46:**629–631.
13. **Barer, M. R., and C. R. Harwood.** 1999. Bacterial viability and culturability. *Adv. Microb. Physiol.* **41:**93–137.
14. **Bernander, R., T. Stokke, and E. Boye.** 1998. Flow cytometry of bacterial cells: comparison between different flow cytometers and different DNA stains. *Cytometry* **31:**29–36.
15. **Bisset, K. A.** 1952. *Bacteria.* E & S Livingstone Ltd., Edinburgh, Scotland.
16. **Bloomfield, S. F., G. S. A. B. Stewart, C. E. R. Dodd, I. R. Booth, and E. G. M. Power.** 1998. The viable but non-culturable phenomenon explained? *Microbiology* **144:**1–3.
17. **Bogosian, G.** 1998. Viable but nonculturable, or dead? *ASM News* **64:**547. (Letter.)
18. **Booth, W.** 1987. Postmortem on Three Mile Island. *Science* **238:**1342–1345.
19. **Bowden, W. B.** 1977. Comparison of two direct-count techniques for enumerating aquatic bacteria. *Appl. Environ. Microbiol.* **33:**1229–1232.

20. **Brayton, P., M. L. Tamplin, A. Huq, and R. R. Colwell.** 1987. Enumeration of *Vibrio cholerae* O1 in Bangladesh waters by fluorescent-antibody direct viable count. *Appl. Environ. Microbiol.* **53:** 2862–2865.
21. **Brayton, P. R., and R. R. Colwell.** 1987. Fluorescent antibody staining method for enumeration of viable environmental *Vibrio cholerae* O1. *J. Microbiol. Methods* **6:**309–314.
22. **Byrd, J. J., and R. R. Colwell.** 1990. Maintenance of plasmids pBR322 and pUC8 in noncultivable *Escherichia coli* in the marine environment. *Appl. Environ. Microbiol.* **56:**2104–2107.
23. **Cano, R. J., and M. Borucki.** 1995. Revival and identification of bacterial spores in 25–40 million-year-old Dominican amber. *Science* **268:**1060–1064.
24. **Caro, A., P. Got, J. Lesne, S. Binard, and B. Baleux.** 1999. Viability and virulence of experimentally stressed nonculturable *Salmonella* Typhimurium. *Appl. Environ. Microbiol.* **65:**3229–3232.
25. **Chowdhury, M. A. R., B. Xu, R. Montilla, J. A. K. Hasan, A. Huq, and R. R. Colwell.** 1995. DFA-DVC: a simplified technique for detection of viable cells of V. cholerae O1 and O139. *J. Microbiol. Methods* **24:**165–170.
26. **Colwell, R. R., J. Kaper, and S. W. Joseph.** 1977. *Vibrio cholerae, Vibrio parahaemolyticus* and other vibrios: occurrence and distribution in Chesapeake Bay. *Science* **198:**394–396.
27. **Colwell, R., P. Brayton, D. Grimes, D. Roszak, S. Huq, and L. Palmer.** 1985. Viable, but non-culturable *Vibrio cholerae* and related pathogens in the environment: implications for release of genetically engineered microorganisms. *Bio/Technology* **3:**817–820.
28. **Colwell, R. R., M. L. Tamplin, P. R. Brayton, A. L. Gauzens, B. D. Tall, D. Harrington, M. M. Levine, S. Hall, A. Huq, and D. A. Sack.** 1990. Environmental aspects of *V. cholerae* in transmission of cholera, p. 327–343. *In* R. B. Sack and Y. Zinnaka (ed.), *Advances in Research on Cholera and Related Diarrhoeas*, 7th ed. K. T. K. Scientific Publishers, Tokyo, Japan.
29. **Colwell, R. R., P. Brayton, A. Huq, B. Tall, P. Harrington, and M. Levine.** 1996. Viable but non-culturable *Vibrio cholerae* O1 revert to a culturable state in the human intestine. *World J. Microbiol. Biotechnol.* **12:**28–31.
30. **Cornax, R., M. A. Morinigo, P. Romero, J. J. Borrego.** 1990. Survival of pathogenic microorganisms in seawater. *Curr. Microbiol.* **20:**293–298.
31. **Crowe, J. H., and A. F. Cooper.** 1971. Cryptobiosis. *Sci. Am.* **225**(6)**:**30–36.
32. **Dawes, E. A.** 1989. Growth and survival of bacteria, p. 67–187. *In* J. S. Poindexter and E. Leadbetter (ed.), *Bacteria in Nature*, vol. 3. Plenum Press, New York, N.Y.
33. **Dawson, M. P., B. A. Humphrey, and K. C. Marshall.** 1981. Adhesion: a tactic in the survival strategy of a marine vibrio during starvation. *Curr. Microbiol.* **6:**195–199.
34. **Dixon, B.** 1998. Viable but nonculturable. *ASM News* **64:**372–373.
35. **Dodd, C. E. R., R. L. Sharman, S. F. Bloomfield, I. R. Booth, and G. S. A. B. Stewart.** 1997. Inimical processes: bacterial self-destruction and sublethal injury. *Trends Food Sci. Technol.* **8:**38–241.
36. **Domingue, G. J., G. M. Ghoniem, K. L. Bost, C. Fermin, and L. G. Human.** 1995. Dormant microbes in interstitial cystitis. *J. Urol.* **158:**1921–1926.
37. **Dupray, E., M. Pommepuy, A. Derrien, M. P. Caprais, and M. Cormier.** 1993. Use of the direct viable count (D.V.C.) for the assessment of survival of *E. coli* in marine environments. *Water Sci. Tech.* **27(3–4):**395–399.
38. **Ehrenberg, C. G.** 1838. Die Infusionsthierchen als volkomene Organismen. Ein Blick in das tiefere organische Leben der Natur. Leipzig.
39. **Felter, R. A., R. R. Colwell, and G. B. Chapman.** 1969. Morphology and round body formation of *Vibrio marinus. J. Bacteriol.* **99:**326–335.
40. **Felter, R. A., S. F. Kennedy, R. R. Colwell, and G. B. Chapman.** 1969. Intracytoplasmic membrane structures in *Vibrio marinus. J. Bacteriol.* **102:**552–560.
41. **Fontana, F.** 1776. Lettre à un de ses amis sur l'ergot et le Tremella. *J. de Physique, l'Abbé Rozier.* **7**:42.
42. **Fry, J. C.** 1990. Direct methods and biomass estimation. *Methods Microbiol.* **22:**41–85.
43. **Gest, H.** 1996. Microorganisms are ubiquitous on Earth—did they also evolve on Mars? *ASM News* **63:**296–297.
44. **Grimes, D. J., and R. R. Colwell.** 1986. Viability and virulence of *Escherichia coli* suspended by membrane chamber in semitropical ocean water. *FEMS Microbiol. Lett.* **34:**161–165.

45. **Grimes, D. J., R. W. Attwell, P. R. Brayton, L. M. Palmer, D. M. Rollins, D. B. Roszak, F. L. Singleton, M. L. Tamplin, and R. R. Colwell.** 1986. Fate of enteric pathogenic bacteria in the estuarine and marine environment. *Microbiol. Sci.* **3:**324–329.
46. **Guelin, A. M., I. E. Mishustina, L. V. Andreev, M. A. Bobyk, and V. A. Lambina.** 1979. Some problems of the ecology and taxonomy of marine microvibrios. *Biol. Bull. Acad. Sci. USSR* **5:**336–340.
47. **Hallock, F. A.** 1960. The life cycle of *Vibrio alternans* (sp. nov.). *Trans. Am. Microsc. Soc.* **79:** 404–412.
48. **Hamamoto, T., and K. Horikoshi.** 1994. Characterization of a bacterium isolated from amber. *Biodiv. Conserv.* **3:**567–572.
49. **Hasan, J. A. K., A. Huq, and R. R. Colwell.** 1991. A method of determination of individual respiring microorganisms and substrate responsive cells in aquatic microcosms. Proc. Abst., p. 212. USA-Japan Cholera Conference, Charlottesville, Va.
50. **Hasan, J. A. K., A. Huq, M. L. Tamplin, R. Siebeling, and R. R. Colwell.** 1994. A novel kit for rapid detection of *V. cholerae* O1. *J. Clin. Microbiol.* **32:**249–252.
51. **Hasan, J. A. K.** 1995. Development and application of rapid test kits for the detection of *V. cholerae* in water, food and clinical samples. Ph.D. dissertation. University of Maryland, College Park.
52. **Hasan, J. A. K., A. Huq, G. B. Nair, S. Garg, A. K. Mukhopadhyay, L. Loomis, D. Bernstein, and R. R. Colwell.** 1995. Development and testing of monoclonal antibody-based rapid immunodiagnostic test kits for direct detection of *Vibrio cholerae* O139 synonym Bengal. *J. Clin. Microbiol.* **33:**2935–2939.
53. **Hattori, T.** 1988. *The Viable Count: Quantitative and Environmental Aspects.* Springer-Verlag, Berlin, Germany.
54. **Hobbie, J. E., R. J. Daley, and S. Jasper.** 1977. Use of nucleopore filters for counting bacteria by fluorescent microscopy. *Appl. Environ. Microbiol.* **33:**1225–1228.
55. **Holmquist, L., and S. Kjelleberg.** 1993a. Changes in viability, respiratory activity and morphology of the marine *Vibrio* sp. strain S14 during starvation of individual nutrients and subsequent recovery. *FEMS Microbiol. Ecol.* **12:**215–224.
56. **Holmquist, L., and S. Kjelleberg.** 1993b. The carbon starvation stimulon in the marine *Vibrio* sp. S14 (CCUG15956) includes three periplasmic space protein responders. *J. Gen. Microbiol.* **139**:209–215.
57. **Huq, A., R. R. Colwell, R. Rahman, A. Ali, M. A. R. Chowdhury, S. Parveen, D. A. Sack, and E. Russek-Cohen.** 1990. Occurrence of *V. cholerae* in the aquatic environment measured by fluorescent antibody and culture method. *Appl. Environ. Microbiol.* **56:**2370–2373.
58. **Huq, A., J. A. K. Hasan, G. Losonsky, V. Diomin, and R. R. Colwell.** 1994. Colonization of professional divers by toxigenic *Vibrio cholerae* O1 and *V. cholerae* non-O1 at dive sites in the United States, Ukraine and Russia. *FEMS Microbiol. Lett.* **120:**137–142.
59. **Huq, A., R. R. Colwell, M. A. R. Chowdhury, B. Xu, S. M. Muniruzzaman, M. S. Islam, M. Alam, M. Yunus, and M. J. Albert.** 1995. Coexistence of *Vibrio cholerae* O1 and O139 Bengal in plankton in Bangladesh. *Lancet* **345:**1249.
60. **Jones, D. M., E. M. Sutcliffe, and A. Curry.** 1991. Recovery of viable but nonculturable *Campylobacter jejuni. J. Gen. Microbiol.* **25:**415–420.
61. **Keilin, D.** 1959. The problem of anabiosis or latent life: history and current concept. *Proc. R. Soc. Biol. B* **150:**149–191.
62. **Kell, D. B., A. S. Kaprelyants, D. Weichart, C. R. Harwood, and M. R. Barer.** 1998. Viability and activity in readily culturable bacteria: a review and discussion of the practical issues. *Antonie Van Leeuwenhoek.* **73:**169–187.
63. **Kell, D. B., H. M. Davey, G. V. Mukamolova, T. V. Votyakova, and A. S. Kaprelyants.** 1995. A summary of recent work on dormancy in nonsporulating bacteria: Its significance for marine microbiology and biotechnology. *J. Mar. Biotechnol.* **3:**24–25.
64. **Kennedy, S., R. Colwell, and G. Chapman.** 1970. Ultrastructure of a psychrophilic marine vibrio. *Can. J. Microbiol.* **16:**1027–1032.
65. **Kepner, R. L., Jr., and J. R. Pratt.** 1994. Use of fluorochromes for direct enumeration of total bacteria in environmental samples. *Microbiol. Rev.* **58:**603–615.
66. **Koch, R.** 1884. An address on cholera and its bacillus. *Br. Med. J.* **2:**403–407, 453–459.

67. **Kogure, K., U. Simidu, and N. Taga.** 1979. A tentative direct microscopic method for counting living marine bacteria. *Can. J. Microbiol.* **25:**415–420.
68. **Kolter, R., D. A. Siegele, and A. Tormo.** 1993. The stationary phase of the bacterial life cycle. *Annu. Rev. Microbiol.* **47:**855–874.
69. **Kondo, K., A. Takade, and K. Amako.** 1994. Morphology of the viable but nonculturable *Vibrio cholerae* as determined by the freeze fixation technique. *FEMS Microbiol. Lett.* **123:**179–184.
70. **Kurath, G., and Y. Morita.** 1983. Starvation survival and physiological studies of a marine *Pseudomonas* sp. *Appl. Environ. Microbiol.* **45:**1206–1211.
71. **Lange, R., and R. Hengge-Aronis.** 1991. Identification of a central regulator of stationary-phase gene expression of *Escherichia coli. Mol. Microbiol.* **5**:49–59.
72. **Lazaro, B., J. Cárcamo, A. Audícana, I. Perales, and A. Fernández-Astorga.** 1999. Viability and DNA maintenance in nonculturable spiral *Campylobacter jejuni* cells after long-term exposure to low temperatures. *Appl. Environ. Microbiol.* **65:**4677–4681.
73. **Linder, K., and J. D. Oliver.** 1989. Membrane fatty acid and virulence changes in the viable but nonculturable state of *Vibrio vulnificus. Appl. Environ. Microbiol.* **55:**2837–2842.
74. **Lloyd, D.** 1993. *Flow Cytometry in Microbiology.* Springer-Verlag, London, United Kingdom.
75. **Losonsky, G. A., J. A. K. Hasan, A. Huq, S. Kaintuch, and R. R. Colwell.** 1994. Serum antibody responses of diverse to waterborne pathogens. *J. Clin. Diagnos. Lab. Immun.* **1**:182–185.
76. **Lowell, P.** 1908. *Mars as the Abode of Life.* The Macmillan Co., New York, N.Y.
77. **McCabe, A.** 1990. The potential significance of microbial activity in radioactive waste disposal. *Experientia* **46:**779–787.
78. **McCann, M. P., J. P. Kidwell, and A. Matin.** 1991. The putative factor KatF has a central role in development of starvation-mediated general resistance in *Escherichia coli. J. Bacteriol.* **173:**4188–4194.
79. **McFeters, G. A., and A. Singh.** 1991. Effects of aquatic environmental stress on enteric bacterial pathogens. *J. Appl. Bacteriol.* **70**(Symp. Suppl.)**:**115S–120S.
80. **McFeters, G. A., F. P. Yu, B. H. Pyle, and P. S. Stewart.** 1995. Physiological assessment of bacteria using fluorochromes. *J. Microbiol. Methods* **21:**1–13.
81. **McKay, D. S., E. K. Gibson, Jr., K. L. Thomas-Keprta, H. Vali, C. S. Romanek, S. J. Clemett, X. D. F. Chillier, C. R. Maechling, and R. N. Zare.** 1996. Search for past life on Mars: possible relic biogenic activity in Martian meteorite ALH84001. *Science* **273:**924–930.
82. **Medema, G. J., F. M. Schets, A. W. van de Giessen, and A. H. Haveljar.** 1992. Lack of colonization of 1 day chicks by viable non-culturable *Campylobacter jejuni. J. Appl. Bacteriol.* **72:**512–516.
83. **Mishustina, I. E., and T. G. Kameneva.** 1981. Bacterial cells of the minimal size in the Barents Sea during the polar night. *Mikrobiologiya* **50:**360–363.
84. **Mitscherlich, E., and E. H. Marth**. 1984. *Microbial Survival in the Environment: Bacteria and Rickettsiae Important in Human and Animal Health.* Springer-Verlag, Berlin, Germany.
85. **Monfort, P., and B. Baleux.** 1991. Distribution and survival of motile *Aeromonas* spp. in brackish water receiving sewage treatment effluent. *Appl. Environ. Microbiol.* **57:**2459–2467.
86. **Monfort, P., and B. Baleux.** 1992. Comparison of flow cytometry and epifluorescence microscopy for counting bacteria in aquatic ecosystems. *Cytometry* **13:**188–192.
87. **Monfort, P., and B. Baleux.** 1999. Bactéries viables non cultivables: réalité et conséquences. *Bull. Soc. Fr. Microbiol.* **14**(3)**:**201–207.
88. **Monfort, P., M.-H. Ratinaud, P. Got, and B. Baleux.** 1995. Apports de la cytométrie en flux et en image en écologie bactérienne des milieux aquatiques. *Océanis* **21:**97–111.
89. **Morita, R. Y.** 1993. Bioavailability of energy and the starvation state, p. 1–23. *In* S. Kjelleberg (ed.), *Starvation in Bacteria.* Plenum Press, New York, N.Y.
90. **Moseley, B. E. B.** 1983. Photobiology and radiobiology of *Micrococcus* (*Deinococcus*) radiodurans. *Photochem. Photobiol. Rev.* **7:**223–234.
91. **Munro, P. M., and R. R. Colwell.** 1996. Fate of *Vibrio cholerae* O1 in seawater microcosms. *Water Res.* **30**(1)**:**47–50.
92. **Needham, J. T.** 1745. *New Microscopical Discoveries.* [Of eels in blighted wheat. pp. 85–89.] London.
93. **Needham, J. T.** 1775. Lettre écrite à l'auteur de ce Receuil. *J. de Physique, l'Abbé Rozier* **5:**226.

94. **New York Times**, May 19, 1995.
95. **Novitsky, J. A., and R. Y. Morita.** 1978. Possible strategy for the survival of marine bacteria under starvation conditions. *Marine Biol.* **48**:289–295.
96. **Nwoguh, C. E., C. R. Harwood, and Barer, M. R.** 1993. Detection of induced beta-galactosidase activity in individual non-culturable cells of pathogenic bacteria by quantitative cytological assay. *Mol. Microbiol.* **17**:545–554.
97. **Oliver, J. D.** 1993. Formation of viable but nonculturable cells, p. 239–272. *In* S. Kjelleberg (ed.), *Starvation in Bacteria.* Plenum Press, New York, N.Y.
98. **Oliver, J. D.** 1995. The viable but non-culturable state in the human pathogen *Vibrio vulnificus. FEMS Microbiol. Lett.* **133**:203–208.
99. **Pitonzo, B. J., P. S. Amy, and M. Rudin.** 1999. Effect of gamma radiation on native endolithic microorganisms from a radioactive waste deposit site. *Radiation Res.* **152**:64–70.
100. **Pitonzo, B. J., P. S. Amy, and M. Rudin.** 1999. Resuscitation of microorganisms after gamma irradiation. *Radiation Res.* **152**:71–75
101. **Pommepuy, M., M. B Utin, A. Derrien, M. Gourmelon, R. R. Colwell, and M. Cormier.** 1996. Retention of enteropathogenicity by viable but nonculturable *Escherichia coli* exposed to seawater and sunlight. *Appl. Environ. Microbiol.* **62**:4621–4626.
102. **Porter, K. G., and Y. S. Fieg.** 1980. The use of DAPI for identifying and counting aquatic microflora. *Limnol. Oceanogr.* **25**:943–948.
103. **Porter, J., D. Deere, R. Pickup, and C. Edwards.** 1996. Fluorescent probes and flow cytometry: new insights into environmental bacteriology. *Cytometry* **23**:91–96.
104. **Postgate, J. R.** 1967. Viability measurements and the survival of microbes under minimum stress. *Adv. Microb. Physiol.* **1**:1–24.
105. **Postgate, J. R., and J. R. Hunter.** 1962. The survival of starved bacteria. *J. Gen. Microbiol.* **29**: 233–263.
106. **Postgate, J. R.** 1969. Viable counts and viability. *Methods Microbiol.* **1**:611–628.
107. **Preyer, W.** 1872. Mitteilung aus dem Vortrag des Herrn Prof. Dr. Preyer. Tagebl. 45 Versamml. Dtsch. Naturs. Aertze. Leipzig. **18**:46.
108. **Preyer, W.** 1891. Ueber dàle Anabiosc. *Biol. Zbl.* **11**:1.
109. **Rahman, I., M. Shahamat, P. A. Kirchman, E. Russek-Cohen, and R. R. Colwell.** 1994. Methionine uptake and cytopathogenicity of viable but nonculturable *Shigella dysenteriae* type 1. *Appl. Environ. Microbiol.* **60**:3573–3578.
110. **Rahman, I., M. Shahamat, M. A. R. Chowdhury, and R. R. Colwell.** 1996. Potential virulence of viable but nonculturable *Shigella dysenteriae* type I. *Appl. Environ. Microbiol.* **62**:115–120.
111. **Ravel, J., R. T. Hill, and R. R. Colwell.** 1994. Isolation of a *Vibrio cholerae* transposon-mutant with an altered viable but nonculturable response. *FEMS Microbiol. Lett.* **120**:57–62.
112. **Roffredi, M.** 1775. Sur l'origine des petits vers ou Anguilles du Bled rachitique. *J. Physique, l' Abbé Rozier.* Paris. **5**:1.
113. **Roffredi, M.** 1775. Seconde lettre ou suite d'observations sur le rachitism du Bled, sur les Anguilles de la colle de farine et sur le grain charbonné. *J. Physique, l'Abbé Rozier.* **5**:197.
114. **Roffredi, M.** 1776. Mémoire pour servir de supplément et d'éclaircissement aux mémoires sur les Anguilles du Bled et de la colle de farine. *J. Physique, l'Abbé Rozier.* **7**:369.
115. **Rollins, D. M., and R. R. Colwell.** 1986. Viable but non-culturable stage of *Campylobacter jejuni* and its role in survival in the natural aquatic environment. *Appl. Environ. Microbiol.* **52**:531–538.
116. **Roszak, D. B., D. J. Grimes, and R. R. Colwell.** 1984. Viable but non-recoverable stage of *Salmonella enteritidis* in aquatic systems. *Can. J. Microbiol.* **30**:334–338.
117. **Roszak, D. B., and R. R. Colwell.** 1987. Metabolic activity of bacterial cells enumerated by direct viable count. *Appl. Environ. Microbiol.* **53**:2889–2983.
118. **Roszak, D. B., and R. R. Colwell.** 1987. Survival strategies of bacteria in the natural environment. *Microbiol. Rev.* **51**:365–379.
119. **Shahamat, M., U. Mai, C. Paszko-Kolva, M. Kessel, and R. Colwell.** 1993. Use of autoradiography to assess viability of *Helicobacter pylori* in water. *Appl. Environ. Microbiol.* **59**:1231–1235.
120. **Shirai, H., M. Nishibuchi, T. Ramamurthy, S. K. Bhattacharya, S. C. Pal, and Y. Takeda.** 1991. Polymerase chain reaction for detection of cholera enterotoxin operon of *V. cholerae*. *J. Clin. Microbiol.* **29**:2517–2521.

121. **Singh, A., M. W. LeChavallier, and G. A. McFeters.** 1985. Reduced virulence of *Yersinia enterocolitica* by copper-induced injury. *Appl. Environ. Microbiol.* **50:**406–411.
122. **Singh, A., R. Yeager, and G. A. McFeters.** 1986. Assessment of in vivo revival, growth, and pathogenicity of *Escherichia coli* strains after copper- and chlorine-induced injury. *Appl. Environ. Microbiol.* **52:**832–837.
123. **Spallanzani, L.** 1769. Nouvelles recherches sur les découvertes microscopiques et sur la géneration des corps organisés, p. 25. *In* Needham (lr.). *Nouvelles Recherches Physiques et Métaphysiques sur la Nature et la Religion*, vol. 1. Lacombe, London.
124. **Spallanzani, Abbé.** 1787. Opuscules de physique animale et végétale (translated from Italian, 1776, by J. Senebier), 2 vol. Paris. [Vol. 2 contains a chapter, "Observations et experiences sur quelques animaux surprenant que l'observateur peut à son gré faire passer de la mort à la vie, p. 203–285.]
125. **Spallanzani, Abbé.** 1803. *Tracts on the Natural History of Animals and Vegetables*, 2 vol. Creech and Constable, Edinburgh, Scotland. (translated from Italian by J. G. Dalyell.) [Second edition, see pp. 119–194: "Observations and experiments on some singular animals which may be killed and revived." An abridged edition in one volume appeared in 1799.]
126. **Stevenson, L. H.** 1978. A case for bacterial dormancy in aquatic systems. *Microb. Ecol.* **4:**127–133.
127. **Strugger, S.** 1949. *Fluoreszensmikroskopie and Mikrobiologie*, p. 151–173. M. V. H. Schaper, Hanover, West Germany.
128. **Valentine, R. C., and J. R. G. Bradfield.** 1954. The urea method for bacterial viability counts with the electron microscope and its relation to other viability counting methods. *J. Gen. Microbiol.* **11:**349–357.
129. **van Leeuwenhoek, A.** 1702. On certain animalcules found in the sediment in gutters of the roofs of houses. Letter 144, vol. 2, p. 207–213. *In* Samuel Hoole (ed.), *The Select Works of Antony van Leeuwenhoek.* H. Fry, London, England.
130. **Wertheim, M.** 1999. Indestructible, p. 43. *N. Sci.* October 23, 1999.
131. **Wheals, A.** 1997. Stationary phase: entry, residence and exit. *SGM Q.* **1997:**16.
132. **Xu, Huai-Shu, N. Roberts, F. L. Singleton, R. W. Attwell, D. J. Grimes, and R. R. Colwell.** 1982. Survival and viability of nonculturable *Escherichia coli* and *Vibrio cholerae* in the estuarine and marine environment. *Microb. Ecol.* **8:**313–323.
133. **Yu, F. P., and G. A. McFeters.** 1994. Physiological responses of bacteria in biofilms to disinfection. *Appl. Environ. Microbiol.* **60:**2462–2466.
134. **Zimmerman, R., R. Iturriaga, and J. Becker-Birck.** 1978. Simultaneous determination of the total number of aquatic bacteria and the number thereof involved in respiration. *Appl. Environ. Microbiol.* **36:**926–935.

INDEX